AS Level and A Level
# Biology

Mary Jones, Richard Fosbery,
Dennis Taylor, Jennifer Gregory

CAMBRIDGE
UNIVERSITY PRESS

CAMBRIDGE UNIVERSITY PRESS
Cambridge, New York, Melbourne, Madrid, Cape Town, Singapore, São Paulo

Cambridge University Press
The Edinburgh Building, Cambridge CB2 2RU, UK

www.cambridge.org
Information on this title: www.cambridge.org/9780521536745

First published 2003
4th printing 2005

Printed in Dubai by Oriental Press

*A catalogue record for this publication is available from the British Library*

ISBN-13  978-0-521-53674-5 paperback
ISBN-10  0-521-53674-X paperback

Produced by Kamae Design, Oxford

Front cover photograph: © Chase Jarvis/Getty Images

**Past examination questions**
Past examination questions and mark schemes are reproduced by permission of the
University of Cambridge Local Examinations Syndicate. Figures: 1.1. Don Fawcett/
Science Photo Library; 5.1 Fig. 14.34, p. 421 from MOLECULAR BIOLOGY OF THE GENE,
Vol. 1, 4th ed. by James D. Watson, Nancy H. Hopkins, Jeffrey W. Roberts, Joan Argetsinger
Steitz, and Alan M. Weiner. Copyright © 1965, 1970, 1976, 1987 by James D. Watson.
Reprinted by permission of Addison Wesley Longman Publishers, Inc.; 8.1 Science
Photo Library.

# Contents

# Part 2: A Level

# Introduction

This book has been produced to help you to do well in your Cambridge International Examinations (CIE) AS and A Level (9700) Biology courses. The book is divided into two parts.

- **Part 1** provides complete coverage of the AS Level syllabus. This is also the first year of study for A Level. The AS material is designed to be accessible to students with a background of O Level or IGCSE Biology.
- **Part 2** covers all the **core** material for the second year of study for A Level. (It should be noted that the book does **not** include the four A2 Options. The chosen Option can be studied from one of the Cambridge Advanced Sciences books *Mammalian Physiology and Behaviour; Microbiology and Biotechnology; Growth, Development and Reproduction; Applications of Genetics.*)

The book does not always follow the order of topics and learning outcomes in the syllabus. The table below provides a guide to syllabus coverage.

|   | Syllabus section | Chapter |
|---|---|---|
| **AS** | A  Cell structure | 1 |
|   | B  Biological molecules | 2 |
|   | C  Enzymes | 3 |
|   | D  Cell membranes and transport | 4 |
|   | E  Cell and nuclear division | 6 |
|   | F  Genetic control | 5 |
|   | G  Transport in multicellular plants Transport in mammals | 10 8 and 9 |
|   | H  Gas exchange [H(d) is covered in chapter 4] | 11 and 12 |
|   | I  Infectious disease | 13 |
|   | J  Immunity | 14 |
|   | K  Ecology | 7 |
| **A Level** | L  Energy and respiration | 15 |
|   | M  Photosynthesis | 16 |
|   | N  Regulation and control | 19 |
|   | O  Inherited change and gene technology [O(m)] can be found in chapter 19] | 17 |
|   | P  Selection and evolution | 18 |

At the end of your course, you will be tested on three sets of Assessment Objectives.

- **Knowledge with understanding.** You are expected to know and understand all of the facts and concepts listed in the syllabus. These are all covered in this book (except the options).
- **Handling information and solving problems.** Questions testing this will expect you to use your knowledge and understanding in an unfamiliar context. A good knowledge and understanding of the topics covered in this book will enable you to apply your knowledge with confidence in new situations.
- **Experimental skills and investigations.** This involves practical work and your examination will test your practical skills. Try to do plenty of practical work. Key information is provided in this book on some practical aspects of the course. Since questions involving an understanding of the use of t- and $\chi^2$ tests may be set in the A Level practical paper, you will find an explanation of the use of $\chi^2$ in Chapter 17 and of the t-test in Appendix 3.

Biology involves a number of technical terms. Each time a new term is introduced it is shown in **bold** and its meaning is explained. The Glossary (page 329) contains definitions of the technical terms used in the book.

Each chapter of the book contains self-assessment questions (SAQs). These are to help you think about what you have read and make sure that you have understood it. Doing the questions will help you to remember what you have read. Answers to the SAQs can be found on pages 300–317.

The book includes some CIE examination questions so that you can familiarise yourself with the style of the questions and practise your skills. Answers for these questions can be found on pages 318 to 328.

This book has been adapted mainly from Biology 1 and Biology 2 from the Cambridge Advanced Science series.

# Acknowledgements

## Photographs

1.1, 2.7, 4.13, 7.5a, Dr Jeremy Burgess/Science Photo Library; 1.3, Alfred Pasieka/Science Photo Library; 1.7, 1.12, 1.22a, 7.5b, A M Page, Royal Holloway College, University of London; 1.8a, 1.8b, 2.26, Claude Nuridsany & Marie Perennou/Science Photo Library; 1.11, 6.14a, 13.1, Eye of Science/Science Photo Library; 1.14, Don Fawcett/Science Photo Library; 1.15, 1.20, 1.21, 1.25, 6.1, 6.2, 8.5, 9.1, 9.4a, 10.6c, 10.13, 14.1, 14.2, ©Biophoto Associates; 1.16, 11.3, Secchi-Leacaque-Roussel-Uclaf/CNRI/Science Photo Library; 1.17, 1.23, Dr Kari Lounatmaa/Science Photo Library; 1.18, 14.6, Dr Gopal Murti/Science Photo Library; 1.19, 2.23b, Bill Longcore/Science Photo Library; 2.9, 4.11c, 10.14e, Geoff Jones; 11.1a, 11.1b, John Adds; 2.12, Peter Gould; 2.13, Tom McHugh/Science Photo Library; 2.20, Dr Arthur Lesk/Science Photo Library; 2.23a, 13.4, Omikron/Science Photo Library; 2.24d, J Gross, Biozentrum/Science Photo Library; 2.24e, Quest/Science Photo Library; 3.8, Simon Fraser/RVI, Newcastle-upon-Tyne/Science Photo Library; 4.9, 14.8, J C Revy/Science Photo Library; 5.10, Professor Oscar Miller/Science Photo Library; 6.11, 6.13, Eric Grave/Science Photo Library; 6.12, Manfred Kage/Science Photo Library; 6.14b, 6.15a, 11.2, 12.8, 12.9, Science Photo Library; 6.15b, James Stevenson/Science Photo Library; 7.5b, 8.12, 10.6b, 10.14c, 10.14d, ©Andrew Syred; 7.7, Vaughan fleming/Science Photo Library; 8.4, 8.10b, 11.1c, Biophoto Associates/Science Photo Library; 8.17, Chris Bonnington Picture Library (©Doug Scott); 9.10, Simon Fraser/Coronary Care Unit, Freeman Hospital, Newcastle-upon-Tyne/Science Photo Library; 10.11, Martyn F Chillmaid/Science Photo Library; 10.14a, Andrew Syred/Science Photo Library; 10.14b, Sinclair Stammers/Science Photo Library; 10.17, Ann Langham, Bethany, Toronto, Canada; 13.2, Ahmed Jadallah/Popperfoto/Reuters; 12.2, GCa-CNRI/Science Photo Library; 12.3, Dr Tony Brain/Science Photo Library; 12.5, Mirror Syndication International; 13.6, NIBSC/Science Photo Library; 13.7, Kwangshin Kim/Science Photo Library; 13.8, T Falise/WHO; 14.14, Hutchison Picture Library; 14.15, Kamal Kishore/Popperfoto/Reuters; 15.14, Dr Keith Porter/Science Photo Library; 15.15, Dr J E Walker, MRC Cambridge; 16.7, 16.8, ©Andrew Syred; 16.10, 19.22, Electron Microscopy Unit, Royal Holloway, University of London; 16.11, 17.3, 19.6a, 19.6b, 19.18, 19.29, 19.42, 19.45, Biophoto Associates; 17.6a, Wayne Hutchison/Holt Studios; 17.6b, G I Bernard/Oxford Scientific Films; 17.6c, John Daniels/Ardea London Ltd; 17.6d, Hans Reinhard/Bruce Coleman; 17.9, Eye of Science/Science Photo Library; 18.2, Popperfoto; 18.3, Jane Burton/Bruce Coleman; 18.6, Dick Roberts/Holt Studios; 18.7, Eric Dragesco/Ardea London Ltd; 18.8, John Durham/Science Photo Library; 18.9a, 18.9b, J L Mason/Ardea London Ltd; 18.12, Wellcome Trust Medical Photographic Library; 18.13 (from top), P Morris/Ardea London Ltd, J B & S Bottomley/Ardea London Ltd; 19.24, St Bartholomew's Hospital/SPL; 19.38, H. Lindscoy/Geoscience Features; 19.47, Nigel Cattlin/Holt Studios.

## Diagrams

6.16, adapted from *Understanding Cancer and its Treatment*, ABPI; 11.4, adapted from *Advanced Biology Principles and Applications, Study Guide* C J Clegg *et al*, 1996, John Murray; 11.6, adapted from *Essentials of Exercise Physiology* McArdle *et al*, 1994, Lea and Febiger; 12.1, adapted from Tetley, *Biological Sciences Review*, May 1990; 12.4, adapted from *Smoking* ASH, 1994; 12.10, from *Coronary Heart Disease Statistics Book*, 1999, British Heart Foundation; 13.5, adapted from Brown, *Inside Science*, *New Scientist*, 18 April 1992; 13.10, 13.9, from the Global TB programme of the World Health Organisation; 14.10, adapted from *Biology of Microorganisms* Brook *et al*, 1994, Prentice-Hall; 14.3, 14.13, adapted from *Medical Immunology for Students* Playfair & Lydyard, 1995, Churchill Livingston; 14.16, from the World Health Organisation internet site; 19.49, Geoff Jones.

## Tables

11.1, 11.2, data from *Human Physiology, Foundations and Frontiers* Schauf *et al*, 1990, Times Mirror/Mosby College Publishing; 12.1, data from *Factsheet* No. 4, ASH; 12.4, data from *Factsheet* No. 1, ASH and *Coronary Heart Disease Statistics* Boaz & Rayner, 1995, British Heart Foundation; 12.3, World Health Organisation MONICA Project, 1989; 13.1, 13.2, 13.3, 13.4, data from World Health Organisation internet site; 13.5, data from *Biology of Microorganisms* Brook *et al*, 1994, Prentice-Hall

## Appendix 3

The example of the use of the t-test is taken from pages 88–90 of *Environmental Biology* by Michael Reiss and Jenny Chapman (Cambridge University Press).

# Part 1
# AS Level

# Cell structure

In the early days of microscopy an English scientist, Robert Hooke, decided to examine thin slices of plant material and chose cork as one of his examples. On looking down the microscope he was struck by the regular appearance of the structure and in 1665 he wrote a book containing the diagram shown in *figure 1.1*.

If you examine the diagram you will see the 'pore-like' regular structures that he called 'cells'. Each cell appeared to be an empty box surrounded by a wall. Hooke had discovered and described, without realising it, the fundamental unit of *all* living things.

Although we now know that the cells of cork are dead, further observations of cells in living materials were made by Hooke and other scientists. However, it was not until almost 200 years later that a general cell theory emerged from the work of two German scientists. In 1838 Schleiden, a botanist, suggested that all plants are made of cells, and a year later Schwann, a zoologist, suggested the same for animals. The **cell theory** states that **the basic unit of structure and function of all living**

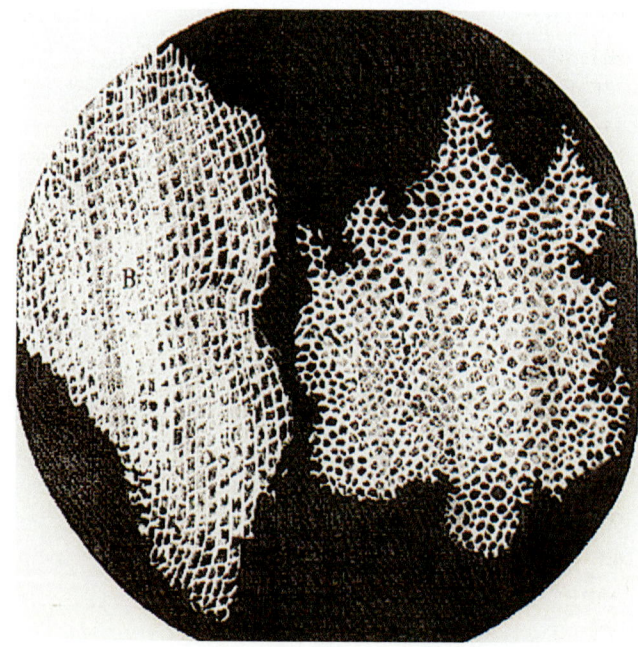

● **Figure 1.1** Drawing of cork cells published by Robert Hooke in 1665.

**organisms is the cell**. Now, over 150 years later, this idea is one of the most familiar and important theories in biology. To it has been added Virchow's theory of 1855 that **all cells arise from pre-existing cells by cell division**.

## Why cells?

A cell can be thought of as a bag in which the chemistry of life is allowed to occur, partially separated from the environment outside the cell. The thin membrane which surrounds all cells is essential in controlling exchange between the cell and its environment. It is a very effective barrier, but also allows a controlled traffic of materials across it in both directions. The membrane is therefore described as **partially permeable**. If it were **freely permeable**, life could not exist because the chemicals of the cell would simply mix with the surrounding chemicals by diffusion.

## Cell biology and microscopy

The study of cells has given rise to an important branch of biology known as **cell biology**. Cells can now be studied by many different methods, but scientists began simply by looking at them, using various types of microscope.

There are two fundamentally different types of microscope now in use: the light microscope and the electron microscope. Both use a form of radiation in order to create an image of the specimen being examined. The **light microscope** uses *light* as a source of radiation, while the **electron microscope** uses *electrons*, for reasons which are discussed later.

### Light microscopy

The 'golden age' of light microscopy could be said to be the nineteenth century. Microscopes had been available since the beginning of the seventeenth century but, when dramatic improvements were made in the quality of glass lenses in the early nineteenth century, interest among scientists became widespread. The fascination of the micro-

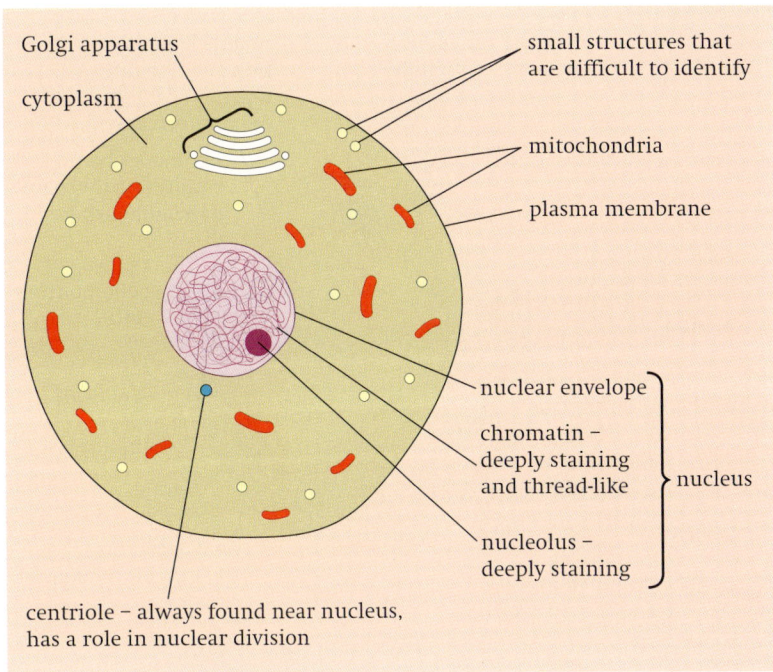

- **Figure 1.2** Structure of a generalised animal cell (diameter about 20 μm) as seen with a very high quality light microscope.

scopic world that opened up in biology inspired rapid progress both in microscope design and, equally importantly, in preparing material for examination with microscopes. This branch of biology is known as **cytology**. By 1900, all the structures shown in *figures 1.2, 1.3* and *1.4*, except lysosomes, had been discovered.

*Figure 1.2* shows the structure of a generalised animal cell and *figure 1.4* the structure of a generalised plant cell as seen with a light microscope. (A generalised cell shows *all* the structures that

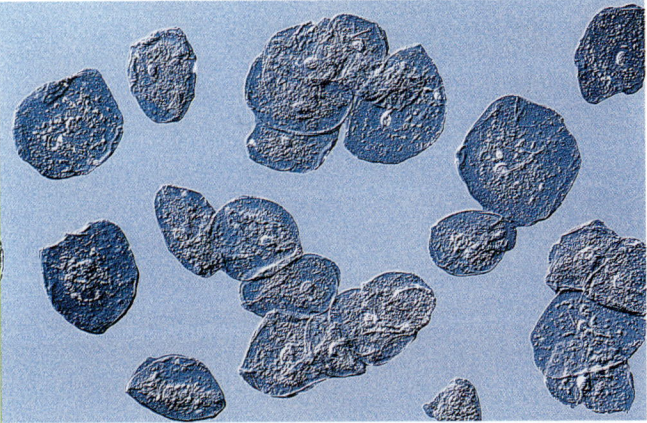

- **Figure 1.3** Cells from the lining of the human cheek (× 300), showing typical animal cell characteristics: a centrally placed nucleus and many organelles such as mitochondria. The cells are part of a tissue known as squamous (flattened) epithelium.

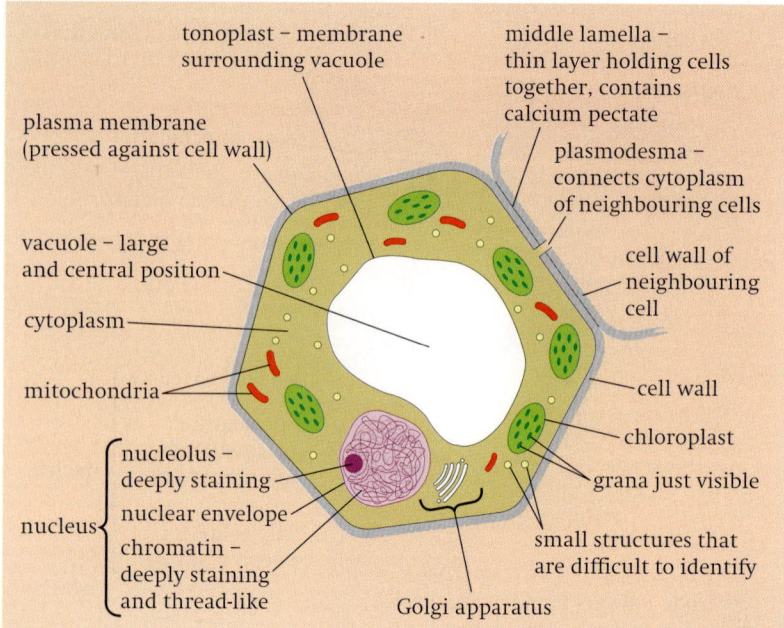

tonoplast – membrane surrounding vacuole

middle lamella – thin layer holding cells together, contains calcium pectate

plasma membrane (pressed against cell wall)

plasmodesma – connects cytoplasm of neighbouring cells

vacuole – large and central position

cell wall of neighbouring cell

cytoplasm

mitochondria

cell wall

chloroplast

nucleolus – deeply staining

grana just visible

nuclear envelope

nucleus

chromatin – deeply staining and thread-like

small structures that are difficult to identify

Golgi apparatus

- **Figure 1.4** Structure of a generalised plant cell (diameter about 40 μm) as seen with a very high quality light microscope.

are typically found in a cell.) *Figure 1.3* shows some *actual* human cells.

## SAQ 1.1

Using *figures 1.2* and *1.4*, name the structures that animal and plant cells have in common and those which are special only to animal or plant cells.

# Animal and plant cells have features in common

In animals and plants each cell is surrounded by a very thin, **plasma** (cell surface) **membrane** which is too thin to be seen with a light microscope. Many of the cell contents are colourless and transparent so they need to be stained to be seen. Each cell has a **nucleus** which is a relatively large structure that stains intensely and is therefore very conspicuous. The deeply staining material in the nucleus is called **chromatin** and is a mass of loosely coiled threads. This material collects together to form visible separate chromosomes during nuclear division (see page 82). It contains **DNA (deoxyribonucleic acid)**, a molecule which contains the instructions that control the

activities of the cell (see chapter 5). Within the nucleus an even more deeply staining area is visible, the **nucleolus**, which is made of loops of DNA from several chromosomes.

The material between the nucleus and the plasma membrane is known as **cytoplasm**. Cytoplasm is an aqueous (watery) material, varying from a fluid to a jelly-like consistency. Many small structures can be seen within it. These have been likened to small organs and hence are known as **organelles**. An organelle can be defined as a **functionally and structurally distinct part of a cell**. Organelles themselves are often surrounded by membranes so that their activities can be separated from the surrounding cytoplasm. This is described as **compartmentalisation**. Having separate compartments is essential for a structure as complex as a cell to work efficiently. Since each type of organelle has its own function, the cell is said to show **division of labour**, a sharing of the work between different specialised organelles.

The most numerous organelles seen with the light microscope are usually **mitochondria** (singular **mitochondrion**). They are only just visible, but extraordinary films of living cells, taken with the aid of a light microscope, have shown that they can move about, change shape and divide. They are specialised to carry out aerobic respiration.

The use of special stains containing silver enabled the **Golgi apparatus** to be detected for the first time in 1898 by Camillo Golgi. The Golgi apparatus is part of a complex internal sorting and distribution system within the cell (see page 14).

# Differences between animal and plant cells

The only structure commonly found in animal cells which is absent from plant cells is the **centriole**. Under the light microscope it appears as a small structure close to the nucleus (*figure 1.2*). It is involved in nuclear division (see page 82).

Individual plant cells are more easily seen with a light microscope than animal cells are because they are usually larger and surrounded by a relatively rigid **cell wall** outside the plasma membrane. The cell wall gives the cell a definite shape. It prevents the cell from bursting when water enters by osmosis, allowing large pressures to develop inside the cell (see page 56). Cell walls may also be reinforced for extra strength. Plant cells are linked to neighbouring cells by means of fine strands of cytoplasm called **plasmodesmata** (singular **plasmodesma**) which pass through pore-like structures in the walls of these neighbouring cells. Movement through the pores is thought to be controlled by their structure.

Apart from a cell wall, mature plant cells differ from animal cells in often possessing a **large central vacuole** and, if the cell carries out photosynthesis, in containing **chloroplasts**. The vacuole is surrounded by a membrane, the **tonoplast**, which controls exchange between the vacuole and the cytoplasm. The fluid in the vacuole is a solution of mineral salts, sugars, oxygen, carbon dioxide, pigments, enzymes and other organic compounds, including some waste products. Vacuoles help to regulate the osmotic properties of cells (the flow of water inwards and outwards) as well as having a wide range of other functions. For example, the pigments which colour the petals of certain flowers and parts of some vegetables, such as the red pigment of beetroots, are sometimes located in vacuoles.

**Chloroplasts** are relatively large organelles which are green in colour due to the presence of chlorophyll. At high magnifications small 'grains', or **grana**, can be seen in them. During the process of photosynthesis light is absorbed by these grana, which actually consist of stacks of membranes. Starch grains may also be visible within chloroplasts. Chloroplasts are found in the green parts of the plant, mainly in the leaves.

### Points to note

- You can think of a plant cell as being very similar to an animal cell but with extra structures.
- Plant cells are often larger than animal cells, although cell size varies enormously.
- Do not confuse the cell *wall* with the plasma *membrane*. Cell walls are relatively thick and physically strong, whereas plasma membranes are very thin. *All* cells have a plasma membrane.
- Vacuoles are not confined to plant cells; animal cells may have small vacuoles, such as phagocytic vacuoles (see page 59), although these are often not permanent structures.

We return to the differences between animal and plant cells as seen using the *electron* microscope on page 15.

# Units of measurement in cell studies

In order to measure objects in the microscopic world, we need to use very small units of measurement which are unfamiliar to most people. According to international agreement, the International System of Units (SI units) should be used. In this system the basic unit of length is the **metre**, symbol **m**. Additional units can be created in multiples of a thousand times larger or smaller, using standard prefixes. For example, the prefix **kilo** means **1000** times. Thus 1 kilometre = 1000 metres. The units of length relevant to cell studies are shown in *table 1.1*.

It is difficult to imagine how small these units are, but, when looking down a microscope and seeing cells clearly, we should not forget how amazingly small the cells *actually* are. *Figure 1.5* shows the sizes of some structures. The smallest

| Fraction of a metre | Unit | Symbol |
|---|---|---|
| one thousandth = 0.001 = 1/1000 = $10^{-3}$ | millimetre | mm |
| one millionth = 0.000 001 = 1/1 000 000 = $10^{-6}$ | micrometre | μm |
| one thousand millionth = 0.000 000 001 = 1/1 000 000 000 = $10^{-9}$ | nanometre | nm |

μ is the Greek letter mu
1 micrometre is a thousandth of a millimetre
1 nanometre is a thousandth of a micrometre

● **Table 1.1** Units of measurement relevant to cell studies.

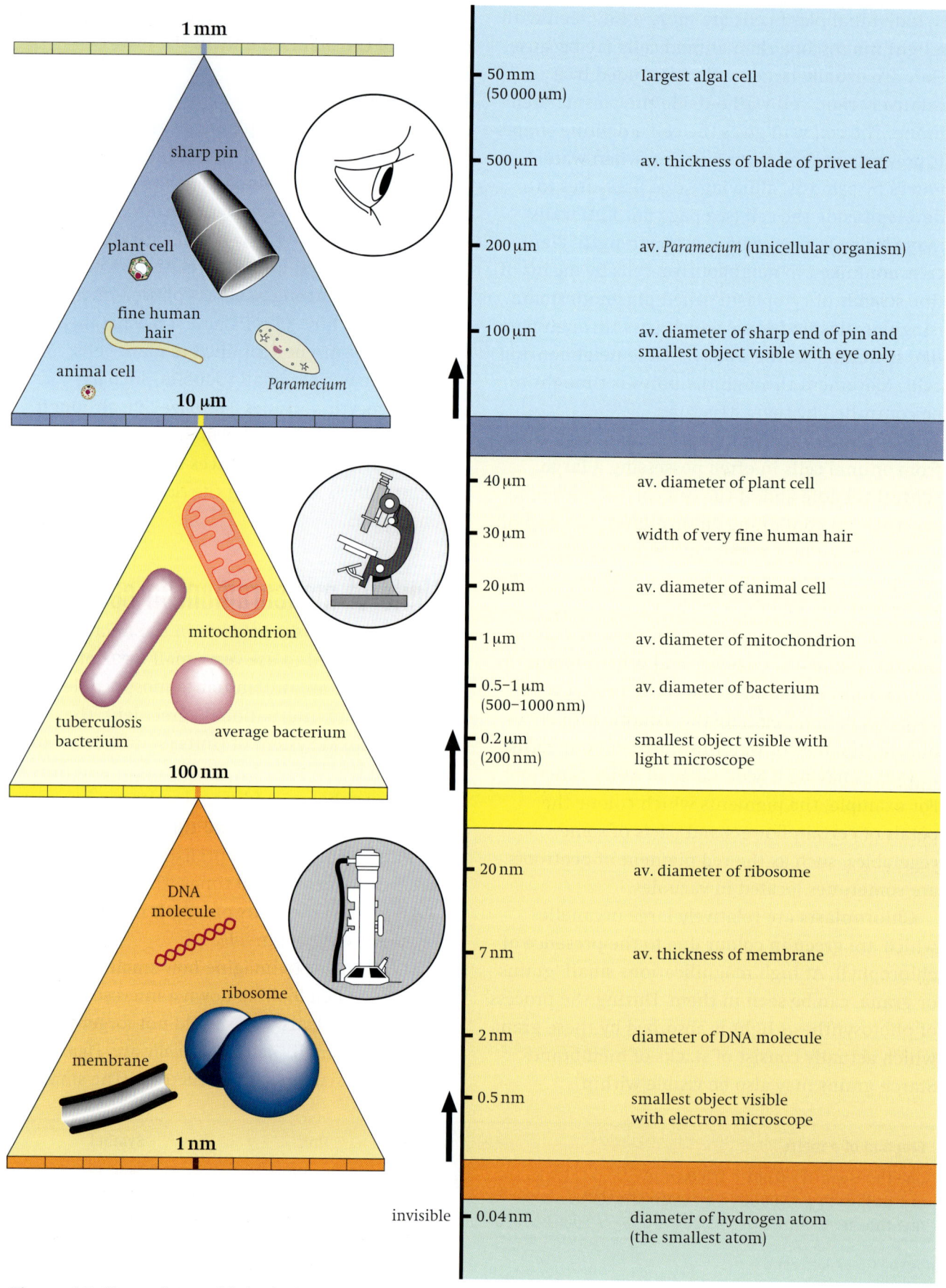

- **Figure 1.5** Sizes of some biological structures.

structure visible with only the human eye is about 50–100 μm in diameter. Your body contains about 60 million million cells, varying in size from about 5 μm to 40 μm. Try to imagine structures like mitochondria, which have an average diameter of 1 μm, or bacteria with an average diameter of 0.5 μm. The smallest cell organelles we deal with in this book, ribosomes, are only about 20 nm in diameter! When we consider processes such as diffusion (chapter 4), it is also helpful to have an appreciation of the distances involved.

## Measuring cells

Cells and organelles can be measured with a microscope by means of an **eye-piece graticule** (see *figure 1.6a*). This is a transparent scale, usually with 100 divisions, which is placed in the microscope eyepiece so that it can be seen at the same time as the object to be measured (see *figure 1.6b*).

The human cheek epithelial cell (see *figure 1.3*) shown superimposed on the scale measures 20 units in diameter. We do not know the actual value of these units until the eyepiece graticule scale is calibrated.

Calibrating the eyepiece graticule scale is done by placing a miniature transparent ruler called a **stage micrometer** scale on the microscope stage and focussing on it. This scale may be etched onto a glass slide or printed on a transparent film. It commonly has subdivisions of 0.1 and 0.01 mm. The images of the two scales can then be superimposed as shown in *figure 1.6c*.

In the figure, 100 eyepiece graticule divisions measure 0.25 mm. Hence, the value of each eyepiece graticule division is

$$\frac{0.25}{100} = 0.0025 \text{ mm}$$

$$\text{or } \frac{0.25 \times 1000}{100} = 2.5 \text{ μm.}$$

The diameter of the cell shown superimposed on the scale in *figure 1.6b* measures 20 eyepiece graticule units and so its actual diameter is

$$20 \times 2.5 \text{ μm} = 50 \text{ μm.}$$

This diameter is greater than that of many human cells because the cell is a flattened epithelial cell.

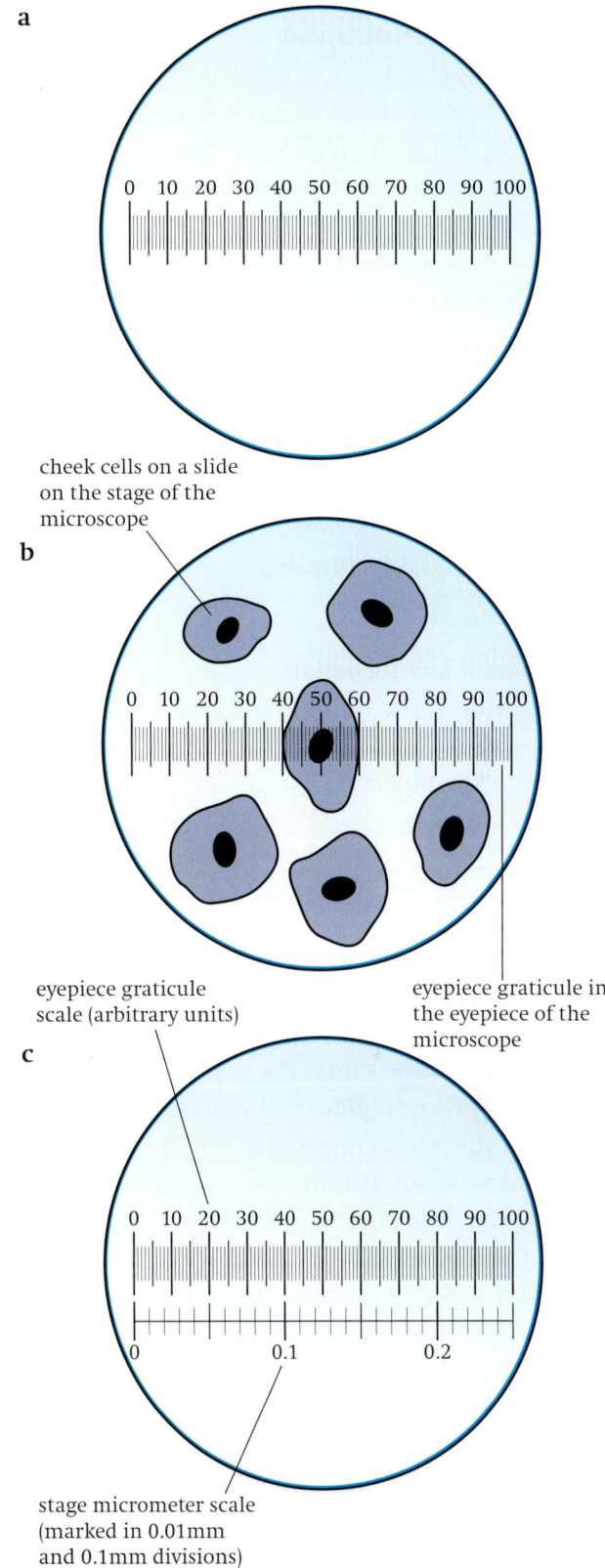

cheek cells on a slide on the stage of the microscope

eyepiece graticule scale (arbitrary units)

eyepiece graticule in the eyepiece of the microscope

stage micrometer scale (marked in 0.01 mm and 0.1 mm divisions)

● **Figure 1.6** Microscopical measurement. Three fields of view seen using a high power (× 40) objective lens. **a** An eyepiece graticule scale. **b** Superimposed images of human cheek epithelial cells and the eyepiece graticule scale. **c** Superimposed images of the eyepiece graticule scale and the stage micrometer scale.

# Electron microscopes

Earlier in this chapter it was stated that by 1900 almost all the structures shown in *figures 1.2* and *1.4* had been discovered. There followed a time of frustration for microscopists because they realised that no matter how much the design of light microscopes improved, there was a limit to how much could ever be seen using light.

In order to understand the problem, it is necessary to know something about the nature of light itself and to understand the difference between **magnification** and **resolution**.

## Magnification and resolution

**Magnification** is the **number of times larger an image is compared with the real size of the object.**

$$\text{magnification} = \frac{\text{size of image}}{\text{actual size of specimen}}$$

*Figure 1.7* shows two photographs of sections through the same group of plant cells. The magnifications of the two photographs are the same. The real length of the central plant cell was about 150 μm. In the photographs, the length appears to be about 60 mm.

To calculate the magnification, it is easiest if we convert all the measurements to the same units, in this case micrometres. 60 mm is 60 000 μm, therefore

$$\text{magnification} = \frac{60\,000}{150}$$

$$= \times 400$$

## SAQ 1.2

**a** Calculate the magnification of the drawing of the animal cell in *figure 1.2*.

**b** Calculate the actual length of the chloroplast in *figure 1.23*.

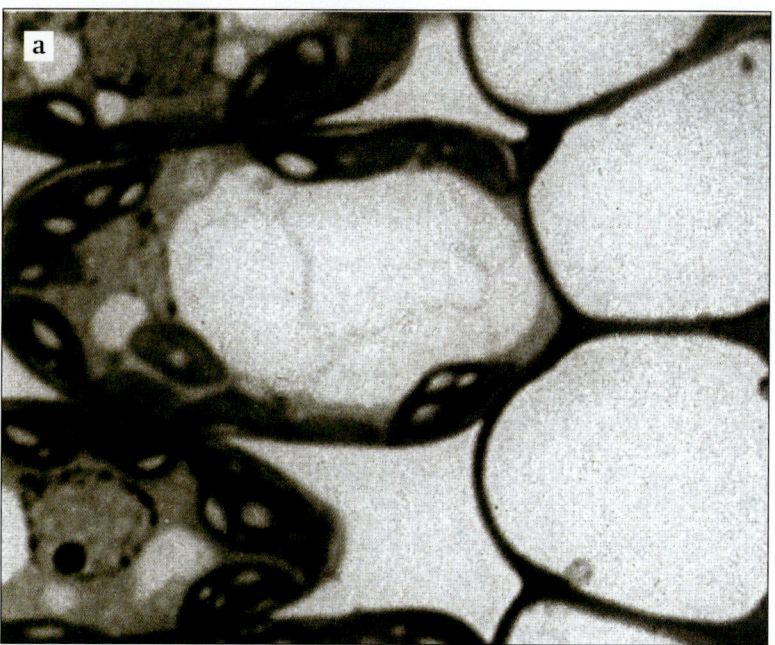

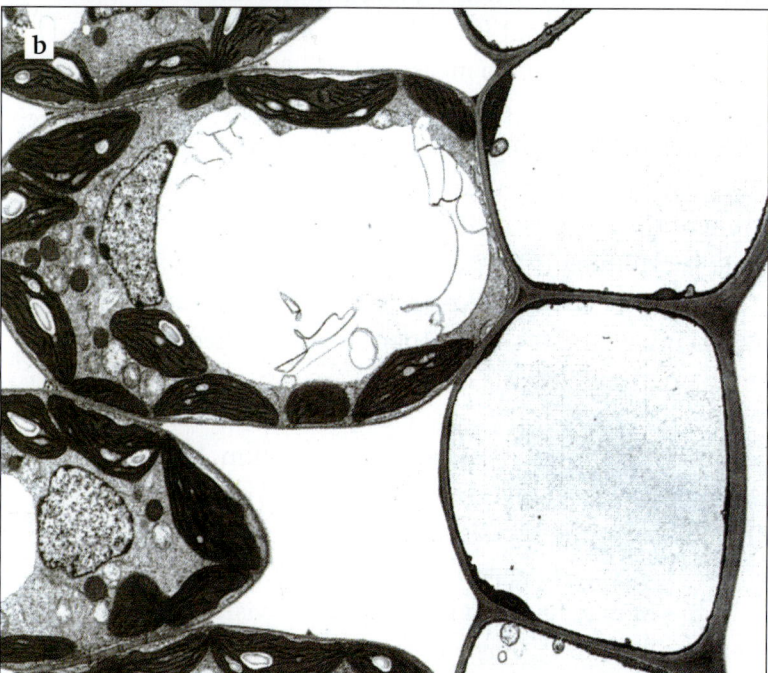

● **Figure 1.7** Photographs of the same plant cells seen **a** with a light microscope, **b** with an electron microscope, both shown at a magnification of about × 400.

Although both photographs in *figure 1.7* are shown at the same magnification, you can see that **b**, the electron micrograph, is much clearer. (An electron micrograph is a picture taken with an electron microscope.) This is because it has greater resolution. **Resolution** is defined as **the ability to distinguish between two separate points**. If the two points cannot be **resolved**, they will be seen as one point. The maximum resolution of a light microscope is 200 nm. This means that if two points or objects are closer together than 200 nm they cannot be distinguished as separate.

It is possible to take a photograph such as *figure 1.7a* and to magnify (enlarge) it, but we see no more *detail*; in other words, we do not improve resolution, even though we often enlarge photographs because they are easier to see when larger. Thus an increase in magnification is not necessarily accompanied by an increase in resolution. With a microscope, magnification up to the limit of resolution can reveal further detail, but any further magnification increases blurring as well as the size of the picture.

## The electromagnetic spectrum

How is resolution linked with the nature of light? One of the properties of light is that it travels in waves. The length of the waves of visible light varies, ranging from about 400 nm (violet light) to about 700 nm (red light). The human eye can distinguish between these different wavelengths, and in the brain the differences are converted to colour differences. (Colour is an invention of the brain!) Some animals can see wavelengths that humans cannot. Bees, for example, can see ultraviolet light. Flowers that to us do not appear to have markings often have ultraviolet markings that guide bees to their nectaries (*figure 1.8*). If you happen to be sharing a dark room with a cobra, the cobra will be able to see *you*, even though you cannot see *it*, because warm bodies give off (radiate) infrared radiation which cobras can see.

The whole range of different wavelengths is called the electromagnetic spectrum. Visible light is only one part of this spectrum. *Figure 1.9* shows some of the parts of the electromagnetic spectrum. The longer the electromagnetic waves, the lower their frequency (all the waves travel at the same speed, so imagine them passing a post: shorter waves pass at higher frequency).

In theory, there is no limit to how short or how long the waves can be. Wavelength changes with energy: the greater the energy,

● **Figure 1.8** The eye of a bee is sensitive to ultraviolet light and can see the guides which lead to the nectaries at the centre of the flower. **a** In normal light, the nectar guides of the *Potentilla* flower cannot be seen by the human eye. **b** In ultraviolet, they appear as dark patches.

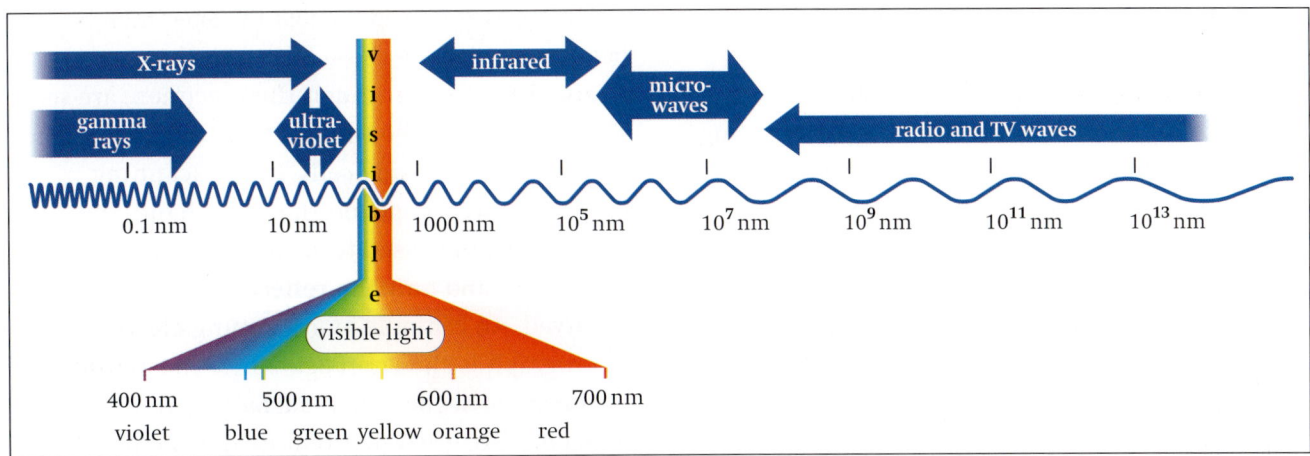

● **Figure 1.9** Diagram of the electromagnetic spectrum (the waves are not drawn to scale). The numbers indicate the wavelengths of the different types of electromagnetic radiation. Visible light is a form of electromagnetic radiation.

the shorter the wavelength (rather like squashing a spring!). Now look at *figure 1.10*, which shows a mitochondrion, some very small cell organelles called ribosomes (see page 13) and light of 400 nm wavelength, the shortest visible wavelength. The mitochondrion is large enough to interfere with the light waves. However, the ribosomes are far too small to have any effect on the light waves. The general rule is that **the limit of resolution is about one half the wavelength of the radiation used to view the specimen**. In other words, if an object is any smaller than half the wavelength of the radiation used to view it, it cannot be seen separately from nearby objects. This means that the best resolution that can be obtained using a microscope that uses visible light (a light microscope) is 200 nm, since the shortest wavelength of visible light is 400 nm (violet light). In practice, this corresponds to a maximum useful magnification of about 1500 times. Ribosomes are approximately 22 nm in diameter and can therefore never be seen using light.

If an object is transparent it will allow light waves to pass through it and therefore will still not be visible. This is why many biological structures have to be stained before they can be seen.

## The electron microscope

Biologists, faced with the problem that they would never see anything smaller than 200 nm using a light microscope, realised that the only solution would be to use radiation of a shorter wavelength than light. If you study *figure 1.9*, you will see that ultraviolet light, or better still X-rays, look like possible candidates. Both ultraviolet and X-ray microscopes have been built, the latter with little success partly because of the difficulty of focussing X-rays. A much better solution is to use electrons. **Electrons** are negatively charged particles which orbit the nucleus of an atom. When a metal becomes very hot, some of its electrons gain so much energy that they escape from their orbits, like a rocket escaping from Earth's gravity. Free electrons behave like electromagnetic radiation. They have a very short wavelength: the greater the energy, the shorter the wavelength. Electrons are a very suitable form of radiation for microscopy for two major reasons. Firstly, their wavelength is

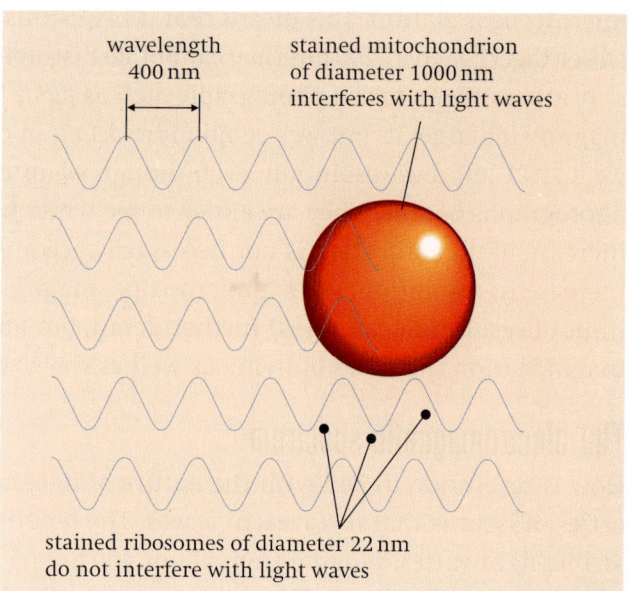

● **Figure 1.10** A mitochondrion and some ribosomes in the path of light waves of 400 nm length.

extremely short (at least as short as that of X-rays); secondly, because they are negatively charged, they can be focussed easily using electromagnets (the magnet can be made to alter the path of the beam, the equivalent of a glass lens bending light).

Electron microscopes were developed during the 1930s and 1940s but it was not until after the Second World War that techniques improved enough to allow cells to be studied with the electron microscope.

### Transmission and scanning electron microscopes

Two types of electron microscope are now in common use. The **transmission electron microscope** was the type originally developed. Here the beam of electrons is passed *through* the specimen before being viewed. Only those electrons that are **transmitted** (pass through the specimen) are seen. This allows us to see thin sections of specimens, and thus to see inside cells. In the **scanning electron microscope**, on the other hand, the electron beam is used to scan the **surfaces** of structures, and only the **reflected** beam is observed. An example of a scanning electron micrograph is shown in *figure 1.11*. The advantage of this microscope is that surface structures can be seen. Also, great depth of field is obtained so that much of the specimen is in focus at the same time. Such a picture would be impossible to obtain with a light microscope, even using the

● **Figure 1.11** False-colour scanning electron micrograph (SEM) of the head of a cat flea (× 100).

same magnification and resolution, because you would have to keep focussing up and down with the objective lens to see different parts of the specimen. The disadvantage of the scanning electron microscope is that it cannot achieve the same resolution as a transmission electron microscope.

### Viewing specimens with the electron microscope

It is not possible to see an electron beam, so to make the image visible the electron beam has to be projected onto a fluorescent screen. The areas hit by electrons shine brightly, giving overall a 'black and white' picture. The stains used to improve the contrast of biological specimens for electron microscopy contain heavy metal atoms which stop the passage of electrons. The resulting picture is therefore similar in principle to an X-ray photograph, with the more dense parts of the specimen appearing blacker. 'False-colour' images are created by processing the standard black and white image using a computer.

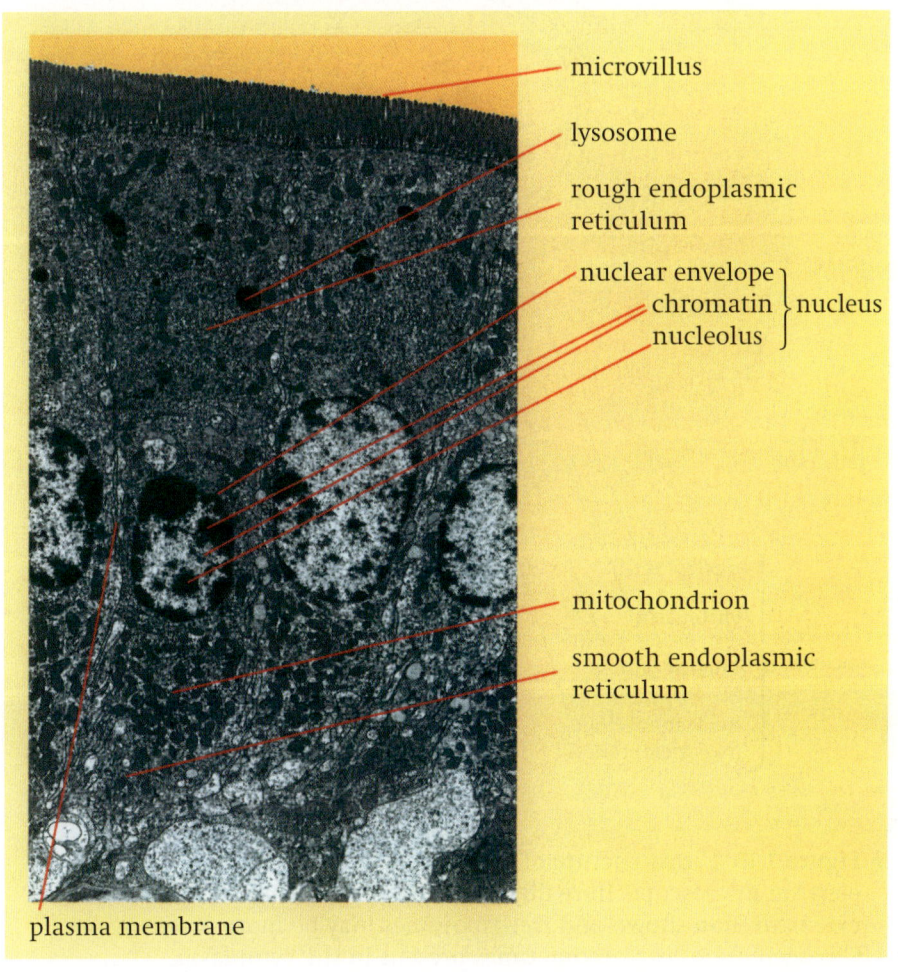

microvillus

lysosome

rough endoplasmic reticulum

nuclear envelope ⎤
chromatin ⎬ nucleus
nucleolus ⎦

mitochondrion

smooth endoplasmic reticulum

plasma membrane

● **Figure 1.12** A representative animal cell as seen with a transmission electron microscope (TEM). The cell is a small intestinal cell from a mouse (× 10 000).

To add to the difficulties of electron microscopy, the electron beam, and therefore the specimen and the fluorescent screen, must be in a vacuum. If electrons collided with air molecules, they would scatter, making it impossible to achieve a sharp picture. Also, water boils at room temperature in a vacuum, so all specimens must be dehydrated before being placed in the microscope. This means that only dead material can be examined. Great efforts are therefore made to try to preserve material in a life-like state when preparing it for the microscope.

**SAQ 1.3**
Explain why ribosomes are not visible using a light microscope.

## Ultrastructure of an animal cell

The 'fine', or detailed, structure of a cell as revealed by the electron microscope is called its **ultrastructure**. *Figure 1.12* shows the appearance of a typical animal cell as seen with an electron microscope and *figure 1.13* is a diagram based on many other such micrographs.

**SAQ 1.4**
Compare *figure 1.13* with *figure 1.2*. Name the structures which can be seen with the electron microscope but not with the light microscope.

## Structure and functions of organelles

Compartmentalisation and division of labour within the cell are even more obvious with an electron microscope than with a light microscope.

We now consider the structure and functions of some of the cell components in more detail.

### Nucleus *(figure 1.14)*

The **nucleus** is the largest cell organelle. It is surrounded by two membranes known as the **nuclear envelope**. The outer membrane of the nuclear envelope is continuous with the endoplasmic reticulum *(figure 1.13)*. The nuclear envelope is conspicuously perforated by the **nuclear pores**. These allow exchange between the nucleus and the cytoplasm, e.g. mRNA and ribosomes leave the

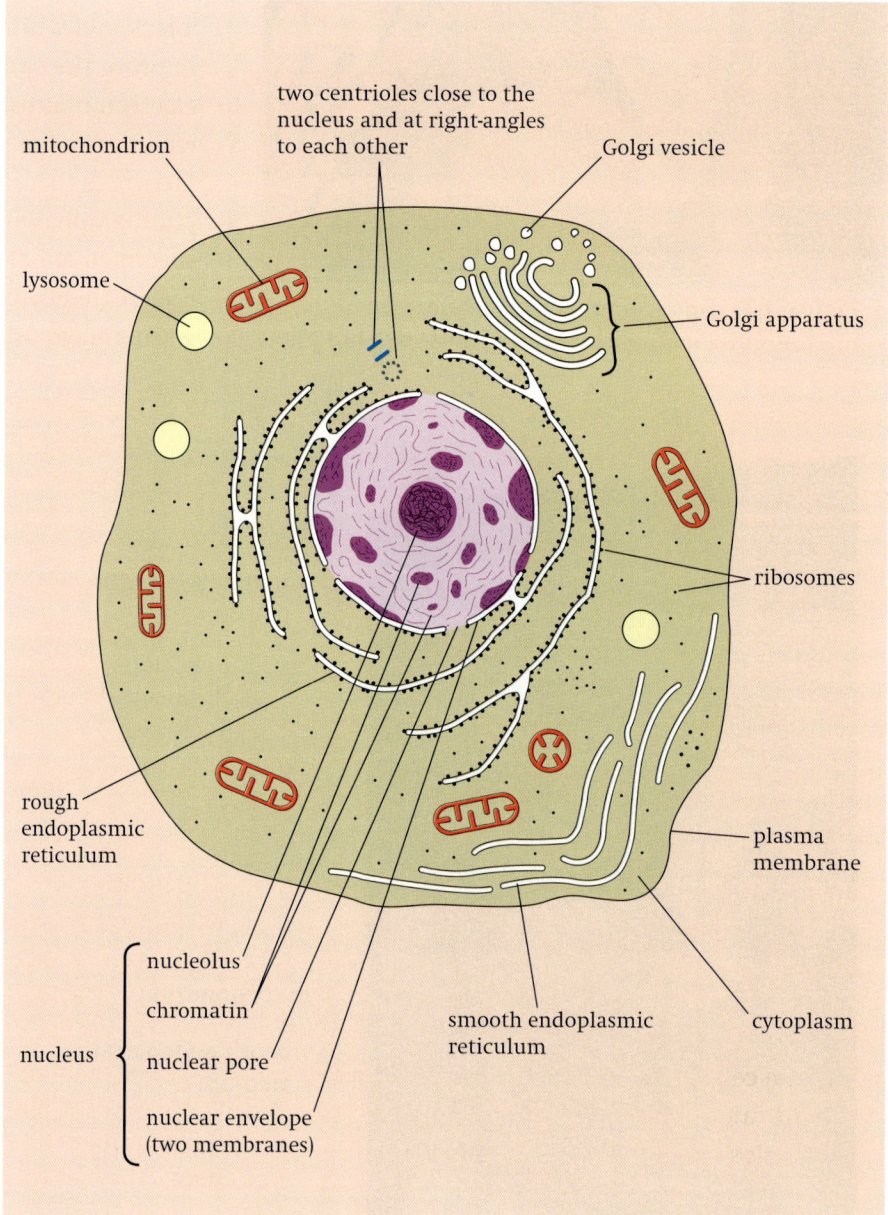

● **Figure 1.13** Ultrastructure of a typical animal cell as seen with an electron microscope. In reality, the endoplasmic reticulum is more extensive than shown and free ribosomes may be more extensive. Glycogen granules are sometimes present in the cytoplasm.

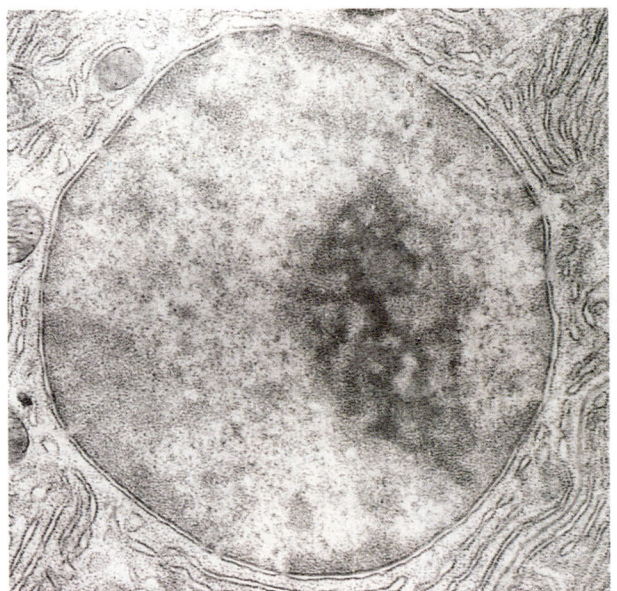

● **Figure 1.14** TEM of the nucleus of a cell from the pancreas of a bat (× 10 000). The circular nucleus displays its double-layered nuclear envelope interspersed with nuclear pores. The nucleolus is more darkly stained. Smooth endoplasmic reticulum is visible in the surrounding cytoplasm.

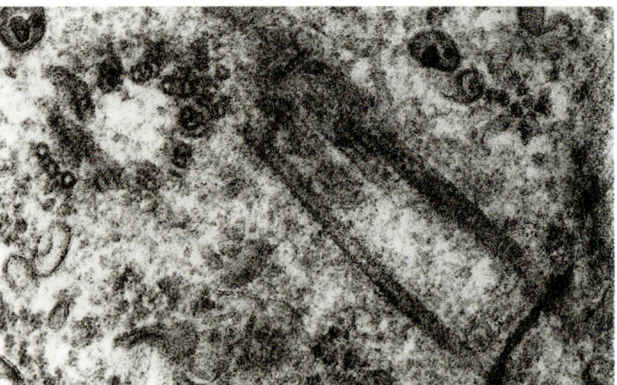

● **Figure 1.15** Centrioles in transverse and longitudinal section (TS and LS) (× 86 000). In TS the nine triplets of microtubules which make up the structure can be clearly seen.

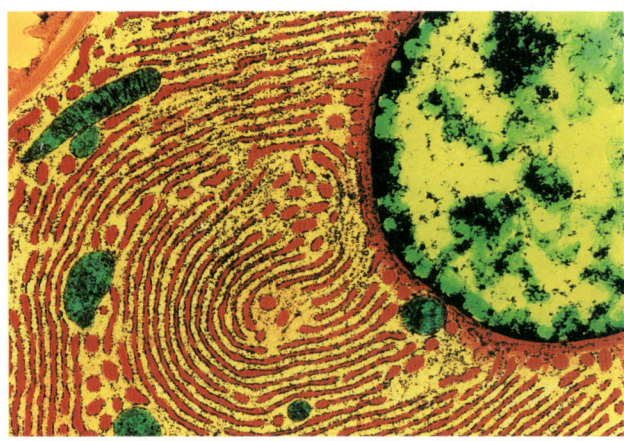

● **Figure 1.16** Coloured TEM of the rough ER (red stripes) covered with ribosomes (black dots) outside the nucleus (green) (× 7000).

nucleus and some hormones and nutrients enter the nucleus through the pores.

Within the nucleus, the chromosomes are in a loosely coiled state known as chromatin (except during nuclear division, see chapter 6). Chromosomes contain DNA which is organised into functional units called genes. Genes control the activities of the cell and inheritance; thus the nucleus controls the cell's activities. Division of the nucleus precedes cell division. Also within the nucleus, the **nucleolus** manufactures ribosomes, using the information in its own DNA.

## Centrioles *(figure 1.15)*

Just outside the nucleus, the extra resolution of the electron microscope reveals that there are really *two* centrioles, not one as it appears under the light microscope (compare with *figure 1.2*). They lie close together at right-angles to each other. A centriole is a hollow cylinder about 0.4 µm long, formed from a ring of microtubules (a kind of cell scaffolding made of protein). These microtubules are used to grow the spindle fibres for nuclear division (see page 82).

## Endoplasmic reticulum and ribosomes *(figure 1.16)*

When cells were first seen with the electron microscope, biologists were amazed to see so much detailed structure. The existence of much of this had not been suspected. This was particularly true of an extensive system of membranes running through the cytoplasm which became known as the **endoplasmic reticulum** (ER).

Attached to the surface of much of the ER are many tiny organelles, now known as **ribosomes**. At very high magnifications these can be seen to consist of two parts, a smaller and a larger subunit. In some areas of the cell, the ER lacks ribosomes and appears smooth. This is called **smooth ER** and is now known to have a different function from ribosome-covered ER, which is called **rough ER**. The membranes form a system of flattened sacs, like sheets, which are called **cisternae**. The space

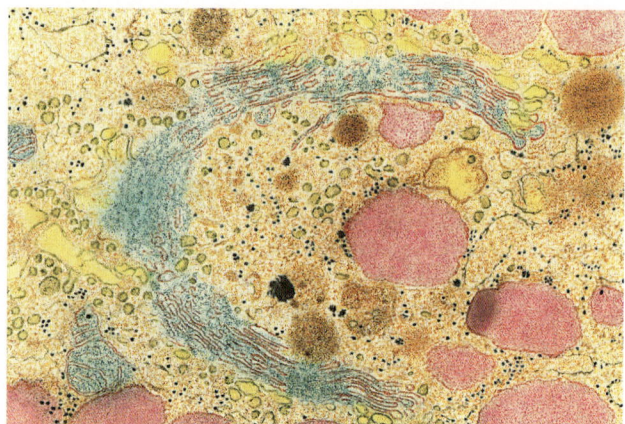

● **Figure 1.17** TEM of a Golgi apparatus (× 30 000). A central stack of saucer-shaped sacs (cisternae) can be seen budding off small Golgi vesicles. These may form secretory vesicles whose contents can be released at the cell surface by exocytosis (for more details see page 59).

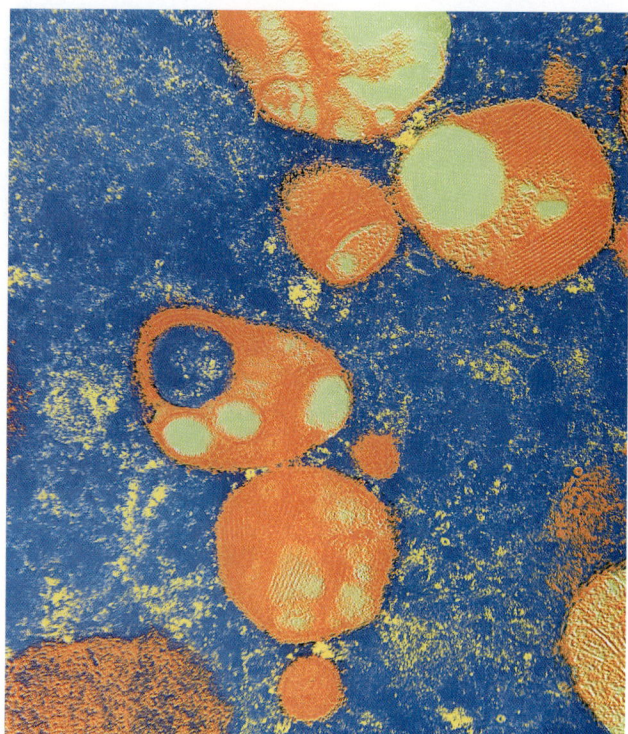

● **Figure 1.18** Lysosomes in a mouse kidney cell (× 55 000). They contain membrane structures in the process of digestion (red) and vesicles (green). Cytoplasm is coloured blue here.

inside the sacs forms a compartment separate from the surrounding cytoplasm. The cisternae can go on to form the Golgi apparatus.

Ribosomes are the sites of protein synthesis (see pages 72–73). They are found free in the cytoplasm as well as on the rough ER. They are very small organelles, only about 22 nm in diameter and are made of **RNA** (ribonucleic acid) and protein.

The proteins that are manufactured on the ribosomes are transported throughout the cell by the rough ER. In contrast, the smooth ER makes lipids (page 28) and steroids (e.g. cholesterol and reproductive hormones).

## Golgi apparatus (figure 1.17)

The Golgi apparatus is a stack of flattened sacs (**cisternae**). The stack is constantly being formed at one end from vesicles which bud off from the ER, and broken down again at the other end to form **Golgi vesicles**.

The apparatus collects, processes and sorts molecules (particularly proteins from the rough ER), ready for transport in Golgi vesicles either to other parts of the cell or out of the cell (secretion). Golgi vesicles are also used to make lysosomes.

## Lysosomes (figure 1.18)

Lysosomes are spherical sacs, surrounded by a single membrane and having no internal structure. They are commonly 0.1–0.5 μm in diameter. They

contain hydrolytic (digestive) enzymes which must be kept separate from the rest of the cell to prevent damage. Lysosomes are responsible for the breakdown (digestion) of unwanted structures, e.g. old organelles or even whole cells, as in mammary glands after lactation (breast feeding). In white blood cells they are used to digest bacteria (see endocytosis, page 59). Enzymes are sometimes released outside the cell, e.g. during replacement of cartilage with bone during development. The heads of sperm contain a special lysosome, the acrosome, for digesting a path to the ovum (egg).

## Mitochondria (figure 1.19)

Mitochondria are slightly larger than lysosomes and are surrounded by two membranes (an envelope). The inner of these is folded to form finger-like **cristae** which project into the interior solution, or **matrix**.

The main function of mitochondria is to carry out the later stages of aerobic respiration. As a result of respiration, they make ATP, the universal energy carrier in cells (see chapter 7). They are also involved in synthesis of lipids (page 28).

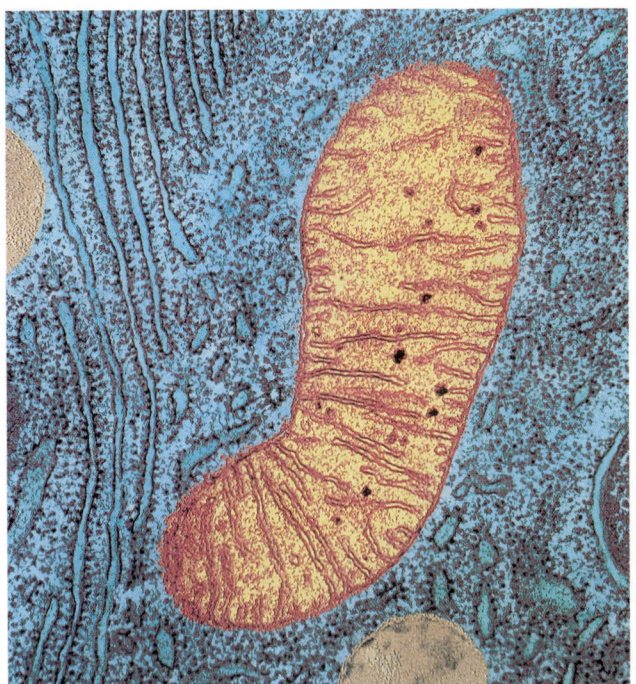

● **Figure 1.19** Mitochondrion (orange) with its double membrane (envelope); the inner membrane is folded to form cristae (× 12 000). Mitochondria are the sites of aerobic cell respiration. Note also the rough ER (turquoise).

● **Figure 1.20** Plasma membrane (× 250 000). At this magnification the membrane appears as two dark lines at the edge of the cell.

## Plasma membrane *(figure 1.20)*

The plasma membrane is extremely thin (about 7 nm). However, at very high magnifications, at least × 100 000, it can be seen to have three layers (**trilaminar appearance**). This consists of two dark lines (heavily stained) either side of a narrow, pale interior. The membrane is partially permeable, controlling exchange between the cell and its environment. Membrane structure is discussed further in chapter 4.

## Cilia *(figure 1.21)*

Some cells have long, thin extensions that can move in a wave-like manner. If there are just a few of these extensions, and they are relatively long,

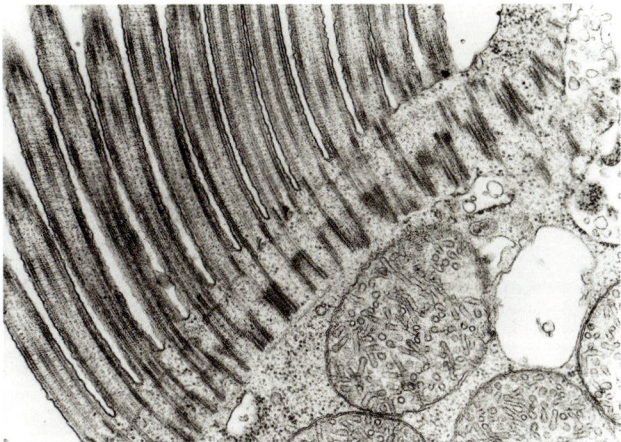

● **Figure 1.21** TEM of cilia in LS (× 18 500).

then they are called **flagella** (singular **flagellum**). If there are many of them, and they are relatively short, then they are called **cilia** (singular **cilium**).

A cilium is usually about 3–4 μm long. It is covered with an extension of the plasma membrane, and it contains microtubules that extend throughout its length. These microtubules arise from a structure called a basal body, in the cytoplasm. The microtubules are arranged in an outer cylinder of 9 pairs, surrounding two central microtubules. Basal bodies are identical in structure to centrioles.

The movement of cilia and flagella is caused by the microtubules, which can slide against each other, causing the whole strucure to bend. Where there are many cilia on a cell, or a group of cells, they all move in a coordinated manner, each slightly out of phase with its neighbour so that the overall effect looks rather like long grass rippling in the wind. As a result, substances around the cell are made to move or – if the cell is not fixed to anything – the cell itself is swept along as the cilia beat.

## Ultrastructure of a plant cell

All the structures found in animal cells are also found in plant cells, except centrioles and – except very rarely – cilia. The appearance of a plant cell as seen with the electron microscope is shown in *figure 1.22a* and a diagram based on many such micrographs in *figure 1.22b*. The relatively thick cell wall and the large central vacuole are obvious, as are the chloroplasts (two

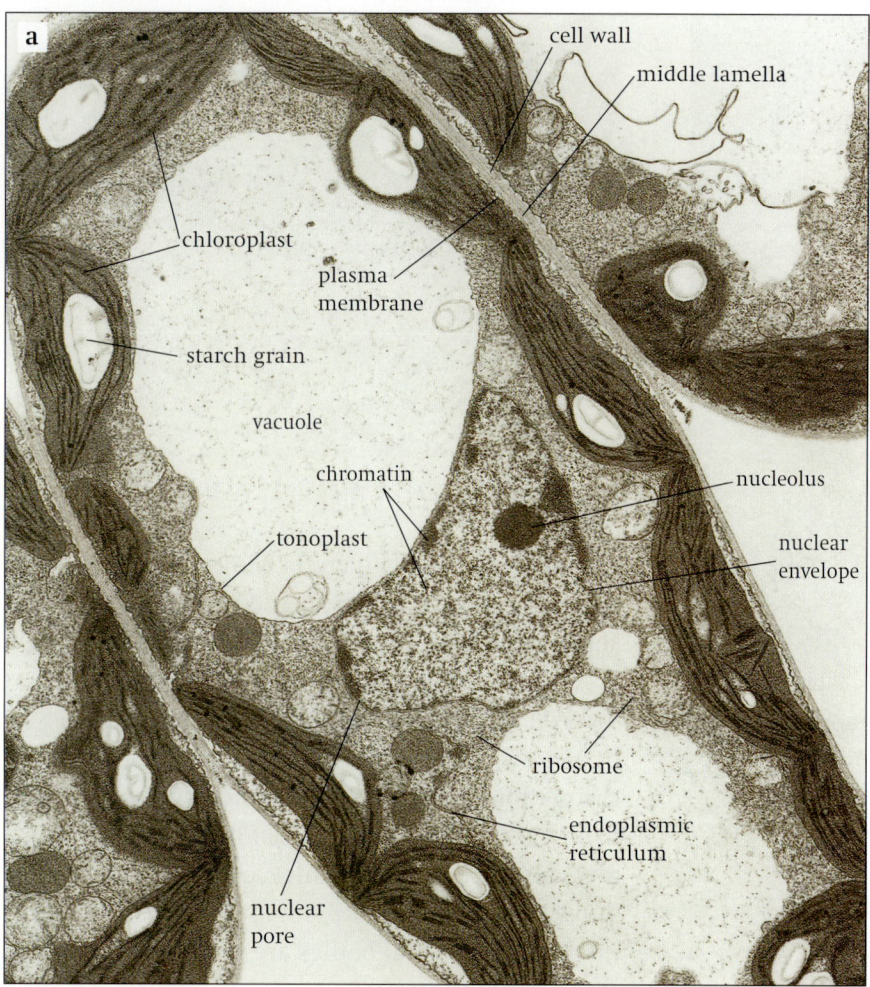

a

cell wall

middle lamella

chloroplast

plasma membrane

starch grain

vacuole

chromatin

nucleolus

tonoplast

nuclear envelope

ribosome

endoplasmic reticulum

nuclear pore

● **Figure 1.22** Appearance of a representative plant cell as seen with an electron microscope. **a** An electron micrograph of a palisade cell from a soya bean leaf (× 5600). **b** A diagram of the ultrastructure of a typical plant cell as seen with the electron microscope. In reality, the ER is more extensive than shown. Free ribosomes may also be more extensive.

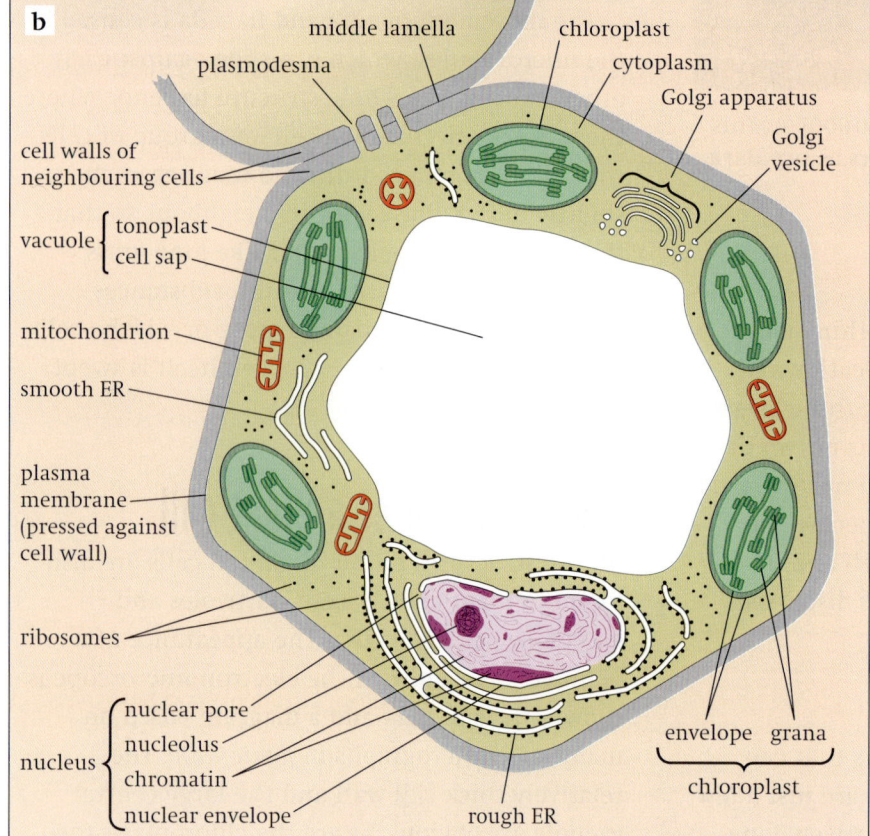

b

middle lamella

plasmodesma

chloroplast

cytoplasm

Golgi apparatus

Golgi vesicle

cell walls of neighbouring cells

vacuole { tonoplast / cell sap

mitochondrion

smooth ER

plasma membrane (pressed against cell wall)

ribosomes

nucleus { nuclear pore / nucleolus / chromatin / nuclear envelope

rough ER

envelope   grana

chloroplast

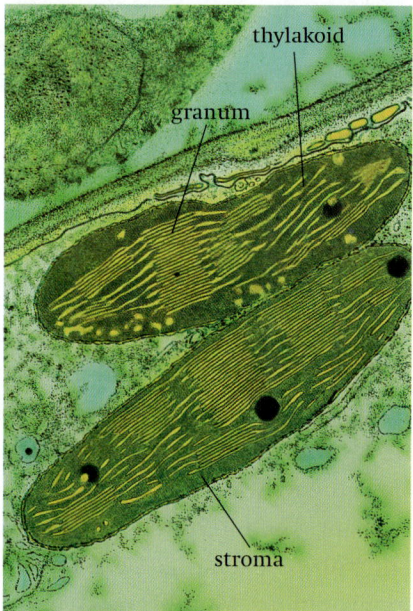

thylakoid

granum

stroma

● **Figure 1.23** Chloroplasts (× 20 000). Parallel flattened sacs (thylakoids) run through the stroma and are stacked in places to form grana. Black circles among the thylakoids are lipid droplets.

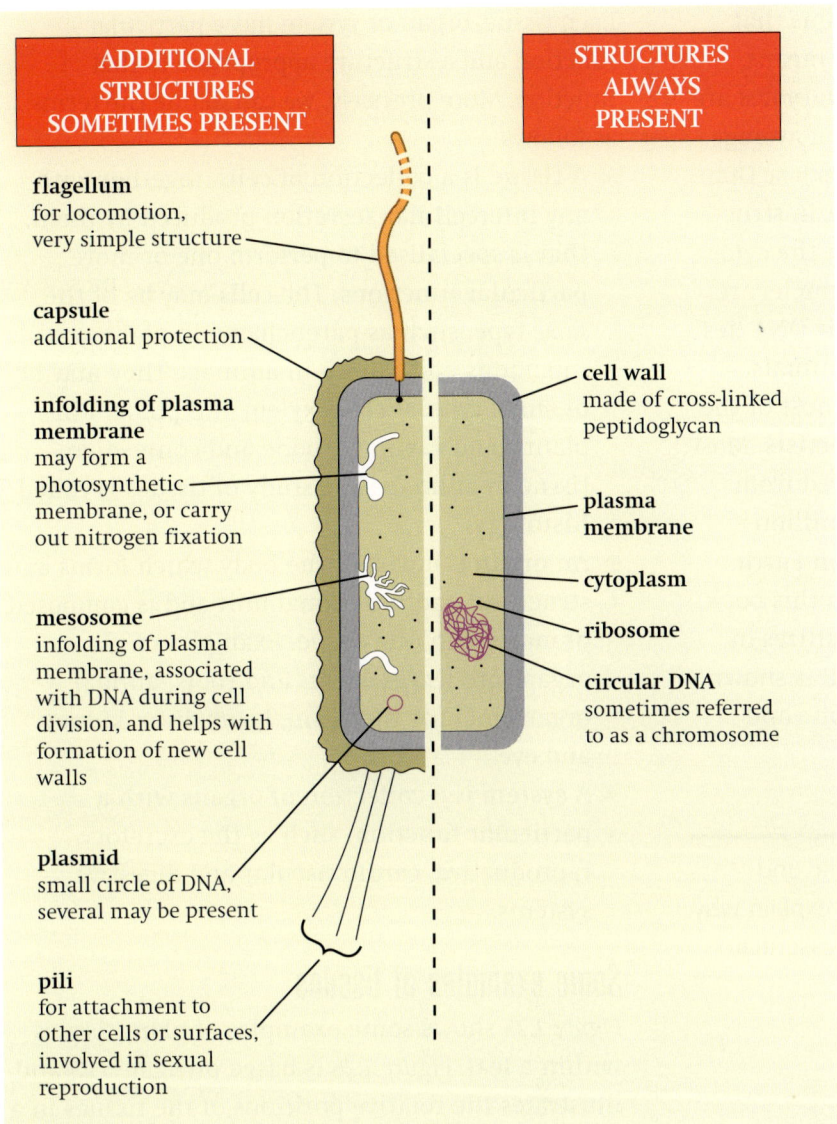

**ADDITIONAL STRUCTURES SOMETIMES PRESENT**

**STRUCTURES ALWAYS PRESENT**

**flagellum**
for locomotion, very simple structure

**capsule**
additional protection

**infolding of plasma membrane**
may form a photosynthetic membrane, or carry out nitrogen fixation

**mesosome**
infolding of plasma membrane, associated with DNA during cell division, and helps with formation of new cell walls

**plasmid**
small circle of DNA, several may be present

**pili**
for attachment to other cells or surfaces, involved in sexual reproduction

**cell wall**
made of cross-linked peptidoglycan

**plasma membrane**

**cytoplasm**

**ribosome**

**circular DNA**
sometimes referred to as a chromosome

● **Figure 1.24** Diagram of a generalised bacterium showing the typical features of a prokaryotic cell.

of which are shown in detail in *figure 1.23*). These structures and their functions have been described on page 5.

*SAQ 1.5*
Compare *figure 1.22b* with *figure 1.4*. Name the structures which can be seen with the electron microscope but not with the light microscope.

# Two fundamentally different types of cell

At one time it was common practice to try to classify *all* living organisms as either animals or plants. With advances in our knowledge of living things, it has become obvious that the living world is not that simple. Fungi and bacteria, for example, are very different from animals and plants, and from each other. Eventually it was realised that there are two fundamentally different types of cell. The most obvious difference between these types is that one *possesses a*

| Prokaryotes | Eukaryotes |
|---|---|
| Average diameter of cell 0.5–5 µm | Cells commonly up to 40 µm diameter and commonly 1000–10 000 times the volume of prokaryotic cells |
| DNA is circular and lies free in the cytoplasm | DNA is not circular and is contained in a nucleus. The nucleus is surrounded by an envelope of two membranes |
| DNA is naked | DNA is associated with protein, forming structures called chromosomes |
| Slightly smaller ribosomes (about 18 nm diameter) | Slightly larger ribosomes (about 22 nm diameter) |
| No ER present | ER present, to which ribosomes may be attached |
| Very few cell organelles; none are surrounded by an envelope of two membranes | Many types of cell organelle present (extensive compartmentalisation and division of labour). Some organelles are bounded by a single membrane, e.g. lysosomes, Golgi apparatus, vacuoles; some are bounded by two membranes (an envelope), e.g. nucleus, mitochondrion; some have no membrane, e.g. ribosomes |
| Cell wall present | Cell wall sometimes present, e.g. in plants |

● **Table 1.2** A comparison of prokaryotic and eukaryotic cells.

*nucleus* and the other does not. Organisms that lack nuclei are called **prokaryotes** (*pro* means before; *karyon* means nucleus). All prokaryotes are now referred to as **bacteria**. They are, on average, about 1000 to 10 000 times smaller in *volume* than cells with nuclei and are much simpler in structure, for example their DNA lies free in the cytoplasm. Organisms whose cells possess nuclei are called **eukaryotes** (*eu* means true). Their DNA lies inside a nucleus. Eukaryotes include **animals**, **plants**, **fungi** and a group containing most of the unicellular eukaryotes known as **protoctists**. Most biologists believe that eukaryotes evolved from prokaryotes, one-and-a-half thousand million years after prokaryotes first appeared on Earth. We mainly study animals and plants in this book, but **all** eukaryotic cells have certain features in common. A generalised prokaryotic cell is shown in *figure 1.24*. A comparison of prokaryotic and eukaryotic cells is given in *table 1.2*.

## SAQ 1.6

List the structural features that prokaryotic and eukaryotic cells have in common. Briefly explain why each of the structures you have listed is essential.

# Tissues and organs

So far we have studied life at the cell level. Some organisms, such as bacteria, consist of one cell only. Many organisms are multicellular, consisting of collections of cells from several hundred to billions in total. One great advantage that multicellular organisms gain over unicellular organisms is greater independence from the environment, but a full discussion of this is outside the scope of this book. In these communities of cells, it is usual for the functions of the organism to be divided among groups of cells which become specialised, both structurally and functionally, for particular roles. We have already seen this distribution of function *within* cells, particularly eukaryotic cells, and have referred to it as 'division of labour'. Usually, specialised cells show division of labour by being grouped into **tissues**; the tissues may be further grouped into **organs** and the organs into **systems**.

Each tissue, organ or system has a particular function and a structure appropriate to that function. More precisely, we can define the terms as follows.

- A **tissue** is a collection of cells, together with any intercellular secretion produced by them, that is specialised to perform one or more particular functions. The cells may be of the *same* type, such as parenchyma in plants and squamous epithelium in animals. They may be of *mixed* type, such as xylem and phloem in plants, and cartilage, bone and connective tissue in animals. The study of tissues is called **histology**.
- An **organ** is a part of the body which forms a structural and functional unit and is composed of more than one tissue. Examples of plant organs are leaves, stems and roots; animal organs include the brain, heart, liver, kidney and eye.
- A **system** is a collection of organs with a particular function, such as the excretory, reproductive, cardiovascular and digestive systems.

## Some examples of tissues

*Figure 1.25* shows some examples of plant tissues within a leaf. *Figure 1.26* is based on *figure 1.25* and illustrates the relative positions of the tissues in a leaf. This is called a **plan diagram**. As its purpose is to show where the different *tissues* are, no individual cells are drawn.

Several kinds of plant tissue are shown in *figure 1.25*. Whereas the palisade mesophyll is a tissue made up of many similar cells, all with the same function, the xylem tissue and phloem tissue are each made of several different types of cells. You can find out about the structure of xylem tissue and phloem tissue in chapter 10.

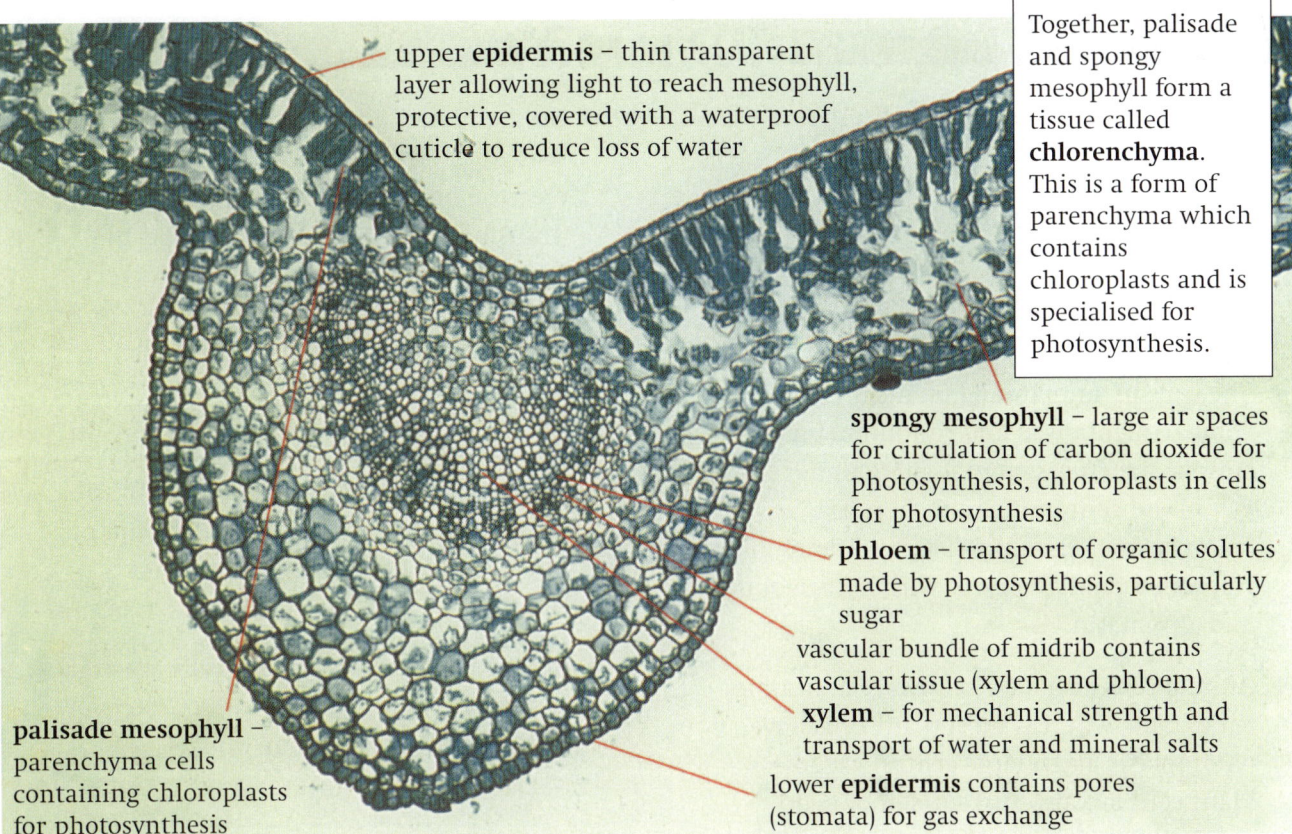

upper **epidermis** – thin transparent layer allowing light to reach mesophyll, protective, covered with a waterproof cuticle to reduce loss of water

Together, palisade and spongy mesophyll form a tissue called **chlorenchyma**. This is a form of parenchyma which contains chloroplasts and is specialised for photosynthesis.

**spongy mesophyll** – large air spaces for circulation of carbon dioxide for photosynthesis, chloroplasts in cells for photosynthesis

**phloem** – transport of organic solutes made by photosynthesis, particularly sugar

vascular bundle of midrib contains vascular tissue (xylem and phloem)

**xylem** – for mechanical strength and transport of water and mineral salts

lower **epidermis** contains pores (stomata) for gas exchange

**palisade mesophyll** – parenchyma cells containing chloroplasts for photosynthesis

● **Figure 1.25** Transverse section through the midrib of a dicotyledonous leaf, *Ligustrum* (privet) (× 50). Tissues are indicated in bold type.

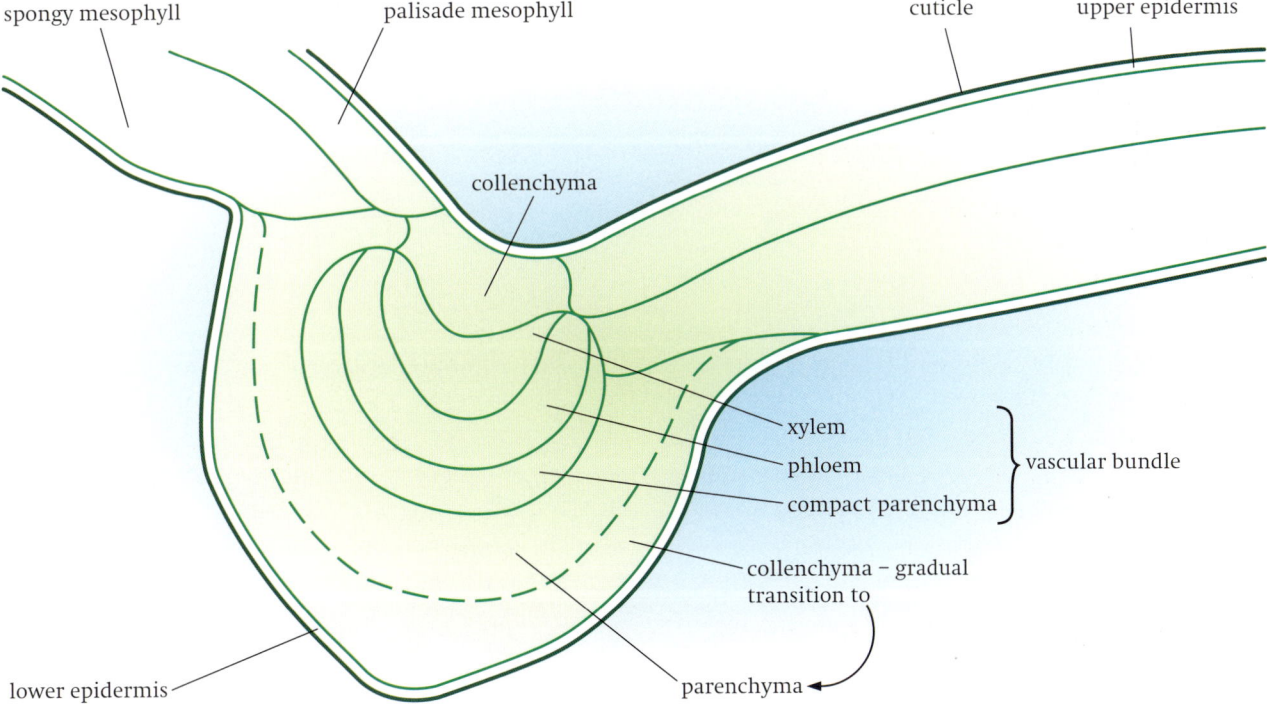

spongy mesophyll

palisade mesophyll

cuticle

upper epidermis

collenchyma

xylem

phloem

compact parenchyma

vascular bundle

collenchyma – gradual transition to

lower epidermis

parenchyma

● **Figure 1.26** A plan diagram of the transverse section through a privet leaf shown in *figure 1.25*. Parenchyma is a tissue made up of unspecialised cells. Collenchyma is made up of cells in which the walls are thickened with extra cellulose, especially at the corners, providing extra strength for support.

# SUMMARY

- All organisms are composed of units called cells.

- All cells are surrounded by a partially permeable membrane that controls exchange between the cell and its environment.

- The cells of animals and plants are compartmentalised and contain many similar structures: plasma membrane; cytoplasm containing mitochondria, endoplasmic reticulum (ER), lysosomes and ribosomes; and a nucleus with a nucleolus and chromatin.

- Animal cells also have centrioles and some-times cilia, whereas most plant cells have chloroplasts and a large central vacuole. Plant cells are also surrounded by rigid cell walls.

- Some of these structures are not visible with the light microscope because of the limit of resolution of light waves.

- Greater detail and smaller structures are seen with electron microscopes which use electron beams transmitted through (transmission electron microscope) or bounced off (scanning electron microscope) the specimen. However, only dead material can be viewed in electron microscopes.

- Prokaryote cells differ from eukaryote cells in being smaller, having free DNA in the cytoplasm, no endoplasmic reticulum or nucleus, few organelles and smaller ribosomes.

- In multicellular organisms, cells are organised into groups called tissues. Groups of different tissues make up organs.

## CHAPTER 2

# Biological molecules

## By the end of this chapter you should be able to:

1 understand the importance in biology of carbohydrates, lipids and proteins;

2 understand that although some important biological molecules are very large, they are made by the relatively simple process of joining together many small repeating subunits;

3 describe the basic structure of the main types of carbohydrates, namely monosaccharides, disaccharides and polysaccharides;

4 describe the structure of the ring forms of alpha- and beta-glucose;

5 describe the formation and breakage of a glycosidic bond, and its significance;

6 describe the structure of the polysaccharides starch, glycogen and cellulose and show how these structures are related to their functions in living organisms;

7 describe the basic structure and properties of triglycerides and phospholipids and relate these structures to their functions in living organisms;

8 distinguish between saturated and unsaturated fatty acids and lipids;

9 describe the structure of amino acids and the way in which peptide bonds are formed and broken;

10 describe the primary structure of polypeptides and proteins and how it affects their secondary and tertiary structure;

11 explain that the quaternary structure of a protein is formed by the combination of two or more polypeptide chains;

12 describe the importance of hydrogen bonds, disulphide bonds, ionic bonds and hydrophobic interactions in maintaining the three-dimensional structure of a protein;

13 discuss the ways in which the structures of haemoglobin and collagen are related to their functions;

14 describe the crucial role that water plays in maintaining life on Earth, both as a constituent of living organisms and as an environment;

15 outline the roles of inorganic ions in living organisms;

16 know how to test for reducing and non-reducing sugars, starch, lipids and proteins.

The study of the structure and functioning of biological molecules now forms an important branch of biology known as **molecular biology**. This is a relatively young science, but the importance of the subject is clear from the relatively large number of Nobel prizes that have been awarded in this field. It has attracted some of the best scientists, even from other disciplines like physics and mathematics.

Molecular biology is closely linked with biochemistry, which looks at the chemical reactions of biological molecules. The sum total of all the biochemical reactions in the body is known as **metabolism**. Metabolism is complex, but it has an underlying simplicity. For example, there are only 20 common amino acids used to make naturally occurring proteins, whereas theoretically there could be millions. Why is there this economy? One possibility is that all the manufacture and interactions of biological molecules must be controlled and regulated and, the more there are, the more complex the control becomes. (Control and regulation by enzymes will be examined in chapter 3.)

Another striking principle of molecular biology is how closely the structures of molecules are related to their functions. This will become clear in this chapter and in chapter 3. Our understanding of how structure is related to function may lead to the creation of a vast range of designer molecules to carry out such varied functions as large-scale industrial reactions and precise targetting of cells in medical treatment.

## The building blocks of life

The four most common elements in living organisms are, in order of abundance, hydrogen, carbon, oxygen and nitrogen. They account for more than 99% of the atoms found in all living things. Carbon is particularly important because carbon atoms can join together to form long chains or ring structures. They can be thought of as the basic skeletons of organic molecules to which groups of other atoms are attached. Organic molecules always contain carbon.

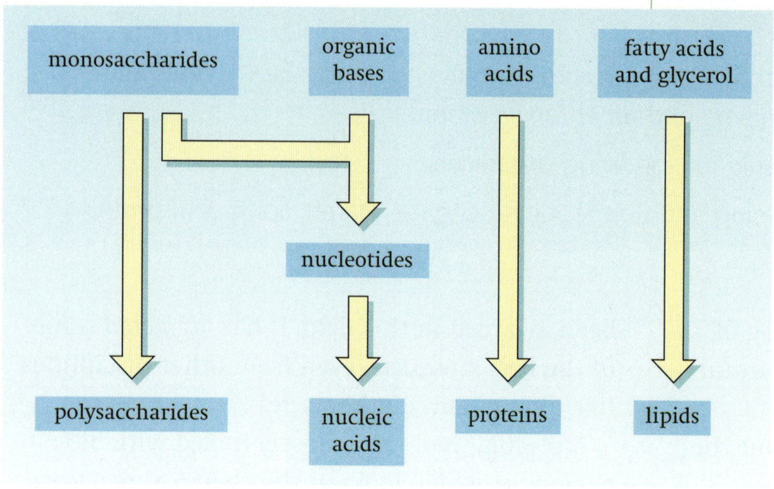

It is believed that, before *life* evolved, there was a period of *chemical* evolution in which thousands of carbon-based molecules evolved from the more simple molecules that existed on the young planet Earth. Such an effect can be artificially created reasonably easily today given similar raw ingredients, such as methane ($CH_4$), carbon dioxide ($CO_2$), hydrogen ($H_2$), water ($H_2O$), nitrogen ($N_2$), ammonia ($NH_3$) and hydrogen sulphide ($H_2S$), and an energy source, for example an electrical discharge. These simple but key biological molecules, which are relatively limited in variety, then act as the building blocks for larger molecules. The main ones are shown in *figure 2.1*.

## Polymers and macromolecules

The term **macromolecule** means 'giant molecule'. There are three types of macromolecule in living organisms, namely polysaccharides, proteins (polypeptides) and nucleic acids (polynucleotides). The prefix *poly* means 'many' and these molecules are **polymers**, that is macromolecules made up of many repeating subunits that are similar or identical to each other and are joined end to end like beads on a string. Making such molecules is relatively easy because the same reaction is repeated many times. This is **polymerisation**.

The subunits from which polysaccharides, proteins and nucleic acids are made are monosaccharides, amino acids and nucleotides respectively as shown in *figure 2.1*. This also shows two types of molecule which, although not polymers, are made up of simpler biochemicals. These are lipids and nucleotides. Natural examples of polymers are cellulose and rubber. There are many examples of industrially produced polymers, such as polyester, polythene, PVC (polyvinyl chloride) and nylon. All these are made up of carbon-based subunits and contain thousands of carbon atoms joined end to end.

We shall now take a closer look at some of the small biological molecules and the larger molecules made from them. (Organic bases and nucleic acids are dealt with in chapter 5).

● **Figure 2.1** The building blocks of life.

# Carbohydrates

All carbohydrates contain the elements carbon, hydrogen and oxygen. The second half of the name comes from the fact that hydrogen and oxygen atoms are present in the ratio of 2:1, as they are in water (*hydrate* refers to water). The **general formula** for a carbohydrate can therefore be written as $C_x(H_2O)_y$.

Carbohydrates are divided into three main groups, namely monosaccharides, disaccharides and polysaccharides.

## Monosaccharides

Monosaccharides are **sugars**. They dissolve easily in water to form sweet solutions (*saccharide* refers to sweet or sugar). Monosaccharides have the general formula $(CH_2O)_n$ and consist of a **single** sugar molecule (*mono* means 'one'). The main types of monosaccharides, if they are classified according to the number of carbon atoms in each molecule, are **trioses** (3C), **pentoses** (5C) and **hexoses** (6C). The names of all sugars end with **-ose**.

### SAQ 2.1

The formula for a hexose is $C_6H_{12}O_6$ or $(CH_2O)_6$. What would be the formula of **a** a triose and **b** a pentose?

### Molecular and structural formulae

The formula for a hexose has been written as $C_6H_{12}O_6$. This is known as the **molecular formula**. It is also useful to show the arrangements of the atoms which can be done by a diagram known as the **structural formula**. *Figure 2.2* shows the structural formula of glucose, a hexose which is the most common monosaccharide.

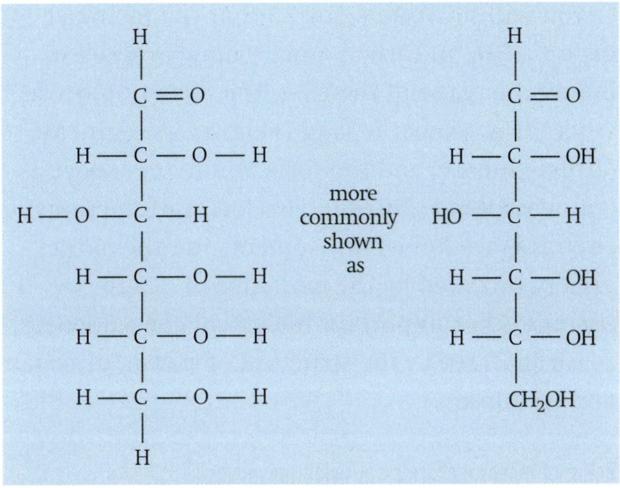

### Ring structures

One important aspect of the structure of pentoses and hexoses is that the chain of

● **Figure 2.2** Structural formula of glucose. –OH is known as a hydroxyl group. There are five in glucose.

carbon atoms is long enough to close up on itself and form a more stable ring structure. This can be illustrated using glucose as an example. When glucose forms a ring, carbon atom number 1 joins to the oxygen on carbon atom number 5 (*figure 2.3*). The ring therefore contains oxygen, and carbon atom number 6 is not part of the ring.

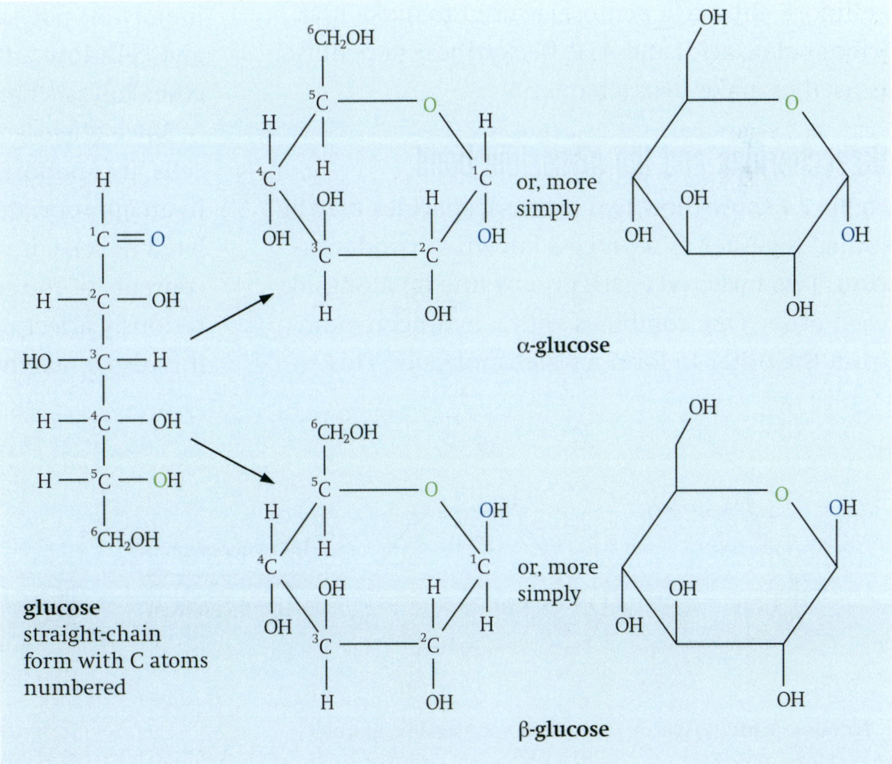

● **Figure 2.3** Structural formulae for the straight-chain and ring forms of glucose. Chemists often leave out the C and H atoms from the structural formula for simplicity.

You will see from *figure 2.3* that the hydroxyl group, –OH, on carbon atom 1 may be **above** or **below** the plane of the ring. The form of glucose where it is below the ring is known as α-**glucose** (**alpha-glucose**) and the form where it is above as β-**glucose** (**beta-glucose**). Two forms of the same chemical are known as **isomers**, and the extra variety provided by the existence of α- and β-isomers has important biological consequences, as we shall see in the structure of starch, glycogen and cellulose.

### Roles of monosaccharides in living organisms

Monosaccharides have two major functions. Firstly, they are commonly used as a source of energy in respiration. This is due to the large number of carbon–hydrogen bonds. These bonds can be broken to release a lot of energy which is transferred to help make ATP (adenosine triphosphate) from ADP (adenosine diphosphate) and phosphate. The most important monosaccharide in energy metabolism is glucose.

Secondly, they are important as building blocks for larger molecules. For example, glucose is used to make the polysaccharides starch, glycogen and cellulose. Ribose (a pentose) is used to make RNA (ribonucleic acid) and ATP. Deoxyribose (a pentose) is used to make DNA (chapter 5).

### Disaccharides and the glycosidic bond

*Figure 2.4* shows how two monosaccharides may be joined together by a process known as **condensation**. Two hydroxyl (–OH) groups line up alongside each other. One combines with a hydrogen atom from the other to form a water molecule. This allows an oxygen 'bridge' to form between the two molecules, holding them together and forming a **disaccharide** (*di* means 'two'). The bridge is called a **glycosidic bond**. In theory any two –OH groups can line up and, since monosaccharides have many –OH groups, there are a large number of possible disaccharides. However, only a few of these are common in nature. Disaccharides, like monosaccharides, are sugars.

The reverse of this kind of condensation is the *addition* of water which is known as **hydrolysis** (*figure 2.4* again). This takes place during the digestion of disaccharides and polysaccharides when they are broken back down to monosaccharides. Like most chemical reactions in cells, hydrolysis and condensation reactions are controlled by enzymes.

## Polysaccharides

Polysaccharides are polymers whose subunits are monosaccharides. They are made by joining many monosaccharide molecules by condensation. Each successive monosaccharide is added by means of a glycosidic bond, as in disaccharides. The final molecule may be several thousand monosaccharide units long, forming a macromolecule. The most important polysaccharides are starch, glycogen and cellulose, all of which are polymers of glucose. Polysaccharides are *not* sugars.

Since glucose is the main source of energy for cells, it is important for living organisms to store it in an appropriate form. If glucose itself accumulated in cells, it would dissolve and make the contents of the cell too concentrated, which would seriously affect its osmotic properties (see page 54). It is also a reactive molecule and would interfere

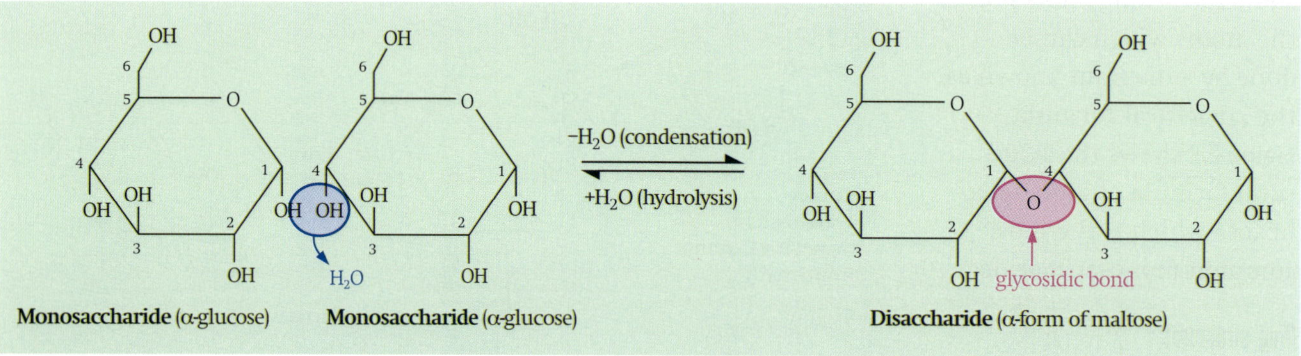

**Monosaccharide (α-glucose)**    **Monosaccharide (α-glucose)**    –H₂O (condensation) / +H₂O (hydrolysis)    **Disaccharide (α-form of maltose)**    glycosidic bond

● **Figure 2.4** Formation of a disaccharide from two monosaccharides by condensation. In this example, the glycosidic bond is formed between carbon atoms 1 and 4 of neighbouring monosaccharides. The process may be repeated many times to form a polysaccharide or reversed by hydrolysis.

## Box 2A Testing for the presence of sugars

If you have a solution that you suspect contains sugar, you can use **Benedict's reagent** to test it. Benedict's reagent is copper(II) sulphate in an alkaline solution and has a distinctive blue colour. If it is added to a **reducing agent** its $Cu^{2+}$ ions will be **reduced** to $Cu^+$ resulting in a change of colour to the red of insoluble copper(I). All monosaccharides and some disaccharides have this effect on Benedict's reagent. This is because they have a —C=O group somewhere in their molecules which can contribute an electron to the copper. They are therefore **reducing sugars**. In the process they themselves become **oxidised**.

reducing sugar + $Cu^{2+}$ → oxidised sugar + $Cu^+$

Add Benedict's reagent to the solution you are testing and heat it in a water bath. If a reducing sugar is present, the solution will gradually turn through green, yellow and orange to brick red as the insoluble copper(I) sulphate forms a precipitate. As long as you use *excess* Benedict's reagent (more than enough to react with all of the sugar present) the intensity of the red colour is related to the concentration of the reducing sugar which you can then estimate. Alternatively you can use a colorimeter to measure subtle differences in colour precisely.

Some disaccharides are *not* reducing sugars, so you would get a negative result from the test as described so far. You must therefore go on to a second stage of the test to be certain whether such a **non-reducing** sugar is present. You need to break non-reducing disaccharides into their constituent monosaccharides, all of which are reducing sugars and *will* react with Benedict's reagent.

Heat the sugar solution with an acid to hydrolyse any glycosidic bonds present. This will release free monosaccharides. Benedict's reagent needs alkaline conditions to work so you need to neutralise the test solution now by adding an alkali such as sodium hydroxide. Add Benedict's reagent and heat as before and look for the colour change. If the solution goes red now but didn't in the first stage of the test, there is non-reducing sugar present. If there is *still* no colour change then there is no sugar of any kind present.

### SAQ 2.2

a Why do you need to use *excess* Benedict's reagent if you want to get an idea of the concentration of a sugar solution?

b Outline how you could use the Benedict's test to estimate the concentration of a solution of a reducing sugar.

### SAQ 2.3

You have a solution which you know contains sugar but you do not know whether it is reducing sugar, non-reducing sugar or a mixture of both. How can you find out?

---

with normal cell chemistry. These problems are avoided by converting it, by condensation reactions, to a storage polysaccharide, which is a convenient, compact, inert and insoluble molecule. This is in the form of starch in plants and glycogen in animals. Glucose can be made available again quickly by an enzyme-controlled reaction.

### SAQ 2.4

What type of chemical reaction would be involved in the formation of glucose from starch or glycogen?

### Starch and glycogen

Starch is a mixture of two substances, **amylose** and **amylopectin**. Amylose is made by many condensations between α-glucose molecules, as shown in *figure 2.4*. In this way a long, unbranching chain of several thousand 1,4 linked glucose molecules is built up. ('1,4 linked' means they are linked between carbon atoms 1 and 4 of successive glucose units.) The chains are curved (*figure 2.5*) and coil up into helical structures like springs, making the final molecule more compact. Amylopectin is also made of many 1,4 linked α-glucose molecules, but the chains are shorter than in amylose, and branch out to the sides. The branches are formed by 1,6 linkages, as shown in *figure 2.6*.

Mixtures of amylose and amylopectin molecules build up into relatively large starch grains which are commonly found in chloroplasts and in storage organs such as potato tubers and the seeds of cereals and legumes (*figure 2.7*). Starch grains are easily seen with a light microscope, especially if stained; rubbing a freshly cut potato tuber on a glass slide and staining with iodine–potassium iodide solution (see *box 2B*) is a quick method of preparing a specimen for viewing.

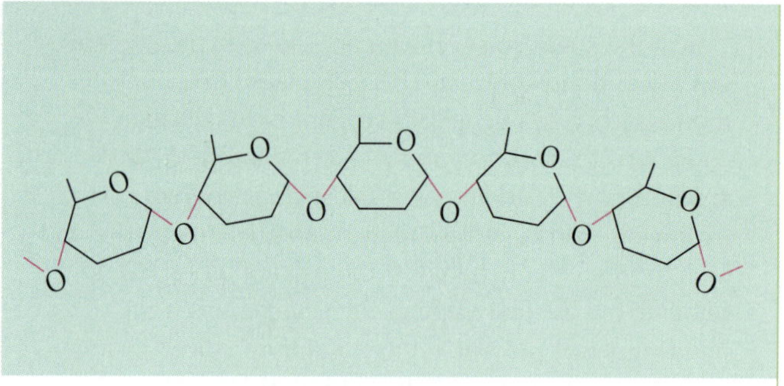

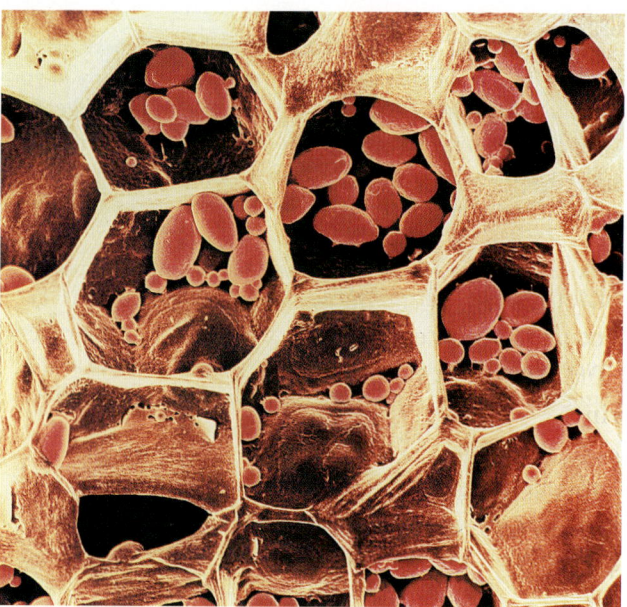

• **Figure 2.5** Arrangement of α-glucose units in amylose. The 1,4 linkages cause the chain to turn and coil. The glycosidic bonds are shown in red and the hydroxyl groups are omitted.

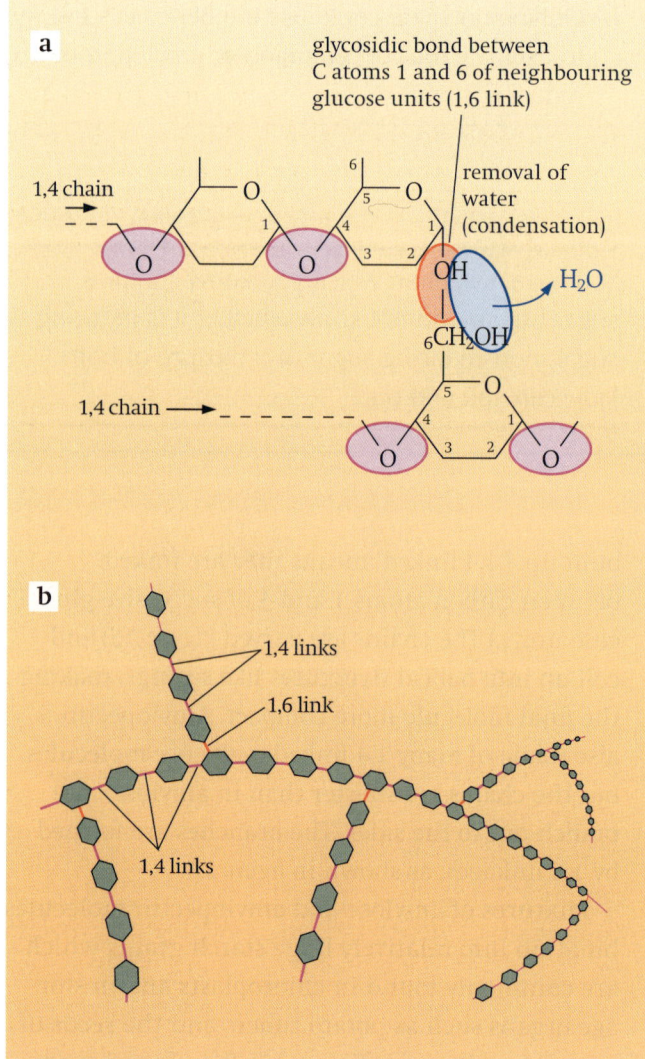

**a**

glycosidic bond between C atoms 1 and 6 of neighbouring glucose units (1,6 link)

1,4 chain

removal of water (condensation)

$H_2O$

$CH_2OH$

1,4 chain

**b**

1,4 links

1,6 link

1,4 links

• **Figure 2.6** Branching structure of amylopectin and glycogen: **a** formation of a 1,6 link, a branchpoint; **b** overall structure of an amylopectin or glycogen molecule. Amylopectin and glycogen only differ in the amount of branching of their glucose chains.

• **Figure 2.7** False-colour SEM of a slice through a raw potato showing starch grains or starch-containing organelles (coloured red) within their cellular compartments (× 200).

**Box 2B  Testing for the presence of starch**

Starch molecules tend to curl up into long spirals. The hole that runs down the middle of this spiral is just the right size for iodine molecules to fit into. The starch-iodine complex that forms has a strong blue-black colour.

So, to test for starch, you use something called 'iodine solution'. In fact, iodine won't dissolve in water, so the 'iodine solution' is actually iodine in potassium iodide solution. This solution is orange-brown. A blue-black colour is quickly produced if it comes into contact with starch.

Starch is never found in animal cells. Instead, a substance with molecules very like those of amylopectin is used as the storage carbohydrate. This is called **glycogen**. Glycogen, like amylopectin, is made of chains of 1,4 linked α-glucose with 1,6 linkages forming branches (*figure 2.6b*). Glycogen molecules tend to be even more branched than amylopectin molecules. Glycogen molecules clump together to form granules, which are visible in liver cells and muscle cells where they form an energy reserve.

**SAQ 2.5**

List five ways in which the molecular structures of glycogen and amylopectin are similar.

## Cellulose

Cellulose is the most abundant organic molecule on the planet due to its presence in plant cell walls and its slow rate of breakdown in nature. It has a structural role, being a mechanically strong molecule, unlike starch and glycogen. However the only difference between cellulose and the latter is that cellulose is a polymer of β-glucose, not α-glucose. Remember that in the β isomer the –OH group on carbon atom 1 projects *above* the ring. In order to form a glycosidic bond with carbon atom 4, where the –OH group is *below* the ring, one glucose molecule must be upside down relative to the other, that is rotated 180°. Thus successive glucose units are linked at 180° to each other, as shown in *figure 2.8*.

This results in a strong molecule because the hydrogen atoms of –OH groups are weakly attracted to oxygen atoms in the same cellulose molecule (the oxygen of the glucose ring) and also to oxygen atoms of –OH groups in neighbouring molecules. These **hydrogen bonds** (see *box 2C*) are individually weak, but so many can form, due to the large number of –OH groups, that collectively they develop enormous strength. Between 60 and 70 cellulose molecules become tightly cross-linked to form bundles called **microfibrils**. Microfibrils are in turn held together in bundles called **fibres** by hydrogen bonding.

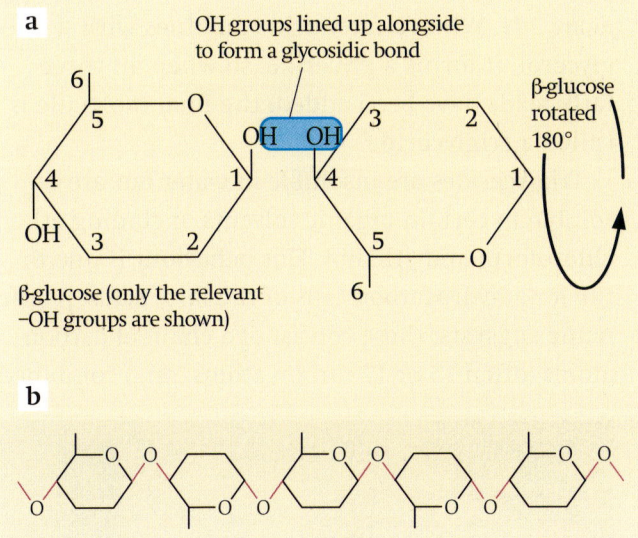

**a** OH groups lined up alongside to form a glycosidic bond

β-glucose (only the relevant –OH groups are shown)

β-glucose rotated 180°

**b**

● **Figure 2.8 a** Two β-glucose molecules lined up to form a 1,4 link; **b** arrangement of β-glucose units in cellulose: glycosidic bonds are shown in red and hydroxyl groups are omitted.

### Box 2C Dipoles and hydrogen bonds

The atoms in molecules are held together because they share electrons with each other. Each shared pair of electrons forms a **covalent bond**. For example, in a water molecule, two hydrogen atoms each share a pair of electrons with an oxygen atom, forming a molecule with the formula $H_2O$.

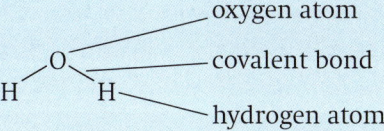

oxygen atom
covalent bond
hydrogen atom

However, the electrons are not shared absolutely equally. The oxygen atom gets slightly more than its fair share, and so has a small negative charge, written $\delta^-$. The hydrogen atoms get slightly less than their fair share, and so have a small positive charge, written $\delta^+$.

$$\delta^+H \quad O^{\delta^-} \quad H^{\delta^+}$$

This unequal distribution of charge is called a **dipole**. In water, the negatively charged oxygen of one molecule is attracted to the positively charged hydrogens of another, and this attraction is called a **hydrogen bond**. It is much weaker than a covalent bond, but still has a very significant effect. You will find out how hydrogen bonds affect the properties of water on pages 37–38.

hydrogen bond

Dipoles occur in many different molecules, particularly wherever there is an –OH, –C=O or >N–H group. Hydrogen bonds can form *between* these groups, as the negatively charged part of one group is attracted to the positively charged part of another. These bonds are very important in the structure and properties of carbohydrates and proteins.

$$-C=O^{\delta^-} \cdots H^{\delta^+}-N<$$

Molecules which have groups with dipoles are said to be **polar**. They are attracted to water molecules, because the water molecules also have dipoles. Such molecules are said to be **hydrophilic** (water-loving), and they tend to be soluble in water. Molecules which do not have dipoles are said to be **non-polar**. They are not attracted to water, and they are **hydrophobic** (water-hating). Such properties make possible the formation of plasma membranes, for example (chapter 4).

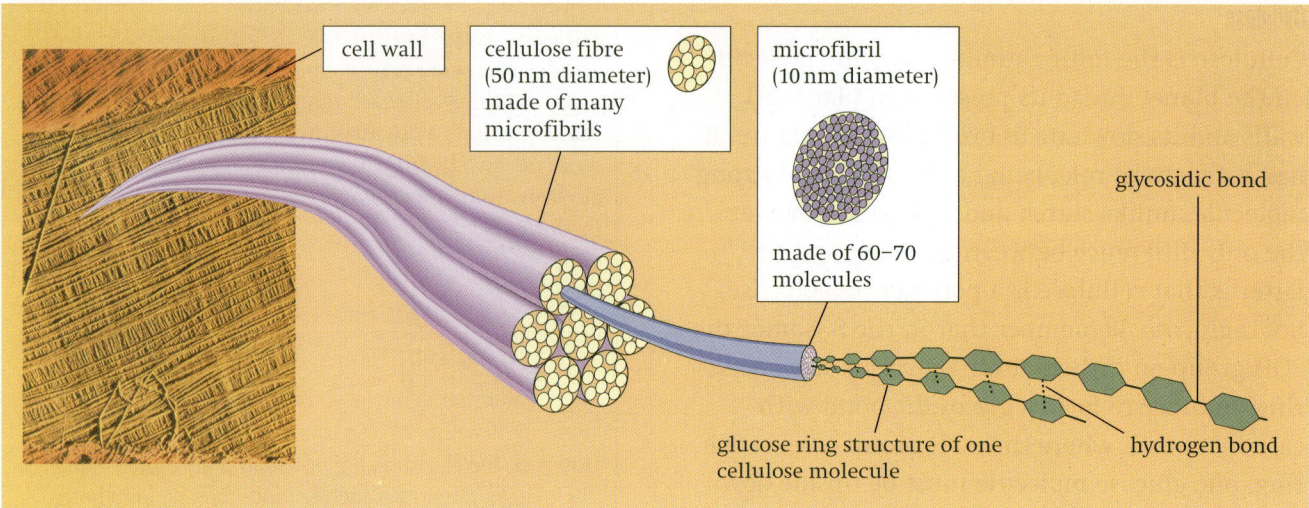

● **Figure 2.9** Structure of cellulose.

A cell wall typically has several layers of fibres, running in different directions to increase strength (*figure 2.9*). Cellulose comprises about 20–40% of the average cell wall; other molecules help to cross-link the cellulose fibres and some form a glue-like matrix around the fibres which further increases strength.

Cellulose fibres have a very high tensile strength, almost equal to that of steel. This means that if pulled at both ends they are very difficult to break, and makes it possible for a cell to withstand the large pressures that develop within it as a result of osmosis (page 56). Without the wall it would burst when in a dilute solution. These pressures help provide support for the plant by making tissues rigid, and are responsible for cell expansion during growth. The arrangement of fibres around the cell helps to determine the shape of the cell as it grows. Despite their strength, cellulose fibres are freely permeable, allowing water and solutes to reach the plasma membrane.

### SAQ 2.6
Make a table to show three ways in which the molecular structures of amylose and cellulose differ.

# Lipids
Lipids are a diverse group of chemicals. The most common type are the **triglycerides**, which are usually known as fats and oils. The main difference between them is that, at room temperature, fats are solid whereas oils are liquid.

## Triglycerides
Triglycerides are made by the combination of three fatty acid molecules with one glycerol molecule. Fatty acids are organic molecules which all have a —COOH group attached to a hydrocarbon tail. Glycerol is a type of alcohol. The triglyceride molecule can be represented diagrammatically as shown in *figure 2.10*. The tails vary in length, depending on the fatty acids used.

Each of the three fatty acid molecules joins to glycerol by a condensation reaction as shown in *figure 2.11*. When a fatty acid combines with glycerol, it forms a glyceride, so when all three fatty acids have been added, the final molecule is called a **triglyceride**.

Triglycerides are insoluble in water but are soluble in certain organic solvents, including ether, chloroform and ethanol. This behaviour is due to the long hydrocarbon tails of the fatty acids. As the name suggests, these consist of a chain of carbon atoms (often 15 or 17 carbon atoms long) combined

● **Figure 2.10** Diagrammatic representation of a triglyceride molecule.

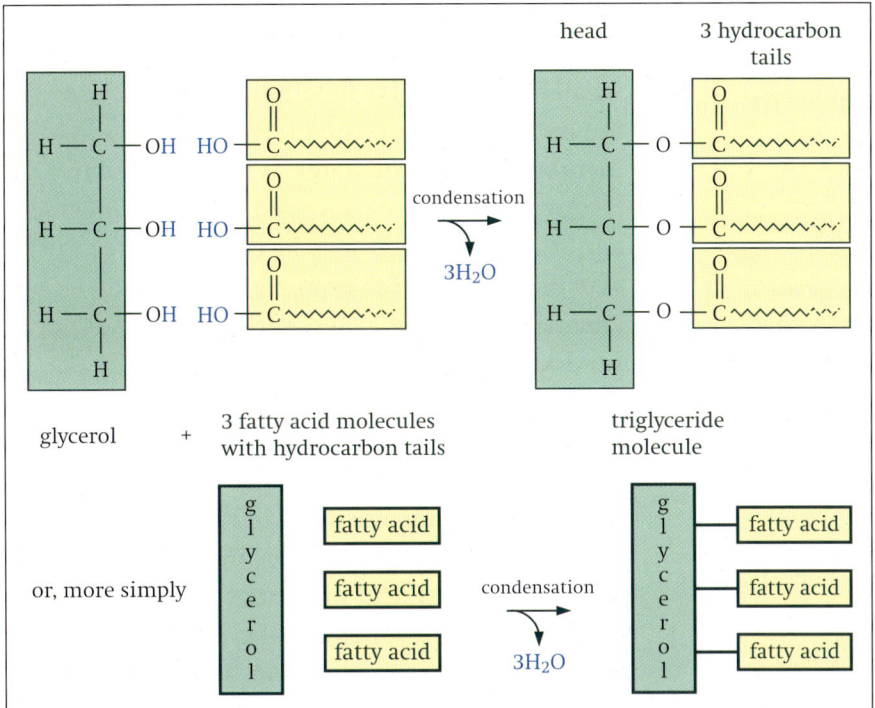

with hydrogen (*figure 2.12*). Unlike water molecules which are polar (see *box 2C*) the fatty acid tails have no uneven distribution of electrical charge. Consequently, they will not mix freely with water molecules. Triglycerides are therefore **non-polar** and **hydrophobic**.

### Saturated and unsaturated fatty acids and lipids

Some fatty acids have double bonds between neighbouring carbon atoms, like this: –C–C=C–C– (*figure 2.12* again). Such fatty acids are described as **unsaturated** (as they do not contain the maximum possible amount of hydrogen) and form unsaturated lipids. Double bonds make fatty acids and lipids melt more easily, for example most oils are unsaturated. If there is more than one double bond, the fatty acid or lipid is described as **polyunsaturated**; if there is only one it is **monounsaturated**. Animal lipids

● **Figure 2.11** Formation of a triglyceride from glycerol and three fatty acid molecules.

saturated fatty acid                    unsaturated fatty acid

● **Figure 2.12** Structure of a saturated and an unsaturated fatty acid. Photographs of models are shown to the right of each structure. In the models hydrogen is white, carbon is black and oxygen is red.

are often saturated and occur as fats, whereas plant lipids are often unsaturated and occur as oils, such as olive oil and sunflower oil.

### Roles of triglycerides

Lipids make excellent **energy reserves** because they are even richer in carbon–hydrogen bonds than carbohydrates. A given mass of lipid will therefore yield more energy on oxidation than the same mass of carbohydrate (it has a higher calorific value), an important advantage for a storage product.

Fat is stored in a number of places in the human body, particularly just below the dermis of the skin and around the kidneys. Below the skin it

● **Figure 2.13**  The desert kangaroo rat uses metabolism of food to provide the water it needs.

also acts as an **insulator** against loss of heat. Blubber, a lipid found in sea mammals like whales, has a similar function, as well as providing buoyancy. An unusual role for lipids is as a **metabolic source of water**. When oxidised in respiration they are converted to carbon dioxide and water. The water may be of importance in very dry habitats. For example, the desert kangaroo rat (*figure 2.13*) never drinks water and survives on metabolic water from its fat intake.

### Phospholipids

Phospholipids are a special type of lipid. Each molecule has the unusual property of having one end which is soluble in water. This is because one of the three fatty acid molecules is replaced by a phosphate group which is polar (see *box 2C*) and can therefore dissolve in water. The phosphate group is **hydrophilic** and makes the head of a phospholipid molecule hydrophilic, though the two remaining tails are still hydrophobic (*figure 2.14*). The biological significance of this will become apparent when we study membrane structure (see chapter 4).

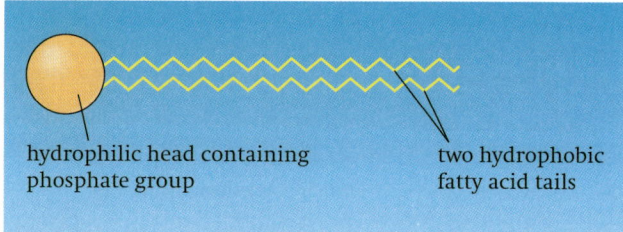

hydrophilic head containing phosphate group

two hydrophobic fatty acid tails

● **Figure 2.14**  Diagrammatic representation of a phospholipid molecule.

# Proteins

Proteins are an extremely important class of molecules in living organisms. More than 50% of the dry mass of most cells is protein. Proteins have many important functions. For example:

- they are essential components of cell membranes;
- the oxygen-carrying pigment haemoglobin is a protein;
- antibodies which attack and destroy invading microorganisms are proteins;
- all enzymes are proteins;

- hair and the surface layers of skin contain the protein keratin;
- collagen, another protein, adds strength to many tissues, such as bone and the walls of arteries.

Despite their tremendous range of functions, all proteins are made from the same basic components. These are **amino acids**.

## Amino acids

*Figure 2.15* shows the general structure of all amino acids and of glycine, the most simple amino acid. They all have a central carbon atom to which is bonded an **amine** group, $-NH_2$, and a **carboxylic acid** group, $-COOH$. It is these two groups which give amino acids their name. The third component bonded to the carbon atom is always a hydrogen atom.

The only way in which amino acids differ from each other is in the remaining, fourth, group of atoms bonded to the central carbon. This is called

amine group      carboxylic acid group

- **Figure 2.15 a** The general structure of an amino acid. **b** Structure of the simplest amino acid, glycine, in which the R group is H, hydrogen. R groups for the 20 naturally occurring amino acids are shown in appendix 1.

the **R group** of which there are many different kinds. There are 20 different amino acids which occur naturally in the proteins of living organisms. (Many others have been synthesised in laboratories.) You can see their molecular formulae in appendix 1. You do not need to remember all of the different R groups! However, it is these R groups that are responsible for the three-dimensional shapes of protein molecules (page 33) and hence their functions.

## The peptide bond

*Figure 2.16* shows how two amino acids can join together. One loses a hydroxyl (–OH) group from its carboxylic acid group, while the other loses a hydrogen atom from its amine group. This leaves a carbon atom of the first amino acid free to bond with the nitrogen atom of the second. The bond is called a **peptide bond**. The oxygen and two hydrogen atoms removed from the amino acids form a water molecule. We have seen this type of reaction, a condensation reaction, in the formation of glycosidic bonds (*figure 2.4*) and in the synthesis of triglycerides (*figure 2.11*).

The new molecule which has been formed, made up of two linked amino acids, is called a **dipeptide**. Any number of extra amino acids could be added to the chain, in a series of condensation reactions. A molecule made up of many amino acids linked together by peptide bonds is called a **polypeptide**. A polypeptide is another example of a polymer and a macromolecule, like polysaccharides. A complete **protein** molecule may contain just one polypeptide chain, or it may have two or more chains which interact with each other.

In living cells, **ribosomes** are the sites where amino acids are linked together to form polypeptides. The reaction is controlled by enzymes. You can read more about this on pages 71–74.

- **Figure 2.16** Amino acids link together by the loss of a molecule of water to form a peptide bond.

Polypeptides can be broken down to amino acids by breaking the peptide bonds. This is a hydrolysis reaction, involving the addition of water (*figure 2.16*), and naturally happens in the stomach and small intestine. Here, protein molecules in food are hydrolysed into amino acids prior to being absorbed into the blood.

## Primary structure

A polypeptide or protein molecule may contain several hundred amino acids linked into a long chain. The types of amino acids contained in the chain, and the sequence in which they are joined, is called the **primary structure** of the protein. *Figure 2.17* shows the primary structure of ribonuclease.

There is an enormous number of different *possible* primary structures. Even a change in one amino acid in a chain made up of thousands may completely alter the properties of the polypeptide or protein.

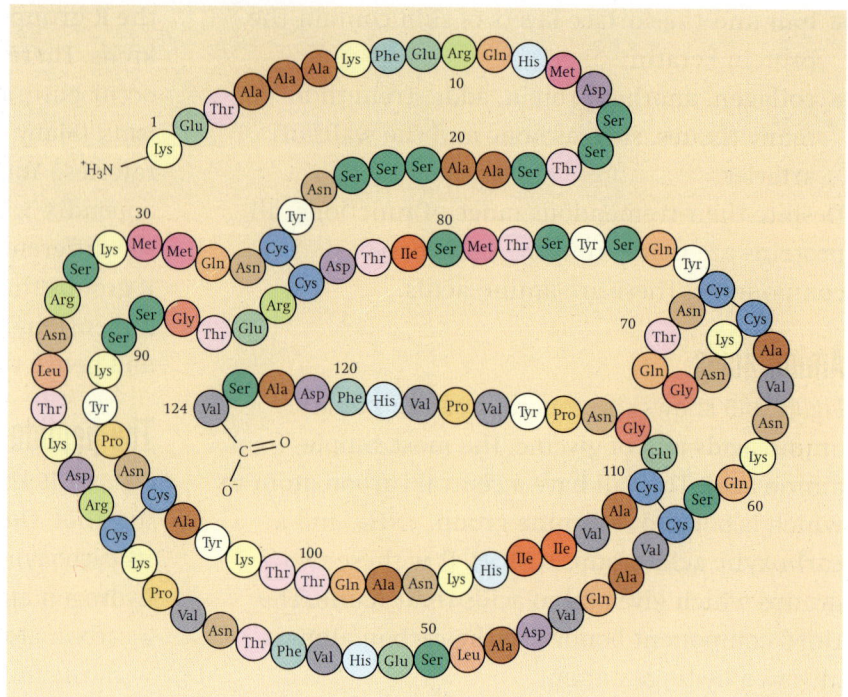

● **Figure 2.17** The primary structure of ribonuclease. Ribonuclease is an enzyme found in pancreatic juice which hydrolyses (digests) RNA (chapter 5). Notice that at one end of the amino acid chain there is an $-NH_3^+$ group, while at the other end there is a $-COO^-$ group. These are known as the amino and carboxyl ends, or the N and C terminals, respectively.

## Secondary structure

The amino acids in a polypeptide chain have an effect on each other even if they are not directly next to each other. A polypeptide chain often coils into an α-**helix** (*figure 2.18a*) due to the attraction between the oxygen of the –CO group of one amino acid and the hydrogen of the –NH group of the amino acid four places ahead of it. This is a result of the polar characteristics of the –CO and –NH groups (*figure 2.19a*) and is another example of hydrogen bonding, (see *box 2C*).

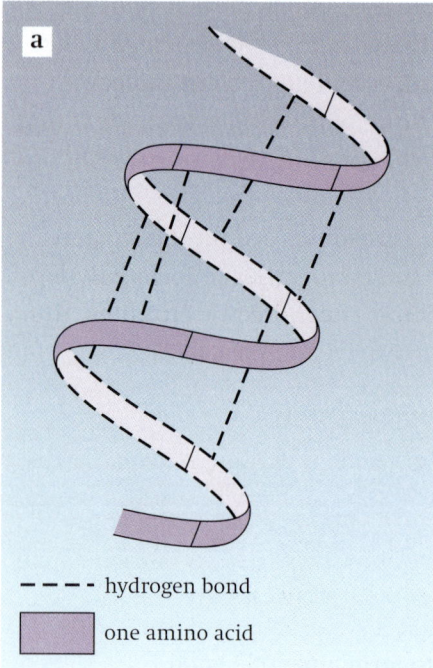

- - - - hydrogen bond

☐ one amino acid

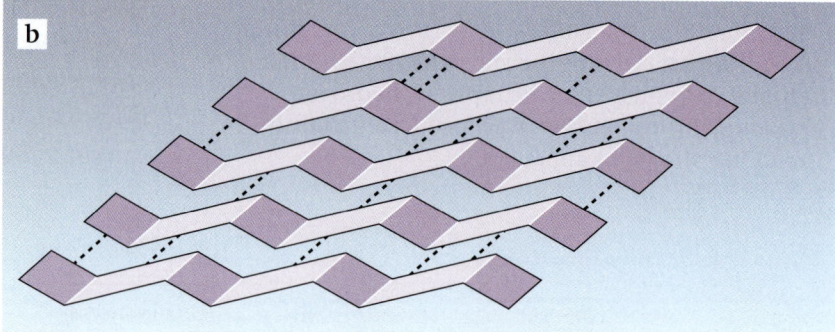

● **Figure 2.18 a** Polypeptide chains often coil into a tightly wound α-helix. **b** Another common arrangement is the β-pleated sheet. Both of these secondary structures are held in shape by hydrogen bonds between the amino acids.

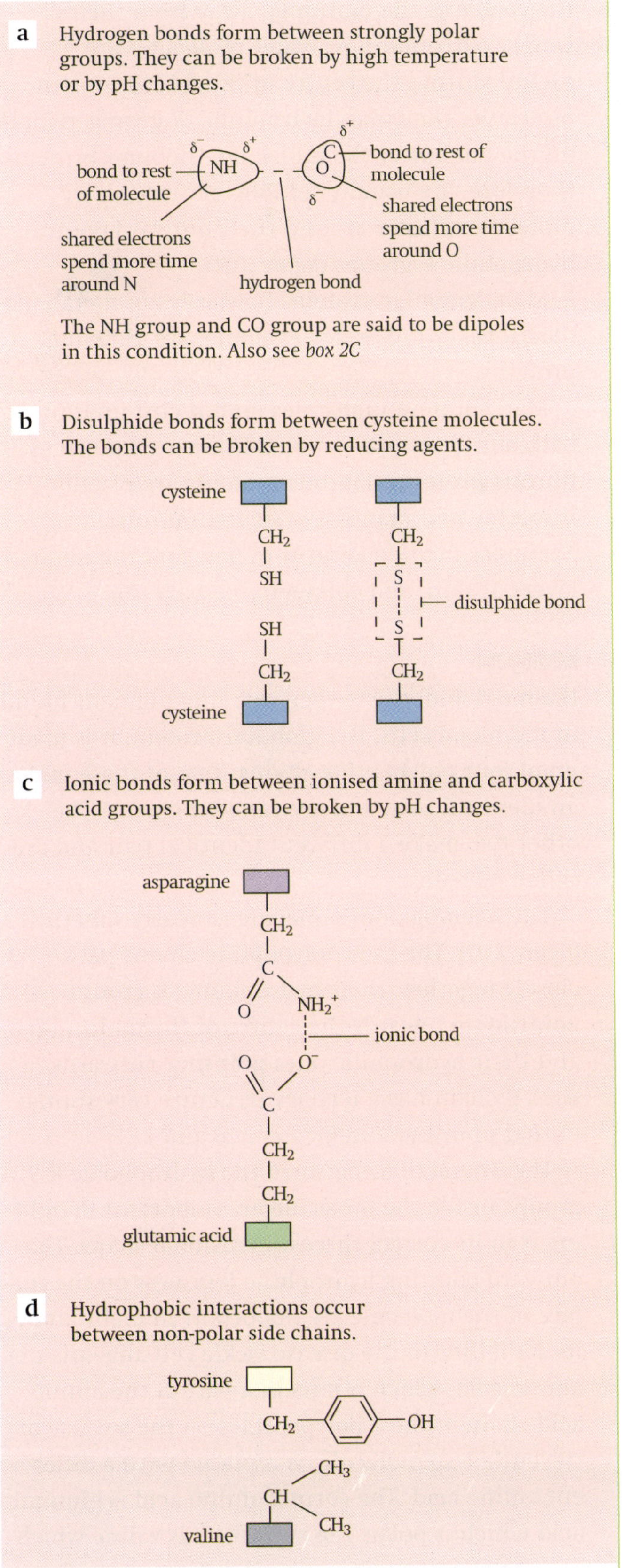

**a** Hydrogen bonds form between strongly polar groups. They can be broken by high temperature or by pH changes.

bond to rest of molecule — NH $\delta^-$ $\delta^+$ $\cdots$ $\delta^+$ C=O $\delta^-$ — bond to rest of molecule

shared electrons spend more time around N

hydrogen bond

shared electrons spend more time around O

The NH group and CO group are said to be dipoles in this condition. Also see *box 2C*.

**b** Disulphide bonds form between cysteine molecules. The bonds can be broken by reducing agents.

cysteine

$CH_2$
SH

SH
$CH_2$

cysteine

$CH_2$
S
S
$CH_2$

— disulphide bond

**c** Ionic bonds form between ionised amine and carboxylic acid groups. They can be broken by pH changes.

asparagine

$CH_2$
C
O    $NH_2^+$
$\cdots\cdots$ — ionic bond
O    $O^-$
C
$CH_2$
$CH_2$

glutamic acid

**d** Hydrophobic interactions occur between non-polar side chains.

tyrosine

$CH_2$ — ⬡ — OH

CH $\cdots$ $CH_3$ / $CH_3$

valine

● **Figure 2.19** The four types of bond which are important in protein secondary and tertiary structure: **a** hydrogen bonds, **b** disulphide bonds, **c** ionic bonds, **d** hydrophobic interactions.

Hydrogen bonds, although strong enough to hold the α-helix in shape, are easily broken by high temperatures and pH changes. As you will see, these effects on proteins have important consequences for living organisms.

Not all proteins coil into an α-helix. Sometimes a much looser, straighter shape is formed, called a β-**pleated sheet** (*figure 2.18b*). Other proteins show no regular arrangement at all. It all depends on which R groups are present and therefore what attractions occur between amino acids in the chain.

## Tertiary structure

In many proteins, the secondary structure itself is coiled or folded. *Figure 2.20* shows the complex way in which a molecule of the protein myoglobin folds.

At first sight, the myoglobin molecule looks like a disorganised tangle, but this is not so. The shape of the molecule is very precise, and held in this exact shape by bonds between amino acids in different parts of the chain. The way in which a protein coils up to form a precise three-dimensional shape is known as its **tertiary structure**.

*Figure 2.19* shows the four types of bonds which help to hold folded proteins in their precise shape.

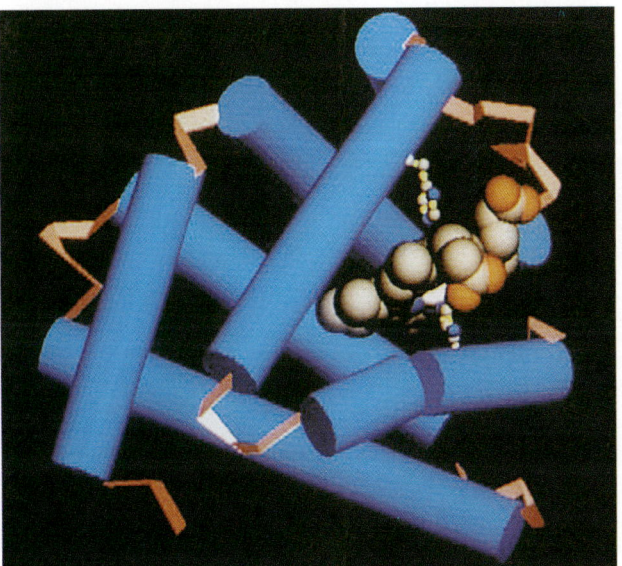

● **Figure 2.20** A computer graphic showing the secondary and tertiary structure of a myoglobin molecule. Myoglobin is the substance which makes meat look red. It is found in muscle, where it acts as an oxygen-storing molecule. The blue sections are α-helixes and are linked by sections of polypeptide chain which are more stretched out – these are shown in red. At the top right is an iron-containing haem group (see page 35).

**Hydrogen bonds** can form between a wide variety of R groups, including those of tryptophan, arginine and asparagine. **Disulphide bonds** form between two cysteine molecules. **Ionic bonds** form between R groups containing amine and carboxyl groups. (Which amino acids have these?) **Hydrophobic interactions** occur between R groups which are non-polar, or hydrophobic.

## Quaternary structure

Many proteins are made up of two or more polypeptide chains. Haemoglobin is an example of this, having four polypeptide chains in each haemoglobin molecule. The association of different polypeptide chains is called the **quaternary structure** of the protein. The chains are held together by the same four types of bond as in the tertiary structure.

## Globular and fibrous proteins

A protein whose molecules curl up into a 'ball' shape, such as myoglobin and haemoglobin, is known as a **globular protein**. In a living organism, proteins may be found in cells, in tissue fluid, or in fluids being transported, such as blood or in the phloem. All these environments contain water. Globular proteins usually curl up so that their non-polar, hydrophobic R groups point into the centre of the molecule, away from their watery surroundings. Water molecules are excluded from the centre of the folded protein molecule. The polar, hydrophilic, R groups remain on the outside of the molecule. Globular proteins, therefore, are usually soluble, because water molecules cluster around their outward-pointing hydrophilic R groups (*figure 2.21*).

Many globular proteins have roles in metabolic reactions. Enzymes, for example, are globular proteins.

Many protein molecules do not curl up into a ball, but form long strands. These are known as **fibrous proteins**. Fibrous proteins are usually insoluble and many have structural roles. Examples include **keratin** in hair and the outer layers of skin, and **collagen** (see page 36).

### Haemoglobin

Haemoglobin, the oxygen-carrying pigment found in red blood cells, is a globular protein. It is made up of four polypeptide chains. Two of these make an identical pair, and are called α chains. The other two make a different identical pair and are called β chains.

The haemoglobin molecule is nearly spherical (*figure 2.22*). The four polypeptide chains pack closely together, their hydrophobic R groups pointing in towards the centre of the molecule and their hydrophilic ones pointing outwards. Each β chain has a tertiary structure very similar to that of myoglobin (*figures 2.20* and *2.22*).

The interactions between the hydrophobic R groups inside the molecule are important in holding it in its correct three-dimensional shape. The outward-pointing hydrophilic R groups on the surface of the molecule are important in maintaining its solubility. In the disease sickle cell anaemia one amino acid, which occurs in a part of the amino acid chain of the β polypeptides on the surface of the curled-up molecule, is replaced with a different amino acid. The correct amino acid is glutamic acid which is polar. The substitute is valine which is non-polar. Having a non-polar R group on the outside of the molecule makes the haemoglobin much less soluble, and causes unpleasant and dangerous symptoms in anyone whose haemoglobin is all of this 'faulty' type (*figure 2.23*).

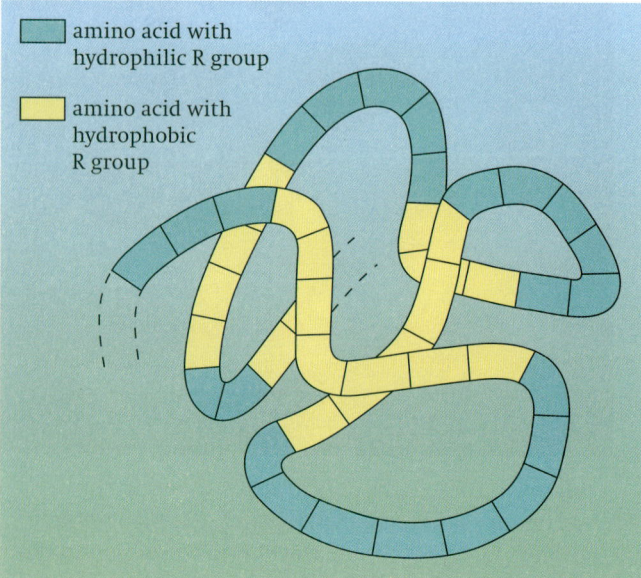

amino acid with hydrophilic R group

amino acid with hydrophobic R group

● **Figure 2.21** A schematic section through part of a globular protein molecule. The polypeptide chain coils up with hydrophilic R groups outside and hydrophobic ones inside, which makes the molecule soluble.

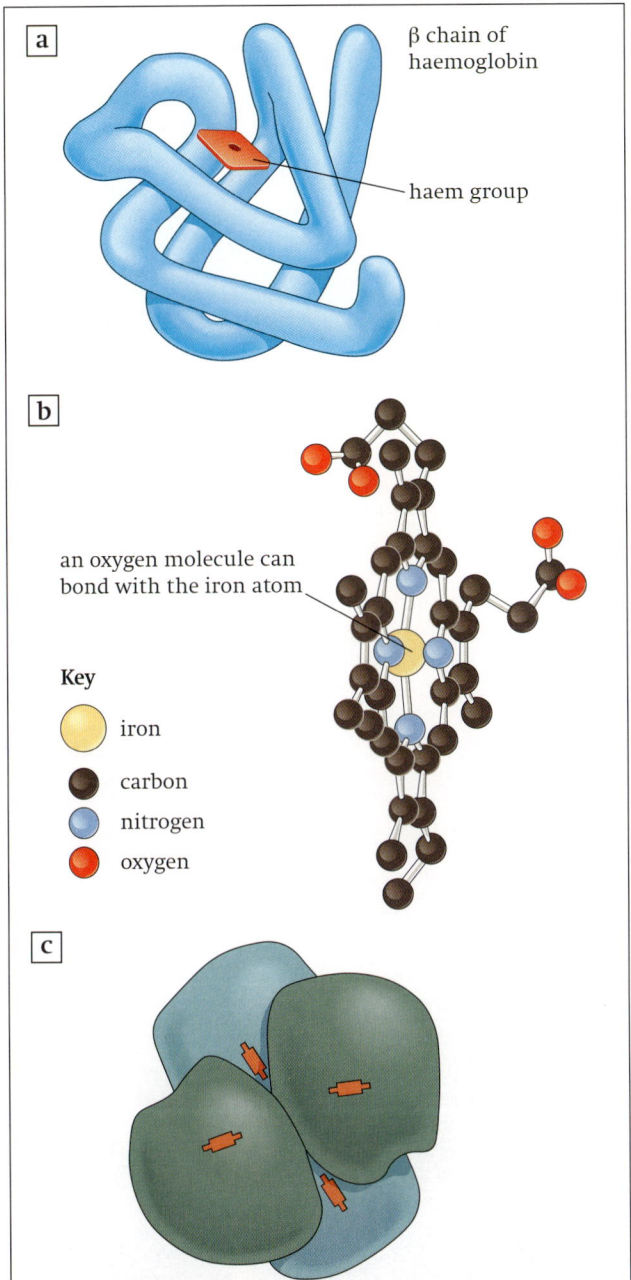

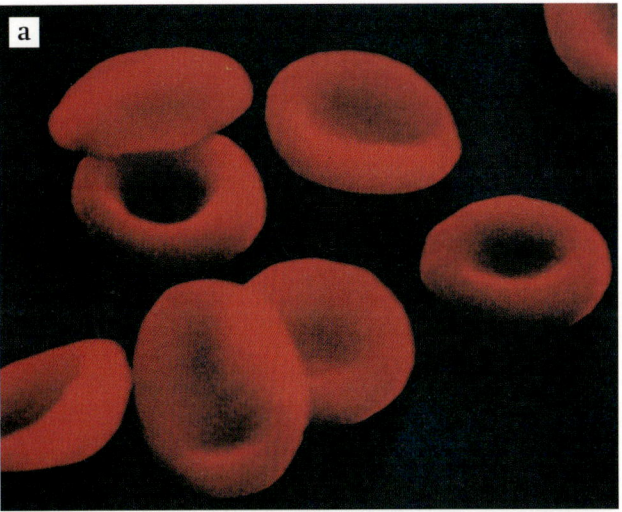

● **Figure 2.23 a** Human red blood cells. Each cell contains about 250 million haemoglobin molecules (× 5500). **b** A scanning electron micrograph of red blood cells from a person with sickle cell anaemia. You can see both normal and sickled cells.

● **Figure 2.22** Haemoglobin. **a** Each haemoglobin molecule contains four polypeptide chains, one of which is shown here. Each polypeptide chain contains a haem group, shown in red. **b** The haem group contains an iron ion which can bond reversibly with an oxygen molecule. **c** The complete haemoglobin molecule is nearly spherical.

Each polypeptide chain contains a **haem group**, shown in *figure 2.22b*. A group like this, which is an important, permanent, part of a protein molecule but is not made of amino acids, is called a **prosthetic group**.

Each haem group contains an iron ion, $Fe^{2+}$. One oxygen molecule, $O_2$, can bind with each iron ion. So a complete haemoglobin molecule, with four haem groups, can carry four oxygen molecules (eight oxygen atoms) at a time.

It is the haem group which is responsible for the colour of haemoglobin. This colour changes depending on whether or not the iron ions are combined with oxygen. If they are, the molecule is known as **oxyhaemoglobin**, and is bright red. If not, the colour is purplish.

## Collagen

Collagen is a fibrous protein that is found in skin, tendons, cartilage, bones, teeth and the walls of blood vessels. It is an important **structural protein**, not only in humans but in almost all animals, and is found in structures ranging from the body wall of sea anemones to the egg cases of dogfish.

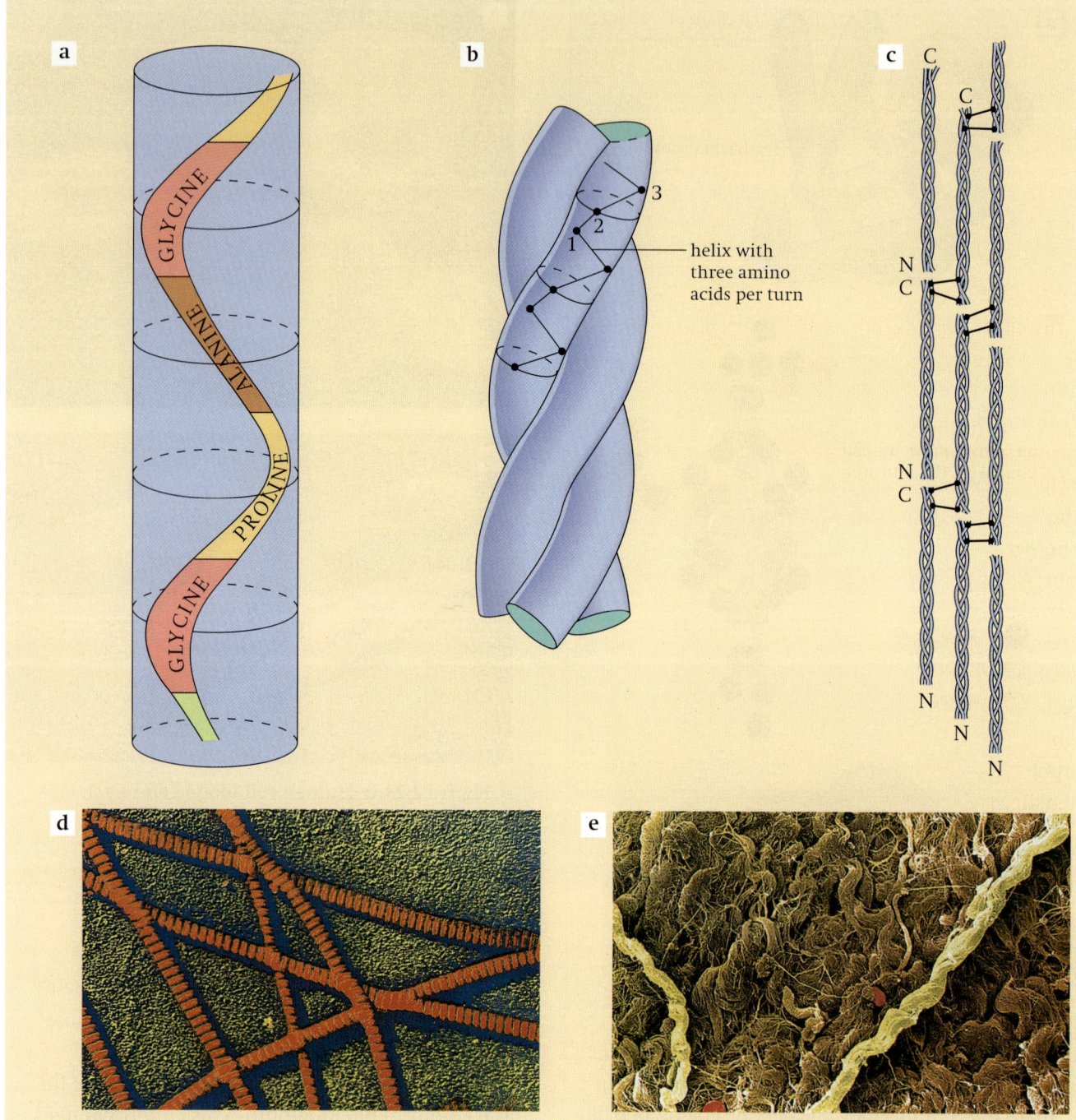

**Figure 2.24** Collagen. The diagrams and photographs begin with the very small and work up to the not-so-small. Thus three of the polypeptide chains shown in **a** make up a collagen molecule shown in **b**; many of these molecules make up a fibril, shown in **c** and **d**; and many fibrils make up a fibre, shown in **e**.

**a** The polypeptides which make up a collagen molecule are in the shape of a stretched-out helix. Every third amino acid is glycine.   **b** Three helixes wind together to form a collagen molecule. These strands are held together by hydrogen bonds.
**c** Many of these triple helixes lie side by side, linked to each other by covalent cross-links between the carboxyl end of one molecule and the amino end of another. Notice that these cross-links are out of step with each other; this gives collagen greater strength.   **d** An SEM of collagen fibrils (× 21 000). Each fibril is made up of many triple helixes lying parallel with one another. The banded appearance is caused by the regular way in which these helixes are arranged, with the staggered gaps between the molecules (shown in **c**) appearing darker.
**e** An SEM of human collagen fibres (× 3000). Each fibre is made up of many fibrils lying side by side. These fibres are large enough to be seen with an ordinary light microscope.

As shown in *figure 2.24*, a collagen molecule consists of three polypeptide chains, each in the shape of a helix. (This is not an α-helix as it is not tightly wound.) The three helical polypeptides then wind around each other to form a three-stranded 'rope'. Almost every third amino acid in each polypeptide is glycine. Its small size allows the three strands to lie close together and so form a tight coil. Any other amino acid would be too large. The three strands are held together by hydrogen bonds.

Each complete, three-stranded molecule of collagen interacts with other collagen molecules running parallel to it. Bonds form between the R groups of lysines in molecules lying next to each other. These cross-links hold many collagen molecules side by side, forming **fibres**. The ends of the parallel molecules are staggered; if they were not, there would be a weak spot running right across the collagen fibre. As it is, collagen has tremendous tensile strength, that is it can withstand large pulling forces. The human Achilles tendon, which is almost all collagen fibres, can withstand a pulling force of 300 N per mm$^2$ of cross-sectional area, about one-quarter the tensile strength of mild steel.

## Box 2E Testing for the presence of proteins

All proteins have several amine, $NH_2$, groups within their molecules. These groups can react with copper ions to form a complex that has a strong purple colour.

The reagent used for this test is called **biuret reagent**. You can use it as two separate solutions – a dilute solution of copper(II) sulphate and a more dilute solution of potassium or sodium hydroxide – which you add in turn to the solution that you suspect might contain protein. A purple colour indicates that protein is present.

Alternatively, you can use a ready-made 'biuret solution' that contains both the copper(II) sulphate solution and the hydroxide ready-mixed. To stop the copper ions reacting with the hydroxide ions and forming a precipitate, this ready-mixed reagent also contains sodium potassium tartrate or sodium citrate.

### SAQ 2.7

Where in a protein molecule are amino groups found?

# Water

Water is arguably the most important biochemical of all. Without water, life would not exist on this planet. It is important for two reasons. Firstly, it is a major component of cells, typically forming between 70 and 95% of the mass of the cell. You yourself are about 60% water. Secondly, it provides an environment for those organisms that live in water. Three-quarters of the planet is covered in water.

Although it is a simple molecule, water has some surprising properties. For example, such a small molecule would exist as a gas at normal Earth temperatures were it not for its special property of hydrogen bonding to other water molecules (see *box 2C*). Also, because it is a liquid, it provides a medium for molecules and ions to mix in and hence a medium in which life could evolve.

The hydrogen bonding of water molecules makes the molecules more difficult to separate and affects the physical properties of water. For example, more energy is needed to break these bonds and convert water from a liquid to a gas than in similar compounds, such as hydrogen sulphide ($H_2S$), which is a gas at normal air temperatures.

## Water as a solvent

Water is an excellent solvent for ions and polar molecules (molecules with an uneven charge distribution such as sugars and glycerol) because the water molecules are attracted to them, collect around and *separate* them (*figure 2.25*). This is what happens when a chemical dissolves in water. Once

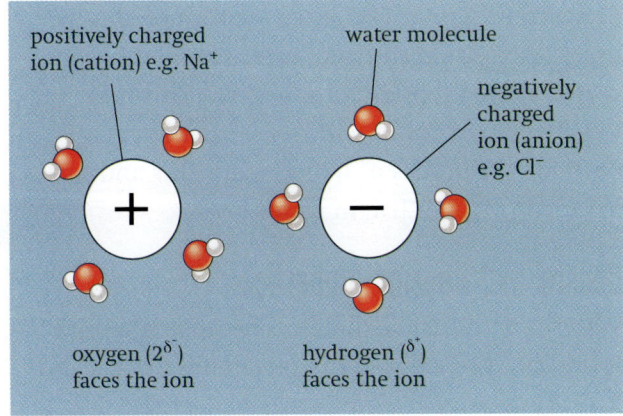

positively charged ion (cation) e.g. $Na^+$

water molecule

negatively charged ion (anion) e.g. $Cl^-$

oxygen ($2^{δ^-}$) faces the ion

hydrogen ($^{δ^+}$) faces the ion

● **Figure 2.25** Distribution of water molecules around ions in a solution.

a chemical is in solution, it is free to move about and react with other chemicals. Most processes in living organisms take place in solution in this way.

By contrast, non-polar molecules such as lipids are insoluble in water and, if surrounded by water, tend to be *pushed together* by the water since the water molecules are attracted to *each other*. This is important, for example, in hydrophobic interactions in protein structure and in membrane structure (see chapter 4) and it increases the stability of these structures.

## Water as a transport medium

Water is the transport medium in the blood, in the lymphatic, excretory and digestive systems of animals, and in the vascular tissues of plants. Here again its solvent properties are essential.

## Thermal properties

As hydrogen bonding restricts the movement of water molecules, a relatively large amount of energy is needed to raise the temperature of water. This means that large bodies of water such as oceans and lakes are slow to change temperature as environmental temperature changes. As a result they are more stable habitats. Due to the high proportion of water in the body internal changes in temperature are also minimised, making it easier to achieve a stable body temperature.

Since a relatively large amount of energy is needed to convert water to a gas, the process of evaporation transfers a correspondingly large amount of energy and can be an effective means of cooling the body, as in sweating and panting. Conversely, a relatively large amount of energy must be transferred from water before it is converted from a liquid to a solid (ice). This makes it less likely that water will freeze, an advantage both for the bodies of living organisms and for organisms which live in water.

## Density and freezing properties

Water is an unusual chemical because the solid form, ice, is less dense than its liquid form. Below 4°C the density of water starts to decrease. Ice therefore floats on liquid water and insulates the water under it. This reduces the tendency for large

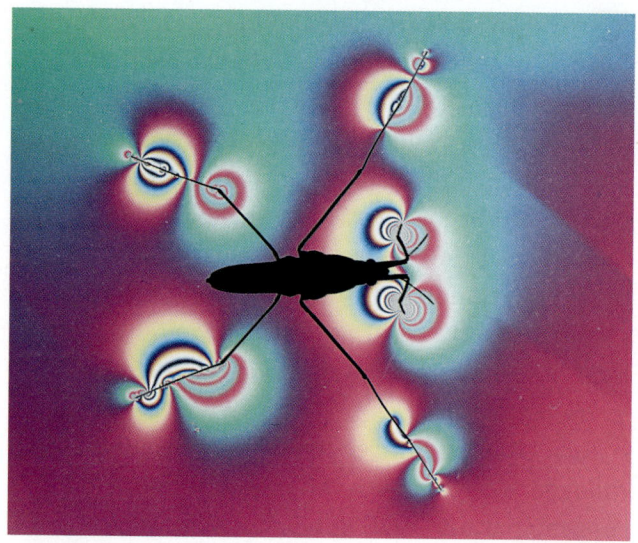

● **Figure 2.26** A pond skater standing on the surface of pond water. This was photographed through an interferometer which shows interference patterns made by the pond skater as it walks on the water's surface. The surface tension of the water means the pond skater never breaks through the surface.

bodies of water to freeze completely, and increases the chances of life surviving in cold conditions.

Changes in density of water with temperature cause currents which help to maintain the circulation of nutrients in the oceans.

## High surface tension and cohesion

Water molecules have very high cohesion, in other words they tend to stick to each other. This is exploited in the way water moves in long, unbroken columns through the vascular tissue in plants (see chapter 10) and is an important property in cells. High cohesion also results in high surface tension at the surface of water. This allows certain small organisms, such as the pond skater, to exploit the surface of water as a habitat, allowing them to settle on or skate over its surface (*figure 2.26*).

### SAQ 2.8
State the property of water that allows each of the following to take place and, in each case, explain its importance.
**a** The cooling of skin during sweating.
**b** The survival of fish in ice-covered lakes.
**c** The ability of insects, such as pond skaters, to walk on water.
**d** The transport of glucose and ions in a mammal.

| Ion | Some roles in living organisms |
|---|---|
| Calcium, $Ca^{2+}$ | Calcium phosphate is an important structural component of bones and teeth. Calcium ions are important in the transmission of electrical impulses across synapses, and in the contraction of muscles. |
| Sodium, $Na^+$ | Sodium ions are involved in the transmission of nerve impulses along neurones. They contribute to the high concentration built up by the loop of Henle in the medulla of the kidney, thus enabling concentrated urine to be excreted so that water is conserved. |
| Potassium, $K^+$ | With sodium, potassium ions are involved in the transmission of nerve impulses along neurones. They contribute to the control of turgidity of guard cells, and thus the opening and closing of stomata. |
| Magnesium, $Mg^{2+}$ | Chlorophyll molecules contain magnesium. Some enzymes that catalyse the breakdown of ATP, called ATPases, have magnesium ions at their active sites. |
| Chloride, $Cl^-$ | With sodium ions, chloride ions contribute to the high concentration built up by the loop of Henle in the medulla of the kidney, thus enabling concentrated urine to be excreted so that water is conserved. Chloride ions help to balance the positive charge of cations such as sodium and potassium, within and around cells. |
| Nitrate, $NO_3^-$ | Plants use the nitrogen from nitrate ions to make amino acids and nucleotides. |
| Phosphate, $PO_4^{3-}$ | Phosphate ions are used for making nucleotides, including ATP. With calcium, they form calcium phosphate, that gives bones their strength. |
| Iron, $Fe^{2+}$ | Haemoglobin molecules contain iron in their prosthetic haem groups. Oxygen binds here for transport in the red blood cells. |

● **Table 2.1** Some functions of eight of the most important ions required by living organisms. Note that many of the terms mentioned are explained in other chapters.

# Inorganic ions

All of the substances described in this chapter – carbohydrates, lipids, proteins and water – are made up of molecules. (A molecule is a group of atoms held together by covalent bonds.) But they are not the only type of substance that is important for the structure and metabolism of living organisms. All living things also need a wide variety of **ions**, as shown in *table 2.1*.

Ions are formed from individual atoms that have gained or lost one or more electrons and are therefore charged negatively or positively. Many are highly soluble in water (*figure 2.25*).

# SUMMARY

◆ Molecular biology is the study of the structure and function of biological molecules.

◆ Many biological molecules are formed from smaller units that bond together. These include carbohydrates, lipids, proteins and nucleic acids. Molecules which are formed from repeating identical or similar subunits are called polymers.

◆ Carbohydrates have the general formula $C_x(H_2O)_y$. Monosaccharides are the smallest carbohydrate units, of which glucose is the most common form. They are important

energy sources in cells and also important building blocks for larger molecules. They may form straight-chain or ring structures and may exist in different isomeric forms. These are important because they bond together in different ways and so affect the structure of polysaccharides, such as starch, glycogen and cellulose. The glycosidic bond forms between monosaccharides by condensation and is broken by hydrolysis. Benedict's reagent can be used to test for reducing and non-reducing sugars.

◆ Starch is formed from some straight and some branched chains of α-glucose molecules and is an energy storage compound in plants. Glycogen is a branched α-glucose chain and is an energy storage compound in animals. Cellulose is a polymer of β-glucose molecules in which the chains are grouped together by hydrogen bonding to form strong fibres that are found in plant cell walls. 'Iodine solution' can be used to test for starch.

◆ Lipids are made from fatty acids and glycerol. They are hydrophobic and do not mix with water. They are energy storage compounds in animals, as well as having other functions such as insulation and buoyancy in marine mammals. Phospholipids have a hydrophilic phosphate head and hydrophobic fatty acid tails. This is important in the formation of membranes. The emulsion test can be used to test for lipids.

◆ Proteins are long chains of amino acids which fold into precise shapes. The sequence of amino acids in a protein, known as its primary structure, determines the way that it folds and hence determines its three-dimensional shape and function.

◆ Many proteins contain areas where the amino acid chain is twisted into an α-helix; this is an example of secondary structure. Further folding produces the tertiary structure. Often, more than one polypeptide associates to form a protein molecule. The association between different polypeptide chains is the quaternary structure of the protein. Tertiary and quaternary structures are held in place by hydrogen, covalent and ionic bonding and hydrophobic interactions.

◆ Proteins may be globular or fibrous. A molecule of a globular protein is roughly spherical. Most globular proteins are soluble and metabolically active. A molecule of a fibrous protein is less folded and forms long strands. Fibrous proteins are insoluble. They often have a structural role. Biuret reagent can be used to test for proteins.

◆ Water is important within bodies where it forms a large part of the mass of the cell. It is also an environment in which organisms can live. As a result of extensive hydrogen bonding, it has unusual properties that are important for life: it is liquid at most temperatures on the Earth's surface; its highest density occurs above its freezing point so that ice floats and insulates water below from freezing air temperatures; it acts as a solvent for ions and polar molecules and causes non-polar molecules to group together; it has a high surface tension which affects the way it moves through narrow tubes and forms a surface on which some organisms can live.

◆ Ions are charged particles, some of which are important in, for example, nerve impulse transmission, excretion from the kidneys and enzyme function. The protein haemoglobin relies on the presence of iron ions in its prosthetic haem groups for its ability to carry oxygen.

## CHAPTER 3

# Enzymes

**By the end of this chapter you should be able to:**

1 explain that enzymes are globular proteins which act as catalysts;

2 explain the way in which enzymes act as catalysts by lowering activation energy;

3 describe examples of enzyme-catalysed reactions;

4 describe methods of following the time-course of an enzyme-controlled reaction;

5 discuss the ways in which temperature, pH, concentration of enzyme, concentration of substrate, and competitive and non-competitive inhibition affect the rate of enzyme-controlled reactions;

6 describe methods of investigating the effects of these factors experimentally.

Enzymes are protein molecules which can be defined as **biological catalysts**. A catalyst is a molecule which speeds up a chemical reaction, but remains unchanged at the end of the reaction. Virtually every metabolic reaction which takes place within a living organism is catalysed by an enzyme. Many enzyme names end in -ase, for example amylase, ATPase.

Enzymes are globular proteins. Like all globular proteins, enzyme molecules are coiled into a precise three-dimensional shape, with hydrophilic R groups (side-chains) on the outside of the molecule ensuring that they are soluble. Enzyme molecules also have a special feature in that they possess an **active site** (*figure 3.1*). The active site of an enzyme is a region, usually a cleft or

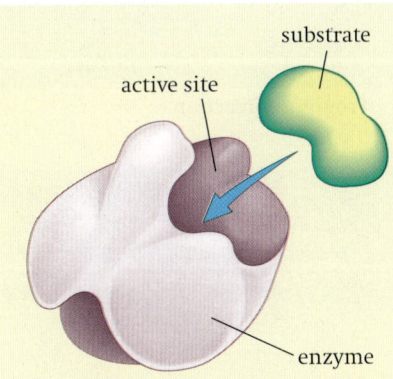 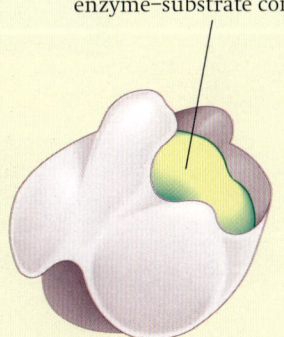 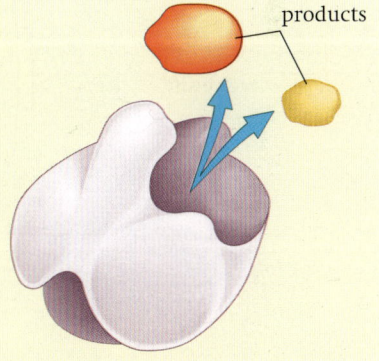

substrate     enzyme–substrate complex     products

active site     enzyme

a   An enzyme has a cleft in its surface called the active site. The substrate molecule has a complementary shape.

b   Random movement of enzyme and substrate brings the substrate into the active site. An enzyme–substrate complex is temporarily formed. The R groups of the amino acids in the active site interact with the substrate.

c   The interaction of the substrate with the active site breaks the substrate apart. The two product molecules leave the active site, leaving the enzyme molecule unchanged and ready to bind with another substrate molecule.

● **Figure 3.1** How an enzyme catalyses the breakdown of a substrate molecule to two product molecules.

depression, to which another molecule or molecules can bind. This molecule is the **substrate** of the enzyme. The shape of the active site allows the substrate to fit perfectly, and to be held in place by temporary bonds which form between the substrate and some of the R groups of the enzyme's amino acids. This combined structure is termed the **enzyme–substrate complex**. A simplified diagram is shown in *figure 3.2*.

Each type of enzyme will usually act on only one type of substrate molecule. This is because the shape of the active site will only allow one shape of molecule to fit. The enzyme is said to be **specific** for this substrate.

The enzyme may catalyse a reaction in which the substrate molecule is split into two or more molecules. Alternatively, it may catalyse the joining together of two molecules, as when making a dipeptide. Interaction between the R groups of the enzyme and the atoms of the substrate can break, or encourage formation of, bonds in the substrate molecule, forming one, two or more **products**.

When the reaction is complete, the product or products leave the active site. The enzyme is unchanged by this process, so it is now available to receive another substrate molecule. The rate at which substrate molecules can bind to the

enzyme's active site, be formed into products and leave can be very rapid. The enzyme catalase, for example, can bind with hydrogen peroxide molecules, split them into water and oxygen and release these products at a rate of $10^7$ molecules per second.

# Enzymes reduce activation energy

As catalysts, enzymes increase the rate at which chemical reactions occur. Most of the reactions which occur in living cells would occur so slowly without enzymes that they would virtually not happen at all.

In many reactions, the substrate will not be converted to a product unless it is temporarily given some extra energy. This energy is called **activation energy** (*figure 3.3a*).

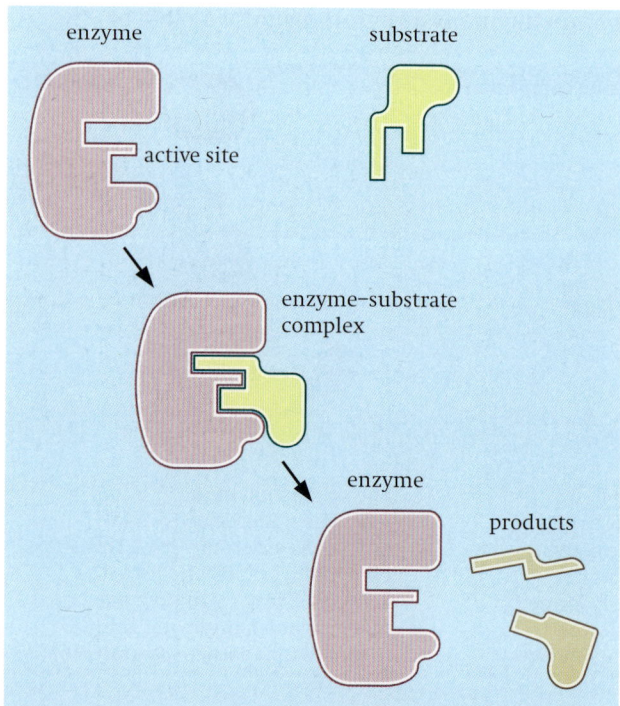

● **Figure 3.2** A simplified diagram of enzyme function.

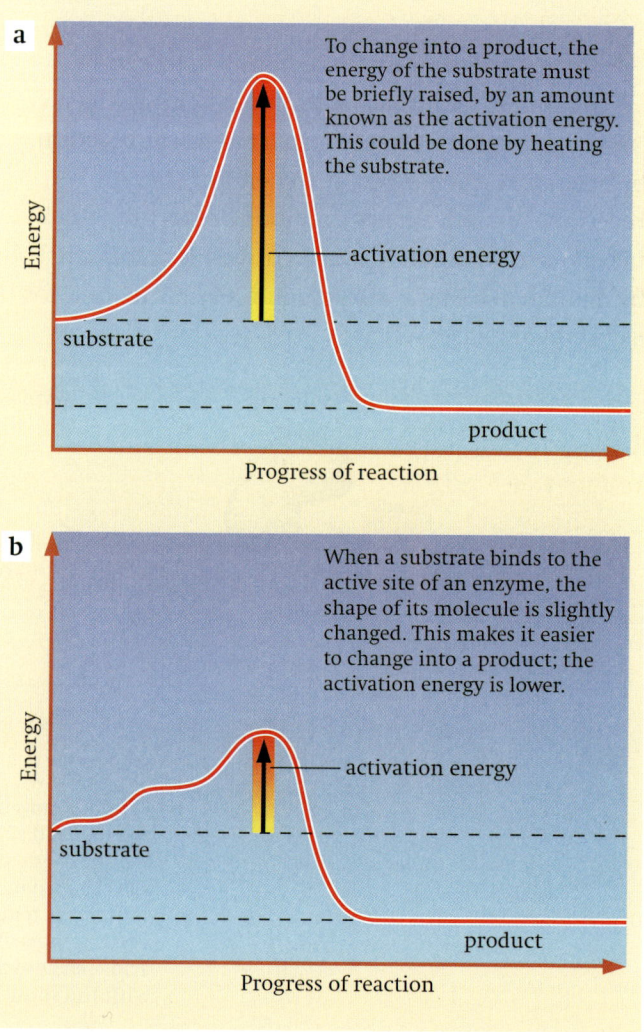

**a** To change into a product, the energy of the substrate must be briefly raised, by an amount known as the activation energy. This could be done by heating the substrate.

**b** When a substrate binds to the active site of an enzyme, the shape of its molecule is slightly changed. This makes it easier to change into a product; the activation energy is lower.

● **Figure 3.3** Activation energy **a** without enzyme, **b** with enzyme.

One way of increasing the rate of many chemical reactions is to increase the energy of the reactants by heating them. You have probably done this on many occasions by heating substances which you want to react together. In the Benedict's test for reducing sugar, for example, you need to heat the Benedict's reagent and sugar solution together before they will react (page 25).

Mammals, such as humans, also use this method of speeding up their metabolic reactions. Our body temperature is maintained at 37 °C, which is usually much warmer than the temperature of the air around us. But even raising the temperature of cells to 37 °C is not enough to give most substrates the activation energy which they need to change into products. We cannot raise body temperature much more than this, as temperatures above about 40 °C begin to cause irreversible damage to many of the molecules from which we are made, especially protein molecules. Enzymes are a solution to this problem because they *decrease* the activation energy of the reaction which they catalyse (*figure 3.3b*). They do this by holding the substrate or substrates in such a way that their molecules can react more easily. Reactions catalysed by enzymes will take place rapidly at a much lower temperature than they would without them.

# The course of a reaction

You may be able to carry out an investigation into the rate at which substrate is converted into product during an enzyme-controlled reaction. *Figure 3.4* shows the results of such an investigation, using the enzyme catalase. This enzyme is found in the tissues of most living things and catalyses the breakdown of hydrogen peroxide into water and oxygen. (Hydrogen peroxide is a toxic product of several different metabolic reactions.) It is an easy reaction to follow as the oxygen that is released can be collected and measured.

The reaction begins very swiftly. As soon as the enzyme and substrate are mixed, bubbles of oxygen are released quickly. A large volume of oxygen is collected in the first minute of the reaction. As the reaction continues, however, the rate at which oxygen is released gradually slows down. The reaction gets slower and slower, until it eventually stops completely.

The explanation for this is quite straightforward. When the enzyme and substrate are first mixed, there is a large number of substrate molecules. At any moment, virtually every enzyme molecule has a substrate molecule in its active site. The rate at which the reaction occurs will depend only on how many enzyme molecules there are, and the speed at which the enzyme can convert the substrate into product, release it, and then bind with another substrate molecule. However, as more and more substrate is converted into product, there are fewer and fewer substrate molecules to bind with enzymes. Enzyme molecules may be 'waiting' for a substrate molecule to hit their active site. As fewer substrate molecules are left, the reaction gets slower and slower, until it eventually stops.

The curve is therefore steepest at the beginning of the reaction: the rate of an enzyme-controlled reaction is always fastest at the beginning. This rate is called the **initial rate of reaction**. You can measure the initial rate of the reaction by calculating the slope of a tangent to the curve, as close to time 0 as possible. An easier way of doing this

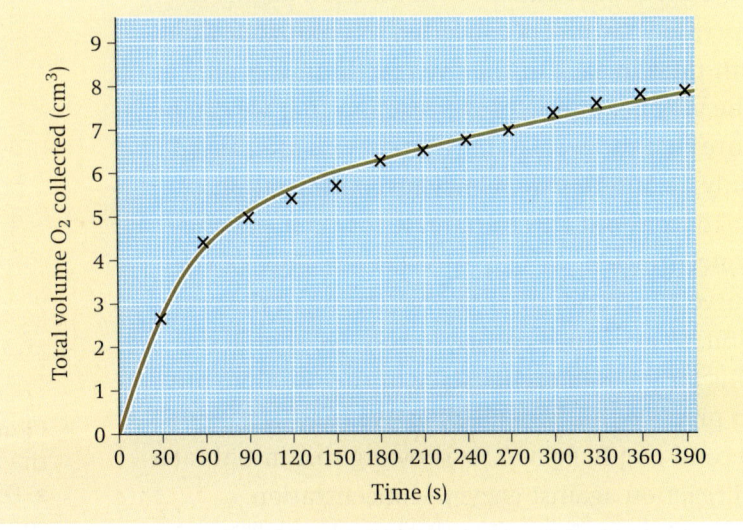

● **Figure 3.4** The course of an enzyme-catalysed reaction. Catalase was added to hydrogen peroxide at time 0. The gas released was collected in a gas syringe, the volume being read at 30 s intervals.

is simply to read off the graph the amount of oxygen given off in the first 30 seconds. In this case, the rate of oxygen production in the first 30 seconds is 2.7 cm$^3$ of oxygen per 30 seconds, or 5.4 cm$^3$ per minute.

## SAQ 3.1

Why is it better to calculate the initial rate of reaction from a curve such as the one in *figure 3.4*, rather than simply measuring how much oxygen is given off in 30 seconds?

## The effect of enzyme concentration

*Figure 3.5a* shows the results of an investigation in which different amounts of catalase were added to the same amount of hydrogen peroxide. You can see that the shape of all five curves is similar. In each case, the reaction begins very quickly (steep curve) and then gradually slows down (curve levels off). Because the amounts of hydrogen peroxide are the same in all five reactions, the total amount of oxygen eventually produced will be the same so, if the investigation goes on long enough, all the curves will meet.

To compare the rates of these five reactions, in order to look at the effect of enzyme concentration on reaction rate, it is fairest to look at the rate *right at the beginning* of the reaction. This is because, once the reaction is under way, the amount of substrate in each reaction begins to vary, as substrate is converted to product at different rates in each of the five reactions. It is only at the very beginning of the reaction that we can be sure that differences in reaction rate are caused only by differences in enzyme concentration.

To work out this initial rate for each enzyme concentration, we can calculate the slope of the curve 30 seconds after the beginning of the reaction, as explained earlier. Ideally, we should do this for an even earlier stage of the reaction, but in practice this is impossible. We can then plot a second graph, *figure 3.5b*, showing this initial rate of reaction against enzyme concentration.

This graph shows that the initial rate of reaction increases linearly. In these conditions, reaction rate is directly proportional to the enzyme concentration. This is just what common sense

says should happen. The more enzyme present, the more active sites will be available for the substrate to slot into. As long as there is plenty of substrate available, the initial rate of a reaction increases linearly with enzyme concentration.

## SAQ 3.2

Sketch the shape of *figure 3.5b* if excess hydrogen peroxide was not available.

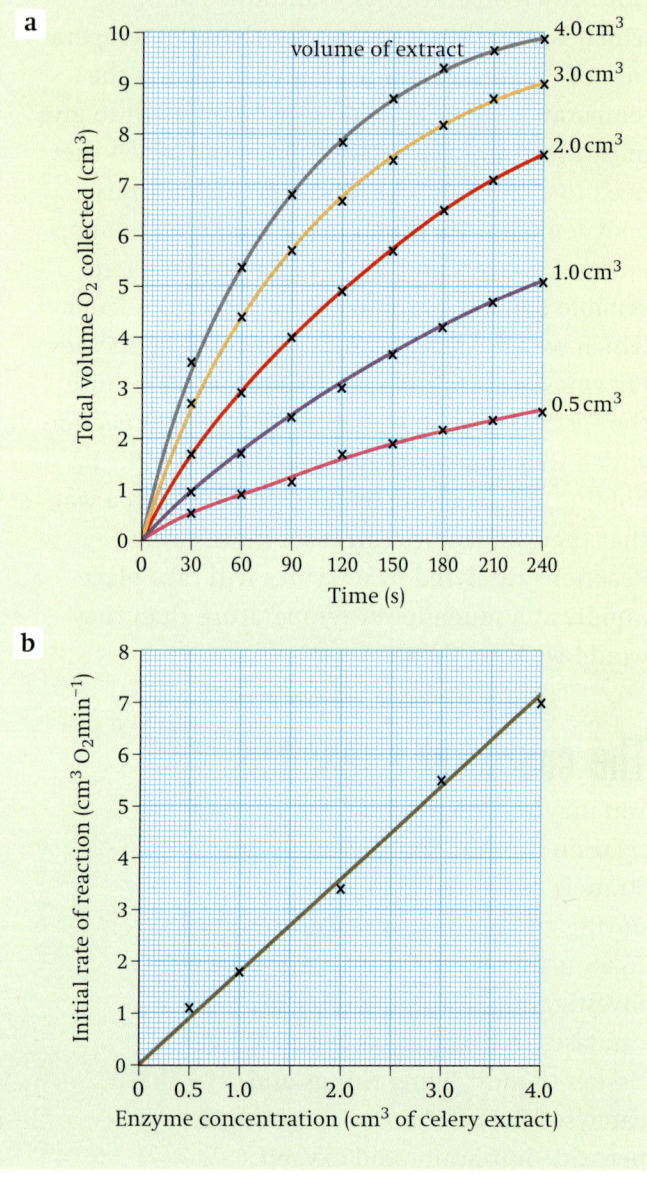

● **Figure 3.5** The effect of enzyme concentration on the rate of an enzyme-catalysed reaction.
**a** Different volumes of celery extract, which contains catalase, were added to the same volume of hydrogen peroxide. Water was added to make the total volume of the mixture the same in each case. **b** The rate of reaction in the first 30 s was calculated for each enzyme concentration.

# Measuring reaction rate

It is easy to measure the rate of the catalase–hydrogen peroxide reaction, because one of the products is a gas, which is released and can be collected. Unfortunately, it is not always so easy to measure the rate of a reaction. If, for example, you wanted to investigate the rate at which amylase breaks down starch, it would be very difficult to observe the course of the reaction because the substrate (starch) and the product (maltose) remain as colourless substances in the reaction mixture.

The easiest way to measure the rate of this reaction is to measure the rate at which starch disappears from the reaction mixture. This can be done by taking samples from the mixture at known times, and adding each sample to some iodine in potassium iodide solution. Starch forms a blue-black colour with this solution. Using a colorimeter, you can measure the intensity of the blue-black colour obtained, and use this as a measure of the amount of starch still remaining. If you do this over a period of time, you can plot a curve of amount of starch remaining against time. You can then calculate the initial reaction rate in the same way as for the catalase–hydrogen peroxide reaction.

## SAQ 3.3

**a** Sketch the curve you would expect to obtain if the amount of starch remaining was plotted against time.

**b** How could you use this curve to calculate the initial reaction rate?

It is even easier to observe the course of this reaction if you mix starch, iodine in potassium iodide solution and amylase in a tube, and take regular readings of the colour of the mixture in this one tube in a colorimeter. However, this is not ideal, because the iodine interferes with the rate of the reaction and slows it down.

## The effect of substrate concentration

*Figure 3.6* shows the results of an investigation in which the amount of catalase was kept constant, and the amount of hydrogen peroxide was varied. Once again, curves of oxygen released against time were plotted for each reaction, and the

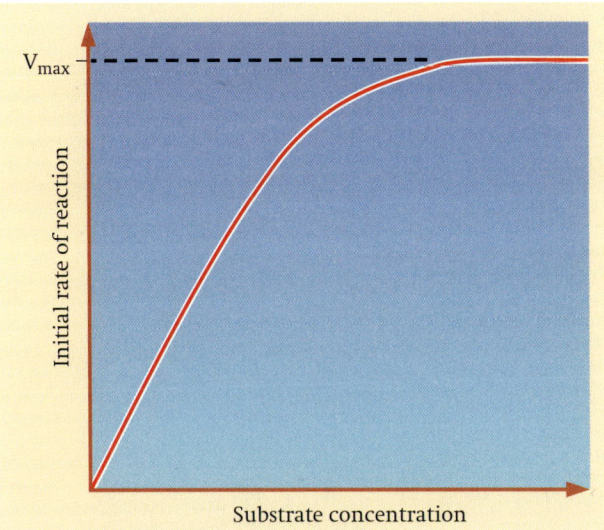

● **Figure 3.6** The effect of substrate concentration on the rate of an enzyme-catalysed reaction.

initial rate of reaction calculated for the first 30 seconds. These initial rates of reaction were then plotted against substrate concentration.

As substrate concentration increases, the initial rate of reaction also increases. Again, this is only what we would expect: the more substrate molecules there are around, the more often an enzyme's active site can bind with one. However, if we go on increasing substrate concentration, keeping the enzyme concentration constant, there comes a point where every enzyme active site is working continuously. If more substrate is added, the enzyme simply cannot work faster; substrate molecules are effectively 'queuing up' for an active site to become vacant. The enzyme is working at its maximum possible rate, known as $V_{max}$.

## Temperature and enzyme activity

*Figure 3.7* shows how the rate of a typical enzyme-catalysed reaction varies with temperature. At low temperatures, the reaction takes place only very slowly. This is because molecules are moving relatively slowly. Substrate molecules will not often collide with the active site, and so binding between substrate and enzyme is a rare event. As temperature rises, the enzyme and substrate molecules move faster. Collisions happen more frequently, so that substrate molecules enter the active site more often. Moreover, when they do collide, they do so with more energy. This makes it easier for bonds to be broken so that the reaction can occur.

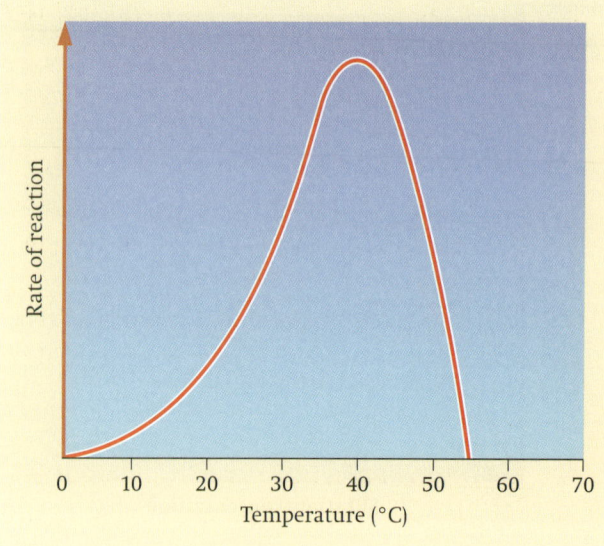

● **Figure 3.7** The effect of temperature on the rate of an enzyme-controlled reaction.

## SAQ 3.4

How could you carry out an experiment to determine the effect of temperature on the rate of breakdown of hydrogen peroxide by catalase?

As temperature continues to increase, the speed of movement of the substrate and enzyme molecules also continues to increase. However, above a certain temperature the structure of the enzyme molecule vibrates so energetically that some of the bonds holding the enzyme molecule in its precise shape begin to break. This is especially true of hydrogen bonds. The enzyme molecule begins to lose its shape and activity and is said to be **denatured**. This is often irreversible.

At first, the substrate molecule fits less well into the active site of the enzyme, so the rate of the reaction begins to slow down. Eventually the substrate no longer fits at all, or can no longer be held in the correct position for the reaction to occur.

The temperature at which an enzyme catalyses a reaction at the maximum rate is called the **optimum temperature**. Most human enzymes have an optimum temperature of around 40 °C. By keeping our body temperatures at about 37 °C, we ensure that enzyme-catalysed reactions occur at close to their maximum rate. It would be dangerous to maintain a body temperature of 40 °C, as even a slight rise above this would begin to denature enzymes.

● **Figure 3.8** Not all enzymes have optimum temperatures of 40 °C. Bacteria and algae living in hot springs such as this one in Yellowstone National Park, USA, are able to tolerate very high temperatures. Enzymes from such organisms are proving useful in various industrial applications.

Enzymes from other organisms may have different optimum temperatures. Some enzymes, such as those found in bacteria which live in hot springs, have much higher optimum temperatures (*figure 3.8*). Some plant enzymes have lower optimum temperatures, depending on their habitat.

## SAQ 3.5

Proteases are used in biological washing powders.
**a** How would a protease remove a blood stain on clothes?
**b** Most biological washing powders are recommended for use at low washing temperatures. Why is this?
**c** Washing powder manufacturers have produced proteases which can work at higher temperatures than 40 °C. Why is this useful?

## pH and enzyme activity

*Figure 3.9* shows how the activity of an enzyme is affected by pH. Most enzymes work fastest at a pH of somewhere around 7, that is in fairly neutral conditions. Some, however, such as the protease pepsin which is found in the acidic conditions of the stomach, have a different optimum pH.

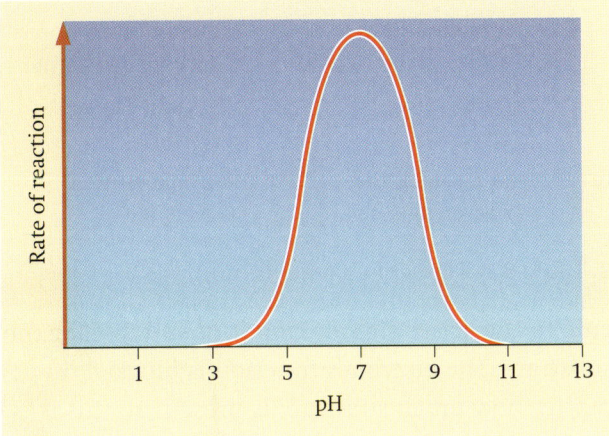

● **Figure 3.9** The effect of pH on the rate of an enzyme-controlled reaction.

pH is a measure of the concentration of hydrogen ions in a solution. The lower the pH, the higher the hydrogen ion concentration. Hydrogen ions can interact with the R groups of amino acids, affecting the way in which they bond with each other and therefore affect their 3D arrangement (see page 33). A pH which is very different from the optimum pH can cause denaturation of an enzyme.

### SAQ 3.6

Trypsin is a protease secreted in pancreatic juice, which acts in the duodenum. If you make up a suspension of milk powder in water, and add trypsin, the enzyme digests the protein in the milk, so that the suspension becomes clear.

How could you carry out an investigation into the effect of pH on the rate of activity of trypsin? (A suspension of 4 g of milk powder in 100 cm³ of water will become clear in a few minutes if an equal volume of a 0.5% trypsin solution is added to it.)

# Enzyme inhibitors

As we have seen, the active site of an enzyme fits one particular substrate perfectly. It is possible, however, for some *other* molecule to bind to an enzyme's active site if it is very similar to the enzyme's substrate. This could **inhibit** the enzyme's function.

If an **inhibitor** molecule binds only briefly to the site there is competition between it and the

substrate for the site. If there is much more of the substrate than the inhibitor present, substrate molecules can easily bind to the active site in the usual way and so the enzyme's function is unaffected. However, if the concentration of the inhibitor rises or the substrate falls, it becomes less and less likely that the substrate will collide with an empty site and the enzyme's function is inhibited. This is therefore known as **competitive inhibition** (*figure 3.10a*). It is said to be **reversible** (not permanent) because it can be reversed by increasing the concentration of the substrate.

An example of competitive inhibition occurs in the treatment of a person who has drunk ethylene glycol. Ethylene glycol is used as antifreeze, and is

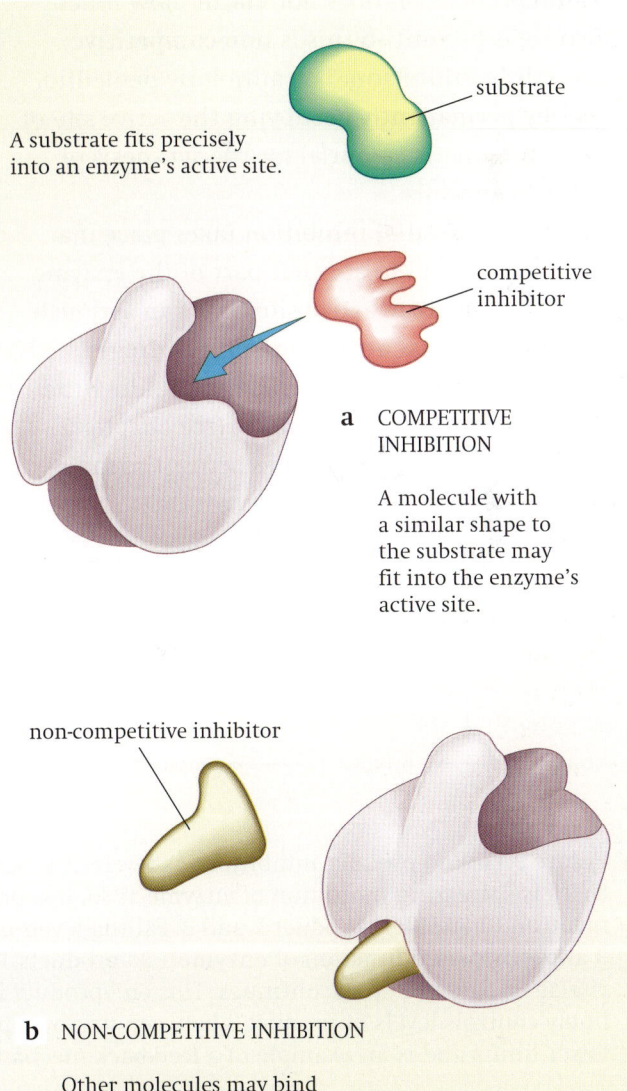

A substrate fits precisely into an enzyme's active site.

substrate

competitive inhibitor

**a** COMPETITIVE INHIBITION

A molecule with a similar shape to the substrate may fit into the enzyme's active site.

non-competitive inhibitor

**b** NON-COMPETITIVE INHIBITION

Other molecules may bind elsewhere on the enzyme, distorting its active site.

● **Figure 3.10** Enzyme inhibition.

sometimes drunk accidentally. Ethylene glycol is rapidly converted in the body to oxalic acid, which can cause irreversible kidney damage. However, the active site of the enzyme which converts ethylene glycol to oxalic acid will also accept ethanol. If the poisoned person is given a large dose of ethanol, the ethanol acts as a competitive inhibitor, slowing down the action of the enzyme on ethylene glycol for long enough to allow the ethylene glycol to be excreted.

Sometimes, the inhibitor can remain permanently bonded with the active site and therefore cause a permanent block to the substrate. No competition occurs as it does not matter how much substrate is present so this is **non-competitive irreversible inhibition**. The antibiotic penicillin works by permanently occupying the active site of an enzyme that is essential for the synthesis of bacterial cell walls.

A different kind of inhibition takes place if a molecule can bind to another part of the enzyme rather than the true active site. This can seriously disrupt the normal arrangement of hydrogen bonds and hydrophobic interactions holding the enzyme molecule in its 3D shape (see chapter 2). The resulting distortion ripples across the mole-

|  | Inhibitor binds to active site on enzyme | Inhibitor binds elsewhere on enzyme |
| --- | --- | --- |
| **Inhibitor binds briefly** | Competitive reversible | Non-competitive reversible |
| **Inhibitor binds permanently** | Non-competitive irreversible | Non-competitive irreversible |

● **Table 3.1** Types of enzyme inhibition.

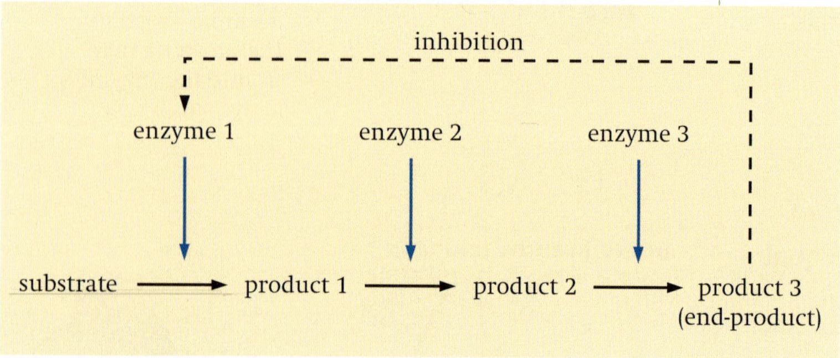

cule to the active site making it unsuitable for the substrate. The enzyme's function is blocked no matter how much substrate is present so this is another type of **non-competitive inhibition** (*figure 3.10b*). It can be **reversible** or **irreversible**, depending on whether the inhibitor bonds briefly or permanently with the enzyme. Digitalis is an example of a non-competitive inhibitor. It binds with the enzyme ATPase, resulting in an increase in the contraction of heart muscle.

Inhibition of enzyme function can be lethal, but in many situations inhibition is essential. For example, metabolic reactions must be very finely controlled and balanced, so no single enzyme can be allowed to 'run wild', constantly churning out more and more product. One way of ensuring that this cannot happen is to use the **end-product** of a chain of reactions as an enzyme inhibitor (*figure 3.11*). As the enzyme converts substrate to product, it is slowed down because the end-product binds to another part of the enzyme and prevents more substrate binding. However, the end-product can lose its attachment to the enzyme and go on to be used elsewhere, allowing the enzyme to reform into its active state. As product levels fall, the enzyme is able to top them up again. This is **end-product inhibition** and is an example of **non-competitive reversible inhibition**.

A summary of the various types of inhibition can be found in *table 3.1*.

● **Figure 3.11** End-product inhibition. As levels of product 3 rise, there is increasing inhibition of enzyme 1. So, less product 1 is made and hence less product 2 and 3. Falling levels of product 3 allow increased function of enzyme 1 so products 1, 2 and 3 rise again and the cycle continues. This end-product inhibition finely controls levels of product 3 between narrow upper and lower limits and is an example of a feed-back mechanism.

# SUMMARY

◆ Enzymes are globular proteins, which act as catalysts by lowering activation energy.

◆ Each enzyme acts on only one specific substrate, because there has to be a perfect match between the shape of the substrate and the shape of the enzyme's active site to form an enzyme–substrate complex.

◆ Anything which affects the shape of the active site, such as high temperature, a change of pH or the binding of a non-competitive inhibitor with the enzyme, will slow down the rate of the reaction.

◆ Competitive inhibitors also slow down the rate of reaction, by competing with the substrate for the active site of the enzyme.

## CHAPTER 4

# Cell membranes and transport

**By the end of this chapter you should be able to:**

1 describe the fluid mosaic model of membrane structure and explain the underlying reasons for this structure;

2 outline the roles of phospholipids, cholesterol, glycolipids, proteins and glycoproteins in membranes;

3 outline the roles of the plasma membrane, and the roles of membranes within cells;

4 describe and explain how molecules can get in and out of cells (cross cell membranes) by the processes of diffusion, facilitated diffusion, osmosis, active transport, endocytosis and exocytosis;

5 describe the effects on animal and plant cells of immersion in solutions of different water potential;

6 describe the features of the gaseous exchange surface of the mammalian lung;

7 describe the features of root hairs that enable the uptake of ions by active transport.

In chapter 1 you saw that *all* living cells are surrounded by a membrane which controls the exchange of materials, such as nutrients and waste products, between the cell and its environment. Although extremely thin, the membrane must be capable of regulating this exchange very precisely. Within cells, particularly eukaryotic cells, regulation of transport across the membranes of organelles is vital. Membranes also have other important functions. For example, they enable cells to receive hormone messages. Therefore it is important to study the structure of membranes if we are to understand how these functions are achieved.

## Phospholipids

An understanding of the structure of membranes depends on an understanding of the structure of phospholipids (see page 30). From phospholipids,

little bags can be formed in which chemicals can be isolated from the external environment. These bags are the membrane-bound compartments that we know as cells and organelles.

*Figure 4.1a* shows what happens if phospholipid molecules are spread over the surface of water. They form a single layer with their heads in the water, because these are polar (hydrophilic), and their tails projecting out of the water, because these are non-polar (hydrophobic). (The term 'polar' refers to the uneven distribution of charge which occurs in some molecules. The significance of this is also discussed on page 27.)

If the phospholipids are shaken up with water they can form stable structures in the water called **micelles** (*figure 4.1b*). Here all the hydrophilic heads face outwards into the water, shielding the hydrophobic tails, which point in towards each other. Alternatively, two-layered structures, called **bilayers**, can form in sheets (*figure 4.1c*). It is now

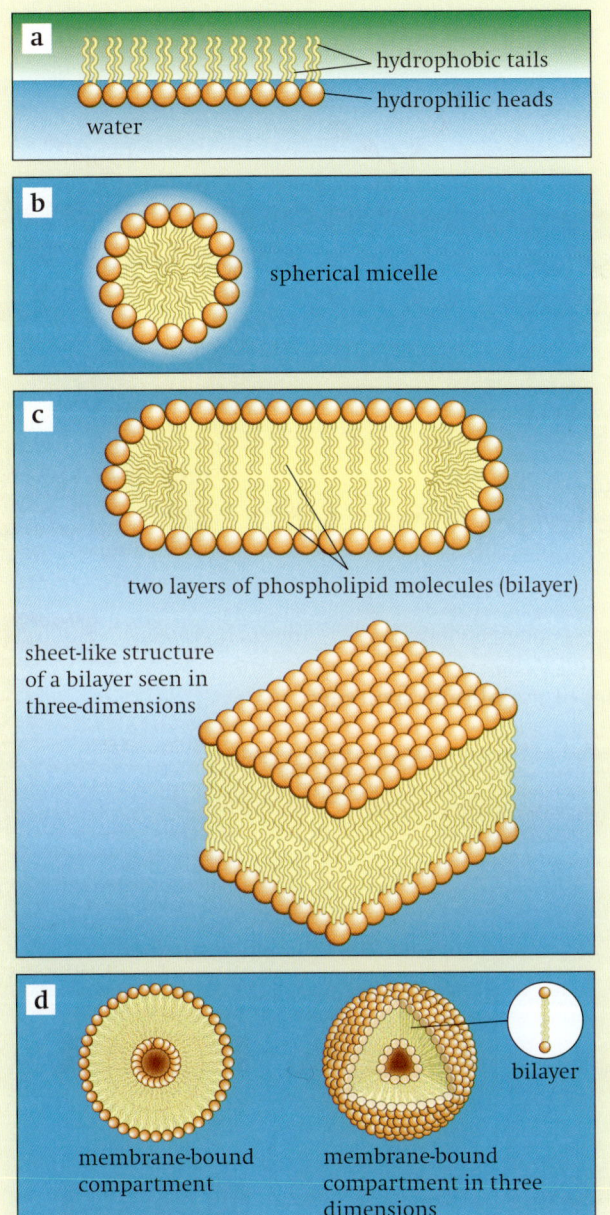

a spread as a single layer of molecules (a monolayer) on the surface of water, b forming micelles surrounded by water, c forming bilayers, d forming membrane-bound compartments.

● **Figure 4.1** Phospholipids in water

known that this phospholipid bilayer is the basic structure of membranes (*figure 4.1d*).

# Structure of membranes

The phospholipid bilayer is visible using the electron microscope at very high magnifications of at least × 100 000 (*figure 1.20*). The double black line visible using the electron microscope is thought

to show the two phospholipid layers. The bilayer (membrane) is about 7 nm wide.

Membranes also contain proteins. *Figures 4.2* and *4.3* show what we imagine a membrane might look like if we could see the individual molecules. This model for the structure of a membrane is known as the **fluid mosaic model**. The word 'fluid' refers to the fact that the individual phospholipid and protein molecules move around within their layer. The word 'mosaic' describes the pattern produced by the scattered protein molecules when the surface of the membrane is viewed from above.

## Features of the fluid mosaic model

■ The membrane is a double layer (**bilayer**) of phospholipid molecules. The individual phospholipid molecules move about by diffusion within their own monolayer.

■ The phospholipid tails point inwards, facing each other and forming a non-polar hydrophobic interior. The phospholipid heads face the aqueous (water-containing) medium that surrounds the membrane.

■ Some of the phospholipid tails are saturated and some are unsaturated. The more unsaturated they are, the more fluid the membrane. This is because the unsaturated fatty acid tails are bent (*figure 2.12*) and therefore fit together more loosely. As temperature decreases membranes become less fluid, but some organisms which cannot regulate their own temperature, such as bacteria and yeasts, respond by increasing the proportion of unsaturated fatty acids in their membranes.

■ Most of the protein molecules float like mobile icebergs in the phospholipid layers, although some are fixed like islands to structures inside the cell and do not move about.

■ Some proteins are embedded in the outer layer, some in the inner layer and some span the whole membrane. They stay in the membrane because they have hydrophobic portions (made from hydrophobic amino acids) which 'sit' among the hydrophobic phospholipid tails. Hydrophilic portions (made from hydrophilic amino acids) face outwards.

■ The total thickness is about 7 nm on average.

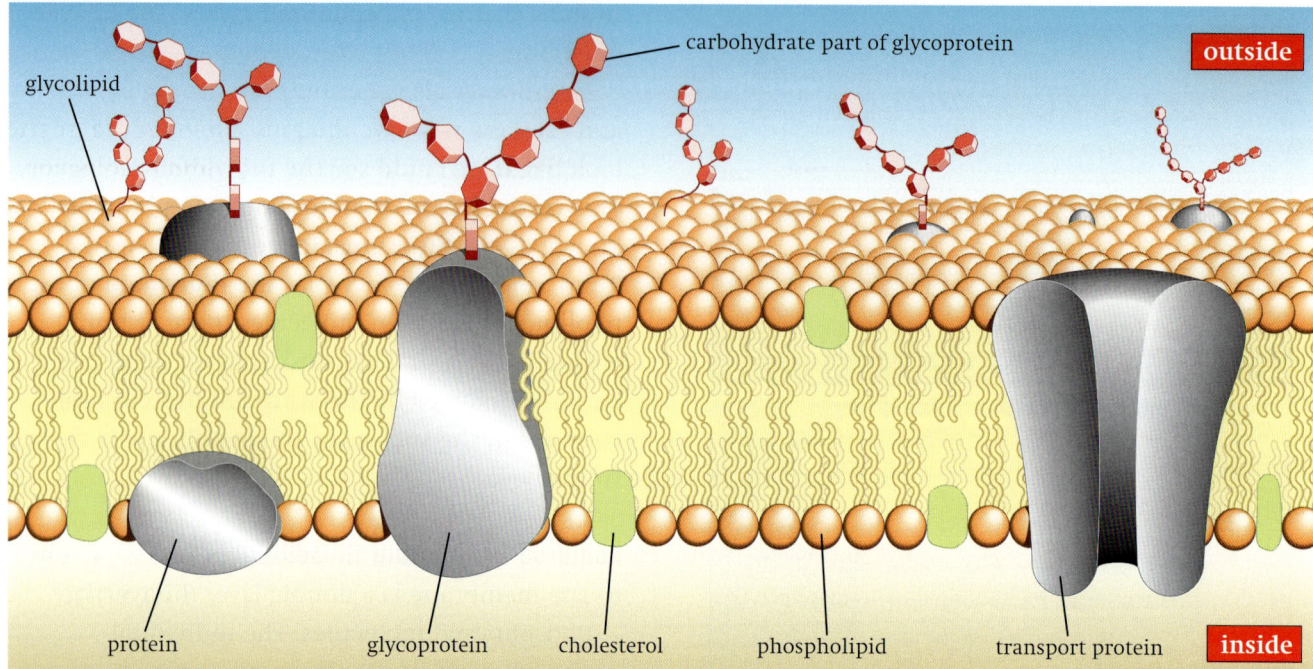

● **Figure 4.2** An artist's impression of the fluid mosaic model of membrane structure.

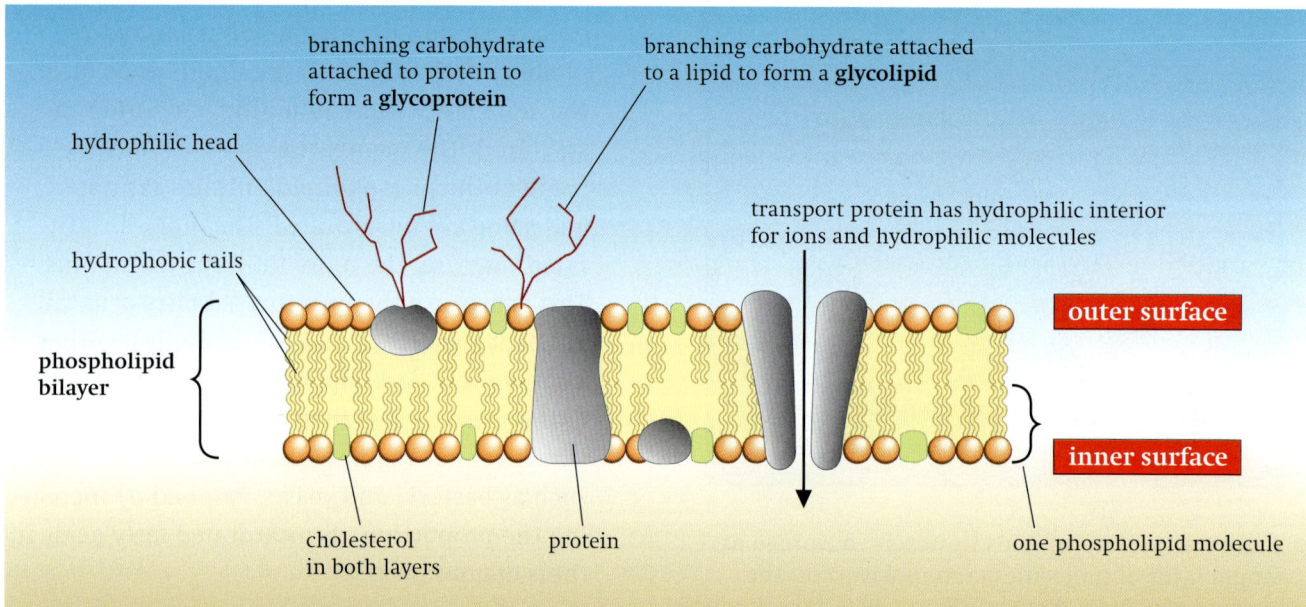

● **Figure 4.3** Diagram of the fluid mosaic model of membrane structure.

■ Many proteins and lipids have short, branching carbohydrate chains attached to the external surface of the membrane, thus forming **glycoproteins** and **glycolipids** respectively.

■ Molecules of cholesterol are also found in the membrane. Their function is described below.

## Roles of the components of cell membranes

We have seen that cell membranes contain several different types of molecule – phospholipids,

cholesterol, proteins, glycolipids and glycoproteins. Each of these has a particular role to play in the overall structure and function of the membrane.

■ **Phospholipids**, as explained on page 50, form the bilayer which is the basic structure of the membrane. Because their tails are non-polar, it is difficult for polar molecules, or ions, to pass through them, so they act as a barrier to most water-soluble substances.

■ **Cholesterol** molecules, like phospholipids, have

hydrophilic heads and hydrophobic tails, so they fit neatly between the phospholipid molecules. They help to regulate the fluidity of the membrane, preventing it from becoming too fluid or too rigid. Cholesterol is also important for the mechanical stability of membranes, as without it membranes quickly break and cells burst open. The hydrophobic regions of cholesterol molecules help to prevent ions or polar molecules from passing through the membrane. This is particulary important in the myelin sheath (made up of many layers of membrane) around nerve cells, where leakage of ions would slow down nerve impulses.

■ **Proteins** have a variety of functions within membranes. Many of them act as **transport proteins**. These provide hydrophilic channels or passageways for ions and polar molecules to pass through the membrane. This is described on pages 54 and 58. There are many different types of protein channel, each specific for a different kind of ion or molecule. Between them, they can control the types of substances that are allowed to enter or leave the cell. Other membrane proteins may be **enzymes**, for example those found in the plasma membranes of the cells on the surface of the small intestine that catalyse the hydrolysis of molecules such as disaccharides (page 24). Proteins also play important roles in the membranes of organelles. For example, in the membranes of mitochondria and chloroplasts, they are involved in the processes of respiration and photosynthesis. (You will find out much more about this if you continue your biology course to A Level.)

■ **Glycolipids** and **glycoproteins**. Many of the lipid molecules on the outer surfaces of plasma membranes, and most of the protein molecules, have short carbohydrate chains attached to them. These 'combination' molecules are known as **glycolipids** and **glycoproteins** respectively. The carbohydrate chains project out into the watery fluids surrounding the cell, where they form hydrogen bonds with the water molecules and so help to stabilise the membrane structure (see page 27). They also act as **receptor molecules**, binding with particular substances such as hormones or neurotransmitters (the chemicals that enable nerve impulses to pass from one nerve cell to another). Different cells have different collections of receptor molecules in their membranes that will bind with particular substances. For example, only certain cells, such as those in the liver and muscles, have receptors for the hormone insulin. When insulin binds with these receptors, it triggers a particular series of chemical reactions in the cell. Other cells, as they do not have insulin receptors, are not affected by insulin.

One group of glycoproteins, known as antigens, are important in allowing cells to recognise each other. Each type of cell has its own type of antigen, rather like countries with different flags.

# Transport across the plasma membrane

A phospholipid bilayer around cells makes a very effective barrier, particularly against the movement of water-soluble molecules and ions. The aqueous contents of the cell are therefore prevented from escaping. However, some exchange between the cell and its environment is essential.

## SAQ 4.1
Suggest three reasons why exchange between the cell and its environment is essential.

There are four basic mechanisms, diffusion, osmosis, active and bulk transport, by which exchange is achieved, which we shall now consider.

## Diffusion and facilitated diffusion

If you open a bottle of perfume in a room, it is not long before molecules of scent spread to all parts of the room (and are detected when they fit into membrane receptors in your nose). This will happen, even in still air, by the process of diffusion. **Diffusion** can be defined as the net movement of molecules (or ions) from a region of their higher concentration to a region of their lower concentration. The molecules move down a **concentration gradient**. It happens because of the natural kinetic energy (energy of movement) possessed by molecules or ions, which makes them move about at random. As a result of diffusion, molecules tend to reach an equilibrium situation where they are evenly spread within a given volume of space.

Some substances have molecules or ions that are able to pass through cell membranes by diffusion. The rate at which a substance diffuses across a membrane depends on a number of factors, including:

- the 'steepness' of the concentration gradient, that is the difference in the concentration of the substance on the two sides of the surface. If there are, for example, many more molecules on one side of a membrane than on the other, then at any one moment more molecules will be moving (entirely randomly) from this side than from the other. The greater the difference in concentration, then the greater the difference in the number of molecules passing in the two directions, and hence the faster the net rate of diffusion.
- temperature. At high temperatures, molecules and ions have much more kinetic energy than at low temperatures. They move around faster, and thus diffusion takes place faster.
- the surface area across which diffusion is taking place. The greater the surface area, then the more molecules or ions can cross it at any one moment, and therefore the faster diffusion can occur.
- the nature of the molecules or ions. Large molecules require more energy to get them moving than small ones do, so substances with large molecules tend to diffuse more slowly than ones with small molecules. Non-polar molecules diffuse more easily through cell membranes than polar ones, as they are soluble in the non-polar phospholipid tails.

The respiratory gases, oxygen and carbon dioxide, cross membranes by diffusion. They are uncharged and non-polar, and so can cross through the phospholipid bilayer directly between the phospholipid molecules. Water molecules, despite being very polar, can diffuse rapidly across the phospholipid bilayer because they too are small enough. However, large polar molecules, such as glucose and amino acids, cannot diffuse through the phospholipid bilayer. Nor can ions such as $Na^+$ or $Cl^-$. These can only cross the membrane by passing through hydrophilic channels created by protein molecules. Diffusion that takes place through these channels is called **facilitated diffusion**. *Facilitate* means 'make easy' or 'make possible', and this is what the protein channels do.

Plasma membranes contain many different types of protein channel, each type allowing only one kind of molecule or ion to pass through it. The movement of the molecules or ions is entirely passive, just as in ordinary diffusion, and net movement into or out of the cell will only take place down a concentration gradient from a high concentration to a low concentration. However, the rate at which this diffusion takes place depends on how many appropriate channels there are in the membrane, and on whether they are open or not. For example, the disease cystic fibrosis is caused by a defect in a protein which should be present in the plasma membranes of certain cells, including those lining the lungs, that normally allows chloride ions to move out of the cells. If this protein is not correctly positioned in the membrane, or if it does not open the chloride channel as and when it should, then the chloride ions cannot move out.

## Osmosis

Osmosis is best regarded as a special type of diffusion involving water molecules only. In the explanations that follow remember that

$$solute + solvent = solution.$$

In a sugar **solution**, for example, the **solute** is sugar and the **solvent** is water.

In *figure 4.4* there are two solutions separated by a **partially permeable membrane**. This is a membrane which allows only certain molecules through, just like membranes in living cells. In the situation shown in *figure 4.4* solution B has a higher concentration of solute molecules than solution A has. Solution B is described as more concentrated than solution A, and solution A as more dilute than solution B.

First imagine the situation if the membrane were *not* present. Because B has the higher

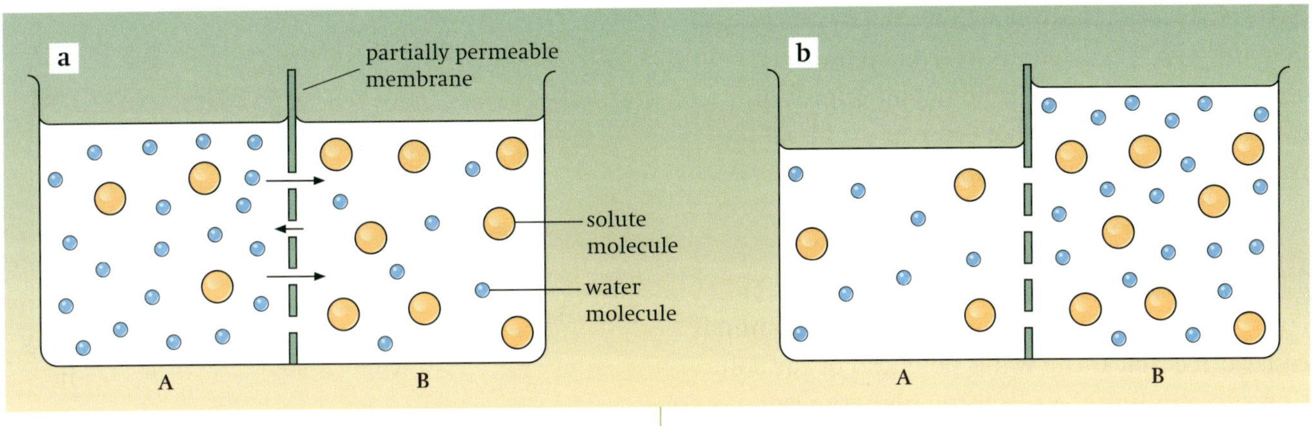

● **Figure 4.4 a** Before osmosis. Two solutions separated by a partially permeable membrane. The solute molecules are too large to pass through the pores in the membrane but the water molecules are small enough. **b** At equilibrium. As the arrows show, more molecules moved from A to B than from B to A so the net movement has been from A to B, raising the level of solution in B and lowering it in A.

concentration of solute molecules, there would be a **net** movement of solute molecules from B to A by diffusion (meaning more solute molecules would pass from B to A in a given time than from A to B). At the same time there would be a net movement of water molecules from A to B by diffusion because solution A has a higher concentration of water molecules than solution B has. Eventually, at equilibrium, the concentrations of water molecules and solute molecules in A would equal that in B.

Now consider the situation where a partially permeable membrane *is* present, as shown in *figure 4.4*. The solute molecules are too large to pass through the membrane, but water molecules can pass easily between solutions A and B, so the only molecules that can diffuse through the membrane are water molecules. There will therefore be a net movement of water molecules from A to B until an equilibrium is reached where solution A has the same concentration of water molecules (and therefore of solute molecules) as solution B. During the process the level of liquid in B will therefore rise and the level in A will fall. The fact that the movement of water molecules *alone*, and not of solute molecules, has brought about the equilibrium is characteristic of osmosis.

## Water potential and solute potential

It is useful to be able to measure the tendency of water molecules to move from one place to another. This tendency is known as **water potential**. The symbol for water potential is the Greek letter psi, ψ. Water *always* moves from a region of higher water potential to a region of lower water potential. It therefore moves down a water potential gradient. Equilibrium is reached when the water potential in one region is the same as in the other. There will then be no net movement of water molecules. We can now define **osmosis** as **the movement of water molecules from a region of higher water potential to a region of lower water potential through a partially permeable membrane**.

In *figure 4.4a*, since water moves from A to B, solution A must have a higher water potential than solution B. Pure water has the highest possible water potential. The effect of solute molecules is therefore to *lower* the water potential.

By convention, the water potential of pure water is set at zero. Since solutes make water potential lower, they make the water potential of solutions *less* than zero, that is negative. The more solute, the more negative (lower) the water potential becomes. The amount that the solute molecules lower the water potential of a solution is called the **solute potential**. Solute potential is therefore always negative. The symbol for solute potential is $\psi_s$.

You should now be able to decide which solution in *figure 4.4a*, A or B, has the lower solute potential. The answer is that B has the lower (more negative) solute potential, and A has the higher solute potential (nearer zero and therefore less negative).

## SAQ 4.2

In *figure 4.4b*, the solutions in A and B are in equilibrium, that is there is no net movement of water molecules now. What can you say about the water potentials of the two solutions?

### Osmosis in animal cells

*Figure 4.5* shows the effect of osmosis on an animal cell. Notice that if the water potential of the solution surrounding the cell is too high, the cell swells and bursts (*figure 4.5a*). If it is too low, the cell shrinks (*figure 4.5c*). This shows one reason why it is important to maintain a constant water potential inside the bodies of animals. In animal cells $\psi = \psi_s$, in other words water potential is equal to solute potential.

## SAQ 4.3

In *figure 4.5*:
**a** which solution has the highest water potential?
**b** which solution has the lowest solute potential?
**c** in which solution is the water potential of the red cell the same as that of the solution?

### Pressure potential

So far, the only factor we have examined which affects water potential is solute potential. It is possible for another factor to come into play, namely pressure. *Figure 4.6* shows a system like

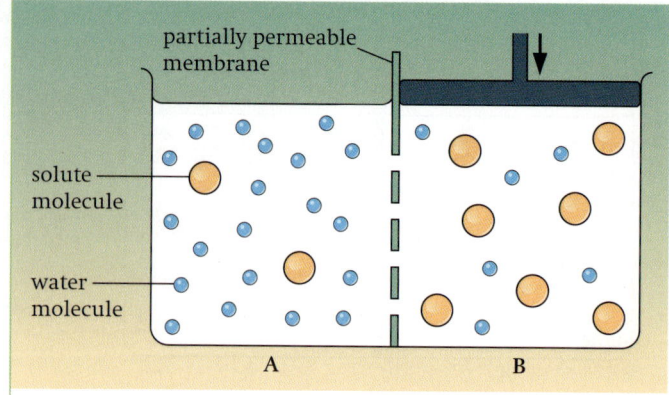

- **Figure 4.6** Two solutions separated by a partially permeable membrane. A piston is added, enabling pressure to be applied to solution **B**.

that in *figure 4.4*, except that a piston has been added, allowing pressure to be applied to solution B. We have seen that, without the piston, there is a net movement of water molecules from A to B. This can be prevented by applying pressure to solution B. The greater the pressure applied, the greater the tendency for water molecules to be forced back from solution B to solution A. Since the tendency of the water molecules to move from one place to another is measured as water potential, it is clear that increasing the pressure *increases* the water potential of solution B. The contribution made by pressure to water potential is known as **pressure potential**, and is given the symbol $\psi_p$. The pressure potential makes the water potential less negative and is therefore positive.

### Osmosis in plant cells

Pressure potential is especially important in plant cells. Unlike animal cells, plant cells are surrounded by cell walls which are very strong and rigid (page 5). Imagine a plant cell being placed in pure water or a dilute solution (*figure 4.7a*). The water or solution has a higher water potential than the plant cell and water therefore enters the cell through its partially permeable plasma membrane by osmosis. Just like the animal cell, the volume of the cell increases but in the plant cell, the protoplast (the living part of the cell inside the cell wall) starts to push against the cell wall and

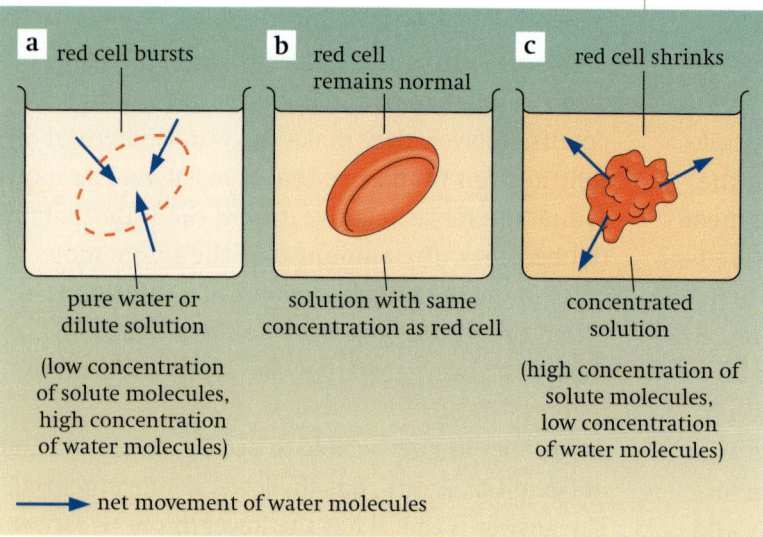

| a red cell bursts | b red cell remains normal | c red cell shrinks |
|---|---|---|
| pure water or dilute solution | solution with same concentration as red cell | concentrated solution |
| (low concentration of solute molecules, high concentration of water molecules) | | (high concentration of solute molecules, low concentration of water molecules) |

→ net movement of water molecules

- **Figure 4.5** Movement of water into or out of red blood cells by osmosis in solutions of different concentration.

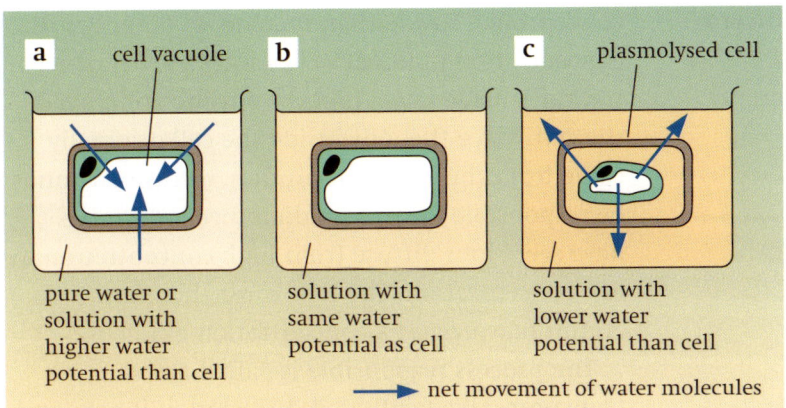

**a** cell vacuole  **b**  **c** plasmolysed cell

pure water or solution with higher water potential than cell

solution with same water potential as cell

solution with lower water potential than cell

→ net movement of water molecules

● **Figure 4.7** Osmotic changes in a plant cell in solutions of different water potential.

pressure starts to build up rapidly. This is the pressure potential and it increases the water potential of the cell until the water potential inside the cell equals the water potential outside the cell, and equilibrium is reached. The cell wall is so inelastic that it takes very little water to enter the cell to achieve this. The cell wall prevents the cell from bursting, unlike the situation when an animal cell is placed in pure water or a dilute solution. When a plant cell is fully inflated with water it is described

as **turgid**. For plant cells, then, water potential is a combination of solute potential and pressure potential. This can be expressed in the following equation:

$$\psi = \psi_s + \psi_p$$

*Figure 4.7c* shows the situation where a plant cell is placed in a solution of lower water potential. An example of the latter would be a concentrated sucrose solution. In such a solution water will *leave* the cell by osmosis. As it does so, the protoplast gradually shrinks until it is exerting no pressure at all on the cell wall. At this point the pressure potential is zero, so the water potential of the cell is equal to its solute potential (see the equation above). As the protoplast continues to shrink it begins to pull away from the cell wall (*figures 4.7c and 4.8*). This process is called **plasmolysis**, and a cell in which it has happened is said to be **plasmolysed**. Both the solute molecules and the water molecules of the external solution can pass through the freely permeable cell wall, and so the external solution remains in contact with the shrinking protoplast. Eventually, as with the animal cell, an equilibrium is reached when the water potential of the cell has decreased to that of the external solution. The point at which pressure potential has just reached zero and plasmolysis is *about* to occur is referred to as **incipient plasmolysis**.

The changes described can easily be observed with a light microscope using strips

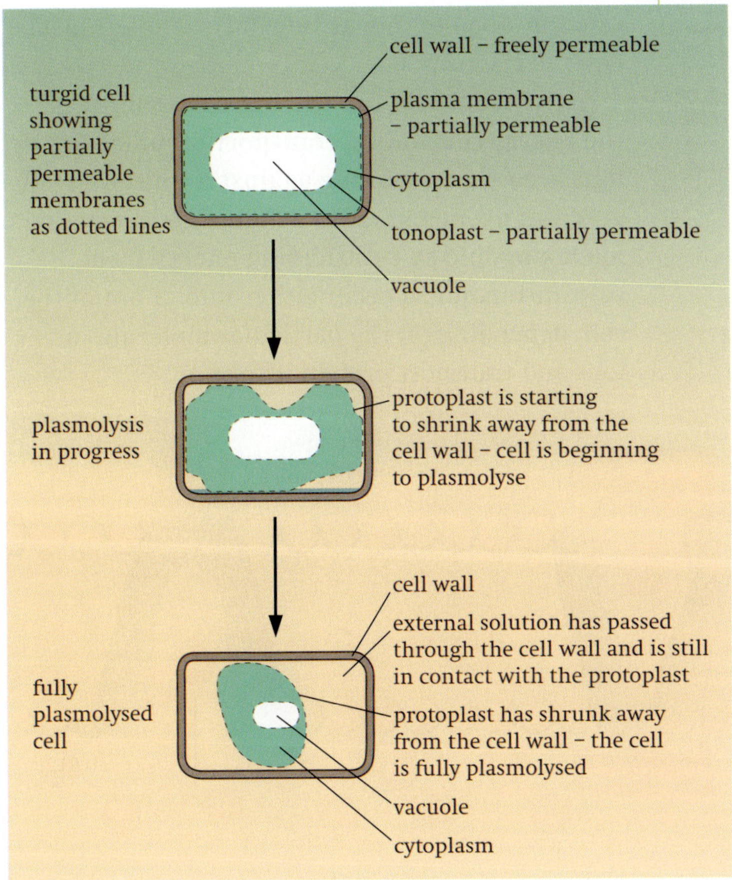

turgid cell showing partially permeable membranes as dotted lines

cell wall – freely permeable

plasma membrane – partially permeable

cytoplasm

tonoplast – partially permeable

vacuole

plasmolysis in progress

protoplast is starting to shrink away from the cell wall – cell is beginning to plasmolyse

fully plasmolysed cell

cell wall

external solution has passed through the cell wall and is still in contact with the protoplast

protoplast has shrunk away from the cell wall – the cell is fully plasmolysed

vacuole

cytoplasm

● **Figure 4.8** How plasmolysis occurs.

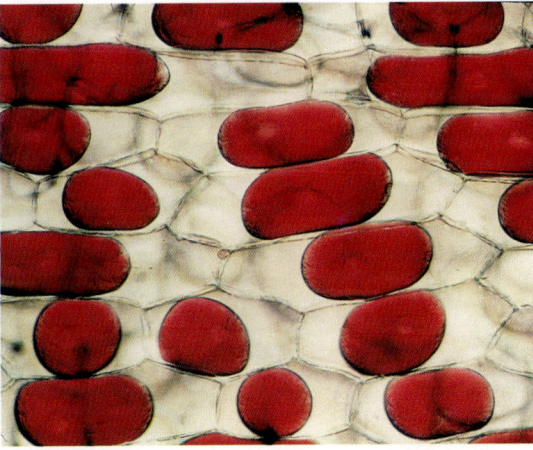

● **Figure 4.9** Light micrograph of red onion cells that have plasmolysed.

of epidermis peeled from rhubarb petioles or from the swollen storage leaves of onion bulbs and placed in a range of sucrose solutions of different concentration (*figure 4.9*).

## SAQ 4.4

Two neighbouring plant cells are shown in the diagram.

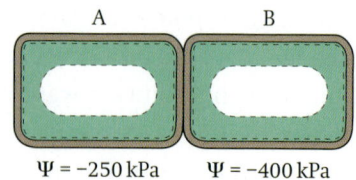

a  In which direction would there be net movement of water molecules?
b  Explain what is meant by *net movement*.
c  Explain your answer to **a**.
d  Explain what would happen if both cells were placed in (i) pure water, (ii) a 1 mol dm$^{-3}$ sucrose solution ($\psi = -3510$ kPa).

## SAQ 4.5

*Figure 4.8* shows a phenomenon called plasmolysis. Why can plasmolysis not take place in an animal cell?

## Active transport

If the concentration of particular ions, such as potassium and chloride, inside cells is measured it is often found that they are 10–20 times more

concentrated inside than outside. In other words, a concentration gradient exists with a lower concentration outside and a higher concentration inside the cell. Since the ions inside the cell originally came from the external solution, diffusion cannot be responsible for this gradient because, as we have seen, ions diffuse from high concentration to low concentration. The ions must therefore accumulate *against* a concentration gradient.

The process responsible is called **active transport**. Like facilitated diffusion, it is achieved by special transport proteins, each of which is specific for a particular type of molecule or ion. However, unlike facilitated diffusion, active transport requires energy because movement occurs *up* a concentration gradient. The energy is supplied by the molecule ATP which is produced during respiration inside the cell. The energy is used to make the transport protein (sometimes called a **carrier protein**) change its 3D shape, transferring the molecules or ions across the membrane in the process (*figure 4.10*). The analogy of a 'kissing gate' can be used to imagine this: there is no direct route through the protein in one step; the 'gate' can only open to the far side once a molecule has entered the near side.

**Active transport** can therefore be defined as **the energy-consuming transport of molecules or ions across a membrane against a concentration gradient** (from a lower to a higher concentration) **made possible by transferring energy from respiration**. It can occur either into or out of the cell, depending on the particular molecules or ions and transport protein involved.

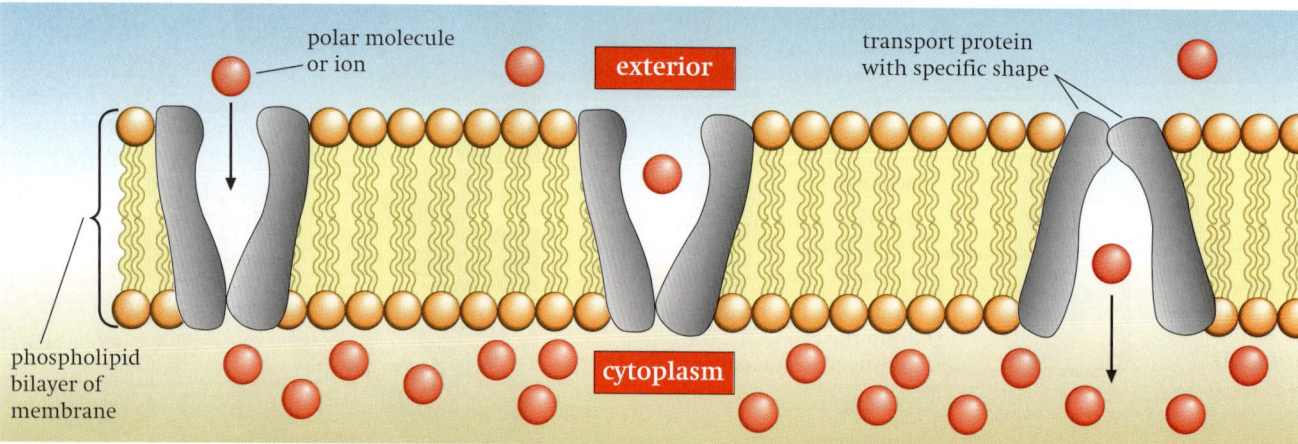

● **Figure 4.10** Changes in the shape of a transport protein during active transport. Here, molecules or ions are being pumped *into* the cell.

Active transport is important in reabsorption in the kidneys where certain useful molecules and ions have to be reabsorbed into the blood after filtration into the kidney tubules. It is also involved in the absorption of some products of digestion from the gut. In plants, active transport is used to load sugar from the photosynthesising cells of leaves into the phloem tissue for transport around the plant (chapter 10), and to load inorganic ions from the soil into root hairs (page 62).

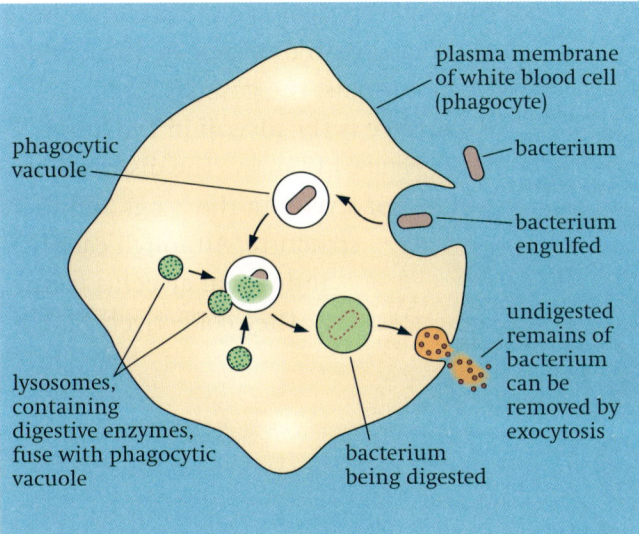

● **Figure 4.11a** Stages in phagocytosis of a bacterium by a white blood cell.

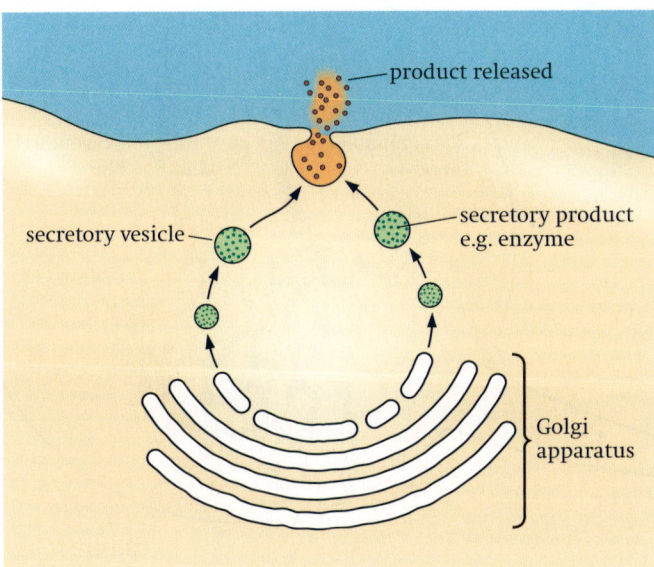

● **Figure 4.11b** Exocytosis in a secretory cell. If the product being secreted is a protein, the Golgi apparatus is often involved in chemically modifying the protein before it is secreted, as in the secretion of digestive enzymes by the pancreas.

## Bulk transport

So far we have been looking at ways in which *individual* molecules or ions cross membranes. Mechanisms also exist for the bulk transport of large quantities of materials into cells (endocytosis) or out of cells (exocytosis).

**Endocytosis** involves the engulfing of the material by the plasma membrane to form a small sac, or 'endocytotic vacuole'. It takes two forms:

■ **phagocytosis** or 'cell eating' – this is the bulk uptake of solid material. Cells specialising in this are called **phagocytes**. The process is called **phagocytosis** and the vacuoles **phagocytic vacuoles**. An example is the engulfing of bacteria by certain white blood cells (*figure 4.11a*). Also see chapter 14.

■ **pinocytosis** or 'cell drinking' – this is the bulk uptake of liquid. The small vacuoles (vesicles) formed are often extremely small, in which case the process is called **micropinocytosis**. The human egg cell takes up nutrients from cells that surround it (the follicle) by pinocytosis.

**Exocytosis** is the reverse of endocytosis and is the process by which materials are removed from cells (*figure 4.11a and b*). It happens, for example, in the secretion of digestive enzymes from cells of the pancreas (*figure 4.11c*). Secretory vesicles carry the enzymes to the cell surface and release their contents. Plant cells use exocytosis to get their cell wall building materials to the outside of the plasma membrane.

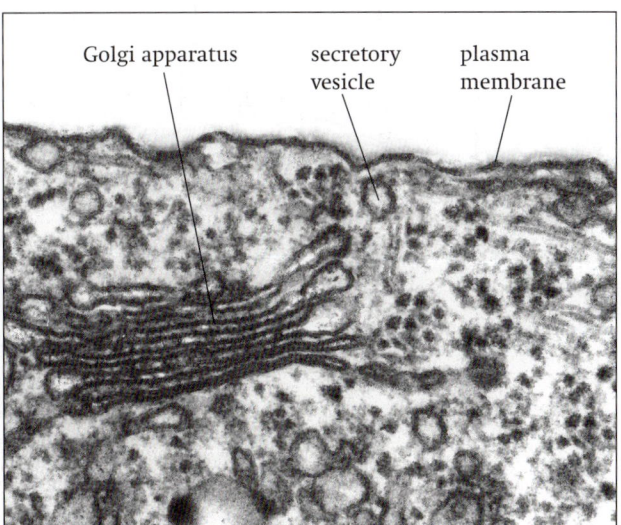

● **Figure 4.11c** EM of pancreatic acinar cell secreting protein.

# Exchange surfaces

We have seen how substances may pass across membranes into and out of *cells* and *organelles* by a variety of passive and active processes. Substances such as gases, water and nutrients also pass into and out of whole *organisms*. We will look at two examples.

# Gaseous exchange in mammalian lungs

Most organisms need a supply of oxygen for respiration. In single-celled organisms, the oxygen simply diffuses from the fluid outside the cell, through the plasma membrane and into the cytoplasm. In a multicellular organism such as a human, however, most of the cells are a considerable distance away from the external environment from which the oxygen is obtained. Multicellular organisms therefore usually have a specialised **gaseous exchange surface** where oxygen from the external environment can diffuse into the body, and carbon dioxide can diffuse out. In humans, the gaseous exchange surface is the **alveoli** in the lungs. *Figure 4.12* shows the distribution of alveoli in the lungs and their structure. Although each individual alveolus is tiny, they collectively have a

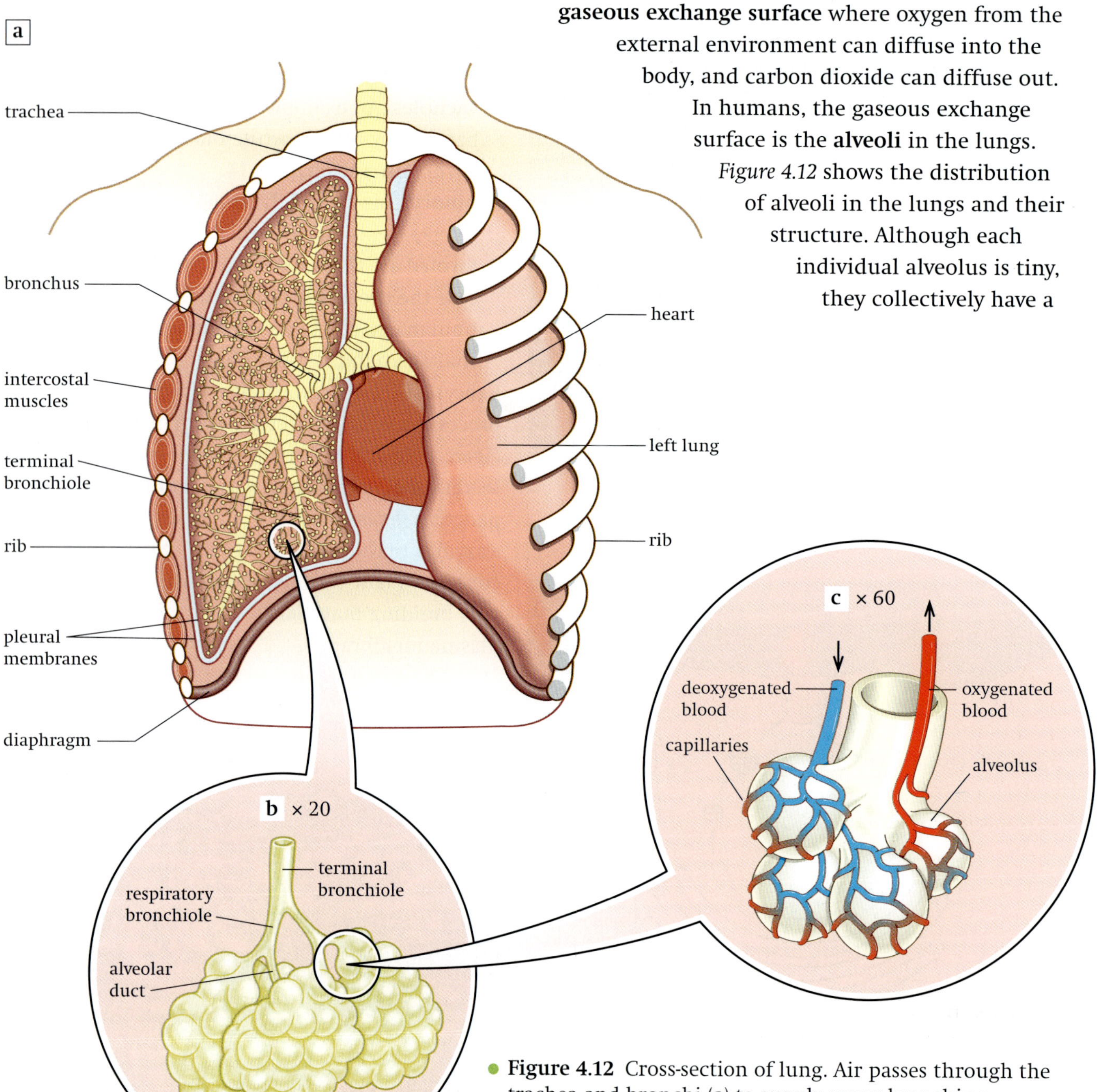

**Figure 4.12** Cross-section of lung. Air passes through the trachea and bronchi (**a**) to supply many branching bronchioles (**b**) which terminate in alveoli (**c**) where gaseous exchange occurs. A gaseous exchange surface of 60–80 $m^2$ fits into a cavity with a capacity of about 5 $dm^3$.

huge surface area, probably totalling around
70 m² in an adult. This increases the number of
oxygen and carbon dioxide molecules that can
diffuse through the surface at any one moment
and so speeds up the rate of gaseous exchange.

The alveoli have extremely thin walls, each
consisting of a single layer of squamous epithelial
cells no more than 0.5 μm thick. Pressed closely
against them are blood capillaries, also with very
thin single-celled walls. The thinness of this
barrier ensures that oxygen and carbon dioxide
molecules can diffuse very quickly across it.

You will remember that diffusion is the net
movement of molecules or ions down a concentration gradient. So, for gaseous exchange to take
place rapidly, a steep concentration gradient must
be maintained. This is done by breathing, and by
the movement of the blood. Breathing brings
supplies of fresh air into the lungs, with a relatively high oxygen concentration and a relatively
low carbon dioxide concentration. Blood is
brought to the lungs with a lower concentration
of oxygen and a higher concentration of carbon
dioxide than the air in the alveoli. Oxygen therefore diffuses down its concentration gradient
from the air in the alveoli to the blood, and
carbon dioxide diffuses down its
concentration gradient in the
opposite direction. The blood is
constantly flowing through and
out of the lungs so, as the
oxygenated blood leaves, more
deoxygenated blood enters to
maintain the concentration
gradients with each new breath.
You can read more about this in
chapters 8, 9 and 11.

## SAQ 4.6

How many times does an oxygen
molecule cross a plasma
membrane as it moves from the air
into a red blood cell?

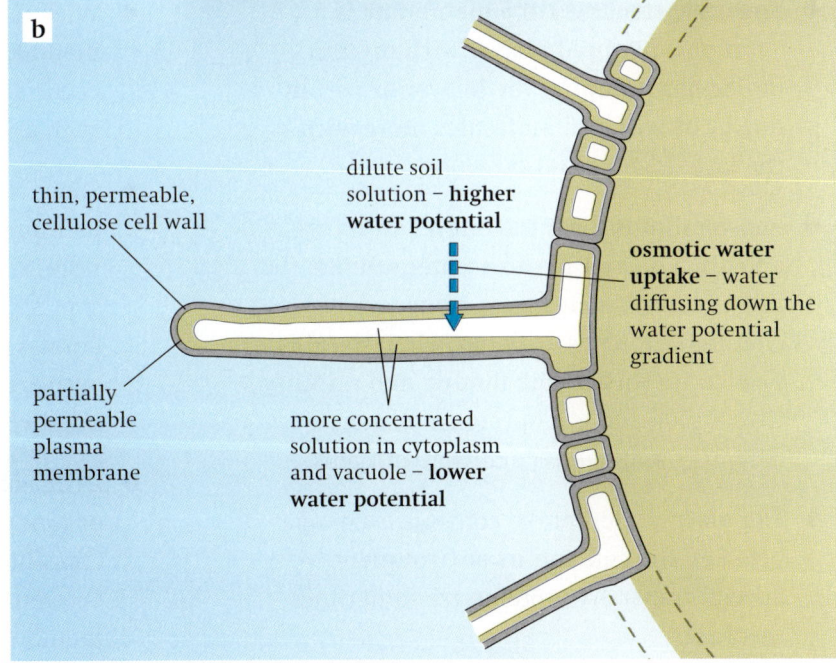

● **Figure 4.13 a** A root of a young radish plant showing the root hairs.
**b** Water uptake by a root hair cell. Mineral ions are also taken up but
by active transport against the concentration gradient via carrier
proteins.

## Uptake of mineral ions in a plant root

A quite different specialised exchange surface is the **root hairs** of a flowering plant. Root hairs are very thin extensions of the cells that make up the outer layer, or **epidermis**, of a root. Each root hair is about 200–250 μm across. There may be thousands of them on each tiny branch of a root, so together they provide an enormous surface area that is in contact with the soil surrounding the root (*figure 4.13*).

Soil is made up of particles of minerals and humus. In between these particles are spaces that are usually filled with air. Unless the soil is extremely dry, there is a thin layer of water coating each soil particle. The root hairs make contact with water, and absorb it by osmosis. The water moves in because there is a lower concentration of solutes in the water in the soil than there is inside the root hair cell. The water potential outside the root hair is therefore higher than the water potential inside it, and water moves passively down the water potential gradient into the cells.

Mineral ions are also absorbed from the soil into root hair cells. If there is a higher concentration of a particular ion outside the root hair cell than inside it, then it can be taken up passively, by facilitated diffusion. More usually, however, the plant requires ions that are present in relatively low concentrations in the soil, while its cells already contain a higher concentration of that ion. In that case, the ions will be taken up by active transport, through carrier proteins, using energy to move them against their concentration gradient. Note that the cell wall provides absolutely no barrier to the diffusion of water molecules or mineral ions – they are all able to pass through it quite freely.

### SAQ 4.7

Alveoli and root hairs are specialised exchange surfaces. What features do they share? How do these features help to increase the rate of exchange of substances across them?

# SUMMARY

◆ The basic structure of a membrane is a 7 nm phospholipid bilayer with protein molecules spanning the bilayer or within one or other layer. Molecules move within the layer (fluid mosaic).

◆ Phospholipids form the bilayer structure and are a barrier to most water-soluble substances. Proteins form transport channels or act as enzymes. Cholesterol is needed for membrane fluidity and stability. Glycolipids and glycoproteins form receptors and cell recognition antigens.

◆ The plasma membrane controls exchange between the cell and its environment. Special transport proteins are sometimes involved.

◆ Within cells, membranes allow compartmentalisation and division of labour to occur, within membrane-bound organelles such as the nucleus, ER and Golgi apparatus.

◆ Some chemical reactions take place on membranes as in photosynthesis and respiration. Membranes also contain receptor sites for hormones and neurotransmitters; possess cell recognition markers, such as antigens; and may contain enzymes, as with microvilli on epithelial cells in the gut.

◆ Diffusion is the net movement of molecules or ions from a region of their higher concentration to one of lower concentration. Oxygen and carbon dioxide cross membranes by diffusion through the phospholipid bilayer. Diffusion of ions and larger polar molecules through membranes is allowed by transport proteins.

◆ Water moves from regions of higher water potential to regions of lower water potential. When this takes place through a partially permeable membrane this diffusion is called osmosis. Pure water has a water potential of zero. Adding solute reduces the water potential by an amount known as the solute potential, which has a negative value. Adding pressure to a solution increases the water potential by an amount known as the pressure potential, which has a positive value.

◆ In dilute solutions, animal cells burst as water moves into the cytoplasm from the solution. In dilute solutions, a plant cell does not burst because the cell wall provides resistance to prevent it expanding. The pressure that builds up is the pressure potential. A plant cell in this state is turgid. In concentrated solutions, animal cells shrink whilst in plant cells the protoplast shrinks away from the cell wall in a process known as plasmolysis.

◆ Some ions and molecules move across membranes by active transport, against the concentration gradient. This needs a carrier protein and ATP to provide energy.

◆ Exocytosis and endocytosis involve the formation of vacuoles to move larger quantities of materials respectively out of, or into, cells by bulk transport. There are two types of endocytosis, namely phagocytosis and pinocytosis.

◆ Multicellular organisms often have surfaces that are specialised to allow exchange of substances to take place between their bodies and the environment. These include exchange surfaces such as alveoli in human lungs and root hairs of plants, which take up water and inorganic ions from the soil.

## CHAPTER 5

# Genetic control of protein structure and function

## By the end of this chapter you should be able to:

1 describe the structures of DNA and RNA, including the importance of base pairing and hydrogen bonding;

2 explain how DNA replicates semi-conservatively during interphase and interpret experimental evidence for this process;

3 know that a gene is part of a DNA molecule, made up of a sequence of nucleotides which codes for the construction of a polypeptide;

4 describe the way in which the nucleotide sequence codes for the amino acid sequence in the polypeptide;

5 describe how transcription and translation take place during protein synthesis, including the roles of messenger RNA, transfer RNA and ribosomes;

6 know that, as enzymes are proteins, their synthesis is controlled by DNA.

If you were asked to design a molecule which could act as the genetic material in living things, where would you start?

One of the features of the 'genetic molecule' would have to be the ability to **carry instructions** – a sort of blueprint – for the construction and behaviour of cells, and the way in which they grow together to form a complete living organism. Another would be the **ability to be copied** perfectly, over and over again, so that whenever the nucleus of a cell divides it can pass on an exact copy of each 'genetic molecule' to the nuclei of each of its 'daughter' cells.

Until the mid 1940s, biologists assumed that the 'genetic molecule' must be a protein. Only proteins were thought to be complex enough to be able to carry the huge number of instructions which would be necessary to make such a complicated

structure as a living organism. But during the 1940s and 1950s a variety of evidence came to light that proved beyond doubt that the 'genetic molecule' was not a protein at all, but DNA.

## The structure of DNA and RNA

DNA stands for **deoxyribonucleic acid** and RNA for **ribonucleic acid**. As we saw in chapter 2, DNA and RNA, like proteins and polysaccharides, are **macromolecules** (page 22). They are also **polymers**, made up of many similar, smaller molecules joined into a long chain. The smaller molecules from which DNA and RNA molecules are made are **nucleotides**. DNA and RNA are therefore **polynucleotides**. They are often referred to simply as nucleic acids.

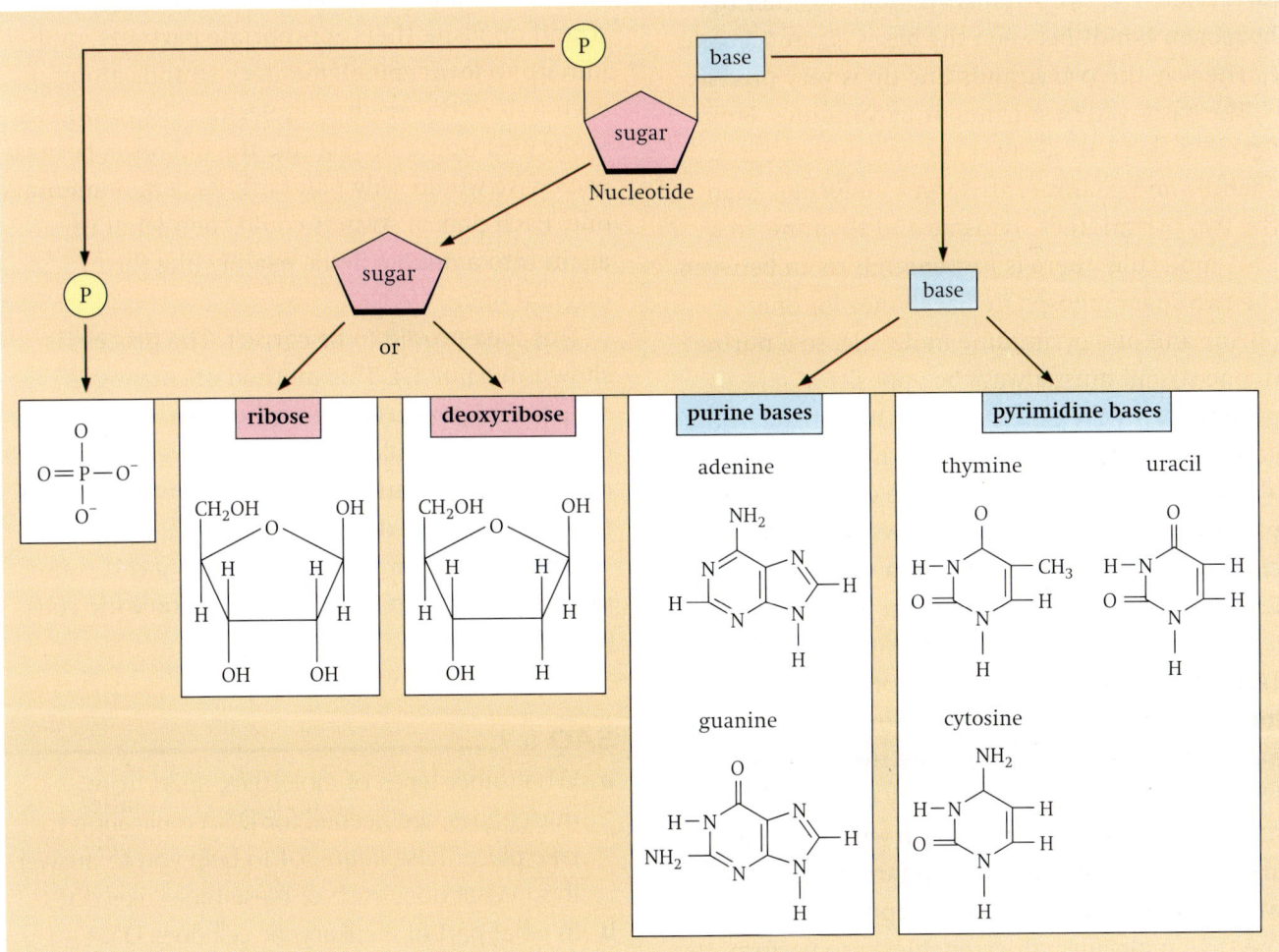

● **Figure 5.1** Nucleotides. A nucleotide is made of a nitrogen-containing base, a pentose sugar and a phosphate group Ⓟ.

## Nucleotides

*Figure 5.1* shows the structure of nucleotides. Nucleotides are made up of three smaller components. These are:

■ a nitrogen-containing base;
■ a pentose sugar;
■ a phosphate group.

There are just five different nitrogen-containing bases found in DNA and RNA. In a DNA molecule there are four: **adenine**, **thymine**, **guanine** and **cytosine**. (Do not confuse adenine with adenosine which is part of the name of ATP – adenosine is adenine with a sugar joined to it; and don't confuse thymine with thiamine, which is a vitamin.) An RNA molecule also contains four bases, but the base thymine is never found. Instead, RNA molecules contain a base called **uracil**. These bases are often referred to by their first letters: **A, T, C, G** and **U**.

The pentose (5-carbon) sugar can be either **ribose** (in RNA) or **deoxyribose** (in DNA). As their

● **Figure 5.2** The components of nucleotides.

names suggest, deoxyribose is almost the same as ribose, except that it has one fewer oxygen atoms in its molecule.

*Figure 5.1* shows the five different nucleotides from which DNA and RNA molecules can be built up. *Figure 5.2* shows the structure of their components in more detail; you do not need to remember these structures, but if you enjoy biochemistry you may find them interesting.

## Polynucleotides

To form the polynucleotides DNA and RNA, many nucleotides are linked together into a long chain. This takes place inside the nucleus, during inter-phase of the cell cycle (page 80).

*Figure 5.3a* shows the structure of part of a polynucleotide strand. In both DNA and RNA it is formed of alternating sugars and phosphates linked together, with the bases projecting sideways.

DNA molecules are made of *two* polynucleotide strands lying side by side running in opposite directions. The two strands are held together by **hydrogen bonds** between the bases (*figure 5.3b* and *c*). The way the two strands line up is very precise.

The bases can be purines or pyrimidines. From *figure 5.2*, you will see that the two purine bases, adenine and guanine, are larger molecules than the two pyrimidines, cytosine and thymine. In a DNA molecule, there is just enough room between the two sugar–phosphate backbones for one purine and one pyrimidine molecule, so a purine in one strand must always be opposite a pyrimidine in the other. In fact, the pairing of the bases is even more precise than this. Adenine always pairs with thymine, while cytosine always pairs with guanine: A with **T**, **C** with **G**. This **complementary base pairing** is a very important feature of polynucleotides, as you will see later.

DNA is often referred to as the 'double helix'. This refers to the 3D shape that DNA molecules form (*figure 5.3d*). The hydrogen bonds linking the bases, and therefore holding the two strands together, can be broken relatively easily. This happens during DNA replication (copying) and also during protein synthesis (manufacture). As we shall see, this too is a very important feature of the DNA molecule, which enables it to perform its role in the cell.

RNA molecules, unlike DNA, remain as *single* strands of polynucleotide and can form a very different 3D structure. We will look at this later in the chapter when we consider protein synthesis.

## DNA replication

We said at the beginning of this chapter that one of the features of the 'genetic molecule' would have to be the **ability to be copied** perfectly many times over.

It was not until 1953 that James Watson and Francis Crick used the results of work by Rosalind Franklin and others to work out the basic structure of the DNA molecule that we have just been looking at. To them, it was immediately obvious how this molecule could be copied perfectly, time and time again.

Watson and Crick suggested that the two strands of the DNA molecule could split apart. New nucleotides could then line up along each strand opposite their appropriate partners, and join up to form complementary strands along each half of the original molecule. The new DNA molecules would be just like the old ones, because each base would only pair with its complementary one. Each pair of strands could then wind up again into a double helix, exactly like the original one.

This idea proved to be correct. The process is shown in *figure 5.4*. This method of copying is called **semi-conservative replication**, because *half* of the original molecule is *kept* (conserved) in each of the new molecules. The experimental evidence for this process is described in *box 5A*.

DNA replication takes place when a cell is not dividing. This is in interphase in eukaryotic cells (chapter 6).

### SAQ 5.1

a What other types of molecules, apart from nucleotides, are needed for DNA replication to take place? (Use *figure 5.4* to help you to answer this.) What does each of these molecules do?
b In what part of a eukaryotic cell does DNA replication take place?

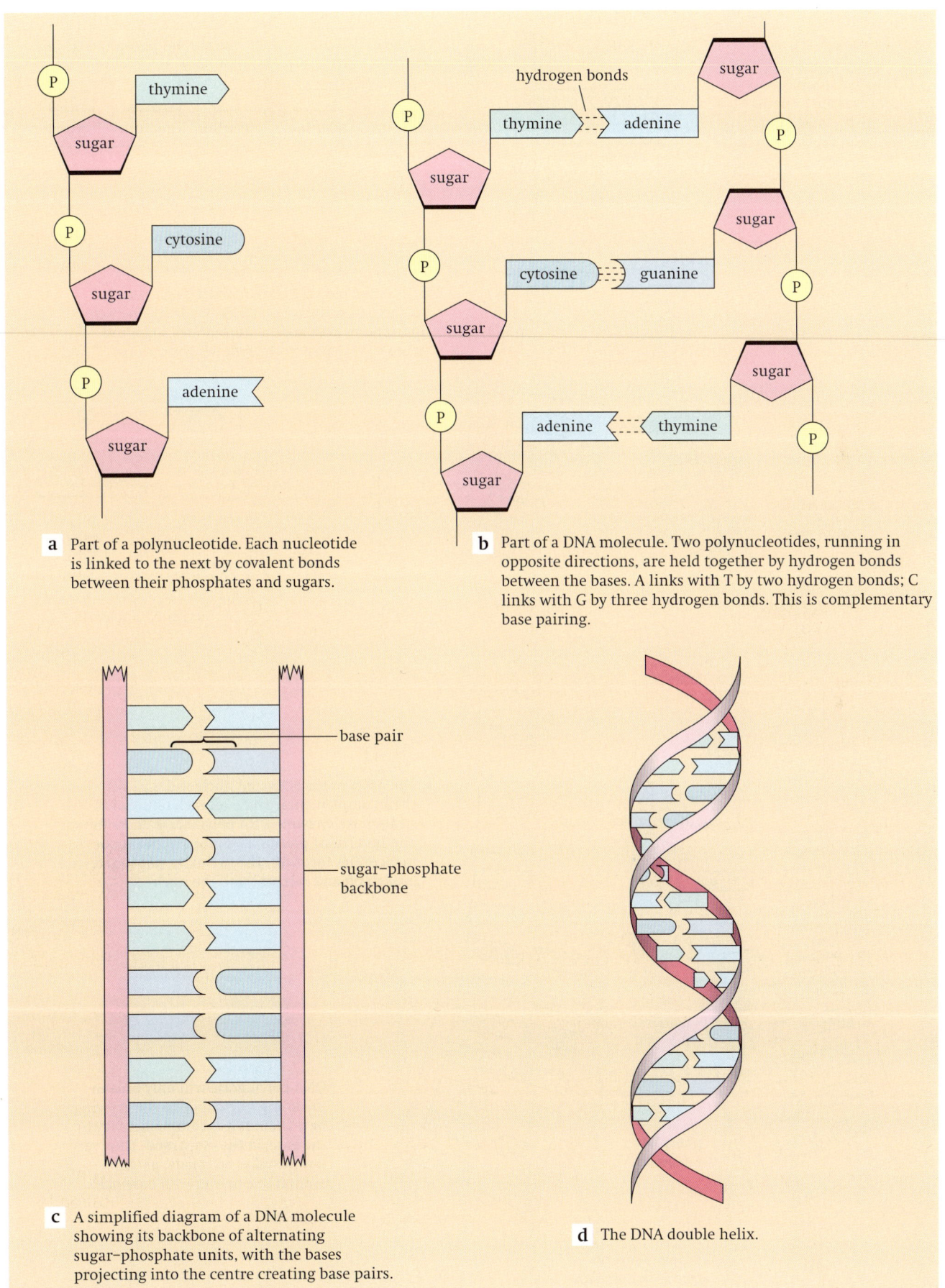

a Part of a polynucleotide. Each nucleotide is linked to the next by covalent bonds between their phosphates and sugars.

b Part of a DNA molecule. Two polynucleotides, running in opposite directions, are held together by hydrogen bonds between the bases. A links with T by two hydrogen bonds; C links with G by three hydrogen bonds. This is complementary base pairing.

c A simplified diagram of a DNA molecule showing its backbone of alternating sugar–phosphate units, with the bases projecting into the centre creating base pairs.

d The DNA double helix.

● **Figure 5.3** The structure of DNA.

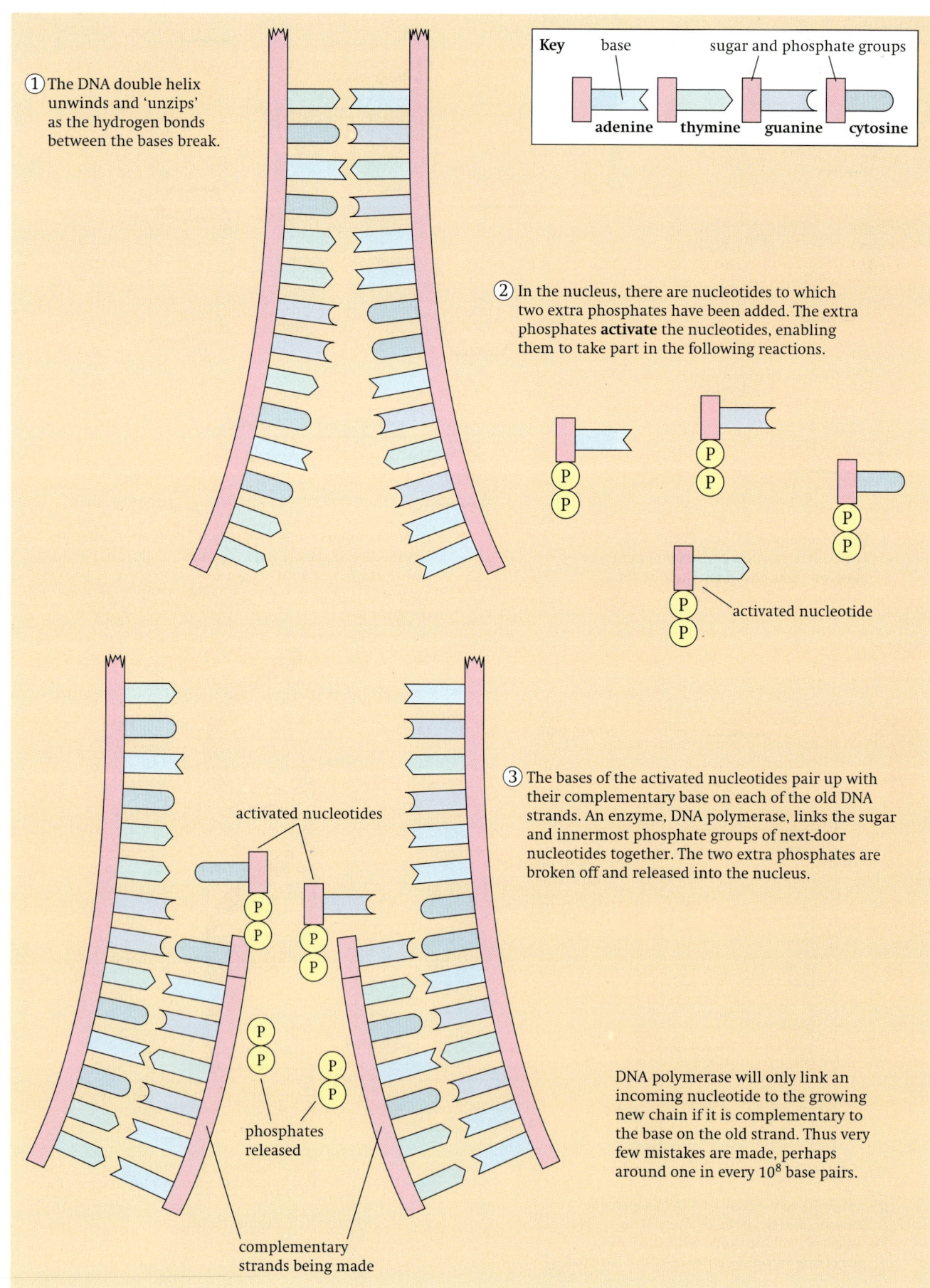

① The DNA double helix unwinds and 'unzips' as the hydrogen bonds between the bases break.

**Key** base sugar and phosphate groups

adenine thymine guanine cytosine

② In the nucleus, there are nucleotides to which two extra phosphates have been added. The extra phosphates **activate** the nucleotides, enabling them to take part in the following reactions.

activated nucleotide

③ The bases of the activated nucleotides pair up with their complementary base on each of the old DNA strands. An enzyme, DNA polymerase, links the sugar and innermost phosphate groups of next-door nucleotides together. The two extra phosphates are broken off and released into the nucleus.

activated nucleotides

phosphates released

DNA polymerase will only link an incoming nucleotide to the growing new chain if it is complementary to the base on the old strand. Thus very few mistakes are made, perhaps around one in every $10^8$ base pairs.

complementary strands being made

● **Figure 5.4** DNA replication.

## Box 5A Experimental evidence for the semi-conservative replication of DNA

In the 1950s, no-one knew exactly how DNA replicated. Three possibilities were suggested:

- **conservative replication**, in which one completely new double helix would be made from the old one (*figure 5.5a*);
- **semi-conservative replication**, in which each new molecule would contain one old strand and one new one (*figure 5.5b*);
- **dispersive replication**, in which each new molecule would be made of old bits and new bits scattered randomly through the molecules (*figure 5.5c*).

Most people thought that semi-conservative replication was most likely, because they could see how it might work. However, to make sure, experiments were carried out in 1958 by Mathew Meselsohn and Franklin Stahl in America.

They used the bacterium *Escherichia coli* (*E. coli* for short), a common, usually harmless, bacterium which lives in the human alimentary canal. They grew populations of the bacterium in a food source that contained ammonium chloride as a source of nitrogen.

The experiment relied on the variation in structure of nitrogen atoms. All nitrogen atoms contain 7 protons, but the number of neutrons can vary. Most nitrogen atoms have 7 neutrons, so their relative atomic mass (the total mass of all the protons and neutrons in an atom) is 14. Some nitrogen atoms have 8 neutrons, so their relative atomic mass is 15. These two forms are said to be **isotopes** of nitrogen.

The nitrogen atoms in the ammonium chloride that Meselsohn and Stahl supplied to the bacteria were the heavy isotope, nitrogen-15 ($^{15}$N). The bacteria used the $^{15}$N

to make their DNA. They were left in it long enough for them to divide many times, so that nearly all of their DNA contained only $^{15}$N atoms, not nitrogen-14 ($^{14}$N). This DNA would be heavier than 'normal' DNA containing $^{14}$N. Some of these bacteria were then transferred to a food source in which the nitrogen atoms were all $^{14}$N. Some were left there for just long enough for their DNA to replicate once – about 50 minutes. Others were left long enough for their DNA to replicate two, three or more times.

DNA was then extracted from each group of bacteria. The samples were placed into a solution of caesium chloride and spun in a centrifuge. The heavier the DNA was, the closer to the bottom of the tube it came to rest.

*Figure 5.6* on page 70 shows the results of Meselsohn and Stahl's experiment.

### SAQ 5.2

Looking at *figure 5.6* on page 70:

**a** Assuming that the DNA has reproduced semi-conservatively, explain why the band of DNA in tube 2 is higher than that in tube 1.

**b** What would you expect to see in tube 2 if the DNA had replicated conservatively?

**c** What would you expect to see in tube 2 if the DNA had replicated dispersively?

**d** Which is the first tube that provides evidence that the DNA has reproduced semi-conservatively and not dispersively? Explain your answer.

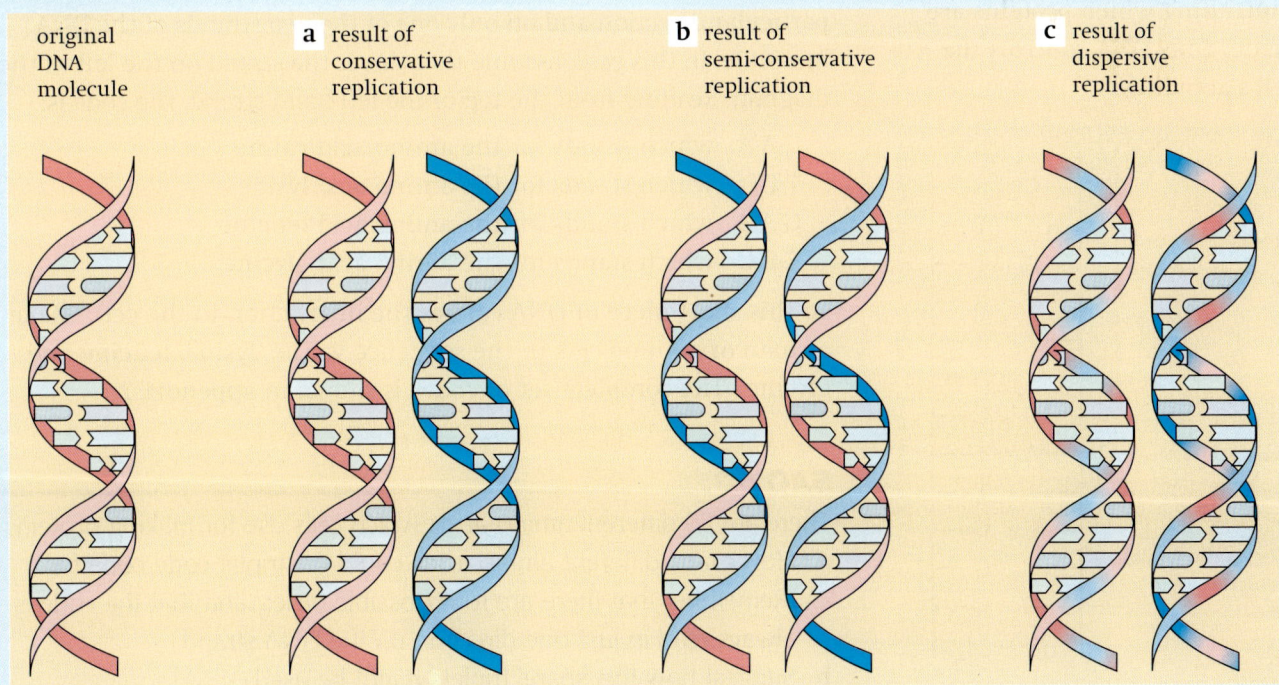

| original DNA molecule | **a** result of conservative replication | **b** result of semi-conservative replication | **c** result of dispersive replication |

● **Figure 5.5** Three suggestions for the method of DNA replication.

**Box 5A continued**

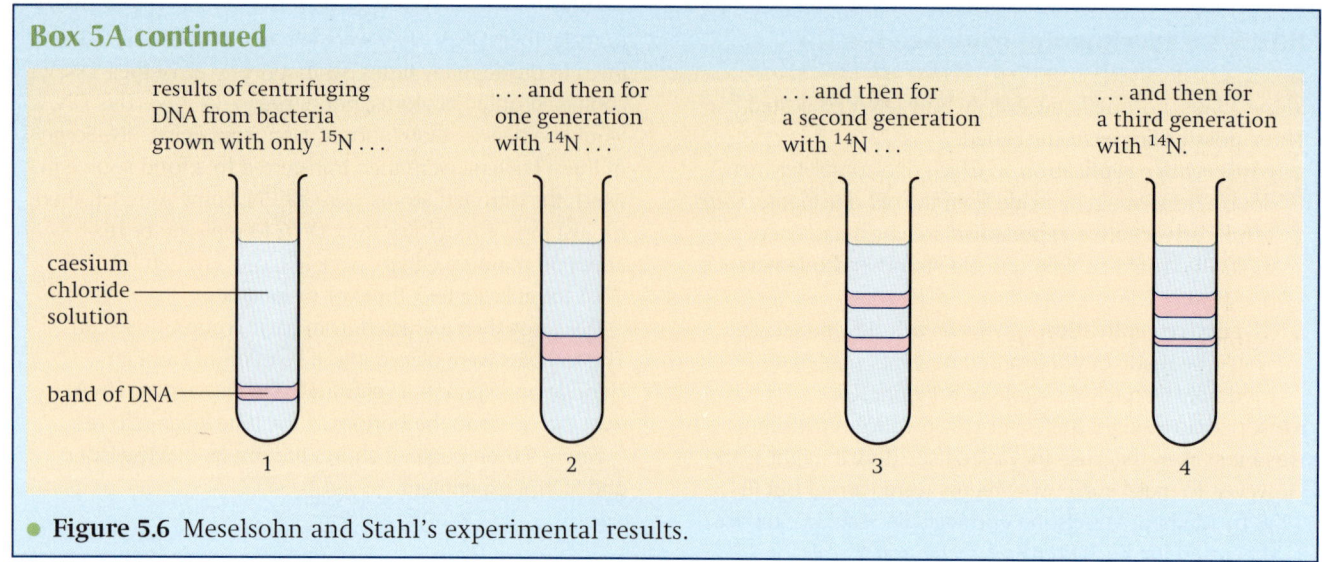

results of centrifuging DNA from bacteria grown with only $^{15}$N ...

... and then for one generation with $^{14}$N ...

... and then for a second generation with $^{14}$N ...

... and then for a third generation with $^{14}$N.

caesium chloride solution

band of DNA

1    2    3    4

● **Figure 5.6** Meselsohn and Stahl's experimental results.

# DNA, RNA and protein synthesis

## DNA controls protein synthesis

How can a single type of molecule like DNA control all the activities of a cell? The answer is very logical. All chemical reactions in cells, and therefore all their activities, are controlled by enzymes. Enzymes are proteins. DNA is a code for proteins, controlling which proteins are made. Thus DNA controls the cell's activities.

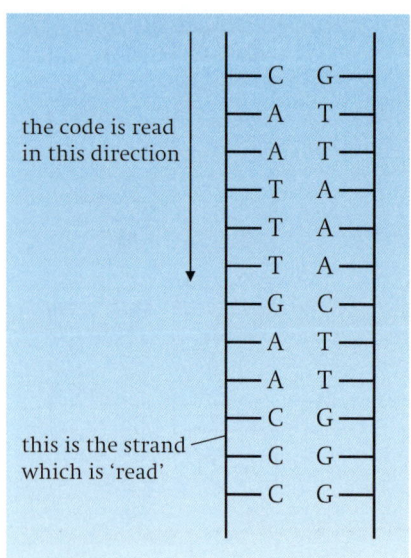

the code is read in this direction

C    G
A    T
A    T
T    A
T    A
T    A
G    C
A    T
A    T
C    G
C    G
C    G

this is the strand which is 'read'

● **Figure 5.7** A length of DNA coding for four amino acids.

Protein molecules are made up of strings of amino acids. The shape and behaviour of a protein molecule depends on the exact sequence of these amino acids, that is its primary structure (page 32). DNA controls protein structure by determining the exact order in which the amino acids join together when proteins are made in a cell.

### The triplet code

The sequence of bases or nucleotides in a DNA molecule is a code for the sequence of amino acids in a polypeptide. *Figure 5.7* shows a very short length of a DNA molecule, just enough to code for four amino acids.

The code is a three-letter, or **triplet**, code. Each sequence of three bases stands for one amino acid. The sequence is always read in one particular direction and on only one of the two strands of the DNA molecule. In this case, assume that this is the strand on the left of the diagram. Reading from the top of the left-hand strand, the code is:

C-A-A   which stands for the amino acid valine
T-T-T   which stands for the amino acid lysine
G-A-A   which stands for the amino acid leucine
C-C-C   which stands for the amino acid glycine

So this short piece of DNA carries the instruction to the cell: 'Make a chain of amino acids in the sequence valine, lysine, leucine and glycine'. The complete set of codes is shown in appendix 2.

### SAQ 5.3

There are 20 different amino acids which cells use for making proteins.
**a** How many different amino acids could the triplet code code for? (Remember that there are four possible bases, and that the code is always read in just one direction on the DNA strand.)
**b** Suggest how the 'spare' triplets might be used.
**c** Explain why it could not be a two–letter code.

## Genes and genomes

DNA molecules can be enormous. The bacterium *E. coli* has just one DNA molecule which is four million base pairs long. There is enough information here to code for several thousand proteins. The total DNA of a human cell is estimated to be about $3 \times 10^9$ base pairs long. However, it is thought that only 3% of this DNA actually codes for protein. The function of the remainder is uncertain.

A part of a DNA molecule which codes for just one polypeptide is called a **gene**. One DNA molecule contains many genes. In humans, it is estimated that there are about 30 000 genes.

The total set of genes in a cell is called the **genome**. The genome is the total information in one cell. Since all cells in the same individual contain the same information, the genome represents the genetic code of that organism.

In 1990, an ambitious project was begun to work out the entire base sequence of the complete human genome. It is called the Human Genome Project, and is being carried out in laboratories all over the world. As there are about 3 billion bases in the human genome, you can see that this is a huge task. It is hoped that this information will help us to identify every human gene and then to find out how at least some of them affect human health.

## Protein synthesis

The code on the DNA molecule is used to determine how the polypeptide molecule is constructed. *Figure 5.8 on pages 72–73 describes the process in detail,* but briefly the process is as follows.

In the nucleus, a complementary copy of the code from a gene is made by building a molecule of a *different* type of nucleic acid, called **messenger RNA (mRNA)**, using one strand of the DNA as a template.

The mRNA leaves the nucleus, and attaches to a **ribosome** in the cytoplasm (page 13).

In the cytoplasm there are molecules of **transfer RNA (tRNA)**. These have a triplet of bases at one end and a region where an amino acid can attach at the other. There are at least 20 different sorts of tRNA molecules, each with a particular triplet of bases at one end and able to attach to a specific amino acid at the other (*figure 5.9*).

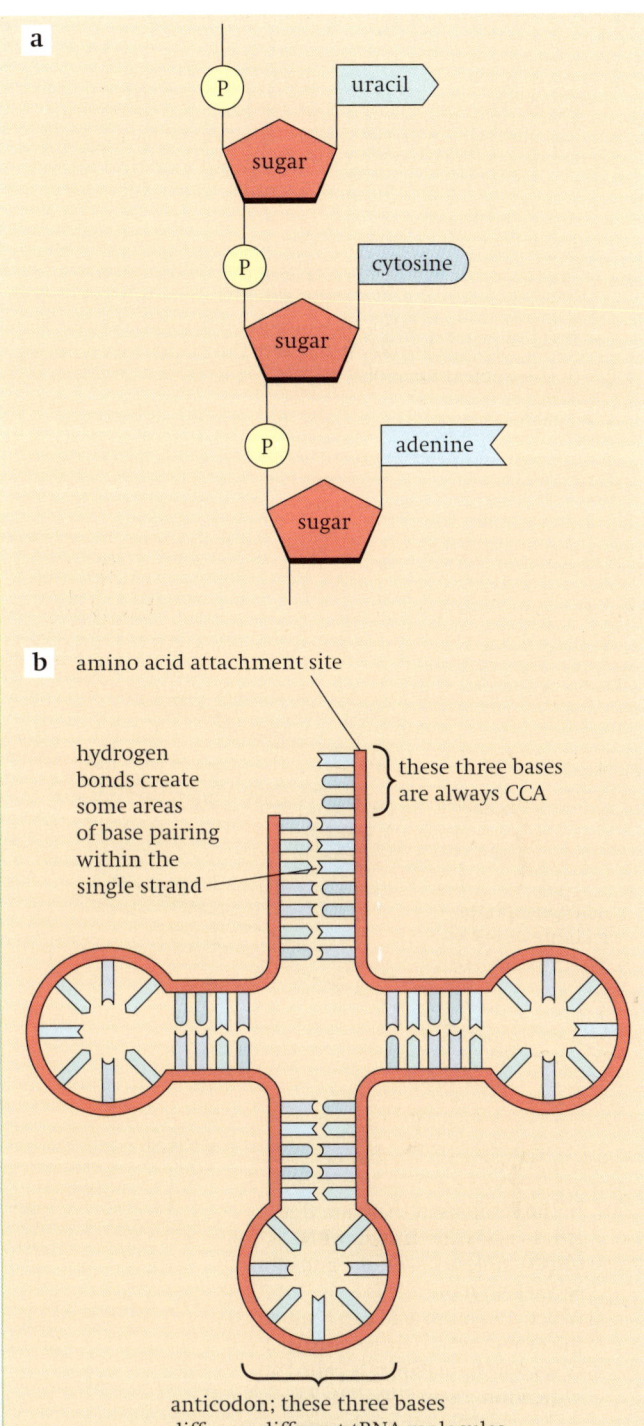

● **Figure 5.9** RNA.
**a** Part of a messenger RNA (mRNA) molecule.
**b** Transfer RNA (tRNA). The molecule is a single-stranded polynucleotide, folded into a clover-leaf shape. Transfer RNA molecules with different anticodons are recognised by different enzymes, which load them with their appropriate amino acid.

### SAQ 5.4

Summarise the differences between the structures of DNA and RNA.

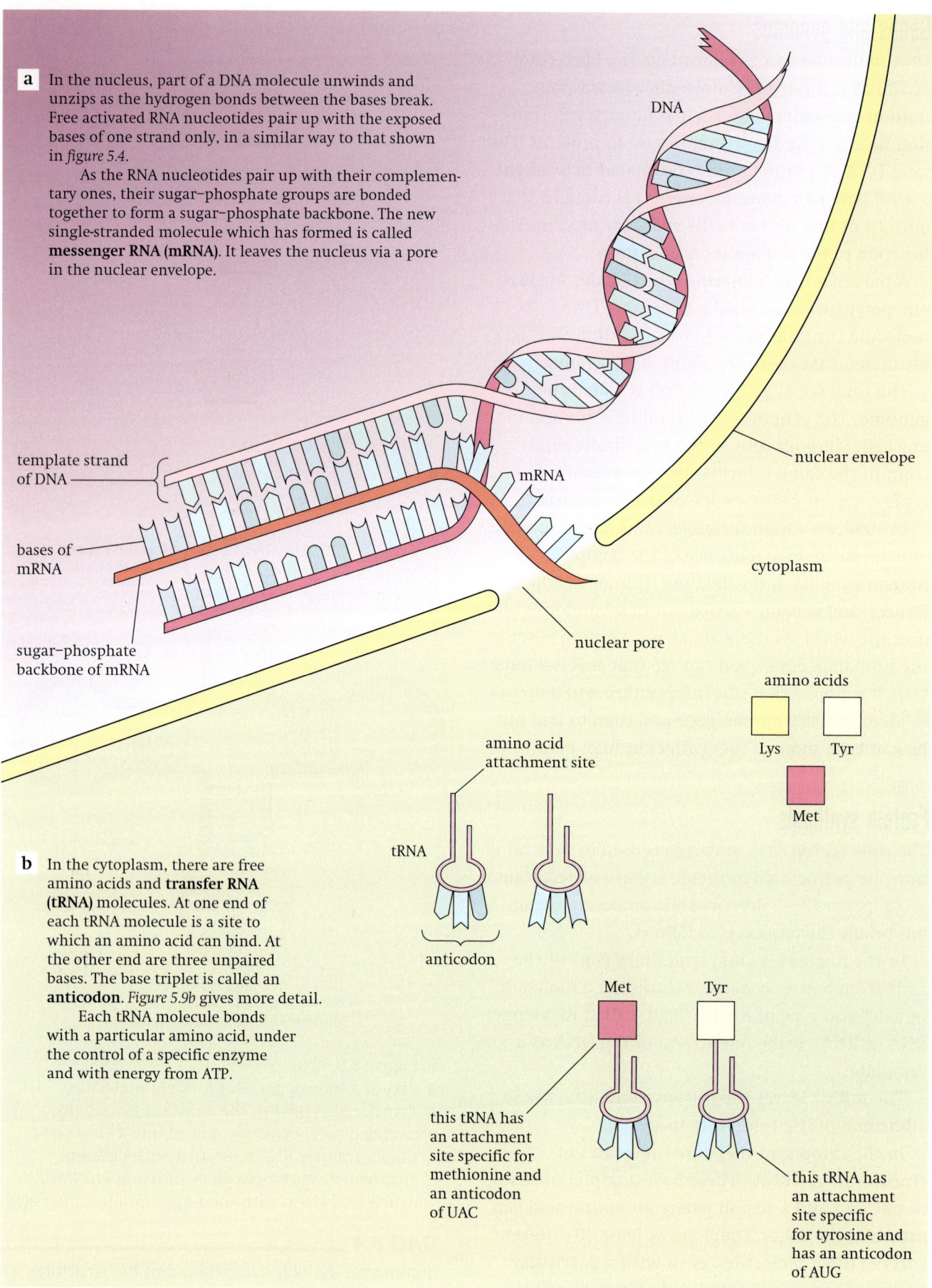

**a** In the nucleus, part of a DNA molecule unwinds and unzips as the hydrogen bonds between the bases break. Free activated RNA nucleotides pair up with the exposed bases of one strand only, in a similar way to that shown in *figure 5.4*.

As the RNA nucleotides pair up with their complementary ones, their sugar–phosphate groups are bonded together to form a sugar–phosphate backbone. The new single-stranded molecule which has formed is called **messenger RNA (mRNA)**. It leaves the nucleus via a pore in the nuclear envelope.

DNA

template strand of DNA

bases of mRNA

sugar–phosphate backbone of mRNA

mRNA

nuclear envelope

cytoplasm

nuclear pore

amino acids

Lys   Tyr

Met

amino acid attachment site

tRNA

anticodon

Met   Tyr

**b** In the cytoplasm, there are free amino acids and **transfer RNA (tRNA)** molecules. At one end of each tRNA molecule is a site to which an amino acid can bind. At the other end are three unpaired bases. The base triplet is called an **anticodon**. *Figure 5.9b* gives more detail.

Each tRNA molecule bonds with a particular amino acid, under the control of a specific enzyme and with energy from ATP.

this tRNA has an attachment site specific for methionine and an anticodon of UAC

this tRNA has an attachment site specific for tyrosine and has an anticodon of AUG

● **Figure 5.8** Protein synthesis – transcription.

**c** Meanwhile, also in the cytoplasm, the mRNA molecule attaches to a ribosome. Ribosomes are made of ribosomal RNA (rRNA) and protein and contain a small and a large subunit. The mRNA binds to the small subunit. Six bases at a time are exposed to the large subunit.

The first three exposed bases, or **codon**, are *always* AUG. A tRNA molecule with the complementary anticodon, UAC, forms hydrogen bonds with this codon. This tRNA molecule has the amino acid methionine attached to it.

**d** A second tRNA molecule bonds with the next three exposed bases. This one brings a different amino acid. The two amino acids are held closely together, and a peptide bond is formed between them. This reaction is catalysed by the enzyme peptidyl transferase, which is found in the small subunit of the ribosome.

**e** The ribosome now moves along the mRNA, 'reading' the next three bases on the ribosome. A third tRNA molecule brings a third amino acid, which joins to the second one. The first tRNA leaves.

**f** The polypeptide chain continues to grow, until a 'stop' codon is exposed on the ribosome. This is UAA, UAC or UGA.

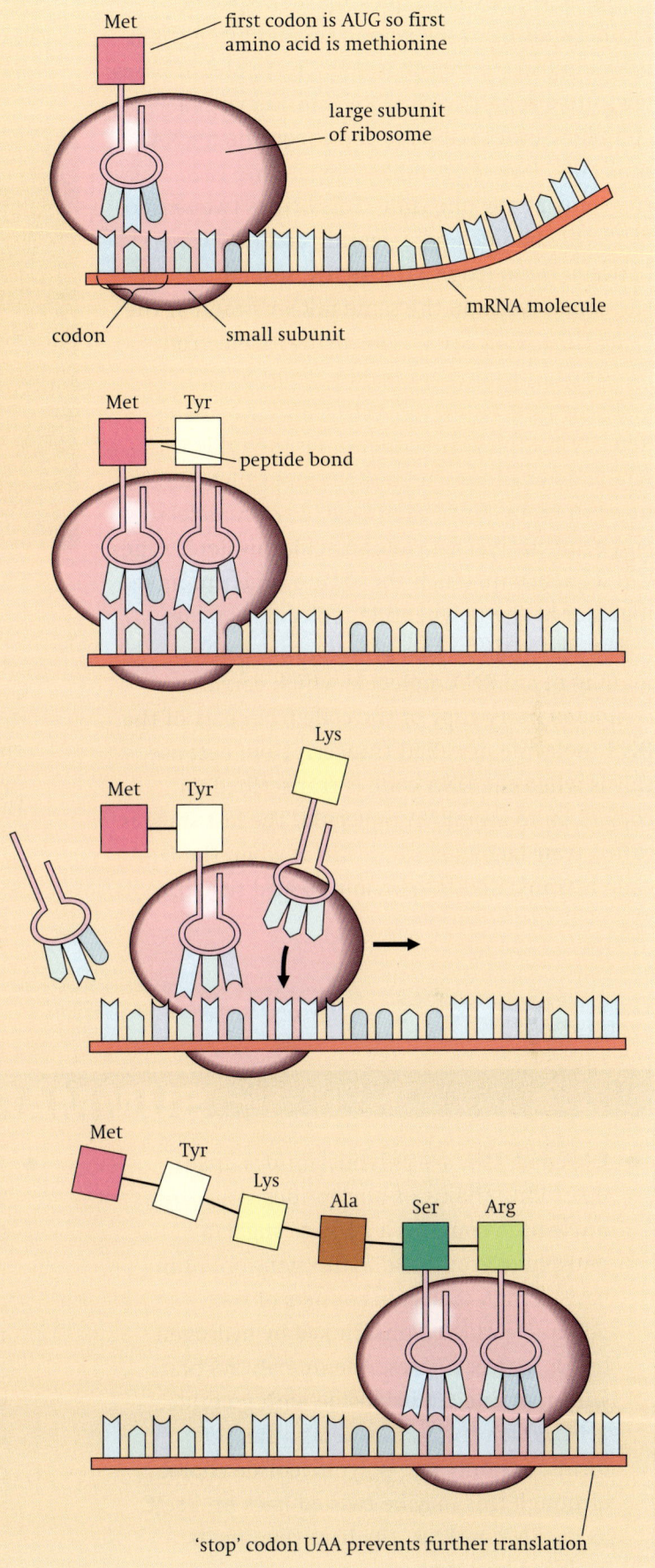

● **Figure 5.8 continued** Protein synthesis – translation.

The tRNA molecules pick up their specific amino acids from the cytoplasm and bring them to the mRNA on the ribosome. The triplet of bases (an **anticodon**) of each tRNA links up with a complementary triplet (a **codon**) on the mRNA molecule. Two tRNA molecules fit onto the ribosome at any one time. This brings two amino acids side by side and a peptide bond is formed between them (page 31). Usually, several ribosomes work on the same mRNA strand at the same time. They are visible, using an electron microscope, as **polyribosomes** (*figure 5.10*).

So the base sequence on the DNA molecule determines the base sequence on the mRNA, which determines which tRNA molecules can link up with them. Since each type of tRNA molecule is specific for just one amino acid, this determines the sequence in which the amino acids are linked together as the polypeptide molecule is made.

The first stage in this process, that is the making of a mRNA molecule which carries a complementary copy of the code from part of the DNA molecule, is called **transcription**, because this is when the DNA code is **transcribed**, or copied, on to an mRNA molecule. The last stage is called **translation**, because this is when the DNA code is **translated** into an amino acid sequence.

*SAQ 5.5*
Draw a simple flow diagram to illustrate the important stages in protein synthesis.

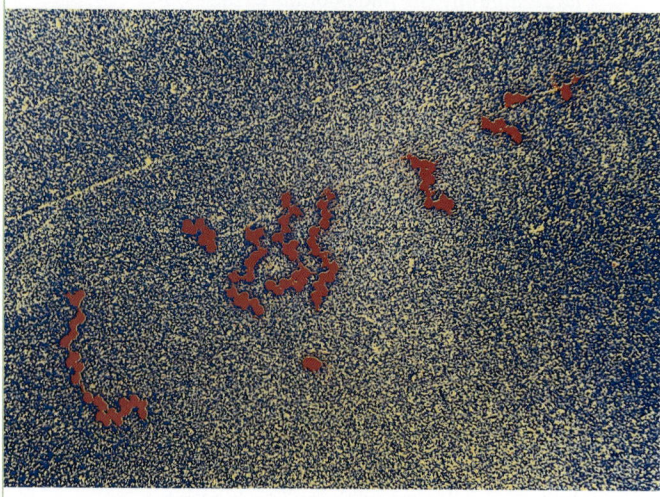

● **Figure 5.10** Protein synthesis in a bacterium. In bacteria, there is no nucleus, so protein synthesis can begin as soon as some mRNA has been made. Here, the long thread running from left to right is DNA. Nine mRNA molecules are being made, using this DNA as a template. Each mRNA molecule is immediately being read by ribosomes, which you can see as red blobs attached along the mRNAs. The mRNA strand at the left hand end is much longer than the one at the right, indicating that the mRNA is being synthesised working along the DNA molecule from right to left. (× 54 000)

# SUMMARY

◆ DNA and RNA are polynucleotides, made up of long chains of nucleotides. A nucleotide contains a pentose sugar, a phosphate group and a nitrogen-containing base. A DNA molecule consists of two polynucleotide chains, linked by hydrogen bonds between bases. Adenine always pairs with thymine, and cytosine with guanine. RNA, which comes in several different forms, has only one polynucleotide chain, although this may be twisted back on itself, as in tRNA. In RNA, the base thymine is replaced by uracil.

◆ DNA molecules replicate during interphase. The hydrogen bonds between the bases break, allowing free nucleotides to fall into position opposite their complementary ones on each strand of the original DNA molecule. Adjacent nucleotides are then linked, through their phosphates and sugars, to form new strands. Two complete new molecules are thus formed from one old one, each new molecule containing one old strand and one new.

◆ The sequence of bases (or nucleotides) on a DNA molecule codes for the sequence of amino acids in a protein (or polypeptide). Each amino acid is coded for by three bases. A length of DNA coding for one complete protein or polypeptide is a gene.

◆ During protein synthesis, a complementary copy of the base sequence on a gene is made, by building a molecule of mRNA against one DNA strand. The mRNA then moves to a ribosome in the cytoplasm.

tRNA molecules with complementary triplets of bases temporarily pair with the base triplets on mRNA, bringing appropriate amino acids. As two amino acids are held side by side, a peptide bond forms between them. The ribosome moves along the mRNA molecule, so that appropriate amino acids are gradually linked together, following the sequence laid down by the base sequence on the mRNA.

## CHAPTER 6

# Nuclear division

## By the end of this chapter you should be able to:

1 explain the need for the production of genetically identical cells within an organism, and hence for precise control of nuclear and cell division;

2 distinguish between haploid and diploid;

3 explain what is meant by homologous pairs of chromosomes;

4 describe how nuclear division comes before cell division and know that replication of DNA takes place during interphase;

5 describe, with the aid of diagrams, the behaviour of chromosomes during the mitotic cell cycle and the associated behaviour of the nuclear envelope, plasma membrane and centrioles;

6 name the main stages of mitosis;

7 explain that as a result of mitosis, growth, repair and asexual reproduction of living organisms is possible;

8 explain why gametes must be haploid and that this is achieved by meiosis (reduction division);

9 explain how cancers are a result of uncontrolled cell division and list factors that can increase the chances of cancerous growth.

All living organisms grow and reproduce. Since living organisms are made of cells, this means that cells must be able to grow and reproduce. Cells reproduce by dividing and passing on their genes (hereditary information) to 'daughter' cells. The process must be very precisely controlled so that no vital information is lost. We have seen how this is achieved at the molecular level (chapter 5). We shall now examine the cellular level, particularly in eukaryotes.

In chapter 1 we saw that one of the most conspicuous structures in eukaryotic cells is the nucleus. Its importance has been obvious ever since it was realised that the nucleus always divides before a cell divides. Each daughter cell therefore contains its own nucleus. This is important because the nucleus controls the cell's activities. It does this through the genetic material DNA. In chapter 5, we saw how DNA is able to act as a set of instructions, or code, for life.

So, nuclear division combined with cell division allows cells, and therefore whole organisms, to reproduce themselves. It also allows multicellular organisms to grow. The cells in your body, for example, are all genetically identical (apart from the gametes); they were all derived from one cell, the zygote, which was the cell formed when two gametes from your parents fused.

## The nucleus contains chromosomes

Just before a eukaryotic cell divides, a number of characteristic thread-like structures gradually become visible in the nucleus. They are easily seen because they stain intensely with particular stains. They were originally termed **chromosomes**

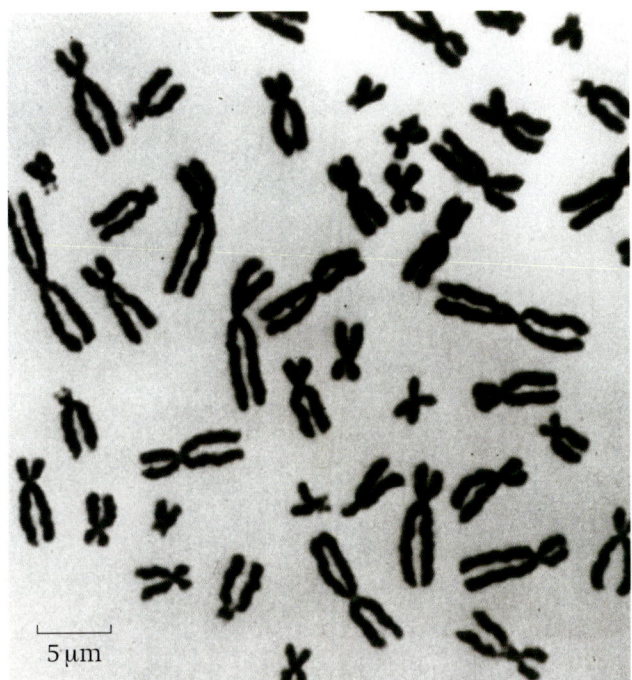

● **Figure 6.1** Photograph of a set of chromosomes in a human male, just before cell division. Each chromosome is composed of two chromatids held at the centromere. Note the different sizes of the chromosomes and positions of the centromeres.

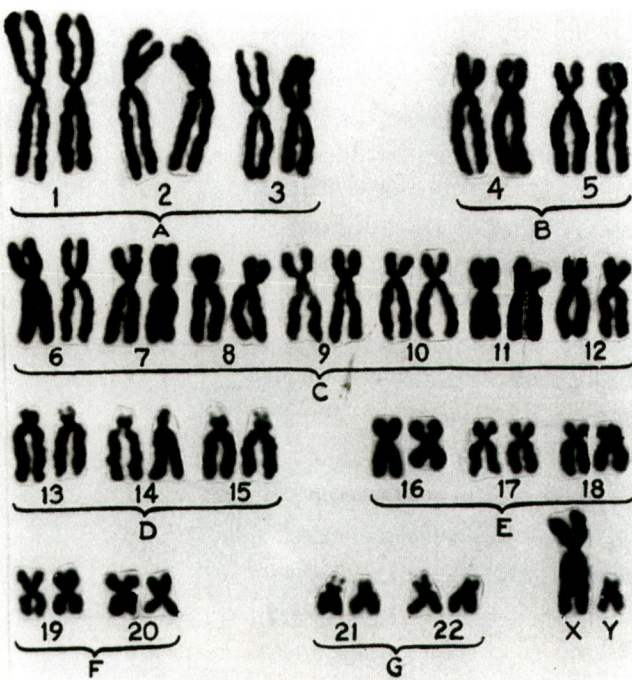

● **Figure 6.2** Karyotype of a human male, prepared from *figure 6.1*. Non-sex chromosomes (autosomes) are placed in the groups A to G. The sex chromosomes (X, female; Y, male) are placed separately.

because *chromo* means 'coloured' and *somes* means 'bodies'. The number of chromosomes is characteristic of the species. For example, in human cells there are 46 chromosomes, and in fruit fly cells there are only 8. *Figure 6.1* shows the appearance of a set of chromosomes in the nucleus of a human cell. *Figure 6.2* shows the same chromosomes rearranged and *figure 6.3* is a diagram of the same chromosomes.

## SAQ 6.1

Look at *figures 6.1, 6.2* and *6.3* and try to decide why the chromosomes are arranged in the particular order shown.

A photograph such as *figure 6.2* is called a **karyotype**. It is prepared by cutting out individual chromosomes from a picture like *figure 6.1* and rearranging them. Note the following.

■ There are matching pairs of chromosomes. These are called **homologous pairs**. Each pair is given a number. In the original zygote, one of each pair came from the mother and one from the father. *Figure 6.3* shows the complete set of 23 chromosomes that derived from just one of

the parents. Accurate and precise nuclear division during growth results in all cells of the body containing the two sets of chromosomes. There is more detail of this later in the chapter.

■ The pairs of chromosomes can be distinguished because each pair has a distinctive banding pattern when stained.

■ Two chromosomes are displayed to one side. These are the **sex chromosomes**, which determine the sex. All the other chromosomes are called **autosomes**. It is conventional to position the two sex chromosomes to one side in a karyotype so that the sex of the organism can be recognised quickly. In humans, females have two X chromosomes, and males have one X and one Y chromosome. The Y chromosome has a portion missing and is therefore smaller than the X chromosome.

Each chromosome has a characteristic set of genes which code for different features. The Human Genome Project is investigating which genes are located on which chromosomes. For example, we now know that the gene for the genetic disease cystic fibrosis is located on chromosome 7.

## Haploid and diploid cells

When animals other than humans are examined, we again find that cells usually contain two sets of chromosomes. Such cells are described as **diploid**. This is represented as **2n**, where n = number of chromosomes in one set of chromosomes.

Not all cells are diploid. As we shall see, gametes have only one set of chromosomes. A cell which contains only one set of chromosomes is described as **haploid**. This is represented as **n**. In humans, therefore, a 2n body cell has 46 chromosomes, and a gamete has 23.

## The structure of chromosomes

Before studying nuclear division, you need to understand a little about the structure of chromosomes. *Figure 6.4* is a simplified diagram of the structure of a chromosome. It can be seen that the chromosome is really a double structure. It is made of two identical structures called **chromatids**. This is because during the period between nuclear divisions, which is known as **interphase**, each DNA molecule in a nucleus makes an identical copy of itself (see chapter 5). Each copy is contained in a chromatid and the two chromatids are held together by a characteristic narrow region called the **centromere**, forming a chromosome. The centromere can be found anywhere along the length of the chromosome, but the position is characteristic for a particular chromosome, as

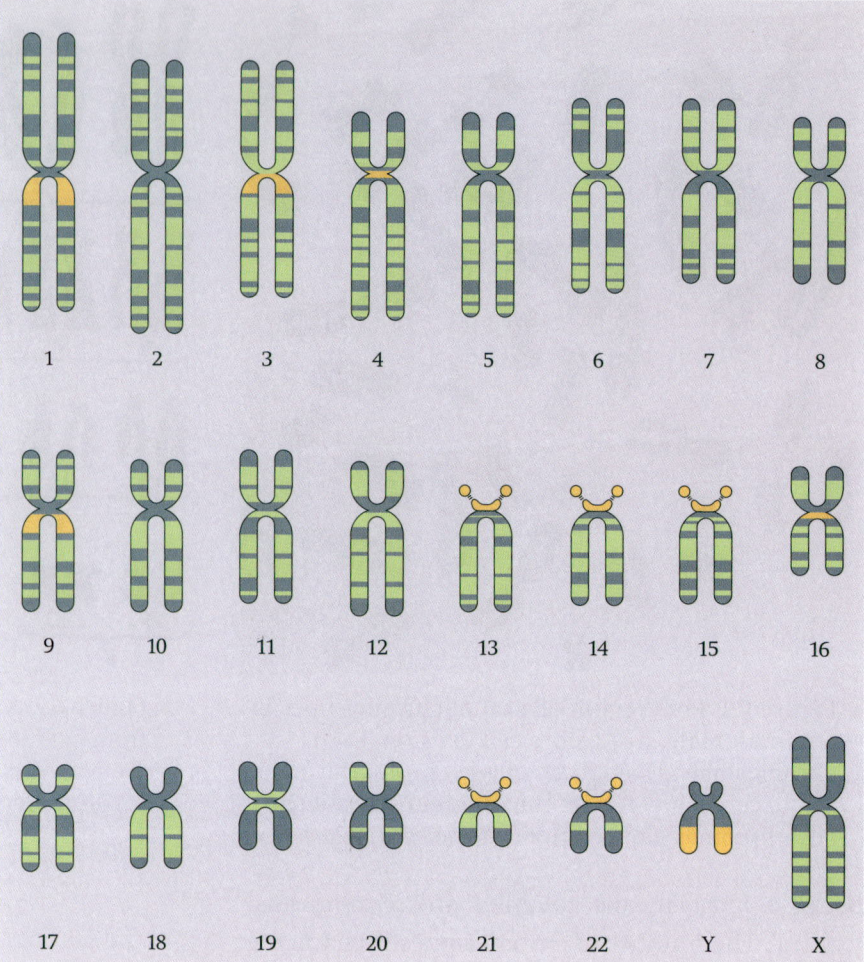

● **Figure 6.3** Diagram showing banding patterns of human chromosomes when stained. Green areas represent those regions that stain with ultraviolet fluorescence staining; orange areas are variable bands. Note that the number of genes is greater than the number of stained bands. Only one chromosome of each pair is shown except for the sex chromosomes which are both shown.

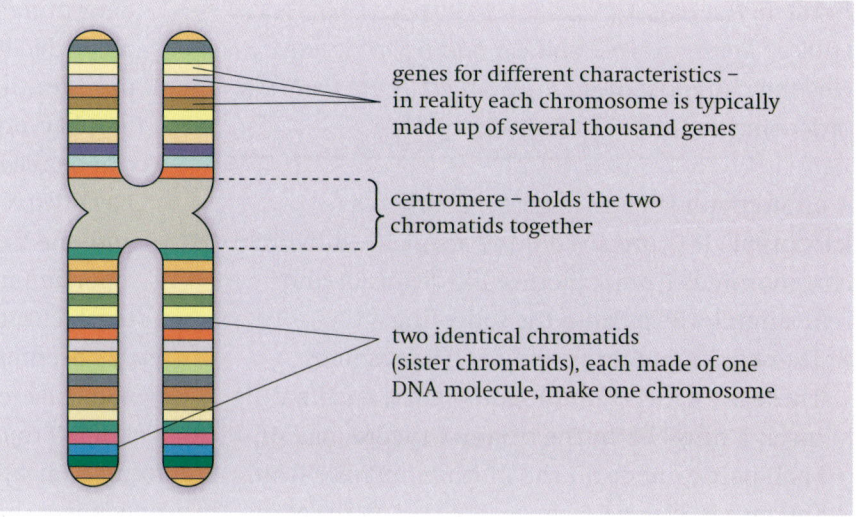

genes for different characteristics – in reality each chromosome is typically made up of several thousand genes

centromere – holds the two chromatids together

two identical chromatids (sister chromatids), each made of one DNA molecule, make one chromosome

● **Figure 6.4** Simplified diagram of the structure of a chromosome.

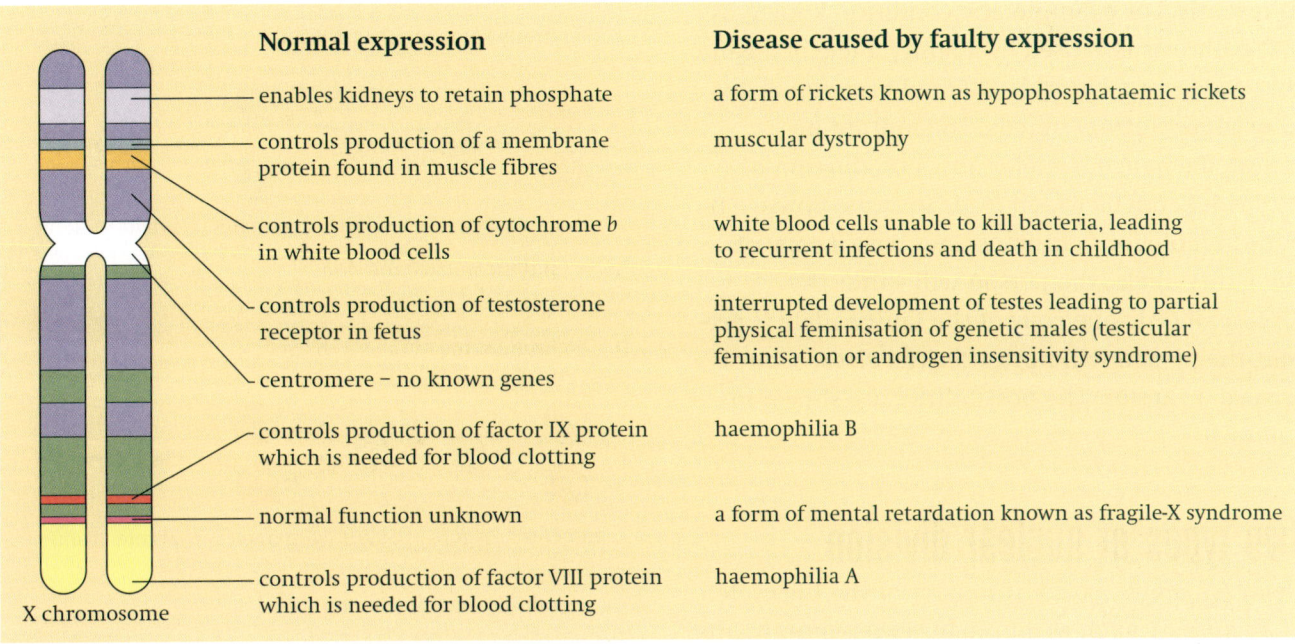

| | Normal expression | Disease caused by faulty expression |
|---|---|---|
| | enables kidneys to retain phosphate | a form of rickets known as hypophosphataemic rickets |
| | controls production of a membrane protein found in muscle fibres | muscular dystrophy |
| | controls production of cytochrome *b* in white blood cells | white blood cells unable to kill bacteria, leading to recurrent infections and death in childhood |
| | controls production of testosterone receptor in fetus | interrupted development of testes leading to partial physical feminisation of genetic males (testicular feminisation or androgen insensitivity syndrome) |
| | centromere – no known genes | |
| | controls production of factor IX protein which is needed for blood clotting | haemophilia B |
| | normal function unknown | a form of mental retardation known as fragile-X syndrome |
| | controls production of factor VIII protein which is needed for blood clotting | haemophilia A |

X chromosome

● **Figure 6.5** Locations of some of the genes on the human female sex chromosome (the X chromosome) showing the effects of normal and faulty expression.

*figures 6.2* and *6.3* show. **Each chromatid contains one DNA molecule.** As you know from chapter 5, DNA is the molecule of inheritance and is made up of a series of genes. Each gene is one unit of inheritance, controlling one characteristic of the organism. The fact that the two DNA molecules in sister chromatids, and hence their genes, are identical is the key to precise nuclear division.

The gene for a particular characteristic is always found at the same position, or **locus** (plural **loci**), on a chromosome. *Figure 6.5* shows a map of some of the genes on the human female sex chromosome which, if faulty, are involved in known genetic diseases.

Each chromosome typically has several hundred to several thousand gene loci, many more than shown in *figure 6.4*. The total number of different genes in humans is thought to be about 30 000.

## Homologous pairs of chromosomes

The word *homologous* means 'similar in structure and composition'. Each member of a homologous pair of chromosomes comes from one of the parents. In humans 23 chromosomes come from the female parent (the **maternal chromosomes**), and 23 from the male parent (the **paternal chromosomes**). There are therefore 23 homologous pairs.

Each member of a pair possesses genes for the same characteristics. They may differ, however, in exactly how they code for those characteristics. A gene controlling a characteristic may exist in different forms (**alleles**) which are expressed

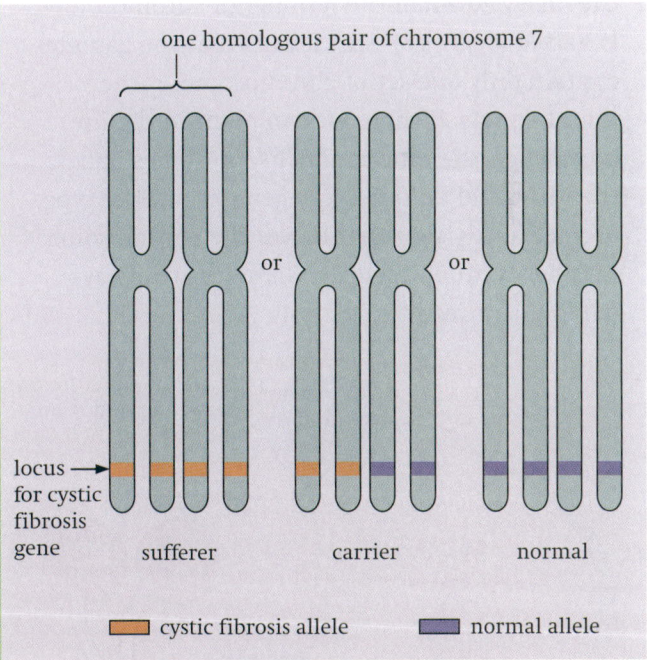

one homologous pair of chromosome 7

or    or

locus for cystic fibrosis gene

sufferer          carrier          normal

■ cystic fibrosis allele          ■ normal allele

● **Figure 6.6** An example of the possible combinations of one gene's alleles on a homologous pair of chromosomes. Note that sister chromatids within each chromosome have identical copies of the gene. The variation occurs between whole chromosomes.

differently. For example, the condition known as cystic fibrosis is caused by a faulty allele of a gene that codes for a chloride channel protein needed to produce normal mucus. The **mutant** or **mutated** (changed) allele causes production of very thick mucus which leads to cystic fibrosis. If both homologous chromosomes have a copy of the faulty allele, the person will suffer the disease; if only one copy of the faulty allele is present the person will not suffer the disease, but is termed a **carrier**. The possibilities are shown in *figure 6.6*.

# Two types of nuclear division

*Figure 6.7* shows a brief summary of the life cycle of an animal, such as a human. Two requirements must be satisfied.

1 **Growth** When a diploid zygote (one cell) grows into a multicellular diploid adult the daughter cells must keep the same number of chromosomes as the parent cell. The type of nuclear division that occurs here is called **mitosis**.

2 **Sexual reproduction** If the life cycle contains sexual reproduction, there must be a point in the life cycle when the number of chromosomes is halved (*figure 6.8*). This means that the gametes contain only one set of chromosomes rather than two sets. If there was no point in the life cycle when the number of chromosomes halved then it would double every generation. The type of nuclear division that halves the chromosome number is called **meiosis**. Gametes are always haploid as a result of meiosis.

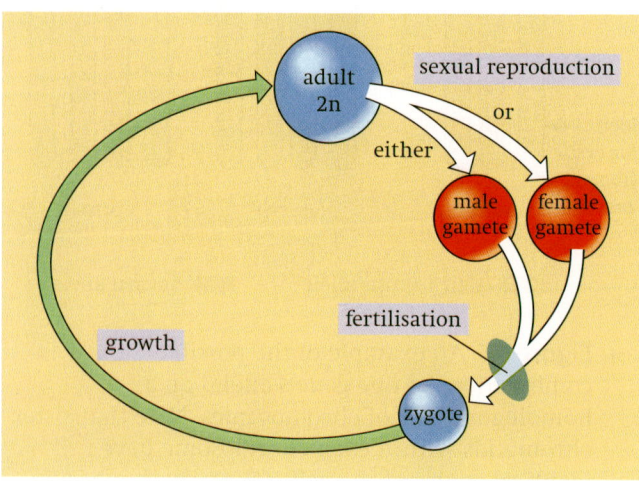

● **Figure 6.7** Outline of the life cycle of an animal.

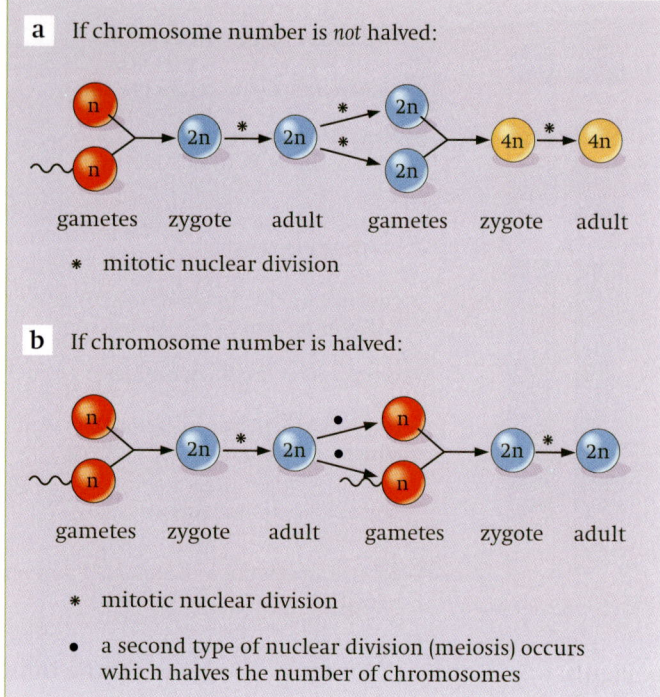

**a** If chromosome number is *not* halved:

gametes  zygote  adult  gametes  zygote  adult

\* mitotic nuclear division

**b** If chromosome number is halved:

gametes  zygote  adult  gametes  zygote  adult

\* mitotic nuclear division

● a second type of nuclear division (meiosis) occurs which halves the number of chromosomes

● **Figure 6.8** A life cycle in which the chromosome number is **a** not halved, **b** halved.

# Mitosis in an animal cell

Mitosis is nuclear division that produces two genetically identical daughter nuclei, each containing the same number of chromosomes as the parent nucleus. A diploid nucleus that divides by mitosis produces two diploid nuclei; a haploid nucleus produces two haploid nuclei. Mitosis, like meiosis, is a form of **nuclear division** and is part of a precisely controlled process called the **cell cycle**.

## The cell cycle

The cell cycle is the period between one cell division and the next. It has three phases, namely **interphase**, **nuclear division** and **cell division**. These are shown in *figure 6.9*.

During interphase the cell grows to its normal size after cell division and carries out its normal functions, synthesising many substances, especially proteins, in the process. At some point during interphase, a signal may be received that the cell should divide again. The DNA in the nucleus replicates so that each chromosome consists of two identical chromatids, each containing one copy of that chromosome's DNA.

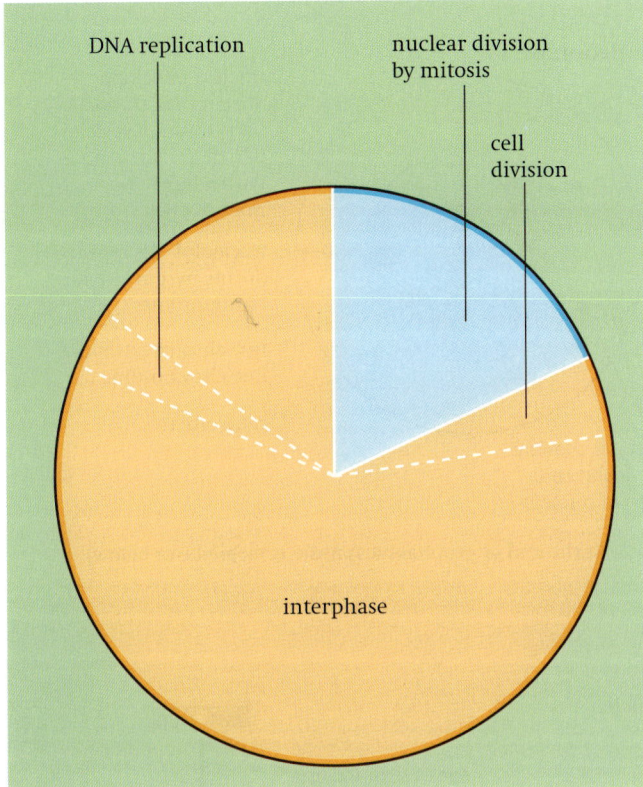

DNA replication

nuclear division
by mitosis

cell
division

interphase

● **Figure 6.9** The mitotic cell cycle.

Nuclear division follows interphase. The whole cell then divides.

The length of the cell cycle is very variable, depending on environmental conditions and cell type. On average, root tip cells of onions divide once every 20 hours; epithelial cells in the human intestine every 10 hours.

In animal cells, cell division involves constriction of the cytoplasm between the two new nuclei, a process called **cytokinesis**. In plant cells it involves the formation of a new cell wall between the two new nuclei.

## Mitosis

The process of mitosis is best described by annotated diagrams as shown in *figure 6.10*. Although in reality the process is continuous, it is shown here divided into four main stages for convenience, like four snapshots from a film. The four stages are called **prophase**, **metaphase**, **anaphase** and **telophase**.

Most nuclei contain many chromosomes, but the diagrams in *figure 6.10* show a cell containing only four chromosomes for convenience (2n = 4). Colours are used to show whether the chromo-

somes are from the female or male parent. An animal cell is used as an example. The behaviour of chromosomes in plant cells is identical. However, plant cells do not contain centrioles and, after nuclear division, a new cell wall must form between the daughter nuclei. It is chromosome behaviour, though, that is of particular interest. *Figure 6.10* summarises the process of mitosis diagrammatically. *Figures 6.11* (animal) and *6.12* (plant) show photographs of the process as seen with a light microscope.

### Biological significance of mitosis

■ The nuclei of the two daughter cells formed have the same number of chromosomes as the parent nucleus and are genetically identical. This allows growth of multicellular organisms from unicellular zygotes. Growth may occur over the entire body, as in animals, or be confined to certain regions, as in the meristems (growing points) of plants.

■ Replacement of cells and repair of tissues is possible using mitosis followed by cell division. Cells are constantly dying and being replaced by identical cells. In the human body, for example, cell replacement is particularly rapid in the skin and in the lining of the gut. Some animals are able to regenerate whole parts of the body, as, for example, the arms of a starfish.

■ Mitosis is the basis of asexual reproduction, the production of new individuals of a species by one parent organism. This can take many forms. For a unicellular organism, such as *Amoeba*, cell division inevitably results in reproduction. For multicellular organisms, new individuals may be produced which bud off from the parent in various ways (*figure 6.13*). This is particularly common in plants, where it is most commonly a form of vegetative propagation in which a bud on part of the stem simply grows a new plant. This eventually becomes detached from the parent and lives independently. The bud may be part of the stem of an overwintering structure such as a bulb or tuber. The ability to generate whole organisms from single cells, or small groups of cells, is becoming important in biotechnology and genetic modification (engineering).

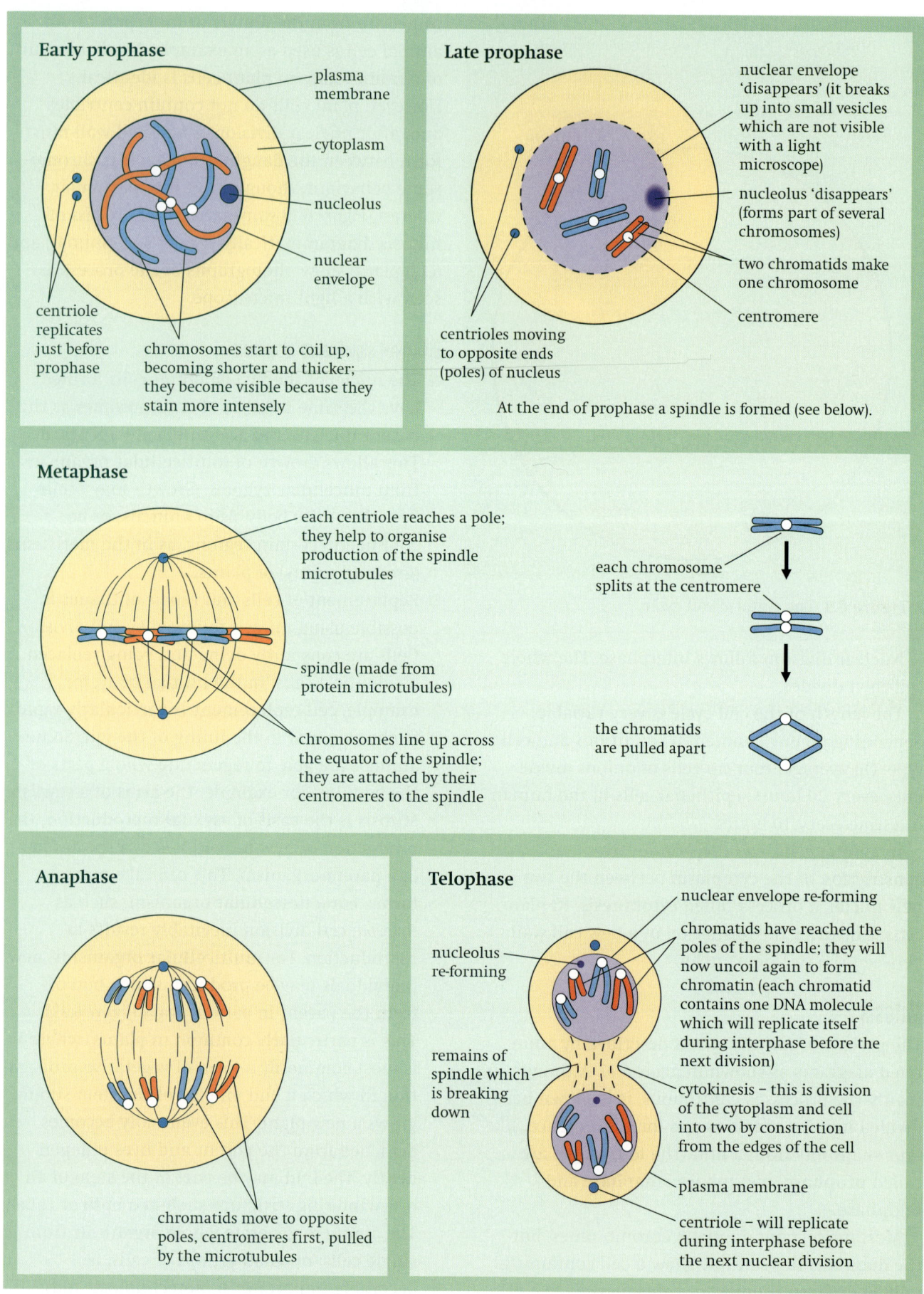

**Early prophase**

plasma membrane

cytoplasm

nucleolus

nuclear envelope

centriole replicates just before prophase

chromosomes start to coil up, becoming shorter and thicker; they become visible because they stain more intensely

**Late prophase**

nuclear envelope 'disappears' (it breaks up into small vesicles which are not visible with a light microscope)

nucleolus 'disappears' (forms part of several chromosomes)

two chromatids make one chromosome

centromere

centrioles moving to opposite ends (poles) of nucleus

At the end of prophase a spindle is formed (see below).

**Metaphase**

each centriole reaches a pole; they help to organise production of the spindle microtubules

spindle (made from protein microtubules)

chromosomes line up across the equator of the spindle; they are attached by their centromeres to the spindle

each chromosome splits at the centromere

the chromatids are pulled apart

**Anaphase**

chromatids move to opposite poles, centromeres first, pulled by the microtubules

**Telophase**

nuclear envelope re-forming

chromatids have reached the poles of the spindle; they will now uncoil again to form chromatin (each chromatid contains one DNA molecule which will replicate itself during interphase before the next division)

nucleolus re-forming

remains of spindle which is breaking down

cytokinesis – this is division of the cytoplasm and cell into two by constriction from the edges of the cell

plasma membrane

centriole – will replicate during interphase before the next nuclear division

● **Figure 6.10** Mitosis and cytokinesis in an animal cell.

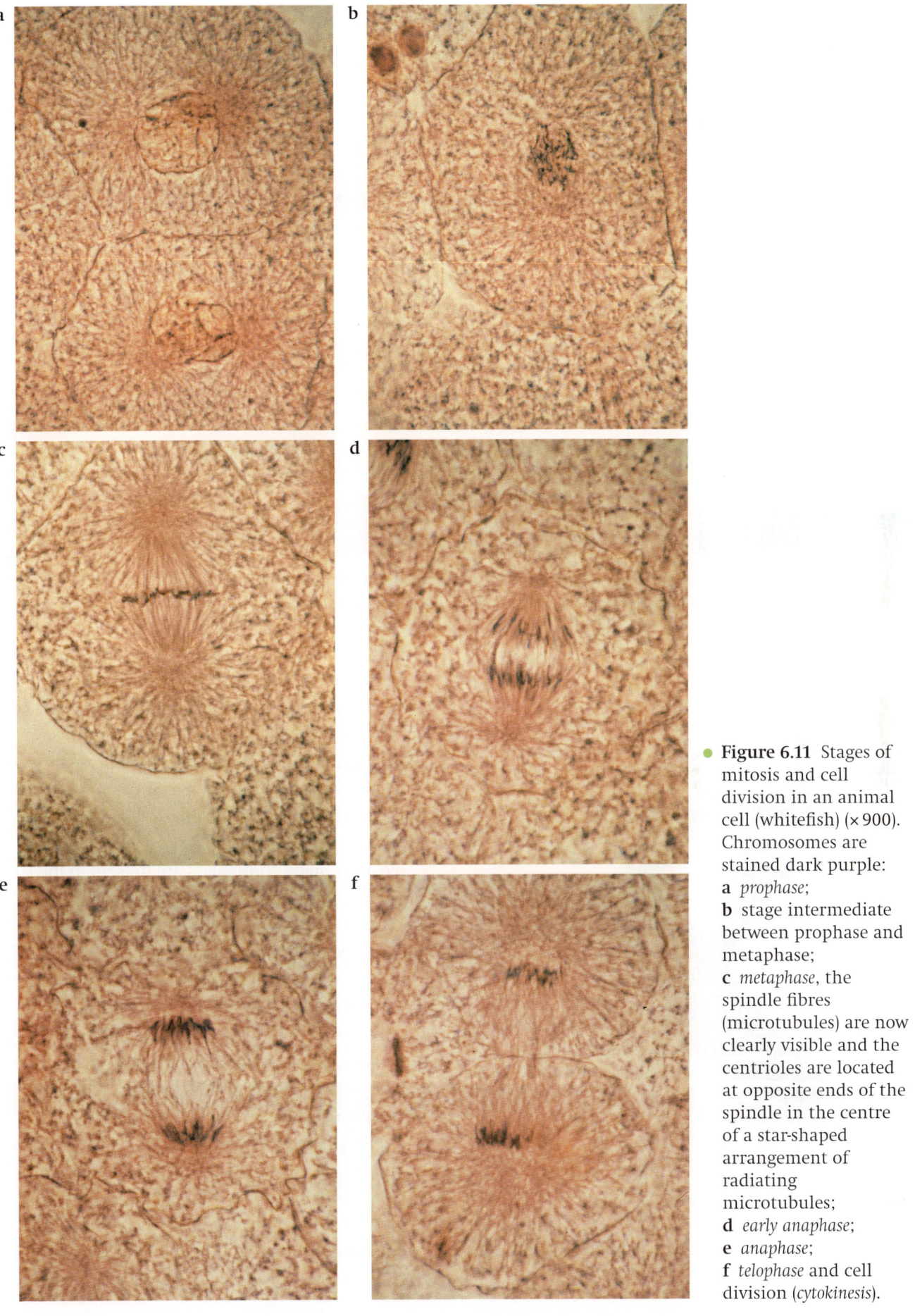

**Figure 6.11** Stages of mitosis and cell division in an animal cell (whitefish) (×900). Chromosomes are stained dark purple:
**a** *prophase*;
**b** stage intermediate between prophase and metaphase;
**c** *metaphase*, the spindle fibres (microtubules) are now clearly visible and the centrioles are located at opposite ends of the spindle in the centre of a star-shaped arrangement of radiating microtubules;
**d** *early anaphase*;
**e** *anaphase*;
**f** *telophase* and cell division (*cytokinesis*).

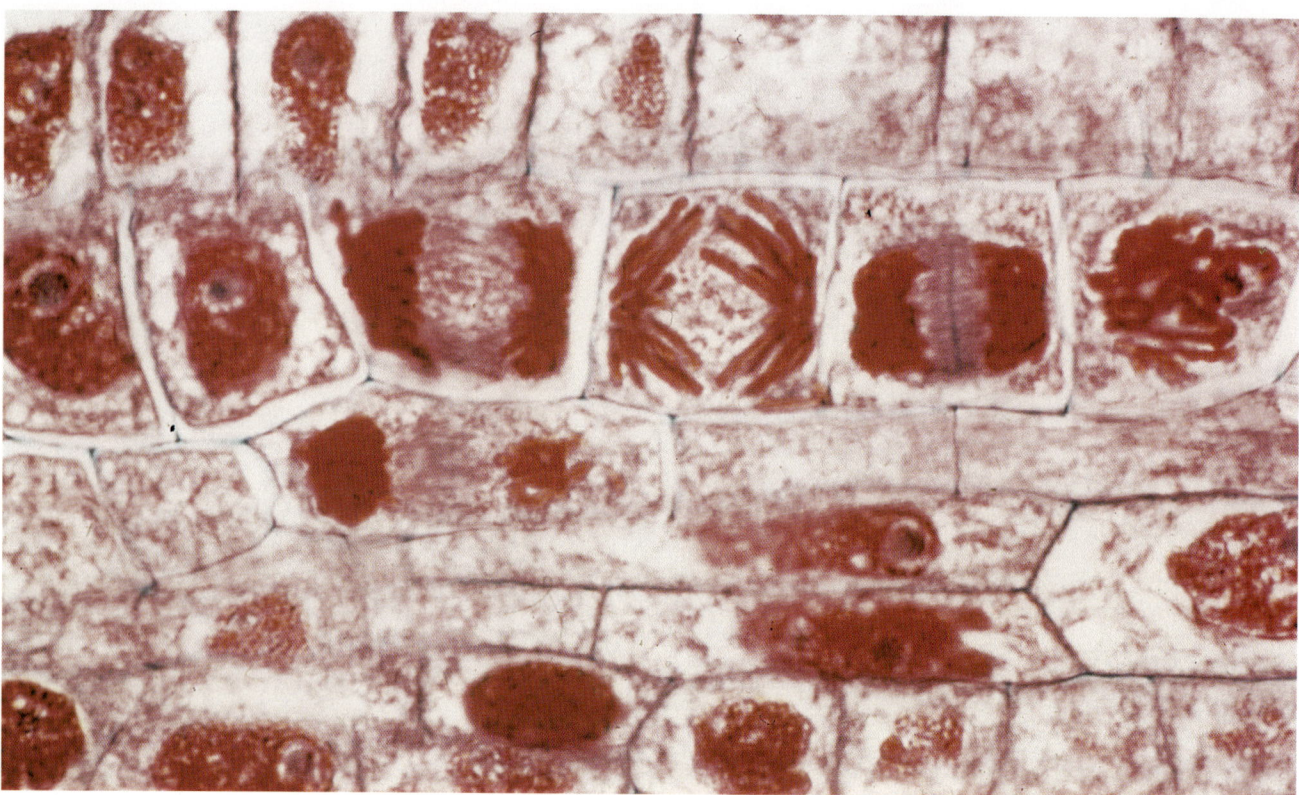

● **Figure 6.12** LS onion root tip showing stages of mitosis and cell division typical of plant cells (× 400). Try to identify the stages based on information given in *figure 6.10*.

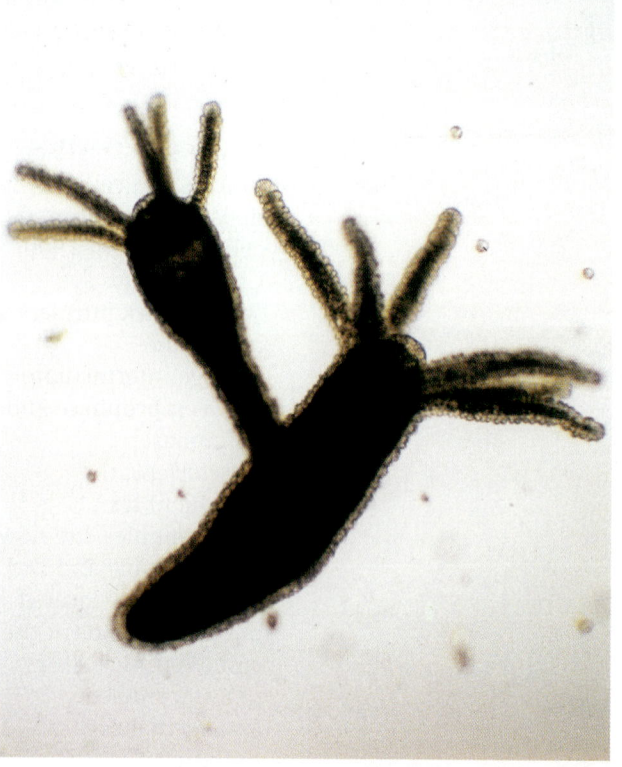

● **Figure 6.13** Asexual reproduction by budding (× 60). *Hydra* lives in fresh water, catching its prey with the aid of its tentacles. The bud growing from its side is genetically identical to the parent and will eventually break free and live independently.

### SAQ 6.2

**a** In the mitotic cell cycle of a human cell:
  (i) how many chromatids are present as the cell enters mitosis?
  (ii) how many DNA molecules are present?
  (iii) how many chromatids are present in the nucleus of each daughter cell after mitosis and cell division?
  (iv) how many chromatids are present in the nucleus of a cell after replication of DNA?

**b** Draw a simple diagram of a cell which contains only one pair of homologous chromosomes (i) at metaphase of mitosis, (ii) at anaphase of mitosis.

**c** What chemical subunits are used to synthesise new DNA molecules during replication of DNA?

**d** Of what elements are these subunits made?

**e** State two functions of centromeres during nuclear division.

**f** Thin sections of adult mouse liver were prepared and the cells stained to show up the chromosomes. In a sample of 75 000 cells examined, nine were found to be in the process of mitosis. Calculate the length of the cell cycle in days in liver cells, assuming that mitosis lasts 1 hour.

# Cancer

Cancer is one of the most common diseases of developed countries, accounting for roughly one in four deaths. Lung cancer alone caused about one in 17 of all deaths in Britain in the 1990s, 1 in 13 deaths in men and 1 in 27 deaths in women. It is the most common form of

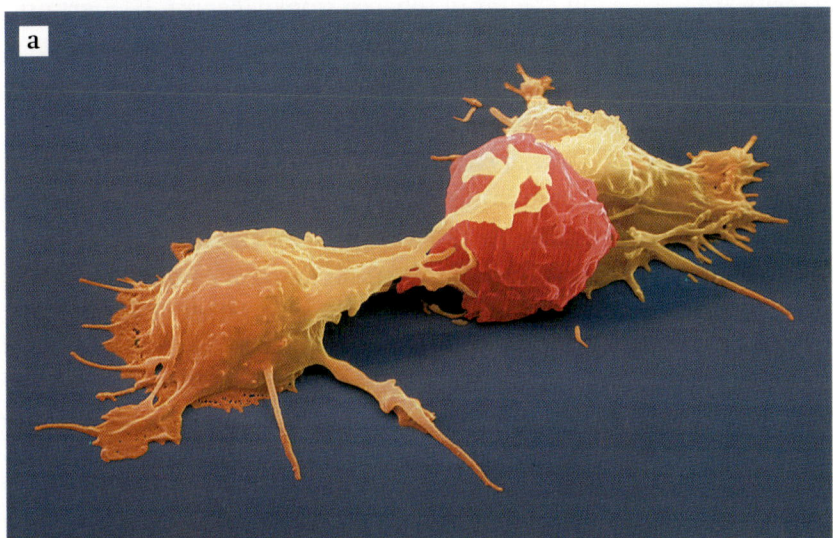

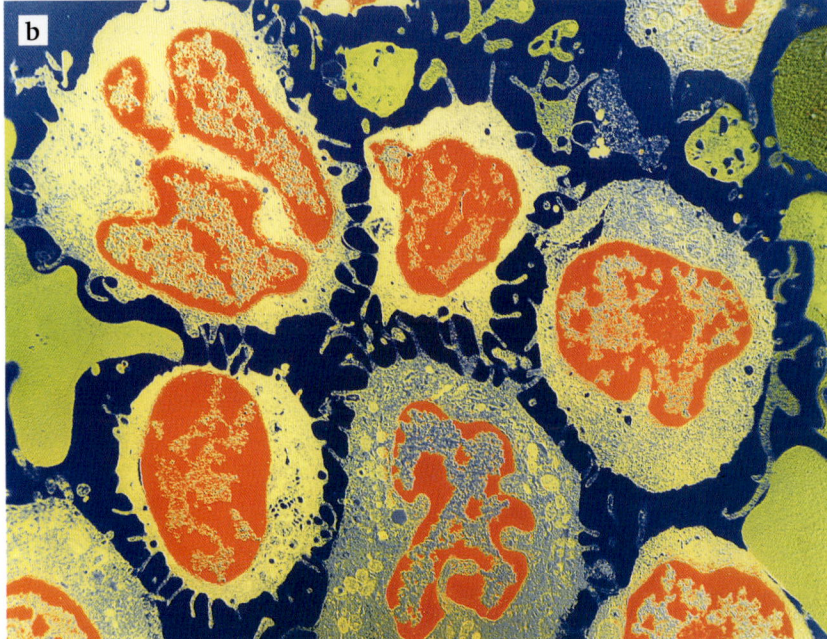

● **Figure 6.14 a** False-colour SEM of a cancer cell (red) and white blood cells (yellow). White blood cells gather at cancerous sites as an immune response. They are beginning to flow around the cancer cell which they will kill using toxic chemicals (× 4500).
**b** False-colour TEM of abnormal white blood cells isolated from the blood of a person suffering from hairy-cell leukaemia. The white blood cells are covered with characteristic hair-like, cytoplasmic projections. Leukaemia is a disease in which the bone marrow and other blood-forming organs produce too many of certain types of white blood cells. These immature or abnormal cells suppress the normal production of white and red blood cells, and increase the sufferer's susceptibility to infection (× 6400).

cancer in men, while breast cancer is the leading form of cancer in women. There are, in fact, more than a hundred different forms of cancer and the medical profession does not think of it as a single disease. Cancers show us the importance of controlling cell division precisely, because cancers are a result of uncontrolled mitosis. Cancerous cells divide repeatedly, out of control, and a **tumour** develops which is an irregular mass of cells. The cells usually show abnormal changes in shape (*figure 6.14*).

## Carcinogens

Cancers are thought to start when changes occur in the genes that control cell division. We have encountered **mutated** genes before when considering different alleles of genes in homologous pairs of chromosomes (page 80). The particular term for a mutated gene that causes cancer is an **oncogene** after the Greek word *onkos* meaning 'bulk' or 'mass'. A change in any gene is called a **mutation**. Mutations are not unusual events, and *most* mutated cells are either crippled in some way that results in their early death or are destroyed by the body's immune system. Since most cells can be replaced, this usually has no detrimental effect on the body. Cancerous cells, however, manage to escape both possible fates, so, although the mutation may originally occur only in one cell, it is passed on to all that cell's descendants. By the time it is detected, a typical tumour usually contains about a thousand million cells.

It is thought that a single mutation cannot be responsible for cancer but that several indepedent rare 'accidents' must all occur in one cell. A factor which brings about any mutation is called a **mutagen** and is described as **mutagenic**. Any agent that causes cancer is called a **carcinogen** and is described as **carcinogenic**. So, some mutagens are carcinogenic.

Some of the factors which can increase mutation rates, and hence the likelihood of cancer, are as follows.

■ *Ionising radiation*
This includes X-rays, gamma rays and particles from the decay of radioactive elements. They cause the formation of damaging ions inside cells which can break DNA strands. Ultraviolet light, although it does not cause the formation of damaging ions, can also damage genes. Depletion of the ozone layer is causing concern because, as a result, more ultraviolet light will penetrate to the Earth's surface and could result in an increase in cases of skin cancer.

■ *Chemicals*
Many different chemicals have been shown to be carcinogenic. About 25% of all cancer deaths in developed countries are due to carcinogens in the tar of tobacco smoke (*figure 6.15*). Certain dyes, such as a group known as the aniline dyes, are also well-known carcinogens. All these chemicals damage DNA molecules.

■ *Virus infection*
Some cancers in animals, including humans, are known to be caused by viruses. Burkitt's lymphoma, the most common cancer in children in certain parts of Africa, is caused by a virus. Another causes a form of leukaemia (cancer of the white blood cells). Papilloma viruses are responsible for some cancers, and include two types that have been linked with cervical cancer, a disease that can be

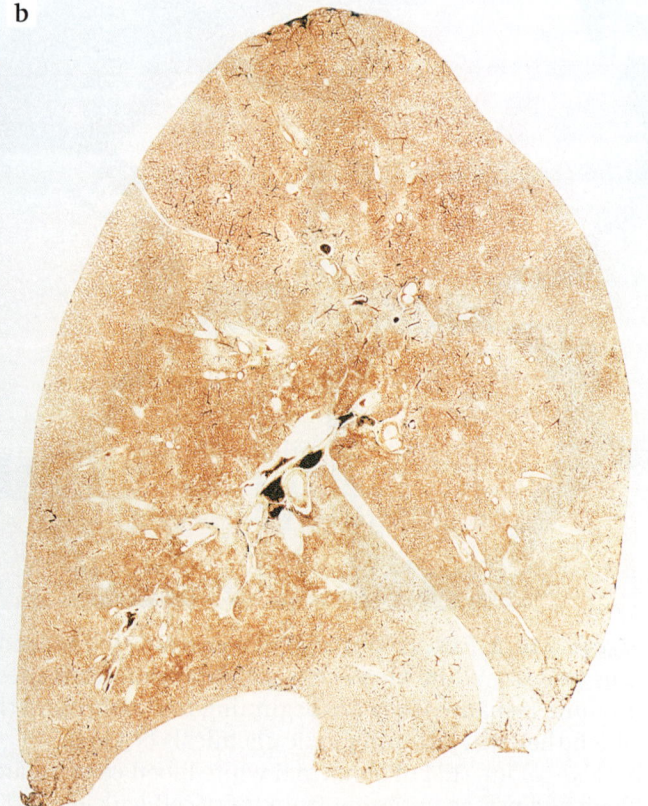

● **Figure 6.15  a** Lung of a patient who died of lung cancer, showing rounded deposits of tumour (bottom, white area). Black tarry deposits throughout the lung show the patient was a heavy smoker.

**b** Section of a healthy human lung. No black tar deposits are visible.

transmitted sexually. Viruses that cause cancer usually carry oncogenes, or regulatory genes that can become oncogenes.

■ *Hereditary predisposition*
Cancer tends to be more common in some families than others, indicating a genetic link. In most cases it is believed that the disease *itself* is not inherited, but susceptibility to the factors that cause the disease is inherited. However, some forms of cancer do appear to be caused by inheritance of a single faulty gene. For example, the inherited form of retinoblastoma, which starts in one or both eyes during childhood and spreads to the brain, causing blindness and then death if untreated, is caused by an error on chromosome 13.

## Benign or malignant?

A small group of tumour cells is called a **primary growth**. There are two types:

■ **benign** tumours, which do not spread from their site of origin, but can compress and displace surrounding tissues, for example warts, ovarian cysts and some brain tumours;

■ **malignant** (cancerous) tumours, which are far more dangerous since they spread throughout the body, invade other tissues and eventually destroy them.

Malignant tumours interfere with the normal functioning of the area where they have started to grow. They may block the intestines, lungs or blood vessels. Cells can break off and spread through the blood and lymphatic system to other parts of the body to form **secondary growths**. The spread of cancers in this way is called **metastasis**. It is the most dangerous characteristic of cancer, since it can be very hard to find secondary cancers and remove them.

The steps involved in the development of cancer are shown in *figure 6.16*.

Note that both benign and malignant tumours involve a huge drain on the body due to the high demand for nutrients that is created by the rapid and continual cell division.

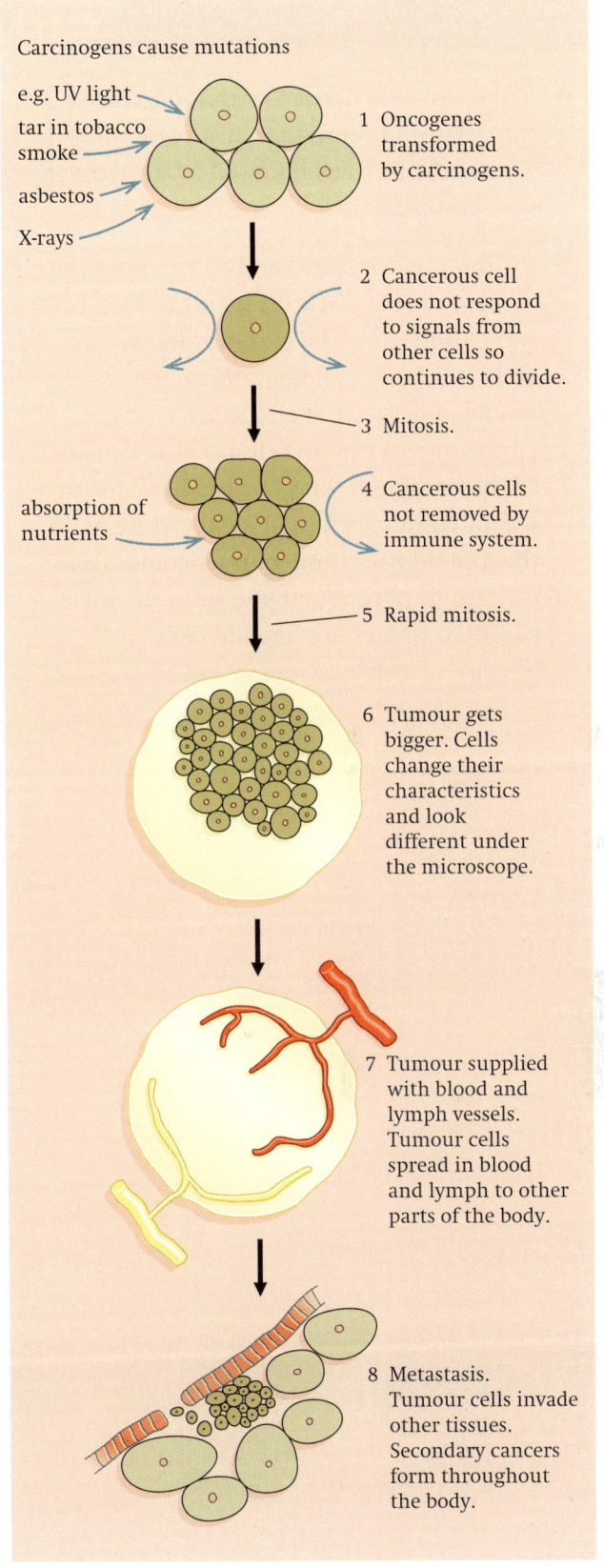

Carcinogens cause mutations

e.g. UV light
tar in tobacco smoke
asbestos
X-rays

1 Oncogenes transformed by carcinogens.

2 Cancerous cell does not respond to signals from other cells so continues to divide.

3 Mitosis.

absorption of nutrients

4 Cancerous cells not removed by immune system.

5 Rapid mitosis.

6 Tumour gets bigger. Cells change their characteristics and look different under the microscope.

7 Tumour supplied with blood and lymph vessels. Tumour cells spread in blood and lymph to other parts of the body.

8 Metastasis. Tumour cells invade other tissues. Secondary cancers form throughout the body.

● **Figure 6.16** Stages in the development of cancer.

# SUMMARY

◆ Cell division is needed so that organisms can grow and reproduce. It involves division of the nucleus followed by division of the cytoplasm.

◆ During nuclear division the chromosomes become visible. Each is seen to be formed of two chromatids. The chromosomes of a body cell can be photographed and arranged in order of size as a karyotype. This shows that each cell has either one set of chromosomes, known as haploid, or two sets of chromosomes, known as diploid. In the diploid condition, one set comes from the female parent and one from the male parent. Gametes are haploid cells.

◆ Body cells divide in the process of mitosis to produce two identical daughter cells whose nuclei contain the same number of chromosomes as the parent cell. This allows growth and repair of a multicellular organism and is the basis of asexual reproduction.

◆ Meiosis halves the number of chromosomes in the nucleus. This prevents chromosome number doubling in each generation of organisms which reproduce sexually.

◆ Cancers are caused by uncontrolled mitosis, possibly as the result of a mutation in a gene or genes which control cell division. Agents which cause cancer are known as carcinogens and include ionising radiation, many chemicals and viruses. Some cancers have a hereditary link.

## CHAPTER 7

# Energy and ecosystems

## By the end of this chapter you should be able to:

1 define the terms *habitat*, *niche*, *population*, *community* and *ecosystem*, and describe examples of each;

2 explain the terms *producer*, *consumer* and *trophic level*, and state examples of these in specific food chains and food webs;

3 describe how energy is transferred through food chains and food webs;

4 explain how energy losses occur along food chains, and understand what is meant by *efficiency* of transfer;

5 describe how nitrogen is cycled within an ecosystem.

So far in this book we have been looking at living things at a very small scale, considering what goes on in organisms in terms of the molecules from which they are built and the structure of their cells. In this chapter we change focus entirely, moving up in scale to think about how whole communities of living organisms interact with each other and with their environment. This branch of biology is called **ecology**.

We will consider two important themes in ecology: the flow of energy through ecological systems and the cycling of materials within those same ecological systems. In order to do this you need to become familiar with some of the specialist terms used in ecology. Ecology has its own set of terms, each with a precise meaning that may differ slightly from the meaning of the word when used in everyday life. Five of these are defined here and you will meet others later in this chapter.

■ A **habitat** is a **place where an organism lives**. The habitat of an oak tree might be the edge of an area of deciduous woodland. The habitat of a leaf-mining caterpillar might be inside a leaf on the oak tree.

■ A **population** is **a group of organisms of the same species, which live in the same place at the same time, and can interbreed with each other**. All of the oak trees in the wood, for example, make up a population of oak trees. However, if the oak trees in a nearby wood can interbreed with those in the first wood, then they too belong to the same population.

■ A **community** is all the **organisms, of all the different species, living in a habitat**. The woodland community includes all the plants – oak trees, ash trees, grasses, hawthorn bushes, bluebells and so on – all the microorganisms and larger fungi, and all the animals which live in the wood.

■ An **ecosystem** is a **relatively self-contained, interacting community of organisms, and the environment in which they live and with which they interact**. Thus the woodland ecosystem includes not only the community of organisms, but also the soil, the water in the rain and streams, the air, the rocks and anything else which is in the wood. As you will see later in this chapter, energy flows into the ecosystem from outside it (as sunlight), flows through the organisms in the ecosystem (as food) and eventually leaves the ecosystem (as heat). Matter, on the other hand – that is, atoms and molecules of substances such as carbon and nitrogen – cycles round an ecosystem, where some atoms are reused over and over again by different organisms.

No ecosystem is entirely self-contained; organisms, energy and matter in one ecosystem do interact with those from other ecosystems. Nevertheless, the concept is a useful one, because it allows you to focus on something of a manageable size.

You can think of ecosystems on different scales. You could consider the surface of a rotting crab apple to be an ecosystem, with its own community of moulds and other organisms, or you could think of the whole hedgerow in which the crab apple tree is growing as an ecosystem.

■ The **niche** of an organism is **its role in the ecosystem**. The niche of an oak tree is as a producer of carbohydrates and other organic substances which provide food for other organisms in the ecosystem. It takes carbon dioxide from the air and returns oxygen to it. Its roots penetrate deeply into the soil, where they take up water and minerals. Water vapour diffuses from its leaves into the air. These leaves provide habitats for myriads of insects and other animals. It is almost impossible to provide a complete description of the niche of any organism, because there are so many ways in which it interacts with other components of the ecosystem of which it is a part.

# Energy flow through organisms and ecosystems

Living organisms need a constant supply of energy to stay alive. At the most basic level, energy is required to drive many of the chemical reactions that take place within living cells. If these **metabolic reactions** stop, then all of the cell's activities stop and it dies. We have already seen that energy is needed for active transport across membranes (page 58) and to achieve cell growth and division. In mammals such as ourselves, large amounts of energy are required to maintain body temperature above that of our surroundings. There are many more such examples of the never-ending demand for energy.

Inside every cell, the immediate source of energy is **ATP** or **adenosine triphosphate**. ATP is the energy 'currency' of a cell. Each cell in every kind of living organism makes its own ATP as and when it needs it. Astonishing amounts of ATP are made every day. You, for example, probably make about 40 kg of ATP inside your cells every day, using it up almost immediately to supply energy for your cells' activities. When energy is required, the ATP is broken down by hydrolysis and its energy used for whatever the cell needs.

So, to keep your cells going you need energy from ATP, and to make ATP you need energy from somewhere else. The source of energy for making ATP is other organic molecules such as carbohydrates, lipids and proteins (see chapter 2). These molecules are high in energy which can be released and used to make ATP as they are broken down in the process of **respiration**. Respiration happens in every living cell, and its whole purpose is to make ATP.

How is energy stored in carbohydrates, lipids and proteins in the first place? It is achieved inside the mesophyll cells of plant leaves. Here, sunlight is captured by chlorophyll in the chloroplasts (page 5) and used to supply energy to drive the reactions of **photosynthesis**. Carbon dioxide from the air and water drawn up from the soil react together to produce carbohydrates and, in the process, energy is transferred from sunlight and converted into chemical energy in the carbohydrate molecules. The plant uses these carbohydrate molecules to make lipids and proteins which also contain some of this energy. When the plant requires energy for its metabolism, it breaks down some of these molecules in respiration and makes ATP.

This, then, is how all living organisms get their energy. Photosynthesis in plants converts sunlight energy to chemical energy in organic molecules. Animals eat plants, obtaining some of this chemical energy in the molecules they take in. Plants and animals then break down these organic molecules in the process of respiration, transferring energy to ATP molecules. The ATP can then itself be broken down to release its energy for use in metabolic reactions. This whole pattern of energy flow is summarised in *figure 7.1*. If you continue your biology studies into the second year, you will cover this subject in more detail.

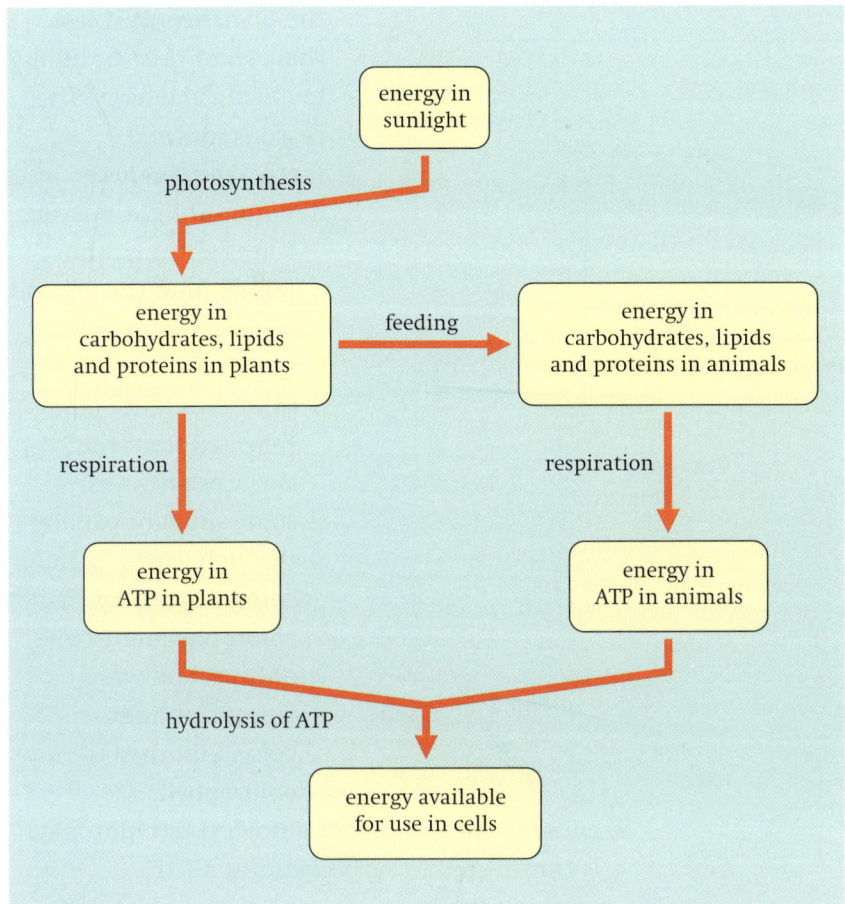

- **Figure 7.1** Energy flow through organisms and ecosystems.

Ultimately, therefore, green plants and other photosynthetic organisms have the essential role of providing the entire input of energy to an ecosystem. They are **producers**. The carbohydrates and other organic chemicals which they synthesise serve as supplies of chemical energy to all of the other organisms in the ecosystem. These other organisms, which include all the animals and fungi, and many of the microorganisms, consume the organic chemicals made by plants. They are **consumers**.

## Food chains and food webs

The way in which energy flows from producer to consumers can be shown by drawing a **food chain**. Arrows in the food chain indicate the direction in which the energy flows. A simple food chain in a deciduous wood could be:

oak tree → winter moth caterpillar → great tit → sparrowhawk

In this food chain, the oak tree is the producer, and the three animals are consumers. The caterpillar is a **primary consumer**, the great tit a **secondary consumer** and the sparrowhawk a **tertiary consumer**. These different positions in a food chain are called **trophic levels**. (*Trophic* means 'feeding'.)

Within this woodland ecosystem, there will be a large number of such food chains. The inter-relationships between many food chains can be drawn as a **food web**. *Figure 7.2* shows a partial food web for such an ecosystem. You can pick out many different food chains within this web.

You may notice that a particular animal does not always occupy the same position in a food chain. While herbivores such as caterpillars and rabbits tend *always* to be herbivores, and therefore always primary consumers, carnivores often feed at several different trophic levels in different food chains. Thus the fox is a primary consumer when it eats a fallen crab apple, a secondary consumer when it eats a rabbit, and a tertiary consumer when it eats a great tit. Animals which regularly feed as both primary and higher-level consumers, such as humans, are known as omnivores.

The food web also shows the importance of a group of organisms called **decomposers**. Most decomposers live in the soil, and their role in an ecosystem is to feed on **detritus** (dead organisms and waste material, such as dead leaves, faeces and urine). You can see that energy from *every* organism in the ecosystem flows into the decomposers. Decomposers include many bacteria, fungi, and also some larger animals such as earthworms. Sometimes, the term 'decomposer' is used only for bacteria and fungi, which feed saprotrophically, while the larger animals are called **detritivores**, meaning 'detritus feeders'. Decomposers are a largely unseen but vitally important group within every ecosystem. You will find out more about their roles in the nitrogen cycle on pages 94–97.

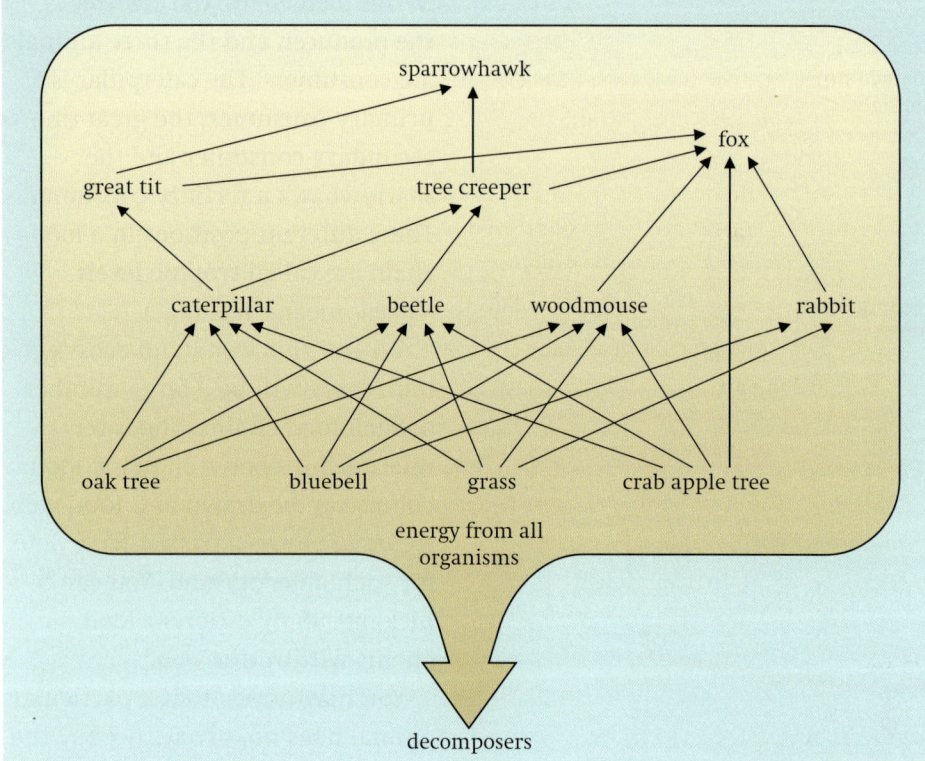

● **Figure 7.2** A food web in oak woodland.

## Energy losses along food chains

Whenever energy is transferred from one form, or from one system, to another some is always lost as heat. As energy passes along a food chain, large losses from the food chain occur at each transfer, both within and between the organisms. *Figure 7.3* shows these losses for a simple food chain.

Of the sunlight falling onto the ecosystem, only a very small percentage is converted by the green plants into chemical energy. In most ecosystems,

the plants convert less than 3% of this sunlight to chemical energy. The reasons for this inefficiency include:

■ some sunlight missing leaves entirely, and falling onto the ground or other nonphoto-synthesising surfaces;
■ some sunlight being reflected from the surfaces of leaves;
■ some sunlight passing through leaves, without being trapped by chlorophyll molecules;
■ only certain wave-lengths of light being absorbed by chlorophyll;
■ energy losses as energy absorbed by chlorophyll is transferred to carbohydrates during photosynthesis.

The chemical potential energy, now in the plants' tissues, is contained in various organic molecules, especially carbohydrates, lipids and proteins. It is from these molecules that the primary consumers in the ecosystem obtain their energy supply. However, in most plants, almost half of the chemical potential energy stored by plants is used by the plants themselves. They release the energy by respiration, using it for purposes such as active

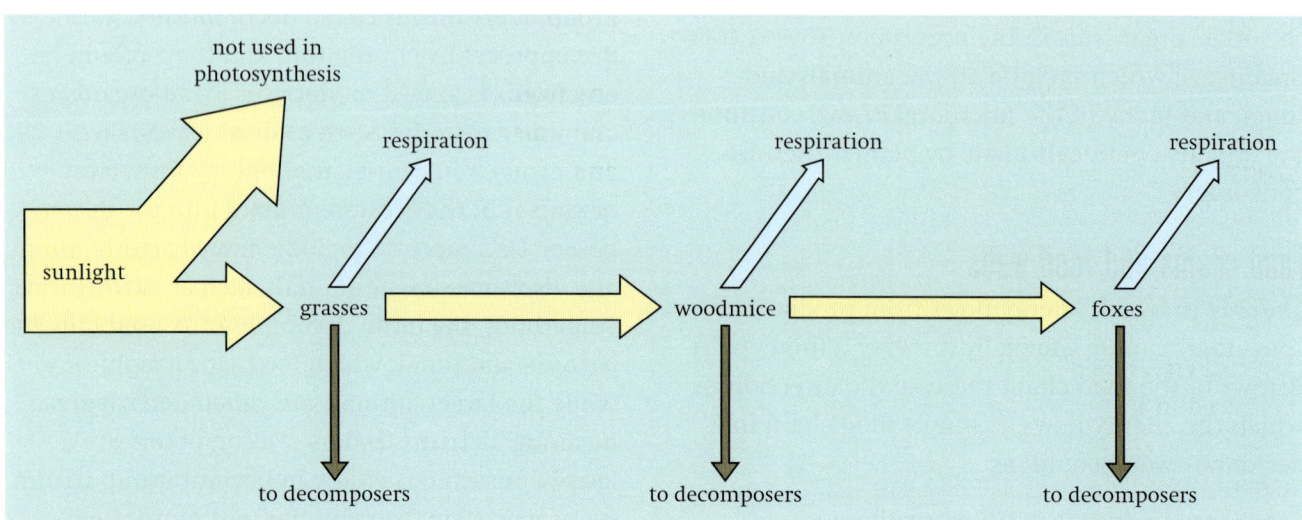

● **Figure 7.3** Energy losses in a food chain.

transport. During these processes, much energy is lost to the environment as heat.

What is left is then available for other organisms, which feed on the plants. Once again, losses occur between the plants and the primary consumers. The reasons for these losses include:

■ not all of the parts of the plants being available to be eaten, such as woody tissues and some roots;

■ not all of the parts of the plants eaten being digestible, so that not all of the molecules can be absorbed and used by the primary consumer;

■ energy losses as heat within the consumer's digestive system, as food is digested.

As a result of the loss of energy during respiration in the plants, and the three reasons listed above, the overall efficiency of transfer of energy from producer to primary consumer is rarely greater than 10%.

Similar losses occur at each trophic level. So, as energy is passed along a food chain, less and less energy is available at each successive trophic level. Food chains rarely have more than four or five links in them because there simply would not be sufficient energy left to support animals so far removed from the original energy input to the producers. If you *can* pick out a five-organism food chain from a food web, you will probably find that the 'top' carnivore also feeds at a lower level in a different food chain.

## SAQ 7.1

Energy losses from mammals and birds tend to be significantly greater than those from other organisms. Suggest why this is so.

### Productivity

The rate at which plants convert light energy into chemical potential energy is called **productivity**, or **primary productivity**. It is usually measured in kilojoules of energy transferred per square metre per year ($kJ\,m^{-2}\,year^{-1}$).

**Gross primary productivity** is the total quantity of energy

converted by plants in this way. **Net primary productivity** is the energy which remains as chemical energy after the plants have supplied their own needs in respiration.

## SAQ 7.2

*Table 7.1* shows some information about energy transfers in three ecosystems.

a Calculate the figures for respiration by plants in the alfalfa field, and the net primary productivity of the young pine forest.

b How much energy is available to the primary consumers in the rain forest?

c Suggest why the gross primary productivity of the rain forest is so much greater than that of the pine forest. (There are many reasons – think of as many as you can.)

d Suggest why the net primary productivity of the alfalfa field is greater than that of the rain forest. (Again, you may be able to think of several reasons.)

# Matter recycling in ecosystems

Living organisms require not only a supply of *energy*, but also a supply of *matter* from which to build their bodies. The elements from which this matter is made are mostly hydrogen, carbon and oxygen, which are contained in all the organic molecules within organisms. Proteins and nucleotides (chapters 2 and 5) also contain nitrogen, and some proteins contain sulphur. Phosphorus is an important component of nucleotides. Other elements are needed in smaller quantities, such as magnesium, calcium, iodine and iron.

Atoms of these elements are used over and over again within an ecosystem, or passed into other

| | Mature rain forest in Puerto Rico | Alfalfa field in USA | Young pine forest in England |
|---|---|---|---|
| Gross primary productivity ($kJ\,m^{-2}\,year^{-1}$) | 188 000 | 102 000 | 51 000 |
| Respiration by plants ($kJ\,m^{-2}\,year^{-1}$) | 134 000 | | 20 000 |
| Net primary productivity ($kJ\,m^{-2}\,year^{-1}$) | 54 000 | 64 000 | |

● **Table 7.1** Energy transfer data.

ecosystems. They are passed from one organism to another, cycling round through the different living organisms, and also through the non-living parts of the ecosystem, such as the air, soil, water and rocks. The principle of recycling can be illustrated by the nitrogen cycle (*figure 7.4*).

## The nitrogen cycle

Nitrogen is an essential element for all living organisms, because of its presence in proteins and nucleic acids. There is a large quantity of nitrogen in the air, which is around 78% $N_2$ gas. However, most organisms cannot use this nitrogen. This is because nitrogen gas exists as molecular nitrogen, in which two nitrogen atoms are linked with a triple covalent bond ($N \equiv N$). In this form, nitrogen is very unreactive. With each breath, you take in around $350 \, cm^3$ of nitrogen gas, but this is completely useless to you. It simply passes in and

out of your body unchanged. Similarly, $N_2$ passes freely in and out of a plant's stomata, with the plant unable to make any use of it.

Before nitrogen can be used by living organisms it must be converted from $N_2$ into some more reactive form, such as ammonia ($NH_3$) or nitrate ($NO_3^-$). This conversion is called **nitrogen fixation**.

### Nitrogen fixation

Nitrogen fixation can take place naturally or synthetically.

### Fixation by living organisms

Only prokaryotes are capable of fixing nitrogen. One of the best-known nitrogen-fixing bacteria is *Rhizobium* (*figure 7.4*). This bacterium lives freely in the soil, and also in the roots of many species of plants, especially leguminous plants (belonging to the pea family, such as peas, beans, clover, alfalfa

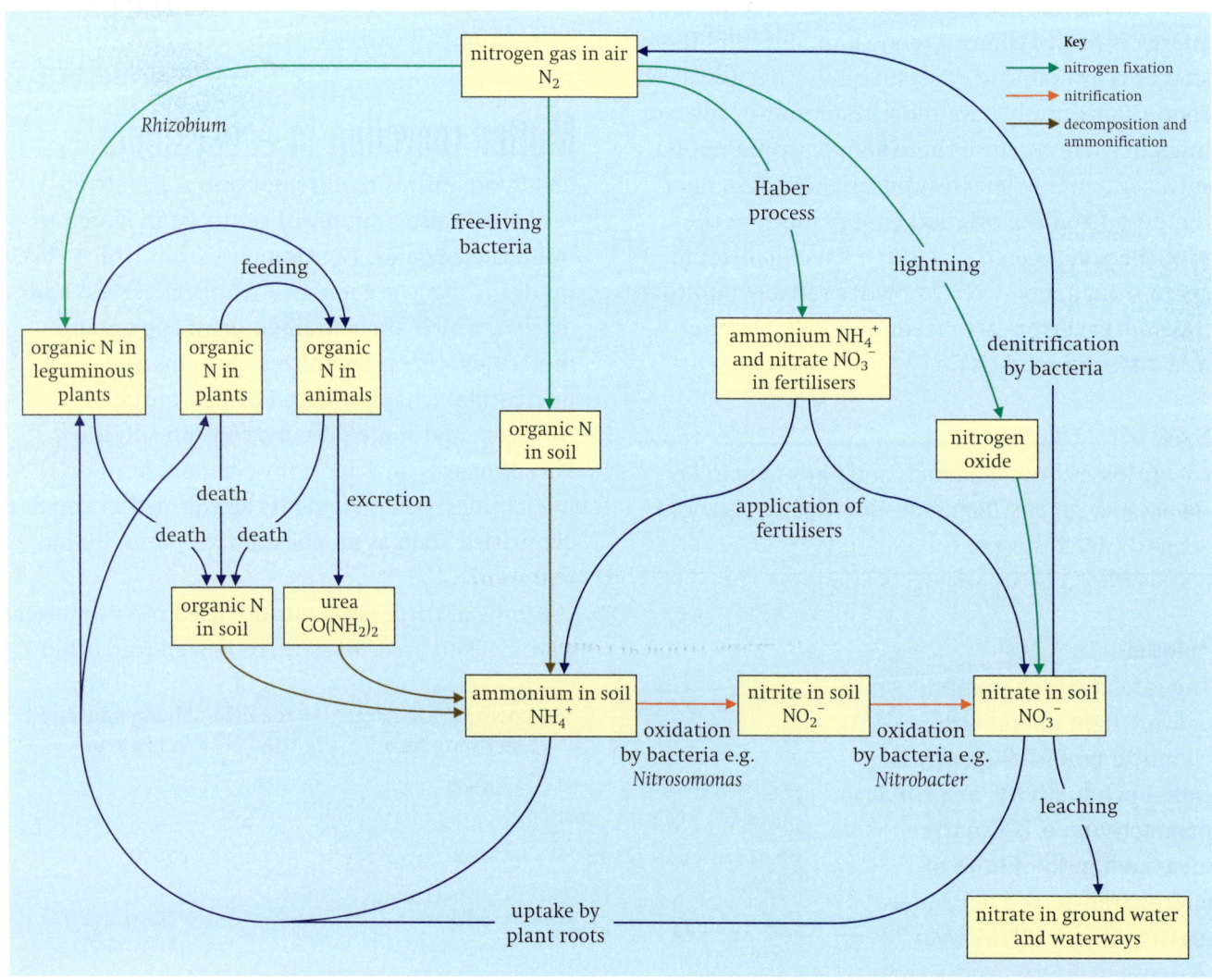

● **Figure 7.4** The nitrogen cycle.

and acacia trees). *Rhizobium* can only fix nitrogen to a very limited extent when living freely in the soil. Most nitrogen fixation by *Rhizobium* occurs when it is living in plant roots. The plant and the bacterium coexist in a rather remarkable way, each benefiting from the presence of the other.

*Rhizobium* is found in most soils. When a leguminous plant germinates, its roots produce proteins called lectins which bind to polysaccharides on the cell surface of the bacteria. The bacteria invade the roots, spreading along the root hairs. They stimulate some of the cells in the root to divide and develop into small lumps or nodules, inside which the bacteria form colonies (*figure 7.5*).

The bacteria fix nitrogen with the help of an enzyme called **nitrogenase**. This enzyme catalyses the conversion of nitrogen gas, $N_2$, to ammonium ions, $NH_4^+$. To do this, it needs:

- a supply of hydrogen;
- a supply of ATP;
- anaerobic conditions, that is the absence of oxygen.

The hydrogen comes from a substance called reduced NADP which is produced by the plant. The ATP comes from the metabolism of sucrose, produced by photosynthesis in the plant's leaves and transported down into the root nodules. Here the sucrose is processed and used in respiration to generate ATP. Anaerobic conditions are maintained through the production, by the plant, of a protein called leghaemoglobin. This molecule has a high affinity for oxygen, and effectively 'mops up' oxygen which diffuses into the nodules.

The relationship between the plant and the bacteria is therefore a very close one. The plant supplies living space, and the conditions required by the bacteria to fix nitrogen. The bacteria supply the plant with fixed nitrogen. This is an example of **mutualism**, in which two organisms of different species live very closely together, each meeting some of the other's needs.

### Fixation in the atmosphere

When lightning passes through the atmosphere, the huge quantities of energy involved can cause nitrogen molecules to react with oxygen, forming nitrogen oxides (*figure 7.4*). These dissolve in rain, and are carried to the ground. In countries where there are frequent thunderstorms, for example many tropical countries, this is a very significant source of fixed nitrogen.

### Fixation by the Haber process

The production of fertilisers containing fixed nitrogen is a major industry. In the Haber process, nitrogen and hydrogen gases are reacted together to produce ammonia. This requires considerable energy inputs, so the resulting fertilisers are not cheap. The ammonia is often converted to ammonium nitrate, which is the most widely used inorganic fertiliser in the world.

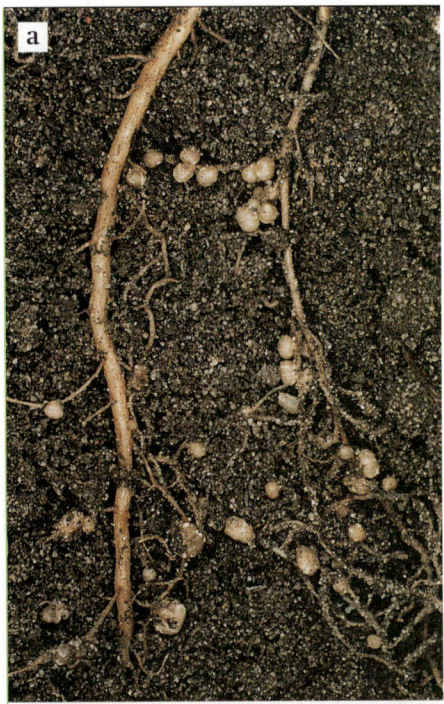

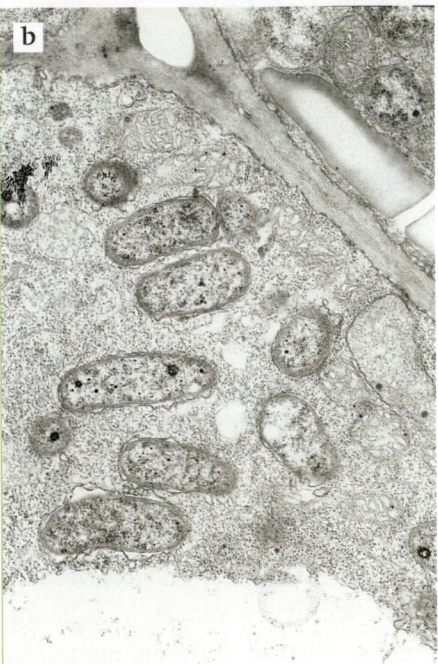

● **Figure 7.5**
**a** Root nodules, containing nitrogen-fixing bacteria, on clover roots (× 1.5).
**b** EM (× 12 000) of part of a cell in a root nodule. The oval structures are nitrogen-fixing bacteria, which are each enclosed by a membrane belonging to the plant cell. You can also see the cell wall, several mitochondria, endoplasmic reticulum and ribosomes in the plant cell.

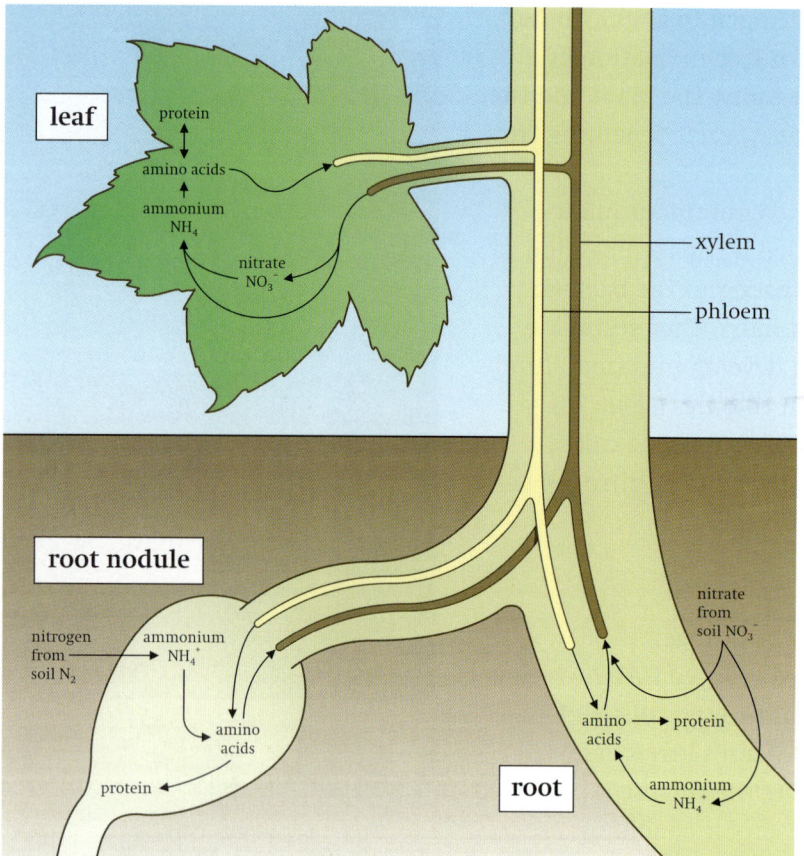

● **Figure 7.6** A summary of nitrogen metabolism and transport in plants.

## Use of fixed nitrogen by plants

In legumes, the fixed nitrogen produced by *Rhizobium* in their root nodules is used to make **amino acids**. These are transported out of the nodules into xylem, distributed to all parts of the plant and used within cells to synthesise proteins (*figure 7.6*).

Other plants rely on supplies of fixed nitrogen in the soil. Their root hairs take up **nitrate ions** by active transport. In many plants, the nitrate is converted in the roots, first to nitrite ($NO_2^-$), then ammonia, and then amino acids which are transported to other parts of the plant in xylem. In other plant species, the nitrate ions are transported, in xylem, to the leaves before undergoing these processes. Again, most of the nitrogen ends up as part of protein molecules in the plant, especially in seeds and storage tissues.

## Assimilation of nitrogen by animals

Animals, including humans, can only use nitrogen when it is part of an organic molecule. Most of our nitrogen supply comes from proteins in the diet, with a small amount from nucleic acids. During digestion, proteins are broken down to amino acids, before being absorbed into the blood and distributed to all cells in the body. Here they are built up again into proteins. Excess amino acids are **deaminated** in the liver, where the nitrogen becomes part of urea molecules. These are excreted in urine.

### Return of nitrate to the soil from living organisms

When an animal or plant dies, the proteins in its cells are gradually broken down to amino acids. This is done by decomposers, especially bacteria and fungi, which produce protease enzymes. The decomposers use some of the amino acids for their own growth, while some are broken down and the nitrogen released as ammonia. Ammonia is also produced from the urea in animal urine. The production of ammonia is called **ammonification**.

Ammonia in the soil is rapidly converted to nitrite ions ($NO_2^-$), and then nitrate ions ($NO_3^-$), by a group of bacteria called **nitrifying bacteria**. They include *Nitrosomonas* and *Nitrobacter* (*figure 7.4*). These bacteria derive their energy from nitrification. In contrast to nitrogen fixation, this

● **Figure 7.7** The greater sundew, *Drosera anglica*, only grows in very wet soils where nitrates are in extremely short supply. The sticky glands on the leaves trap insects. The leaves then curl over, and digest and absorb nutrients, including amino acids, from the insect's body.

only occurs freely in well-aerated soils. Boggy soils are therefore often short of nitrates. Some plants have become adapted to growing in such soils by supplementing their nitrogen intake using animal protein. These carnivorous plants trap insects, whose proteins are digested and absorbed by the plant (*figure 7.7*).

### Denitrification

**Denitrifying bacteria** provide themselves with energy by reversing nitrogen fixation and converting nitrate to nitrogen gas, which is returned to the air. They are common in places such as sewage treatment plants, compost heaps and wet soils. This brings the nitrogen cycle full circle.

## SUMMARY

◆ Energy flows through ecosystems. All energy enters an ecosystem as sunlight, and is converted to chemical energy in organic molecules during photosynthesis in producers. Energy is transferred along food chains as one organism feeds on another. Energy is lost at each transfer within and between organisms, mostly as heat produced during respiration. This results in a decrease in biomass and energy in successive trophic levels.

◆ Matter cycles around ecosystems. Nitrogen from the air is fixed by bacteria, some of which live freely in the soil and some, especially *Rhizobium*, which live in root nodules of leguminous plants. Fixed nitrogen, in the form of nitrates, is taken up by plants and used to synthesise amino acids and proteins, on which animals feed. Decomposers convert dead organisms and their waste products to ammonia, which is then converted to nitrite and nitrate by nitrifying bacteria. Denitrifying bacteria complete the cycle, converting inorganic nitrogen compounds to nitrogen gas.

## CHAPTER 8

# The mammalian transport system

## By the end of this chapter you should be able to:

1 explain why multicellular animals need transport mechanisms;

2 describe the structure of arteries, veins and capillaries, and relate their structure to their functions;

3 recognise micrographs of arteries, veins and capillaries;

4 describe the functions of tissue fluid, and its formation from blood plasma;

5 describe the functions of lymph, and its formation from tissue fluid;

6 describe the composition of blood;

7 outline the functions of white blood cells;

8 describe the role of haemoglobin in the transport of oxygen and carbon dioxide;

9 describe and explain the oxygen dissociation curve for haemoglobin;

10 describe and explain the effects of raised carbon dioxide concentrations on the haemoglobin dissociation curve (the Bohr effect);

11 describe and explain the differences between oxygen dissociation curves for haemoglobin, fetal haemoglobin and myoglobin, and explain the significance of these differences;

12 describe and explain the increase in red blood cell count at high altitude.

hy do humans have a blood system? The answer is fairly obvious even to a non-scientist: our blood system transports substances such as nutrients and oxygen around the body. However, there are many organisms which either have much less complex transport systems or do not have any kind of transport system at all. Before looking in detail at the human transport system, it is worth briefly considering why some organisms can manage without one.

A quick survey of some organisms which have very simple transport systems, or even none at all, will provide an important clue. Table 8.1 lists six kinds of organisms, and gives a brief summary of the type of transport system that each has.

*SAQ 8.1*

From *table 8.1*, state whether each of the following factors appears to be important in deciding whether or not an organism needs an efficient transport system. In each case, identify the information in the table which led you to your answer.

a Size

b The surface area to volume ratio

c Level of activity

All living cells require a supply of nutrients, such as glucose. Most living cells also need a constant supply of oxygen. There will also be waste products, such as carbon dioxide, to be disposed of. Very small organisms, such as *Paramecium*, can

| Type of organism | Single-celled | Cnidarians (jellyfish and sea anemones) | Insects | Green plants | Fish | Mammals |
|---|---|---|---|---|---|---|
| Size range | all microscopic | some microscopic, some up to 60 cm | less than 1 mm to 13 cm | 1 mm to 150 m | 12 mm to 10 m | 35 mm to 34 m |
| Example | *Paramecium* | sea anemone | locust | *Pelargonium* | goldfish | human |
| Level of activity | move in search of food | jellyfish swim slowly; anemones are sedentary and move very slowly | move actively; many fly | no movement of whole plant; parts such as leaves may move slowly | move actively | move actively |
| Type of transport system | no specialised transport system | no specialised transport system | blood system with pumps | xylem and phloem make up transport system; no pump | blood system with pump | blood system with pump |

● **Table 8.1** Different transport systems.

meet their requirements for the supply of nutrients and oxygen, and the removal of waste products, by means of **diffusion** (see page 53). The very small distances across which substances have to diffuse means that the speed of supply or removal is sufficient for their needs. These tiny organisms have a large surface area compared to their total volume, so there is a relativity large area of membrane across which gases can diffuse in and out of their bodies.

Even larger organisms, such as cnidarians, can manage by diffusion alone. Their body is made up of just two layers of cells, so every cell is within a very small distance of the water in which these organisms live and with which they exchange materials. They, too, have relatively large surface area to volume ratios. Moreover, cnidarians are not very active animals, so their cells do not have large requirements for glucose or oxygen, nor do they produce large amounts of waste products.

Diffusion, slow though it is, is quite adequate to supply their needs.

Larger, more active, organisms such as insects, fish and mammals, cannot rely on diffusion alone. Cells, often deep within their bodies, are metabolically very active, with requirements for rapid supplies of nutrients and oxygen, and with relatively large amounts of waste products to be removed. These organisms have well-organised transport systems, with pumps to keep fluid moving through them. Plants, although large, are less metabolically active than these groups of animals and, as you will see in chapter 10, have evolved a very different type of transport system, with no obvious pump to keep fluids moving.

# The cardiovascular system

*Figure 8.1* shows the general layout of the main transport system of mammals, that is the blood system or **cardiovascular system**. It is made up of a pump, the **heart**, and a system of interconnecting tubes, the **blood vessels**. The blood always remains within these vessels, and so the system is known as a **closed** blood system.

If you trace the journey of the blood around the body, beginning in the left ventricle of the heart, you will find that the blood travels twice through the heart on one complete 'circuit'. Blood is pumped out of the left ventricle into the **aorta** (*figure 8.2*), and travels from there to all parts of the body except the lungs. It returns to the right side of the heart in the **vena cava**, and is then pumped out of the right ventricle into the **pulmonary arteries**, which carry it to the lungs. The final part of the journey is along the **pulmonary veins**, which return it to the left side of the heart. This combination of pulmonary circulation and systemic circulation makes a **double circulatory system**.

## SAQ 8.2

*Figure 8.3* shows the general layout of the circulatory system of a fish.

**a** How does this differ from the circulatory system of a mammal?

**b** Suggest the possible advantages of the design of the mammalian circulatory system over that of a fish.

The detailed structure and functions of the heart will be looked at in chapter 9. We now look at the rest of the system.

The vessels making up the blood system are of three main types. *Figure 8.4* shows these

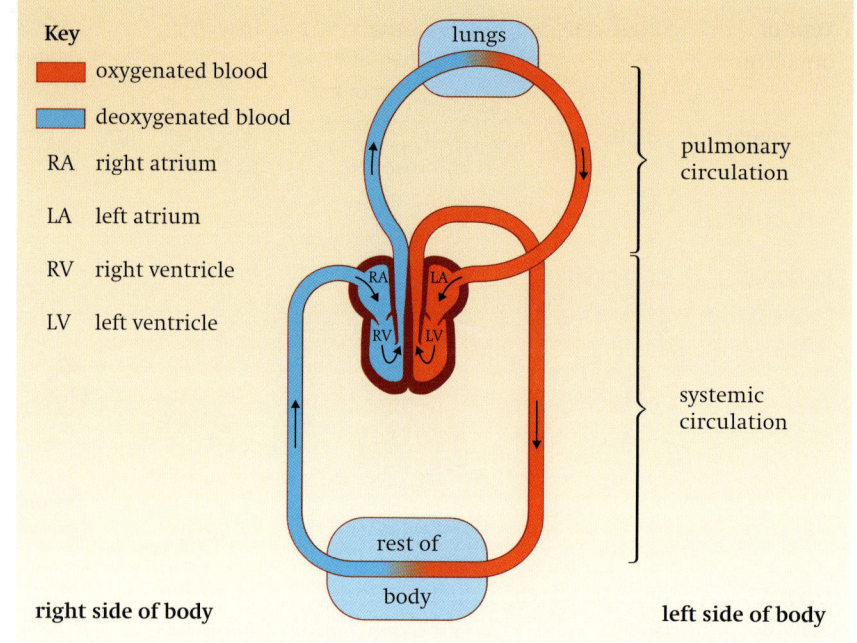

● **Figure 8.1** The general plan of the mammalian transport system, viewed as though looking at someone facing you. It is a closed double circulatory system.

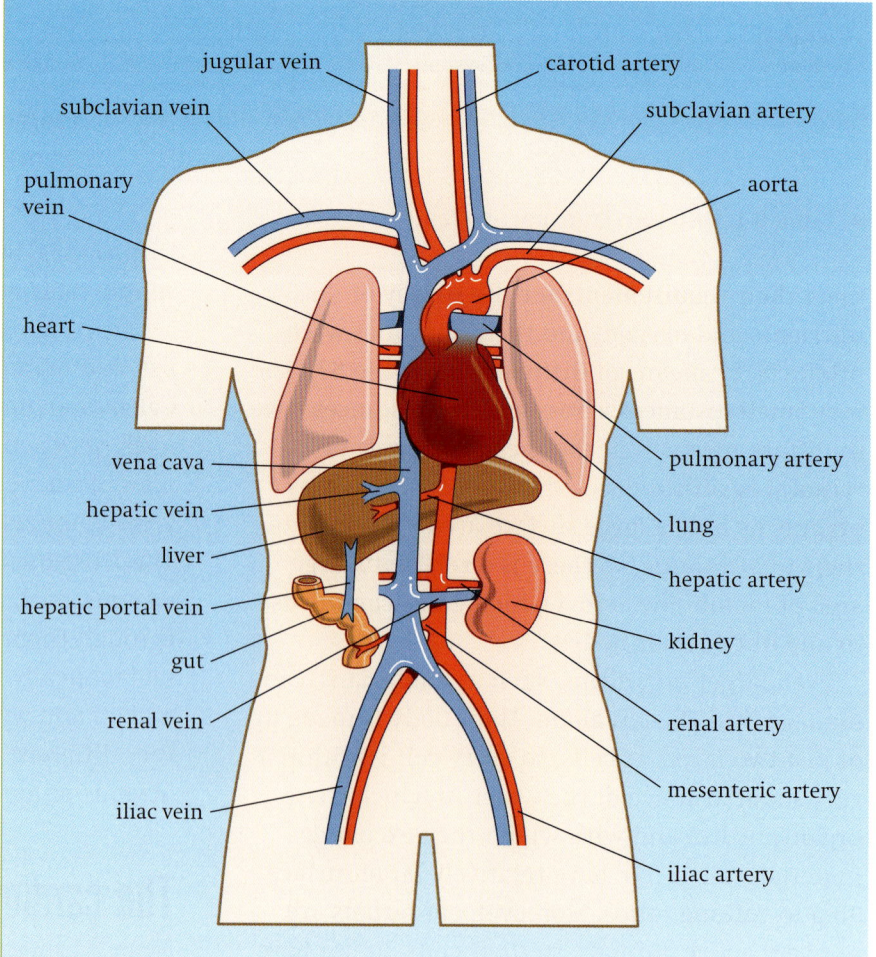

● **Figure 8.2** The positions of some of the main blood vessels in the human body.

vessels in transverse section. Vessels carrying blood *away* from the heart are known as **arteries**, while those carrying blood *towards* the heart are **veins**. Linking arteries and veins, taking blood close to almost every cell in the body, are tiny vessels called **capillaries**.

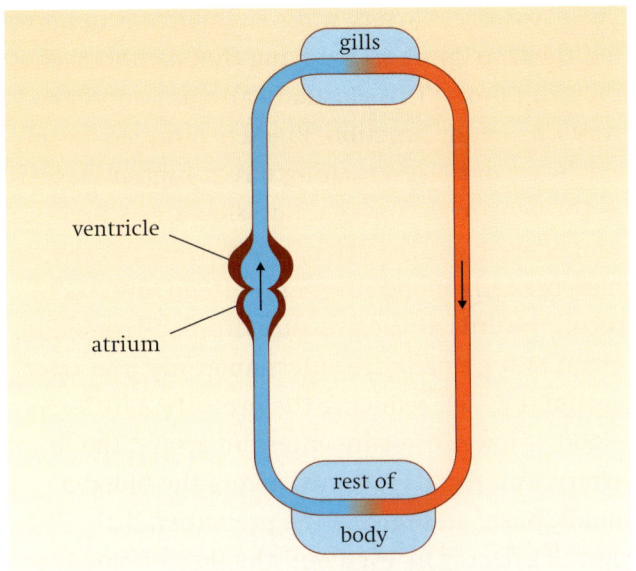

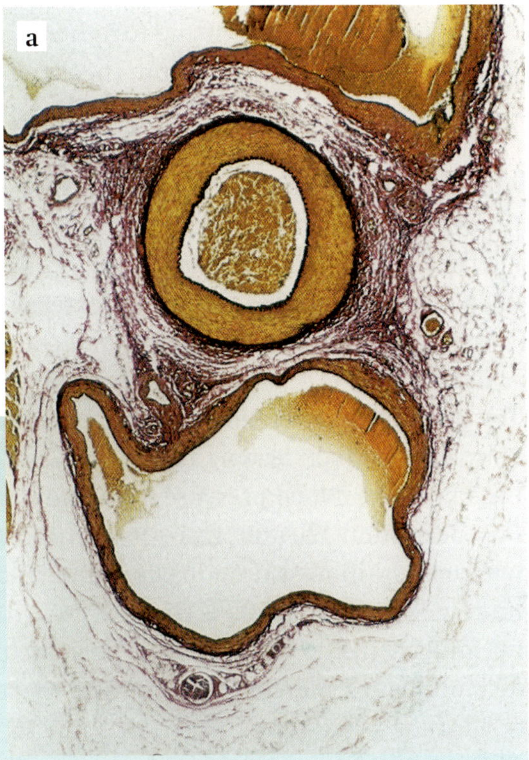

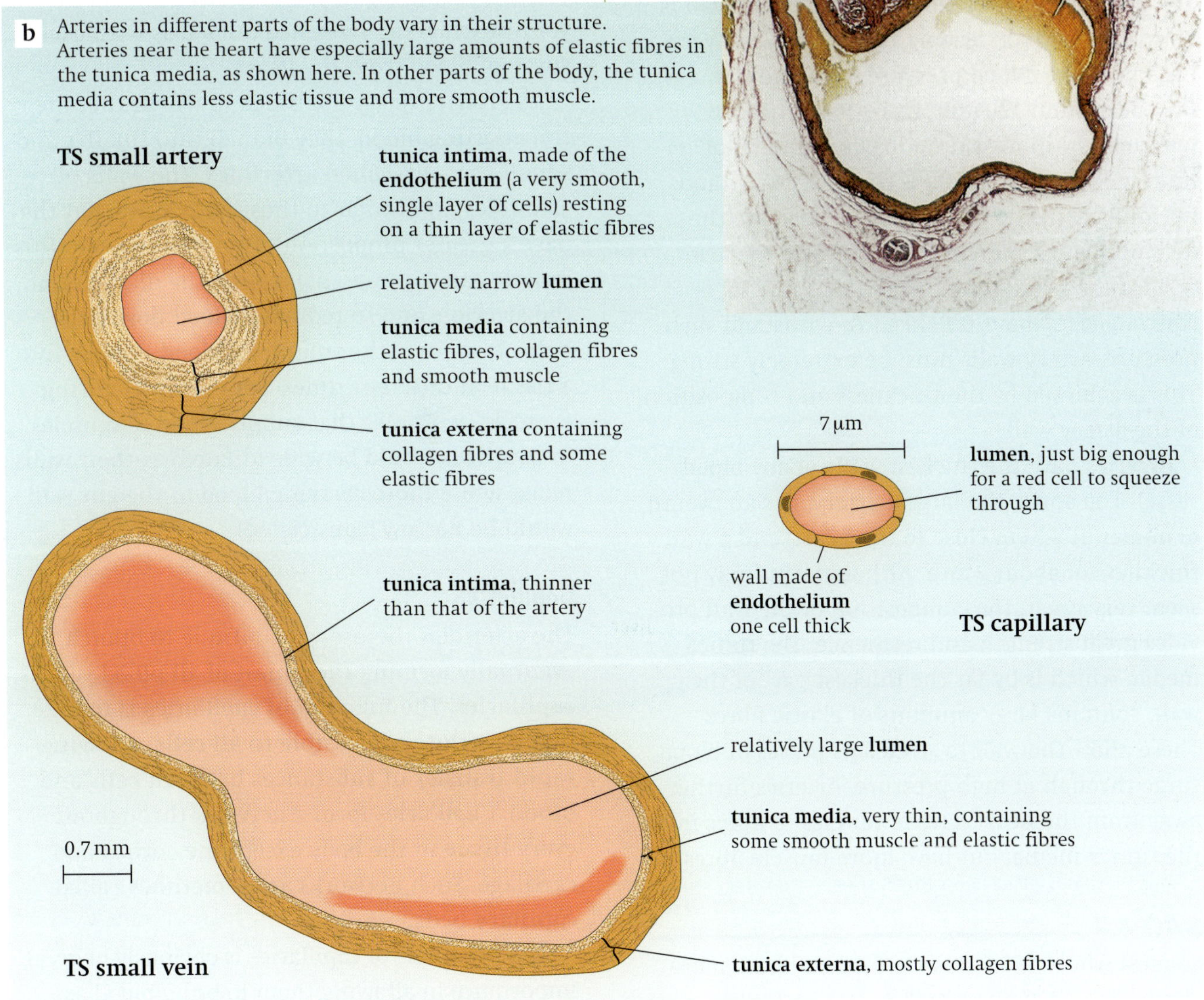

ventricle

atrium

gills

rest of

body

● **Figure 8.3** The general plan of the transport system of a fish.

**a**

**b** Arteries in different parts of the body vary in their structure. Arteries near the heart have especially large amounts of elastic fibres in the tunica media, as shown here. In other parts of the body, the tunica media contains less elastic tissue and more smooth muscle.

**TS small artery**

**tunica intima**, made of the **endothelium** (a very smooth, single layer of cells) resting on a thin layer of elastic fibres

relatively narrow **lumen**

**tunica media** containing elastic fibres, collagen fibres and smooth muscle

**tunica externa** containing collagen fibres and some elastic fibres

7 μm

**lumen**, just big enough for a red cell to squeeze through

wall made of **endothelium** one cell thick

**TS capillary**

**tunica intima**, thinner than that of the artery

relatively large **lumen**

**tunica media**, very thin, containing some smooth muscle and elastic fibres

0.7 mm

**TS small vein**

**tunica externa**, mostly collagen fibres

● **Figure 8.4   a** Micrograph of an artery and a vein in TS (× 15).
**b** The structure of arteries, veins and capillaries.

## Arteries

The function of arteries is to **transport blood, swiftly and at high pressure, to the tissues**.

The structure of the wall of an artery enables it to perform this function efficiently. Arteries and veins both have walls made up of three layers:

- an inner **endothelium** (lining tissue), made up of a layer of flat cells fitting together like jigsaw pieces (squamous epithelium: this layer is very smooth, minimising friction with the moving blood and rests on elastic fibres;
- a middle layer called the **tunica media** ('middle coat'), containing smooth muscle, collagen and elastic fibres (*tunica intima*);
- an outer layer called the **tunica externa** ('outer coat'), containing elastic fibres and collagen fibres.

The distinctive characteristic of an artery wall is its strength. Blood leaving the heart is at a very high pressure. Blood pressure in the human aorta may be around 120 mm Hg, or 16 kPa. (Blood pressure is still measured in the old units of mm Hg even though kPa is the SI unit. It stands for 'millimetres of mercury', and refers to the distance which mercury is pushed up the arm of a U-tube as shown on page 148. 1 mm Hg is equivalent to about 0.13 kPa.) To withstand such pressure, artery walls must be extremely strong. This is achieved by the thickness and composition of the artery wall.

Arteries have the thickest walls of any blood vessel. The aorta, the largest artery, has an overall diameter of 2.5 cm close to the heart, and a wall thickness of about 2 mm. Although this may not seem very great, the composition of the wall provides great strength and resilience. The tunica media, which is by far the thickest part of the wall, contains large amounts of elastic fibres. These allow the wall to stretch as pulses of blood surge through at high pressure. Arteries further away from the heart have fewer elastic fibres in the tunica media, but have more muscle fibres.

### SAQ 8.3

Suggest why arteries close to the heart have more elastic fibres in their walls than arteries further away from the heart.

The elasticity of artery walls is important in allowing them to 'give', so reducing the likelihood that they will burst. This elasticity also has another very important function. Blood is pumped out of the heart in pulses, rushing out at high pressure as the ventricles contract, and slowing as the ventricles relax. The artery walls stretch as the high-pressure blood surges into them, and then recoil inwards as the pressure drops. Therefore, as blood at high pressure enters an artery, the artery becomes wider, reducing the pressure a little. As blood at lower pressure enters an artery, the artery wall recoils inwards, giving the blood a small 'push' and raising the pressure a little. The overall effect is to 'even out' the flow of blood. However, the arteries are not entirely effective in achieving this: if you feel your pulse in your wrist, you can feel the artery, even at this distance from your heart, being stretched outwards with each surge of blood from the heart.

As arteries reach the tissue to which they are transporting blood, they branch into smaller and smaller vessels, called **arterioles**. The walls of arterioles are similar to those of arteries, but they have a greater proportion of smooth muscle. This muscle can contract, narrowing the diameter of the arteriole and so reducing blood flow. This helps to control the volume of blood flowing into a tissue at different times. For example, during exercise, arterioles that supply blood to muscles in your legs would be wide (dilated) as their walls relax, while those carrying blood to the gut wall would be narrow (constricted).

## Capillaries

The arterioles themselves continue to branch, eventually forming the tiniest of all blood vessels, **capillaries**. The function of capillaries is **to take blood as close as possible to all cells, allowing rapid transfer of substances between cells and blood**. Capillaries form a network throughout every tissue in the body except the cornea and cartilage. Such networks are sometimes called **capillary beds**.

The small size of capillaries is obviously of great importance in allowing them to bring blood as closely as possible to each group of cells in the body. A human capillary is approximately 7 μm in diameter, about the same size as a red blood cell (*figure 8.5*). Moreover, their walls are extremely thin, made up of a single layer of endothelial cells. As red blood cells carrying oxygen squeeze through a capillary, they are brought to within as little as 1 μm of the cells outside the capillary which need the oxygen.

## SAQ 8.4

Suggest why there are no blood capillaries in the cornea of the eye. How might the cornea be supplied with its requirements?

In most capillaries, there are tiny gaps between the individual cells that form the endothelium. As we shall see later in this chapter, these gaps are important in allowing some components of the blood to seep through into the spaces between the cells in all the tissues of the body. These components form tissue fluid.

By the time blood reaches the capillaries, it has already lost a great deal of the pressure originally supplied to it by the contraction of the ventricles. As blood enters a capillary from an arteriole, it may have a pressure of around 35 mm Hg or 4.7 kPa; by the time it reaches the far end of the capillary, the pressure will have dropped to around 10 mm Hg or 1.3 kPa.

## Veins

As blood leaves a capillary bed, the capillaries gradually join with one another, forming larger vessels called **venules**. These join to form **veins**. The function of veins is **to return blood to the heart**.

By the time blood enters a vein, its pressure has dropped to a very low value. In humans, a typical value for venous blood pressure is about 5 mm Hg or less. This very low pressure means that there is no need for veins to have thick walls. They have the same three layers as arteries, but the tunica media is much thinner, and has far fewer elastic fibres and muscle fibres.

The low blood pressure in veins creates a problem: how can this blood be returned to the heart? The problem is perhaps most obvious if you

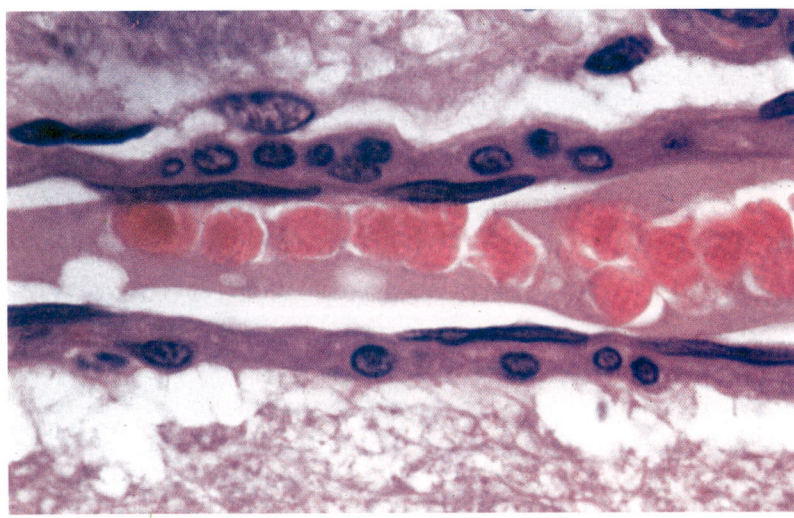

● **Figure 8.5** Micrograph of a blood capillary containing red blood cells, which are stained pink. The cells with dark purple nuclei are the endothelium of the capillary wall.

consider how blood can return from your legs. Unaided, the blood in your leg veins would sink and accumulate in your feet. However, many of the veins run within, or very close to, several leg muscles. Whenever you tense these muscles, they squeeze inwards on the veins in your legs, temporarily raising the pressure within them.

This in itself would not help to push the blood back towards the heart; blood would just squidge up and down as you walked. To keep the blood flowing in the right direction, veins contain half-moon valves, or **semilunar valves**, formed from their endothelium (*figure 8.6*). These valves allow blood to move towards the heart, but not away

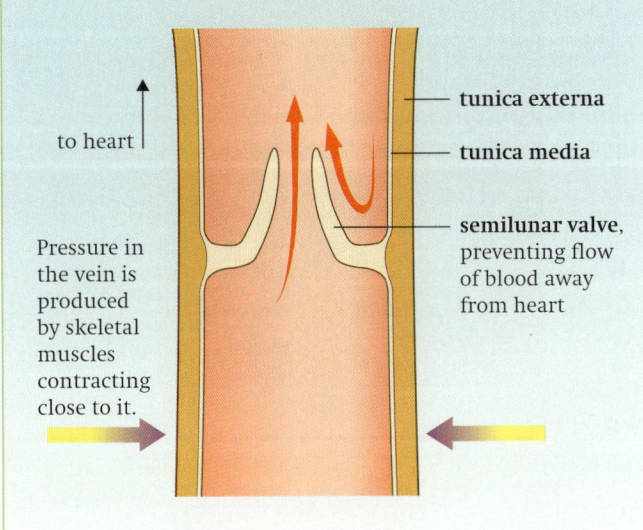

to heart

Pressure in the vein is produced by skeletal muscles contracting close to it.

tunica externa

tunica media

semilunar valve, preventing flow of blood away from heart

● **Figure 8.6** Longitudinal section of a small vein.

from it. Thus, when you contract your leg muscles, the blood in the veins is squeezed *up* through these valves, but cannot pass *down* through them.

### SAQ 8.5

Suggest reasons for each of the following.
a Normal venous pressure in the feet is about 25 mm Hg. When a soldier stands at attention the blood pressure in his feet rises very quickly to about 90 mm Hg.
b When you breathe in, that is when the volume of the thorax increases, blood moves through the veins towards the heart.

### SAQ 8.6

Construct a table comparing the structure of arteries, veins and capillaries. Include both similarities and differences, and give reasons for the differences which you describe.

### SAQ 8.7

Using *figure 8.7*, describe and explain how blood pressure varies in different parts of the circulatory system.

# Blood plasma and tissue fluid

Blood is composed of cells floating in a pale yellow liquid called **plasma**. Blood plasma is mostly water, with a variety of substances dissolved in it. These solutes include nutrients, such as glucose, and waste products, such as urea, that are being transported from one place to another in the body. They also include protein molecules, called **plasma proteins**, that remain in the blood all the time.

As blood flows through capillaries within tissues, some of the plasma leaks out through the gaps between the cells in the walls of the capillary, and seeps into the spaces between the cells of the tissues. Almost one-sixth of your body consists of spaces between your cells. These spaces are filled with this leaked plasma, which is known as **tissue fluid**.

Tissue fluid is almost identical in composition to blood plasma. However, it contains far fewer protein molecules than blood plasma, as these are too large to escape easily through the tiny holes in the capillary endothelium. Red blood cells are much too large to pass through, so tissue fluid does not contain these, but some white blood cells can squeeze through, and move freely around in tissue fluid. *Table 8.2* shows the sizes of the molecules of some of the substances in blood plasma, and the relative ease with which they pass from capillaries into tissue fluid.

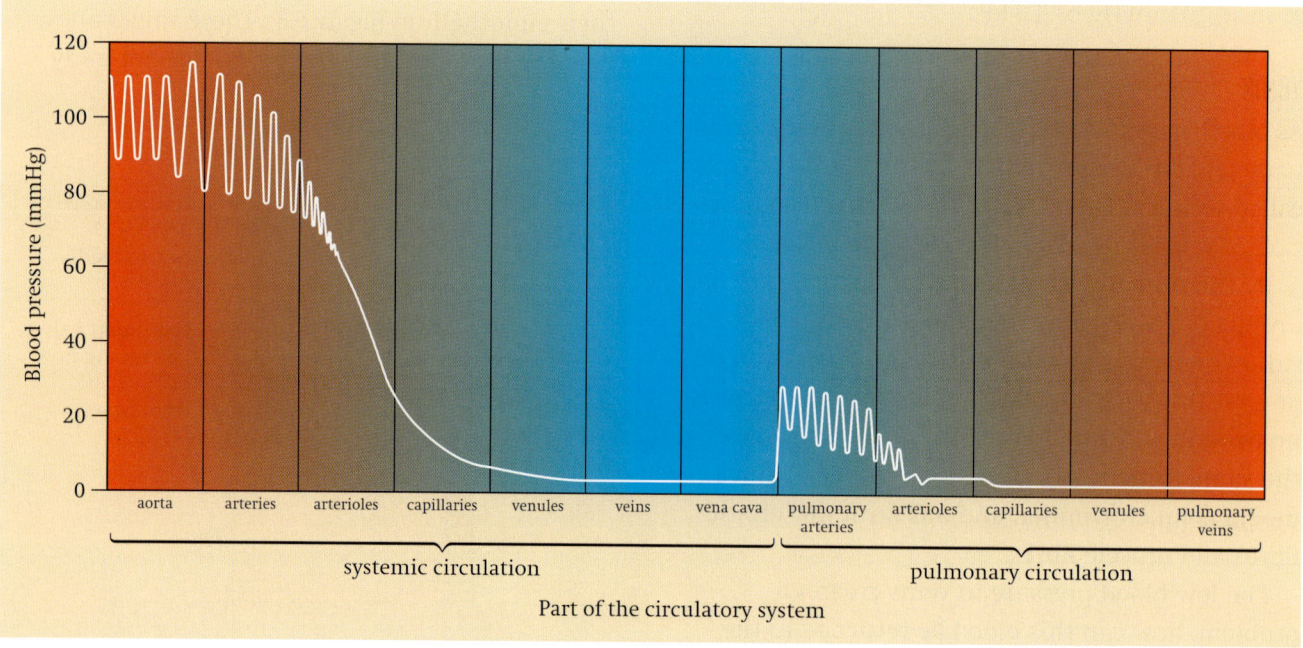

● **Figure 8.7** Blood pressure in different regions of the human circulatory system.

| Substance | Relative molecular mass | Permeability |
|---|---|---|
| water | 18 | 1.00 |
| sodium ions | 23 | 0.96 |
| urea | 60 | 0.8 |
| glucose | 180 | 0.6 |
| haemoglobin | 68 000 | 0.01 |
| albumin | 69 000 | 0.000 01 |

The permeability to water is given a value of 1. The other values are given in proportion to that of water.

● **Table 8.2** Relative permeability of capillaries in a muscle to different substances.

### SAQ 8.8

Use the information in *table 8.2* to answer the following.

**a** How does relative molecular mass of a substance appear to correlate with the permeability of capillary walls to this substance?

**b** In a respiring muscle, would you expect the net diffusion of glucose to be *from* the blood plasma to the muscle cells, or vice versa? Explain your answer.

**c** Albumin is the most abundant plasma protein. Suggest why it is important that capillary walls should not be permeable to albumin.

The amount of fluid which leaves the capillary to form tissue fluid is the result of two opposing pressures. Particularly at the arterial end of a capillary bed, the blood pressure inside the capillary is enough to push fluid out into the tissue. However, we have seen that water moves by osmosis from regions of low solute concentration to regions of high solute concentration (page 55). Since tissue fluid lacks the high concentrations of proteins that exist in plasma, the imbalance leads to osmotic movement of water back into capillaries from tissue fluid. The net result of these competing processes is that fluid tends to

flow *out* of capillaries into tissue fluid at the *arterial* end of a capillary bed and *into* capillaries from tissue fluid near the *venous* end of a capillary bed. Overall, however, rather more fluid flows out of capillaries than into them, so that there is a net loss of fluid from the blood as it flows through a capillary bed.

Tissue fluid forms the immediate environment of each individual body cell. It is through tissue fluid that exchanges of materials between cells and the blood occur. Within our body, many processes take place to maintain the composition of tissue fluid at a constant level, to provide an optimum environment in which cells can work. These processes contribute to the overall process of **homeostasis**, that is the maintenance of a constant internal environment, and include the regulation of glucose concentration, water, pH, metabolic wastes and temperature.

## Lymph

About 90% of the fluid that leaks from capillaries eventually seeps back into them. The remaining 10% is collected up and returned to the blood system by means of a series of tubes known as **lymph vessels** or **lymphatics**.

Lymphatics are tiny, blind-ending vessels, which are found in almost all tissues of the body. The end of one of these vessels is shown in *figure 8.8*. Tissue fluid can flow into the lymphatic through

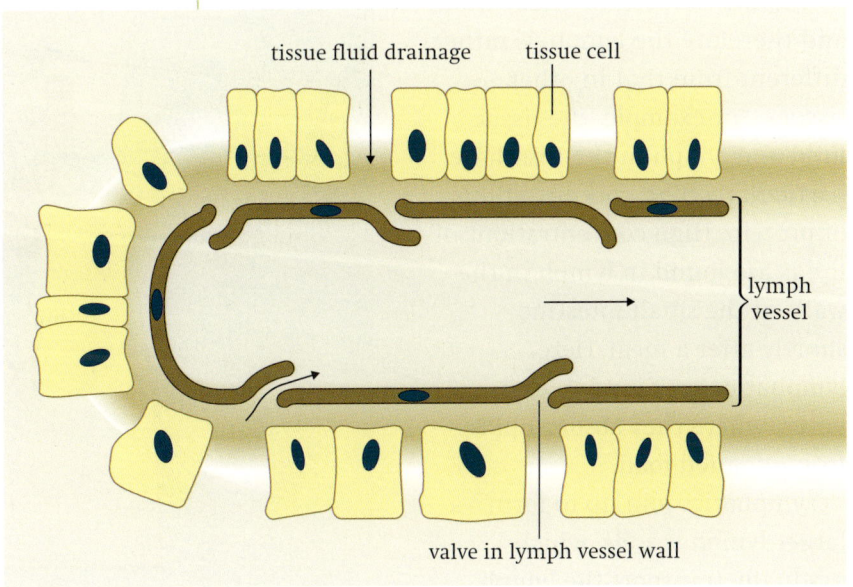

● **Figure 8.8** Drainage of tissue fluid into a lymph vessel.

tiny valves, which allow it to flow in but not out. These valves are wide enough to allow large protein molecules to pass through. This is very important, as such molecules are too big to get into blood capillaries, and so cannot be taken away by the blood. If your lymphatics did not take away the protein in the tissue fluid between your cells, you could die within 24 hours. If the protein concentration and rate of loss from plasma are not in balance with the concentration and rate of loss from tissue fluid, there can be a build up of tissue fluid, called **oedema**.

## SAQ 8.9

a  We have seen that capillary walls are not very permeable to plasma proteins. Suggest where the protein in tissue fluid has come from.

b  The disease kwashiorkor is caused by a diet which is very low in protein. The concentration of proteins in blood plasma becomes much lower than usual. One of the symptoms of kwashiorkor is oedema. Suggest why this is so. (You will need to think about water potential.)

The fluid inside lymphatics is called **lymph**. It is virtually identical to tissue fluid: it has a different name more because it is in a different place than because it is different in composition.

In some tissues, the tissue fluid, and therefore the lymph, is rather different from that in other tissues. For example, the tissue fluid and lymph in the liver have particularly high concentrations of protein. High concentrations of lipids are found in lymph in the walls of the small intestine shortly after a meal. Here, lymphatics are found in each villus, where they absorb lipids from digested food.

Lymphatics join up to form larger lymph vessels, which gradually transport the lymph back to the large veins which run

just beneath the collarbone, the **subclavian veins** (*figure 8.9*). As in veins, the movement of fluid along the lymphatics is largely caused by the contraction of muscles around the vessels, and kept going in the right direction by valves. Lymph vessels also have smooth muscle in their walls, which can contract to push the lymph along. Lymph flow is very slow, and only about $100\,\text{cm}^3$ per hour flows through the largest lymph vessel, the thoracic duct, in a resting human. This contrasts with the flow rate of blood of around $80\,\text{cm}^3$ per second.

At intervals along lymph vessels, there are **lymph nodes**. These are involved in protection against disease. Bacteria and other unwanted particles are removed from lymph by some types of white blood cells as the lymph passes through a node, while other white blood cells within the nodes secrete **antibodies**. For more detail, see chapter 14.

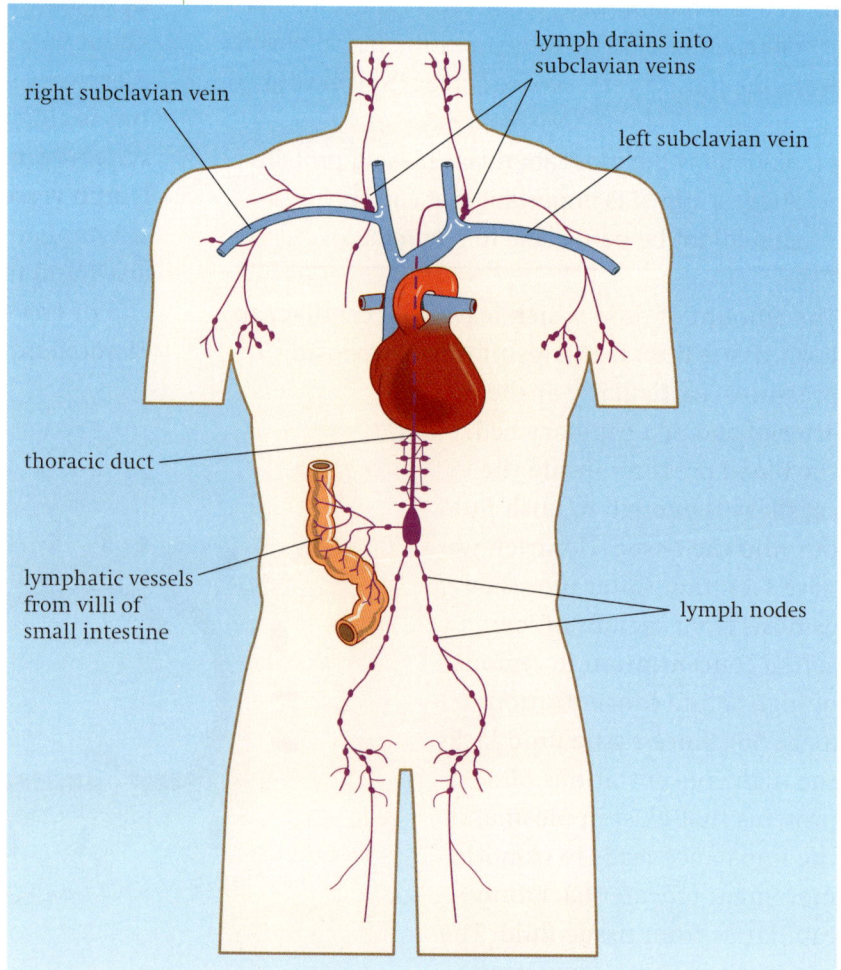

● **Figure 8.9** The human lymphatic system.

# Blood

You have about 5 dm³ of blood in your body, weighing about 5 kg. Suspended in the blood plasma, you have around $2.5 \times 10^{13}$ red blood cells, $5 \times 10^{11}$ white blood cells and $6 \times 10^{12}$ platelets (*figure 8.10*).

## Red blood cells

Red blood cells are also called **erythrocytes**, which simply means 'red cells'. Their red colour is caused by the pigment **haemoglobin**, a globular

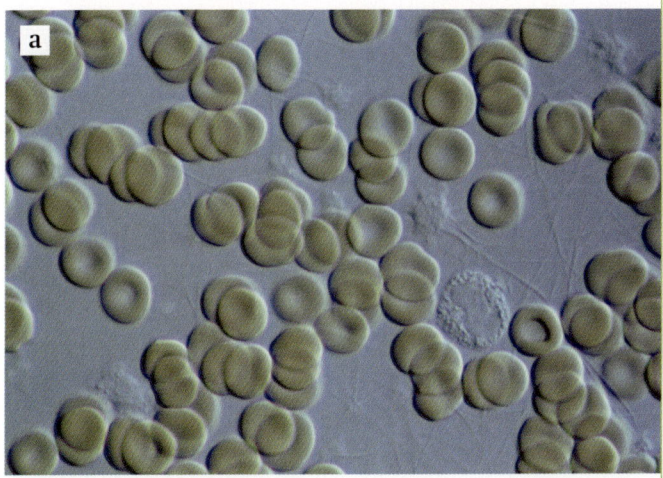

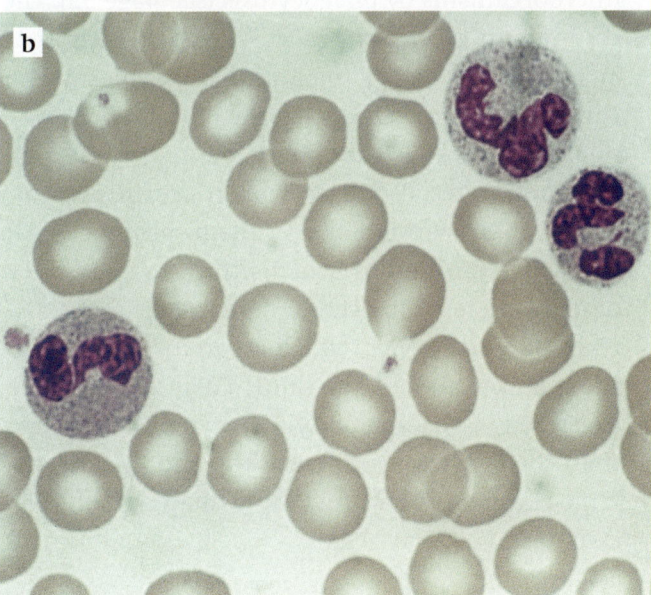

● **Figure 8.10** Micrographs of human blood.
**a** This photograph of unstained blood is taken with an interference contrast light microscope. Most of the cells are red blood cells. You can also see a white cell just to the right of centre (× 900).
**b** This photograph is taken with a normal light microscope. The blood has been stained so that the nuclei of the white cells, and the platelets, are purple (× 1400).

protein (see pages 34–35). The main function of haemoglobin is to transport oxygen from lungs to respiring tissues. This function is described in detail on pages 109–110.

A person's first red blood cells are formed in the liver, while still a fetus inside the uterus. By the time a baby is born, the liver has stopped manufacturing red blood cells. This function has been taken over by the bone marrow. This continues, at first in the long bones such as the humerus and femur, and then increasingly in the skull, ribs, pelvis and vertebrae, throughout life. Red blood cells do not live long; their membranes become more and more fragile and eventually rupture within some 'tight spot' in the circulatory system, often inside the spleen.

### SAQ 8.10

Assuming that you have $2.5 \times 10^{13}$ red blood cells in your body, that the average life of a red blood cell is 120 days, and that the total number of red blood cells remains constant, calculate how many new red blood cells must be made, on average, in your bone marrow each day.

The structure of a red blood cell (*figure 8.11*) is unusual in three ways.

■ **Red blood cells are very small**. The diameter of a human red blood cell is about 7 μm, compared with the diameter of an 'average' liver cell of 40 μm. This small size means that no

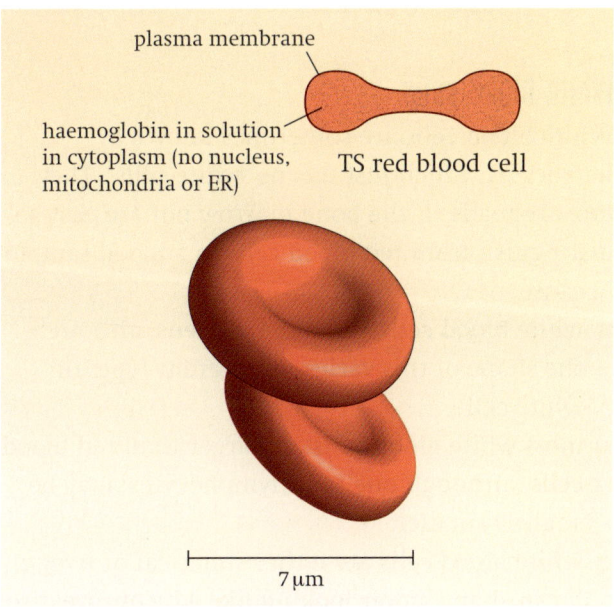

plasma membrane

haemoglobin in solution in cytoplasm (no nucleus, mitochondria or ER)

TS red blood cell

7 μm

● **Figure 8.11** Red blood cells.

haemoglobin molecule within the cell is very far from the cell's plasma membrane, and can therefore quickly exchange oxygen with the fluid outside the cell. It also means that capillaries can be only 7 μm wide and still allow red blood cells to squeeze through them, so bringing oxygen as close as possible to cells which require it.

■ **Red blood cells are shaped like a biconcave disc.** The 'dent' in each side of a red blood cell, like its small size, increases the amount of surface area in relation to the volume of the cell, giving it a large surface area to volume ratio. This large surface area means that oxygen can diffuse quickly into or out of the cell.

■ **Red blood cells have no nucleus, no mitochondria and no endoplasmic reticulum.** The lack of these organelles means that there is more room for haemoglobin, so maximising the amount of oxygen which can be carried by each red blood cell.

### SAQ 8.11

Which of these functions could, or could not, be carried out by a red blood cell? In each case, briefly justify your answer.

**a** Protein synthesis    **b** Cell division
**c** Lipid synthesis       **d** Active transport

## White blood cells

White blood cells are sometimes known as **leucocytes**, which just means 'white cells'. They, too, are made in the bone marrow but are easy to distinguish from red blood cells in a blood sample because:

■ white blood cells all have a nucleus, although the shape of this varies in different types of white cell;

■ most white blood cells are larger than red blood cells, although one type, lymphocytes, may be slightly smaller;

■ white blood cells are either spherical or irregular in shape, never looking like a biconcave disc (*figures 8.10* and *8.12*).

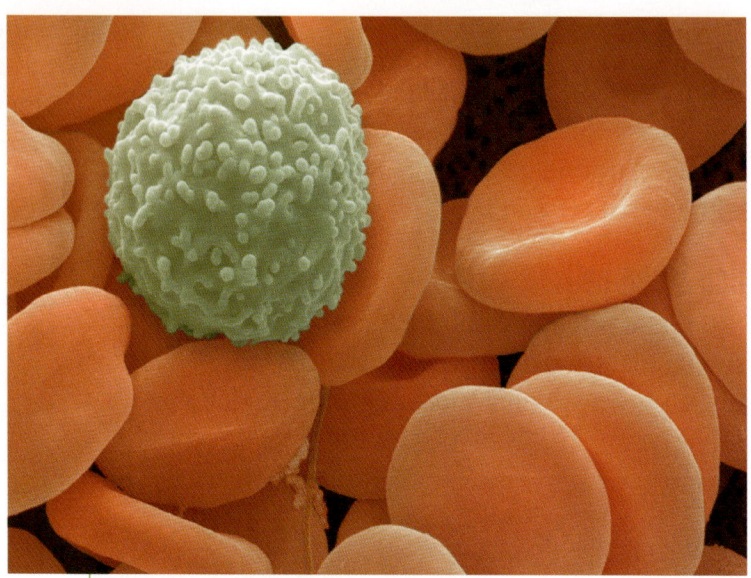

● **Figure 8.12** False-colour SEM of one human white blood cell amongst many red (× 2500). Note the distinctive surface texture. Compare with *figure 6.14*.

There are many different kinds of white blood cell, with a wide variety of functions, although all are concerned with fighting disease. They can be divided into two main groups.

**Phagocytes** are cells that destroy invading microorganisms by phagocytosis (see page 59). The commonest type of phagocytes can be recognised by their lobed nuclei and granular cytoplasm. The three white blood cells with the dark purple nuclei in *figure 8.10b* are phagocytes.

**Lymphocytes** also destroy microorganisms, but not by phagocytosis. Some of them secrete chemicals called **antibodies**, which attach to and destroy the invading cells. There are different types of lymphocytes, which act in different ways, though they all look the same. Their activities are described in chapter 14. Lymphocytes are smaller than the main type of phagocyte, and they have a large round nucleus and only a small amount of cytoplasm.

## Haemoglobin

A major role of the cardiovascular system is to transport oxygen from the gas exchange surfaces of the alveoli in the lungs (see page 60) to tissues all over the body. Body cells need a constant supply of oxygen, in order to be able to carry out aerobic respiration.

Oxygen is transported around the body inside red blood cells in combination with the protein **haemoglobin** (*figure 2.2*).

As we saw in chapter 2, each haemoglobin molecule is made up of four polypeptides each containing one haem group. Each haem group can combine with one oxygen molecule, $O_2$. Overall, then, each haemoglobin molecule can combine with four oxygen molecules (eight oxygen atoms).

$$Hb + 4O_2 \rightleftharpoons HbO_8$$

haemoglobin    oxygen    oxyhaemoglobin

## SAQ 8.12

In an adult healthy human, the amount of haemoglobin in $1\,dm^3$ of blood is about $150\,g$.

**a** Given that $1\,g$ of pure haemoglobin can combine with $1.3\,cm^3$ of oxygen at body temperature, how much oxygen can be carried in $1\,dm^3$ of blood?

**b** At body temperature, the solubility of oxygen in water is approximately $0.025\,cm^3$ of oxygen per $cm^3$ of water. Assuming that blood plasma is mostly water, how much oxygen could be carried in $1\,dm^3$ of blood if we had no haemoglobin?

## The haemoglobin dissociation curve

A molecule whose function is to transport oxygen from one part of the body to another must be able not only to *pick up* oxygen at the lungs, but also to *release* oxygen within respiring tissues. Haemoglobin performs this task superbly.

To investigate how haemoglobin behaves, samples are extracted from blood and exposed to different concentrations, or **partial pressures**, of oxygen. The amount of oxygen which combines with each sample of haemoglobin is then measured. The maximum amount of oxygen with which a sample can possibly combine is given a

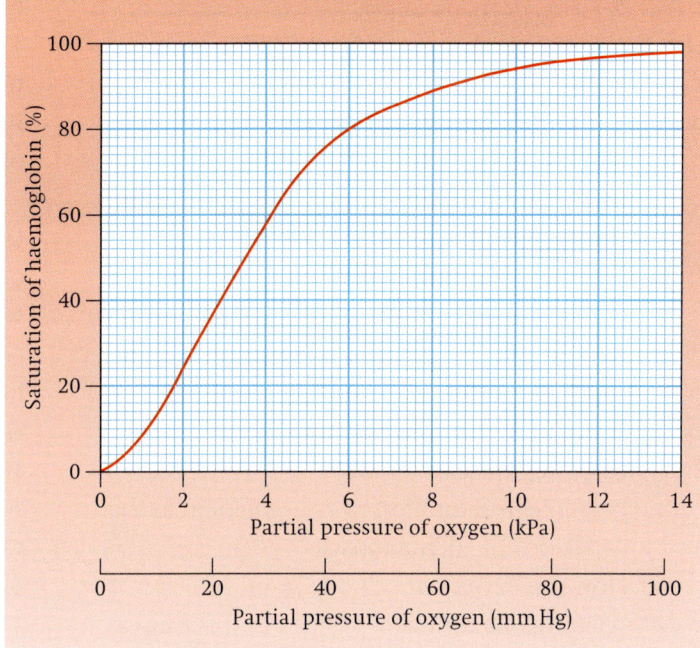

● **Figure 8.13** The haemoglobin dissociation curve.

value of 100%. A sample of haemoglobin which has combined with this maximum amount of oxygen is said to be **saturated**. The amounts with which identical samples combine at lower oxygen partial pressures are then expressed as a percentage of this maximum value. *Table 8.3* shows a series of results from such an investigation.

The percentage saturation of each sample can be plotted against the partial pressure of oxygen to obtain the curve shown in *figure 8.13*. This is known as a **dissociation curve**.

It shows that at low partial pressures of oxygen, the percentage saturation of haemoglobin is very low, that is the haemoglobin is combined with only a very little oxygen. At high partial pressures of oxygen, the percentage saturation of haemoglobin is very high: it is combined with large amounts of oxygen.

| Partial pressure of oxygen (kPa) | 1 | 2 | 3 | 4 | 5 | 6 | 7 | 8 | 9 | 10 | 11 | 12 | 13 | 14 |
|---|---|---|---|---|---|---|---|---|---|---|---|---|---|---|
| Saturation of haemoglobin (%) | 8.5 | 24.0 | 43.0 | 57.5 | 71.5 | 80.0 | 85.5 | 88.0 | 92.0 | 94.0 | 95.5 | 96.5 | 97.5 | 98.0 |

● **Table 8.3** The varying ability of haemoglobin to carry oxygen.

## SAQ 8.13

Use the dissociation curve in *figure 8.13* to answer these questions.

**a** (i) The partial pressure of oxygen in the alveoli of the lungs is about 12 kPa. What is the percentage saturation of haemoglobin in the capillaries in the lungs?

(ii) If one gram of fully saturated haemoglobin is combined with 1.3 cm$^3$ of oxygen, how much oxygen will one gram of haemoglobin in the capillaries in the lungs be combined with?

**b** (i) The partial pressure of oxygen in an actively respiring muscle is about 2 kPa. What is the percentage saturation of haemoglobin in the capillaries of such a muscle?

(ii) How much oxygen will one gram of haemoglobin in the capillaries of this muscle be combined with?

Consider the haemoglobin within a red blood cell in a capillary in the lungs. Here, where the partial pressure of oxygen is high, this haemoglobin will be 95–97% saturated with oxygen, that is almost every haemoglobin molecule will be combined with its full complement of eight oxygen atoms. In an actively respiring muscle, on the other hand, where the partial pressure of oxygen is low, the haemoglobin will be about 20–25% saturated with oxygen, that is the haemoglobin is carrying only a quarter of the oxygen which it is capable of carrying. This means that haemoglobin coming from the lungs carries a lot of oxygen; as it reaches a muscle it releases around three-quarters. This released oxygen diffuses out of the red blood cell and into the muscle where it can be used in respiration.

### The S-shaped curve

The shape of the haemoglobin dissociation curve can be explained by the behaviour of a haemoglobin molecule as it combines with or loses oxygen molecules.

Oxygen molecules combine with the iron atoms in the haem groups of a haemoglobin molecule. You will remember that each haemoglobin molecule has four haem groups. When an oxygen molecule combines with one haem group, the whole haemoglobin molecule is slightly distorted. The distortion makes it easier for a second oxygen molecule to combine with a second haem group. This in turn makes it easier for a third oxygen molecule to combine with a third haem group. It is then still easier for the fourth and final oxygen molecule to combine.

The shape of the curve reflects this behaviour. Up to an oxygen partial pressure of around 2 kPa, on average only one oxygen molecule is combined with each haemoglobin molecule. Once this oxygen molecule is combined, however, it becomes successively easier for the second and third oxygen molecules to combine, so the curve rises very steeply. Over this part of the curve, a *small* change in the partial pressure of oxygen causes a *very large* change in the amount of oxygen which is carried by the haemoglobin.

### The Bohr shift

The behaviour of haemoglobin in picking up oxygen at the lungs, and readily releasing it when in conditions of low oxygen partial pressure, is exactly what is needed. But, in fact, it is even better at this than is shown by the dissociation curve in *figure 8.13*. This is because the amount of oxygen it carries is affected not only by the partial pressure of *oxygen*, but also by the partial pressure of *carbon dioxide*.

Carbon dioxide is continually produced by respiring cells. It diffuses from the cells and into blood plasma, from where some of it diffuses into the red blood cells.

In the cytoplasm of red blood cells there is an enzyme, **carbonic anhydrase**. This enzyme catalyses the following reaction:

$$CO_2 + H_2O \underset{\text{carbonic anhydrase}}{\rightleftharpoons} H_2CO_3$$

carbon dioxide    water             carbonic acid

The carbonic acid dissociates:

$$H_2CO_3 \rightleftharpoons H^+ + HCO_3^-$$

carbonic acid    hydrogen ion    hydrogencarbonate ion

Haemoglobin readily combines with these hydrogen ions, forming **haemoglobinic acid, HHb**. In so doing, it releases the oxygen which it is carrying.

The net result of this reaction is two-fold:

■ The haemoglobin 'mops up' the hydrogen ions which are formed when carbon dioxide dissolves and dissociates. A high concentration of hydrogen ions means a low pH; if the hydrogen ions were left in solution the blood would be very acidic. By removing the hydrogen ions from solution, haemoglobin helps to maintain the pH of the blood close to neutral. It is acting as a **buffer**.

- The presence of a high partial pressure of carbon dioxide causes haemoglobin to release oxygen. This is called the **Bohr effect**, after Christian Bohr who discovered it in 1904. It is exactly what is needed. High concentrations of carbon dioxide are found in actively respiring tissues, which need oxygen; these high carbon dioxide concentrations cause haemoglobin to release its oxygen even more readily than it would otherwise do.

If a dissociation curve is drawn for haemoglobin at a high partial pressure of carbon dioxide, it looks like the lower curve in *figure 8.14*. At each partial pressure of oxygen, the haemoglobin is less saturated than it would be at a low partial pressure of carbon dioxide. The curve therefore lies below, and to the right of, the 'normal' curve.

## Carbon dioxide transport

The description of the Bohr effect above explains one way in which carbon dioxide is carried in the blood. One product of the dissociation of dissolved carbon dioxide is hydrogencarbonate ions, $HCO_3^-$. These are initially formed in the cytoplasm of the red blood cell, because this is where the enzyme carbonic anhydrase is found. Most of them then diffuse out of the red blood cell into the blood plasma, where they are carried in solution. About 85% of the carbon dioxide transported by the blood is carried in this way.

Some carbon dioxide, however, does not dissociate, but remains as carbon dioxide molecules. Some of these simply dissolve in the blood plasma; about 5% of the total is carried in this form. Others diffuse into the red blood cells, but instead of undergoing the reaction catalysed by carbonic anhydrase, combine directly with the terminal amine groups ($-NH_2$) of some of the haemoglobin molecules. The compound formed is called **carbamino-haemoglobin**. About 10% of the carbon dioxide is carried in this way (*figure 8.15*).

When blood reaches the lungs, the reactions described above go into reverse. The relatively low concentration of carbon dioxide in the alveoli compared with that in the blood causes carbon dioxide to diffuse from the blood into the air in the alveoli, stimulating the carbon dioxide of carbamino-haemoglobin to leave the red blood cell, and hydrogencarbonate and hydrogen ions to recombine to form carbon dioxide molecules once more. This leaves the haemoglobin molecules free to combine with oxygen, ready to begin another circuit of the body.

## Fetal haemoglobin

A developing fetus obtains its oxygen not from its own lungs, but from its mother's blood. In the placenta, the mother's blood is brought very close to that of the fetus, allowing diffusion of various substances from mother to fetus or vice versa.

Oxygen arrives at the placenta in combination with haemoglobin, inside the mother's red blood cells. The partial pressure of oxygen in the blood vessels in the placenta is relatively low,

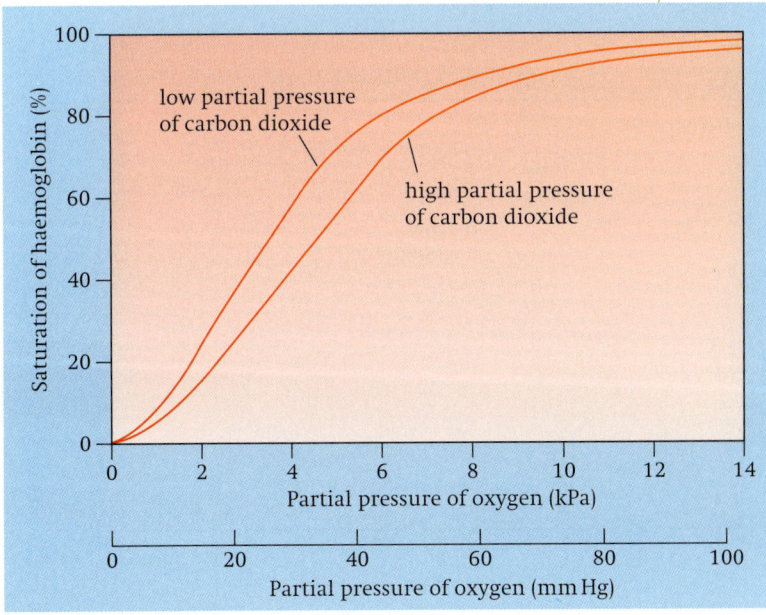

● **Figure 8.14** Dissociation curves for haemoglobin at two different partial pressures of carbon dioxide. The shift of the curve to the right when the haemoglobin is exposed to higher carbon dioxide concentration is called the Bohr effect.

because the fetus is respiring. The mother's haemoglobin therefore releases some of its oxygen, which diffuses from her blood into the fetus's blood.

The partial pressure of oxygen in the fetus's blood is only a little lower than that in its mother's blood. However, the haemoglobin of the fetus is different from its mother's haemoglobin. Fetal haemoglobin combines more readily with oxygen than adult haemoglobin does; thus, the fetal haemoglobin will pick up oxygen which the adult haemoglobin has dropped. Fetal haemoglobin is said to have a **higher affinity** for oxygen than adult haemoglobin.

A dissociation curve for fetal haemoglobin (*figure 8.16*) shows that, at each partial pressure of oxygen, fetal haemoglobin is slightly more saturated than adult haemoglobin. The curve lies *above* the curve for adult haemoglobin.

# Myoglobin

Myoglobin, like haemoglobin, is a red pigment which combines reversibly with oxygen. It is not found in the blood, but inside cells in some tissues of the body, especially in muscle cells. The red colour of meat is largely caused by myoglobin.

Each myoglobin molecule is made up of only one polypeptide, rather than the four in a haemoglobin molecule. It has just one haem group, and can combine with just one oxygen molecule. However, once combined, the oxymyoglobin molecule is very stable, and will not release its oxygen unless the partial pressure

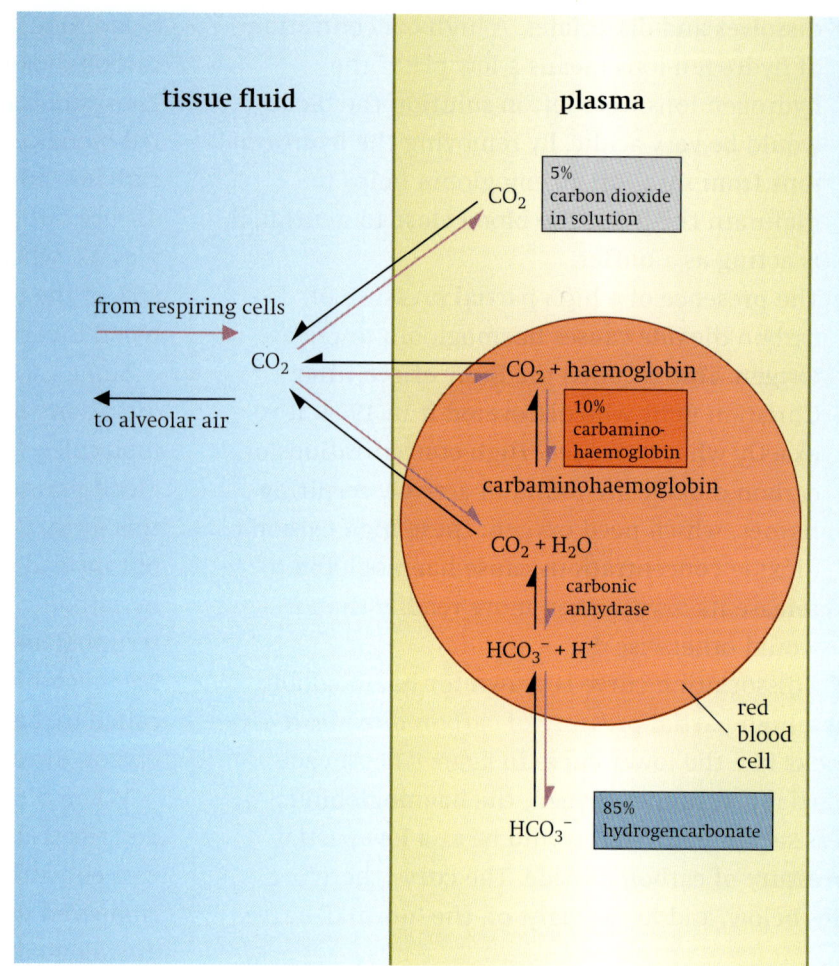

● **Figure 8.15** Carbon dioxide transport in the blood. The blood carries carbon dioxide partly as undissociated carbon dioxide in solution in the plasma, partly as hydrogencarbonate ions in solution in the plasma and partly combined with haemoglobin in the red blood cells.

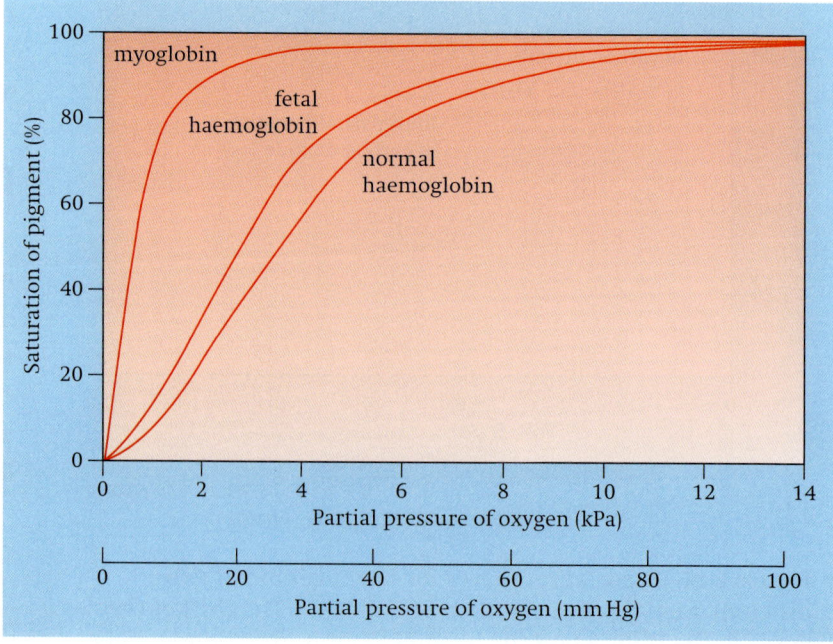

● **Figure 8.16** Dissociation curves for myoglobin and fetal haemoglobin.

of oxygen around it is very low indeed. The curve in *figure 8.16* shows this in comparison with haemoglobin. At each partial pressure of oxygen, myoglobin has a higher percentage saturation with oxygen than haemoglobin does.

This means that myoglobin acts as an **oxygen store**. At the normal partial pressures of oxygen in a respiring muscle, haemoglobin releases its oxygen. Some of this oxygen is picked up and held tightly by myoglobin in the muscle. The myoglobin will not release this oxygen unless the oxygen concentration in the muscle drops very low, that is unless the muscle is using up oxygen at a faster rate than the haemoglobin in the blood can supply it. The oxygen held by the myoglobin is a reserve, to be used only in conditions of particularly great oxygen demand.

# Problems with oxygen transport

The efficient transport of oxygen around the body can be impaired by many different factors. We consider some here and in chapter 12.

## Carbon monoxide

Despite its almost perfect design as an oxygen-transporting molecule, haemoglobin does have one property which can prove very dangerous. It combines very readily, and almost irreversibly, with carbon monoxide.

Carbon monoxide, CO, is formed when a carbon-containing compound burns incompletely. Exhaust fumes from cars contain significant amounts of carbon monoxide, as does cigarette smoke. When such fumes are inhaled, the carbon monoxide readily diffuses across the walls of the alveoli, into blood, and into red blood cells. Here it combines with the haem groups in the haemoglobin molecules, forming **carboxyhaemoglobin**.

Haemoglobin combines with carbon monoxide 250 times more readily than it does with oxygen. Thus, even if the concentration of carbon monoxide in air is much lower than the concentration of oxygen, a high proportion of haemoglobin will combine with carbon monoxide. Moreover, carboxyhaemoglobin is a very stable compound; the carbon monoxide remains combined with the haemoglobin for a long time.

The result of this is that even relatively low concentrations of carbon monoxide, as low as 0.1% of the air, can cause death by asphyxiation. Treatment of carbon monoxide poisoning involves administration of a mixture of pure oxygen and carbon dioxide: high concentrations of oxygen to favour the combination of haemoglobin with oxygen rather than carbon monoxide, and carbon dioxide to stimulate an increase in the breathing rate.

Cigarette smoke contains up to 5% carbon monoxide (see pages 150–151). If you breathed in 'pure' cigarette smoke for any length of time, you would die of asphyxiation. As it is, even smokers who inhale also breathe in some normal air, diluting the carbon monoxide levels in their lungs. Nevertheless, around 5% of the haemoglobin in a regular smoker's blood is permanently combined with carbon monoxide. This considerably reduces its oxygen-carrying capacity.

## High altitude

We obtain our oxygen from the air around us. At sea level, the partial pressure of oxygen in the atmosphere is just over 20 kPa, and the partial pressure of oxygen in an alveolus in the lungs is about 13 kPa. If you look at the oxygen dissociation curve for haemoglobin in *figure 8.13*, you can see that at this partial pressure of oxygen haemoglobin is almost completely saturated with oxygen.

If, however, a person climbs up a mountain to a height of 6500 metres (about 21 000 feet), then the air pressure is much less. The partial pressure of oxygen in the air is only about 10 kPa, and in the lungs about 5.3 kPa. You can see from *figure 8.13* that this will mean that the haemoglobin will become only about 70% saturated in the lungs. Less oxygen will be carried around the body, and the person may begin to feel breathless and ill.

### SAQ 8.14

Mount Everest is nearly 9000 m high. The partial pressure of oxygen in the alveoli at this height is only about 2.5 kPa. Explain what effect this would have on the supply of oxygen to body cells if a person climbed to the top of Mount Everest without a supplementary oxygen supply.

If someone climbs steadily, over a period of just a few days, from sea level to a high altitude, the body does not have enough time to adjust to this drop in oxygen availability, and the person may suffer from **altitude sickness**. The symptoms frequently begin with an increase in the rate and depth of breathing, and a general feeling of dizziness and weakness. These symptoms can be easily reversed by going down to a lower altitude. Some people, however, can quickly become very ill indeed. The arterioles in their brain dilate, increasing the amount of blood flowing into the capillaries, so that fluid begins to leak from them into the brain tissues. This can cause disorientation. Fluid may also leak into the lungs, preventing them from functioning properly. Acute altitude sickness can be fatal, and a person suffering from it must be brought down to low altitude immediately, or given oxygen.

However, if the body is given plenty of time to adapt, then most people can cope well at altitudes up to at least 5000 metres. In 1979, two mountaineers climbed Mount Everest without oxygen, returning safely despite experiencing hallucinations and feelings of euphoria at the summit.

As the body gradually acclimatises to high altitude, a number of changes take place. Perhaps the most significant of these is that the number of red blood cells increases. Whereas red blood cells normally make up about 40–50% of the blood, after a few months at high altitude this rises to as much as 50–70%. However, this does take a long time to happen, and there is almost no change in the number of red blood cells for at least two or three weeks at high altitude.

### SAQ 8.15

Explain how an increase in the number of red blood cells can help to compensate for the lack of oxygen in the air at high altitude.

### SAQ 8.16

Athletes often prepare themselves for important competitions by spending several months training at high altitude. Explain how this could improve their performance.

● **Figure 8.17** Most high-altitude climbers, such as here on Mount Everest, breathe using oxygen carried in cylinders.

People who live permanently at high altitude, such as in the Andes or Himalayas, show a number of adaptations to their low-oxygen environment. It seems that they are not genetically different from people who live at low altitudes, but rather that their exposure to low oxygen partial pressures from birth encourages the development of these adaptations from an early age. They often have especially broad chests, providing larger lung capacities than normal. The heart is often larger than in a person who lives at low altitude, especially the right side that pumps blood to the lungs. They also have more haemoglobin in their blood than usual, so increasing the efficiency of oxygen transport from lungs to tissues.

# SUMMARY

◆ Large, active organisms such as mammals need a transport system, in which fluid driven by a pump carries oxygen and nutrients to all tissues, and removes waste products from them. Mammals have a double circulatory system.

◆ Blood is carried away from the heart in arteries, passes through tissues in capillaries, and is returned to the heart in veins. Blood pressure drops gradually as it passes along this system.

◆ Arteries have thick, elastic walls, to allow them to withstand high blood pressures and to smooth out the pulsed blood flow. Capillaries are only just wide enough to allow the passage of red blood cells, and have very thin walls to allow efficient and rapid transfer of materials between blood and cells. Veins have thinner walls than arteries and possess valves, to help blood at low pressure flow back to the heart.

◆ Plasma leaks from capillaries to form tissue fluid. This is collected into lymphatics and returned to the blood in the subclavian veins.

◆ Red blood cells carry oxygen in combination with haemoglobin. Haemoglobin picks up oxygen at high partial pressures of oxygen in the lungs, and releases it at low partial pressures of oxygen in respiring tissues. It releases oxygen more easily when carbon dioxide concentration is high. Myoglobin and fetal haemoglobin have a higher affinity for oxygen than adult haemoglobin, so they can take oxygen from adult haemoglobin. Myoglobin acts as an oxygen store in muscle.

◆ Carbon dioxide is mostly carried as hydrogencarbonate ions in blood plasma, but also in combination with haemoglobin in red blood cells and dissolved as carbon dioxide molecules in blood plasma.

◆ White blood cells aid in defence against disease.

◆ At high altitudes, the partial pressure of oxygen is so low that altitude sickness can be caused which can be fatal. The body can adapt to gradual changes, however, by producing more red blood cells and haemoglobin.

## CHAPTER 9

# The mammalian heart

**By the end of this chapter you should be able to:**

1 describe the external and internal structure of the human heart;

2 describe the cardiac cycle, and interpret graphs showing pressure changes during this cycle;

3 explain the reasons for the difference in thickness of the atrial and ventricular walls, and of the left and right ventricular walls;

4 describe and explain the functioning of the atrio-ventricular valves, and of the semilunar valves in the aorta and pulmonary artery;

5 explain the role of the sinoatrial node in initiating heart beat, and the roles of the atrio-ventricular node and Purkyne tissue in coordinating the actions of the different parts of the heart.

The heart of an adult human has a mass of around 300 g, and is about the size of your fist (*figure 9.1*). It is a bag of muscle, filled with blood. *Figure 9.2* shows the appearance of a human heart, looking at it from the front of the body.

The muscle of which the heart is made is called **cardiac muscle**. *Figure 9.3* shows the structure of this type of muscle. It is made of interconnecting cells, whose plasma membranes are very tightly joined together. This close contact between the muscle cells allows waves of electrical excitation

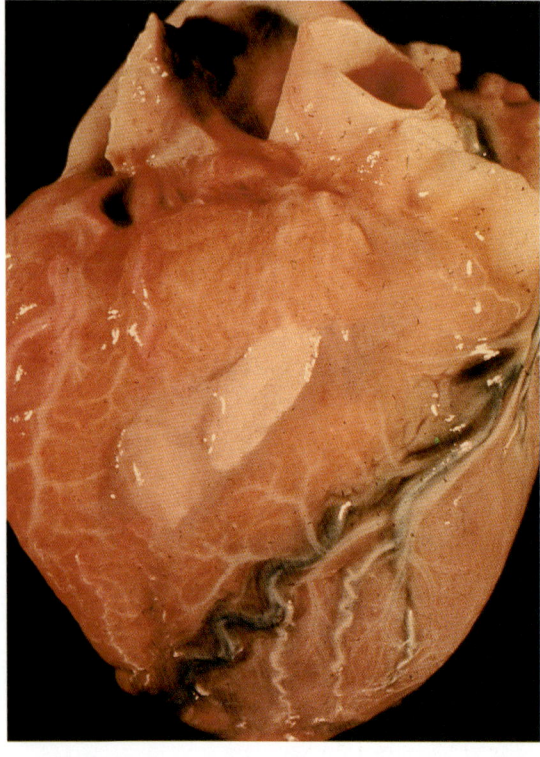

● **Figure 9.1** A human heart.

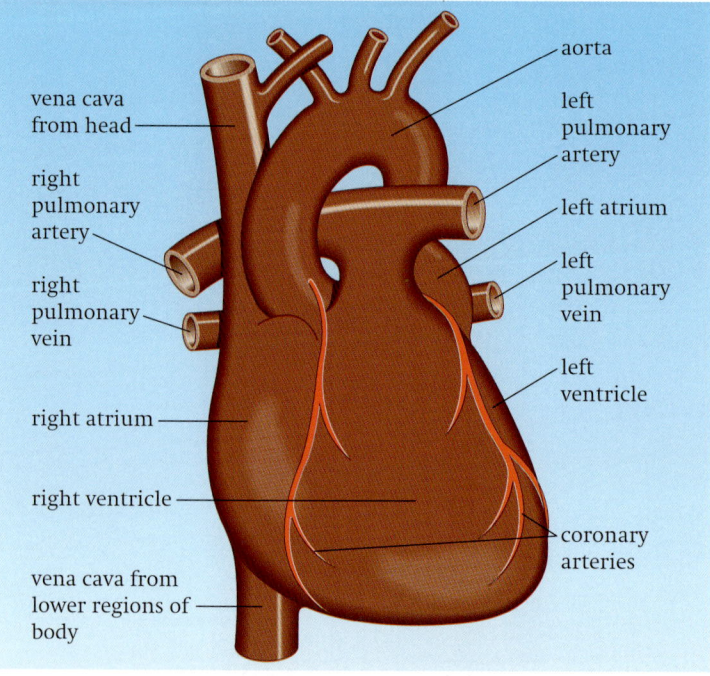

vena cava from head

right pulmonary artery

right pulmonary vein

right atrium

right ventricle

vena cava from lower regions of body

aorta

left pulmonary artery

left atrium

left pulmonary vein

left ventricle

coronary arteries

● **Figure 9.2** A human heart, seen from the front.

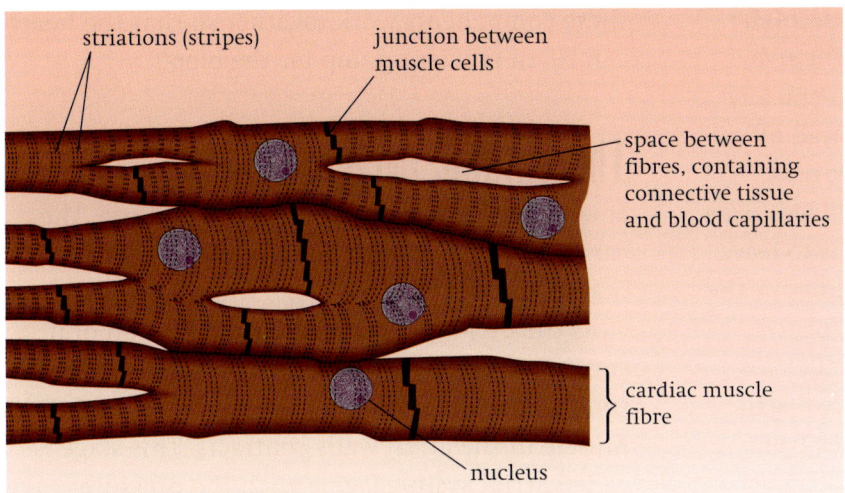

● **Figure 9.3** Cardiac muscle, as it appears under the high power of a light microscope (× 650).

to pass easily between them, which is a very important feature of cardiac muscle, as you will see later.

*Figure 9.2* also shows the blood vessels which carry blood into and out of the heart. The large, arching blood vessel is the largest artery, the **aorta**, with branches leading upwards towards the head and the main flow doubling back downwards to the rest of the body. The other blood vessel leaving the heart is the **pulmonary artery**. This, too, branches very quickly after leaving the heart, into two arteries taking blood to the right and left lungs. Running vertically on the right-hand side of the heart are the two large veins, the **venae cavae**,

one bringing blood downwards from the head and the other bringing it upwards from the rest of the body. The **pulmonary veins** bring blood back to the heart from the left and right lungs.

On the surface of the heart, the **coronary arteries** can be seen (*figures 9.1* and *9.2*). These branch from the aorta, and deliver oxygenated blood to the walls of the heart itself.

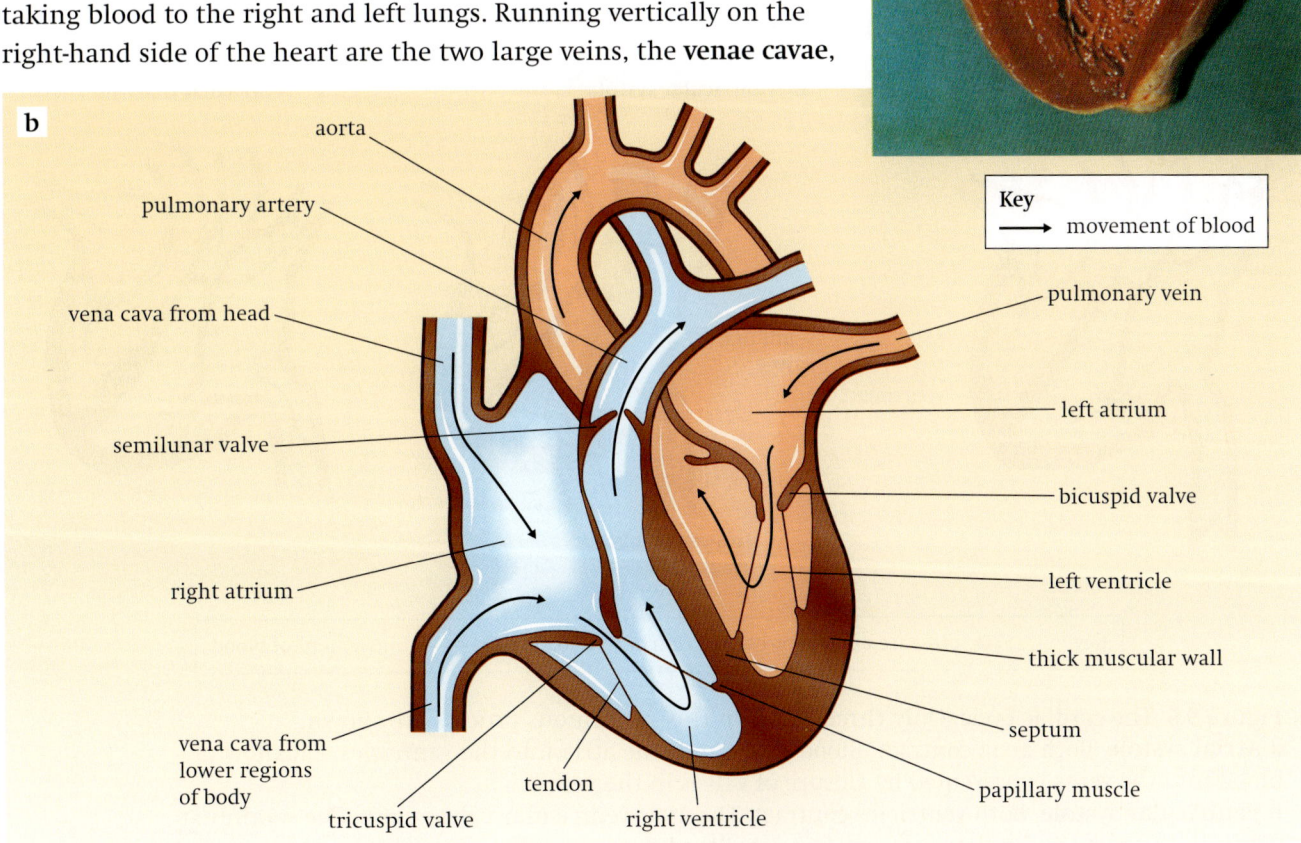

● **Figure 9.4** Vertical sections through a human heart. **a** The heart has been cut through the left atrium and ventricle only in this photograph, whilst **b** shows both sides of the heart.

If the heart is cut open vertically (*figure 9.4*) it can be seen to contain four chambers. The two chambers on the left of the heart are completely separated from those on the right by a wall of muscle called the **septum**. Blood cannot pass through this septum; the only way for blood to get from one side of the heart to the other is to leave the heart, circulate around either the lungs or the rest of the body, and then return to the heart.

The upper chamber on each side is called an **atrium** (or sometimes an **auricle**). The two atria receive blood from the veins. You can see from *figure 9.4* that blood from the venae cavae flows into the right atrium, while blood from the pulmonary veins flows into the left atrium.

The lower chambers are **ventricles**. Blood flows into the ventricles from the atria, and is then squeezed out into the arteries. Blood from the left ventricle flows into the aorta, while blood from the right ventricle flows into the pulmonary arteries.

The atria and ventricles have valves between them, which are known as the **atrio-ventricular valves**. The one on the left is the **mitral** or **bicuspid** valve, and the one on the right is the **tricuspid valve**. We will now consider how all of

these components work together so that the heart can be an efficient pump for the blood.

# The cardiac cycle

Your heart beats around 70 times a minute. The **cardiac cycle** is the sequence of events which makes up one heart beat.

As the cycle is continuous, a description of it could begin anywhere. We will begin with the time when the heart is filled with blood, and the muscle in the atrial walls contracts. This stage is called **atrial systole** (*figure 9.5a*). The pressure developed by this contraction is not very great, because the muscular walls of the atria are only thin, but it is enough to force the blood in the atria down through the atrio-ventricular valves into the ventricles. The blood from the atria does not go back into the pulmonary veins or the venae cavae, because these have semilunar valves to prevent backflow.

About 0.1 second after the atria contract, the ventricles contract. This is called **ventricular systole** (*figure 9.5b*). The thick, muscular walls of the ventricles squeeze inwards on the blood,

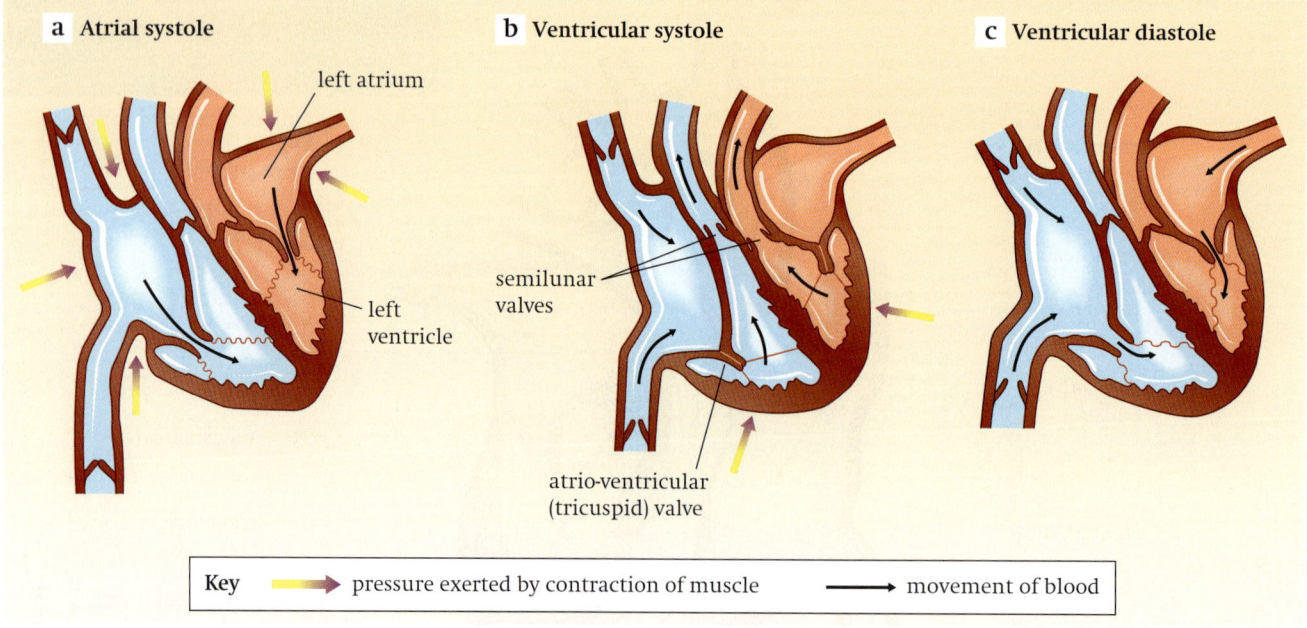

a **Atrial systole**

left atrium

left ventricle

b **Ventricular systole**

semilunar valves

atrio-ventricular (tricuspid) valve

c **Ventricular diastole**

| Key | → pressure exerted by contraction of muscle | → movement of blood |

● **Figure 9.5** The cardiac cycle. Only three stages in this continuous process are shown.
**a Atrial systole** Both atria contract. Blood flows from the atria into the ventricles. Backflow of blood into the veins is prevented by closure of valves in the veins.
**b Ventricular systole** Both ventricles contract. The atrio-ventricular valves close. The semilunar valves in the aorta and pulmonary artery open. Blood flows from the ventricles into the arteries.
**c Ventricular diastole** Atria and ventricles relax. Blood flows from the veins through the atria and into the ventricles.

increasing its pressure and pushing it out of the heart. As soon as the pressure in the ventricles becomes greater than the pressure in the atria, this pressure difference pushes the atrio-ventricular valves shut, preventing blood from going back into the atria. Instead, the blood rushes upwards into the aorta and the pulmonary artery, pushing open the semilunar valves in these vessels as it does so.

Ventricular systole lasts for about 0.3 second. The muscle then relaxes, and the stage called **ventricular diastole** begins (*figure 9.5c*). As the muscle relaxes, the pressure in the ventricles drops. The high-pressure blood which has just been pushed into the arteries would flow back into the ventricles, but for the presence of the semilunar valves, which snap shut as the blood fills their cusps.

During diastole, as the whole of the heart muscle relaxes, blood from the veins flows into the two atria. The blood is at a very low pressure, but the thin walls of the atria are easily distended, providing very little resistance to the blood flow. Some of the blood trickles downwards into the ventricles, through the atrio-ventricular valves. The atrial muscle then contracts, to push blood forcefully down into the ventricles, and the whole cycle begins again.

*Figure 9.6* shows how the atrio-ventricular and semilunar valves work.

The walls of the ventricles are much thicker than the walls of the atria, because the ventricles need to develop much more force when they contract. Their contraction has to push the blood out of the heart and around the body. For the right ventricle, the force required is relatively small, as the blood goes only to the lungs, which are very close to the heart. The left ventricle, however, has to develop sufficient force to push blood around all the rest of the body. Therefore, the thickness of the muscular wall of the left ventricle is much greater than that of the right.

● **Figure 9.6** How the heart valves function.

**a** The atrio-ventricular (bicuspid and tricuspid) valves. During atrial systole (black arrow) the pressure of blood is higher in the atrium than in the ventricle and so forces the valve open. During ventricular systole (white arrow) the pressure of blood is higher in the ventricle than in the atrium. The pressure of the blood pushes up against the cusps of valve, pushing it shut. Contraction of the papillary muscles, attached to the valve by tendons, prevents the valve from being forced inside-out.

**b** The semilunar valves in the aorta and pulmonary arteries. During ventricular systole (white arrow) the pressure of the blood forces the valves open. During ventricular diastole (black arrows) the pressure of blood in the arteries is higher than in the ventricles. The pressure of the blood pushes into the cusps of the valves, squeezing them shut.

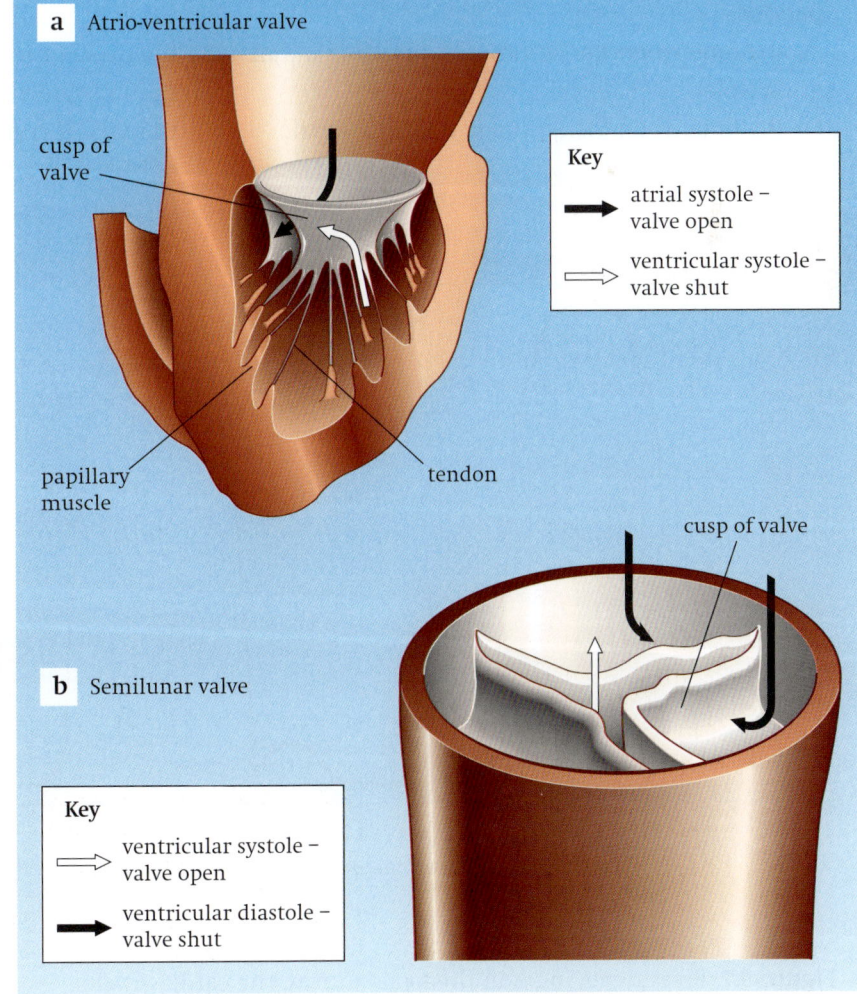

## SAQ 9.1

*Figure 9.7* shows the pressure changes in the left atrium, left ventricle and aorta throughout two cardiac cycles. Make a copy of this diagram.

**a** (i) How long does one heart beat (one cardiac cycle) last?
   (ii) What is the heart rate represented on this graph, in beats per minute?
**b** The contraction of muscles in the ventricle wall causes the pressure inside the ventricle to rise. When the muscles relax, the pressure drops again. On your copy of the diagram, mark the following periods:
   (i) the time when the ventricle is contracting (ventricular systole);
   (ii) the time when the ventricle is relaxing (ventricular diastole).
**c** The contraction of muscles in the wall of the atrium raises the pressure inside it. This pressure is also raised when blood flows into the atrium from the veins, while the atrial walls are relaxed. On your copy of the diagram, mark the following periods:
   (i) the time when the atrium is contracting (atrial systole);
   (ii) the time when the atrium is relaxing (atrial diastole).

**d** The atrio-ventricular valves open when the pressure of the blood in the atria is greater than that in the ventricles. They snap shut when the pressure of the blood in the ventricles is greater than that in the atria. On your diagram, mark the point at which these valves will open and close.
**e** The opening and closing of the semilunar valves in the aorta depends in a similar way on the relative pressures in the aorta and ventricles. On your diagram, mark the point at which these valves will open and close.
**f** The right ventricle has much less muscle in its walls than the left ventricle, and only develops about one-quarter of the pressure developed on the left side of the heart. On your diagram, draw a line to represent the probable pressure inside the right ventricle over the 1.3 seconds shown.

# Control of the heart beat

Cardiac muscle differs from the muscle in all other areas of the body in that it is **myogenic**. This means that it *naturally* contracts and relaxes: it does not need to receive impulses from a nerve to make it contract. If cardiac muscle cells are cultured in a warm, oxygenated solution containing nutrients, they contract and relax rhythmically, all by themselves.

However, the individual heart muscle cells cannot be allowed to contract at their own natural rhythms. If they did, parts of the heart would contract out of sequence with other parts; the cardiac cycle would become disordered, and the heart would stop working as a pump. The heart has its own built-in controlling and coordinating system which prevents this happening.

The cardiac cycle is initiated in a specialised patch of muscle in the wall of the right atrium, called the **sinoatrial node**. It is often

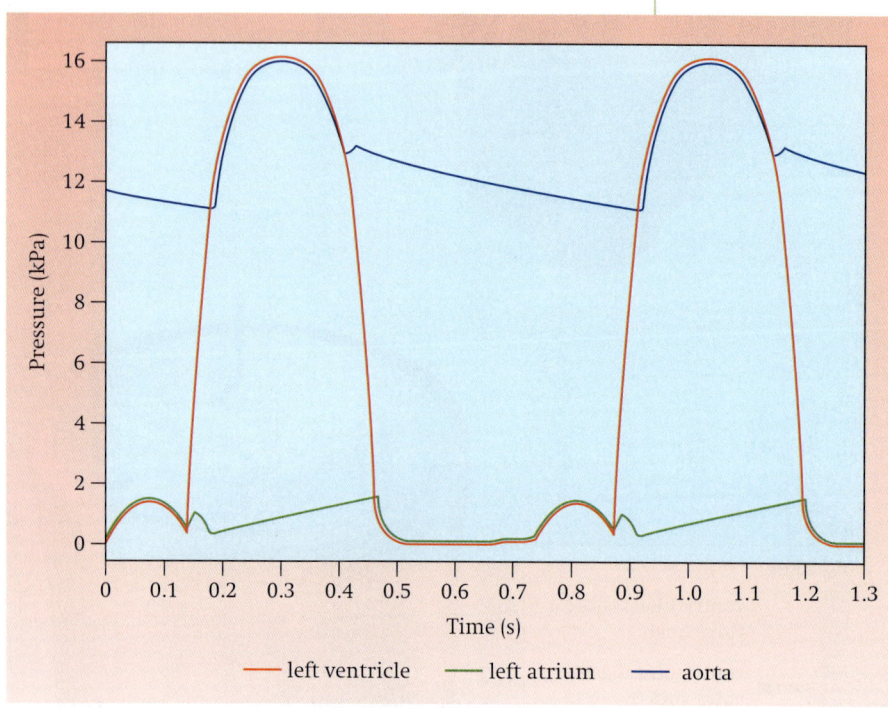

● **Figure 9.7** Pressure changes in the heart during the cardiac cycle.

called the **SAN** for short, or **pacemaker**. The muscle cells of the SAN set the rhythm for all the other cardiac muscle cells. Their natural rhythm of contraction is slightly faster than the rest of the heart muscle. Each time they contract, they set up a wave of electrical activity, which spreads out rapidly over the whole of the atrial walls. The cardiac muscle in the atrial walls responds to this excitation wave by contracting, at the same rhythm as the SAN. Thus, all the muscle in both atria contracts almost simultaneously.

As we have seen, the muscles of the ventricles do not contract until *after* the muscles of the atria. (You can imagine what would happen if they all contracted at once.) This delay is caused by a feature of the heart that briefly delays the excitation wave in its passage from the atria to the ventricles.

There is a band of fibres between the atria and ventricles which does not conduct the excitation wave. Thus, as the wave spreads out from the SAN over the atrial walls, it cannot pass into the ventricle walls. The only route through is via a patch of conducting fibres, situated in the septum, known as the **atrio-ventricular node**, or AVN (*figure 9.8*). The AVN picks up the excitation wave as it spreads across the atria and, after a delay of about 0.1 second, passes it on to a bunch of conducting fibres, called the **Purkyne tissue**, which runs down the septum between the ventricles. This transmits the excitation wave very

rapidly down to the base of the septum, from where it spreads outwards and upwards through the ventricle walls. As it does so, it causes the cardiac muscle in these walls to contract, from the bottom up, so squeezing blood upwards and into the arteries.

In a healthy heart, therefore, the atria contract and then the ventricles contract from the bottom upwards. Sometimes, this coordination of contraction goes wrong. The excitation wave becomes chaotic, passing through the ventricular muscle in all directions, feeding back on itself and restimulating areas it has just left. Small sections of the cardiac muscle contract while other sections are relaxing. The result is **fibrillation**, in which the heart wall simply flutters, rather than contracting as a whole and then relaxing as a whole. Fibrillation is almost always fatal, unless treated instantly. Fibrillation may be started by an electric shock, or by damage to large areas of muscle in the walls of the heart.

## Electrocardiograms

It is relatively easy to detect and record the waves of excitation flowing through heart muscle. Electrodes can be placed on the skin over opposite sides of the heart, and the electrical potentials generated recorded with time. The result, which is essentially a graph of voltage against time, is an **electrocardiogram (ECG)**.

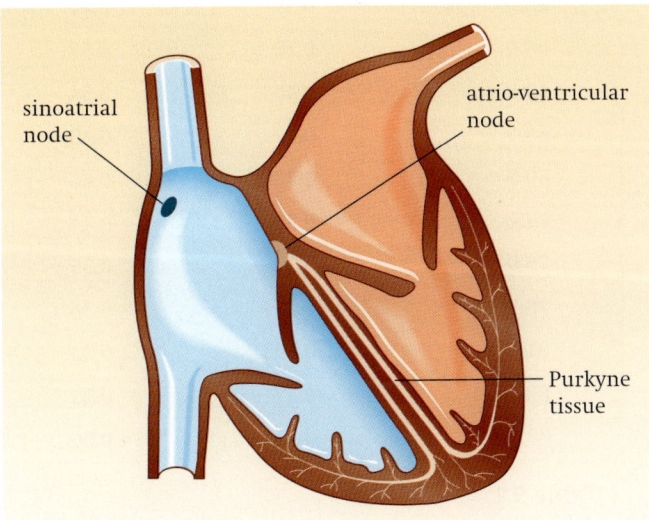

● **Figure 9.8** Vertical section of the heart to show the positions of the sinoatrial node and the atrioventricular node.

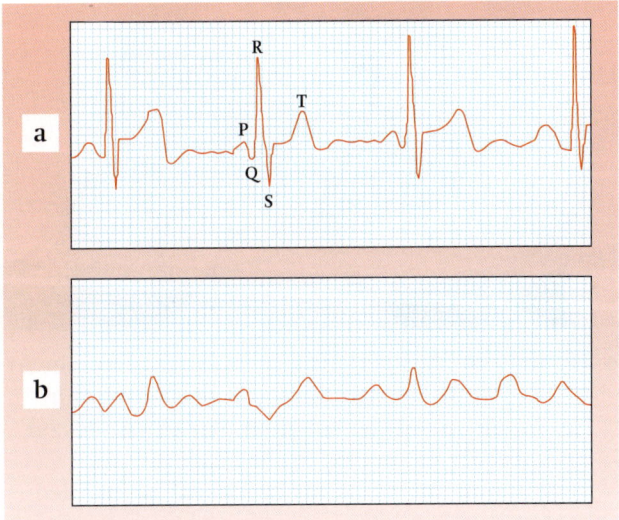

● **Figure 9.9** Electrocardiograms.
a Normal. Notice the regular, repeating pattern.
b Fibrillation.

*Figure 9.9a* shows a normal electrocardiogram. The part labelled **P** represents the wave of excitation sweeping over the atrial walls. The parts labelled **Q**, **R** and **S** represent the wave of excitation in the ventricle walls. The **T** section indicates the recovery of the ventricle walls.

*Figure 9.9b* shows an electrocardiogram from a fibrillating heart. There is no obvious regular rhythm at all. A patient in intensive care in hospital may be connected to a monitor which keeps track of the heart rhythm (*figure 9.10*). If fibrillation begins, a warning sound brings a resuscitation team running. They will attempt to shock the heart out of its fibrillation by passing a strong electric current through the chest wall. This usually stops the heart completely for up to 5 seconds, after which it often begins to beat again in a controlled way. This treatment has to be carried out within one minute of fibrillation beginning to have any real chance of success.

We consider the combined functions of heart and lungs and how they may be improved in chapter 11.

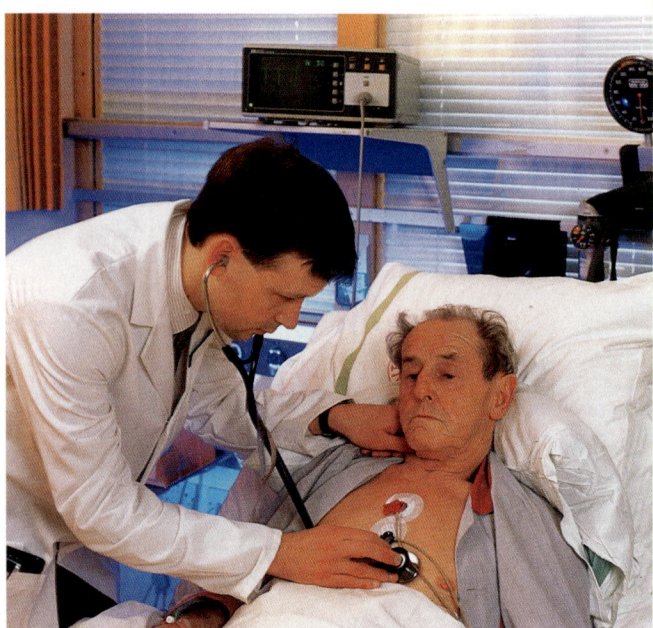

● **Figure 9.10** This patient is recovering from a heart attack in the coronary care unit of a hospital in Newcastle-upon-Tyne, England. The electrodes on his chest are connected to the heart monitor behind him, on which his heart beat is recorded.

## SAQ 9.2

*Figure 9.9a* shows a normal ECG. The paper on which the ECG was recorded was running at a speed of $25 \, mm \, s^{-1}$.

**a** Calculate the heart rate in beats per minute.

**b** The time interval between **Q** and **T** is called the **contraction time**.
　(i) Suggest why it is given this name.
　(ii) Calculate the contraction time from this ECG.

**c** The time interval between **T** and **Q** is called the **filling time**.
　(i) Suggest why it is given this name.
　(ii) Calculate the filling time from this ECG.

**d** An adult male recorded his ECG at different heart rates. The contraction time and filling time were calculated from the ECGs. The results are shown in *table 9.1*.
　(i) Suggest how the man could have increased his heart rate for the purposes of the experiment.
　(ii) Present these results as a line graph, drawing both curves on the same pair of axes.
　(iii) Comment on these results.

| Heart rate (beats per minute) | Contraction time (s) | Filling time (s) |
|---|---|---|
| 39.5 | 0.37 | 1.14 |
| 48.4 | 0.39 | 0.82 |
| 56.6 | 0.39 | 0.66 |
| 58.0 | 0.38 | 0.60 |
| 60.0 | 0.38 | 0.57 |
| 63.8 | 0.40 | 0.54 |
| 68.2 | 0.42 | 0.45 |
| 69.8 | 0.38 | 0.46 |
| 73.2 | 0.38 | 0.44 |
| 75.0 | 0.38 | 0.39 |
| 78.9 | 0.38 | 0.36 |
| 81.1 | 0.39 | 0.33 |
| 85.7 | 0.37 | 0.32 |
| 88.2 | 0.39 | 0.30 |

● **Table 9.1**

# SUMMARY

◆ The human heart, like that of all mammals, has two atria and two ventricles. Blood enters the heart by the atria and leaves from the ventricles. A septum separates the right side of the heart, which contains deoxygenated blood, from the left side, which contains oxygenated blood.

◆ Semilunar valves in the veins and in the entrances to the aorta and pulmonary artery, and atrio-ventricular valves, prevent backflow of blood.

◆ The heart is made of cardiac muscle and is myogenic. The sinoatrial node sets the pace of contraction for the muscle in the heart. Excitation waves spread from the SAN across the atria, causing their walls to contract. A non-conducting barrier prevents these excitation waves from spreading directly into the ventricles, thus delaying their contraction. The excitation wave travels to the ventricles via the atrio-ventricular node (AVN) and the Purkyne tissue, which runs down through the septum before spreading out into the walls of the ventricles. Both sides of the heart contract and relax at the same time. The contraction phase is called systole, and the relaxation phase diastole. One complete cycle of contraction and relaxation is known as the cardiac cycle.

◆ If damaged, the heart can contract in an uncoordinated way, known as fibrillation, which is potentially fatal.

◆ The pattern of contraction can be monitored by producing an electrocardiogram (ECG).

## CHAPTER 10

# Transport in multicellular plants

## By the end of this chapter you should be able to:

1 explain why plants need a transport system;

2 describe the distribution of xylem and phloem tissue in roots, stems and leaves;

3 describe the way in which water is absorbed into a root, and the pathway followed by water from root to leaf;

4 explain the mechanisms by which water moves from roots to leaves;

5 define and describe *transpiration*, and explain the effects of environmental factors on its rate;

6 explain how translocation of organic materials occurs in plants;

7 describe and compare the structure of xylem vessels, phloem sieve tube elements and companion cells;

8 relate these structures to their functions;

9 recognise xylem vessels, sieve tube elements and companion cells in light micrographs;

10 describe how the leaves of xerophytes are adapted to reduce water loss by transpiration.

Plant cells, like animal cells, need a regular supply of oxygen and nutrients. However, their requirements differ from those of animals in several ways, both in the nature of the nutrients and gases required and the rate at which these need to be supplied. Some of the particular requirements of plant cells are as follows:

- **Carbon dioxide** Photosynthetic plant cells require a supply of carbon dioxide during daylight.

- **Oxygen** All plant cells require a supply of oxygen for respiration, but cells which are actively photosynthesising produce more than enough oxygen for their own needs. Cells which are not photosynthesising need to take in oxygen from their environment, but they do not respire at such a high rate as mammals and therefore do

not need such a rapid oxygen supply.

- **Organic nutrients** Some plant cells make many of their own organic food materials, such as glucose, by photosynthesis. However, many plant cells do not photosynthesise and need to be supplied with organic nutrients from photosynthetic cells.

- **Inorganic ions and water** All plant cells require a range of different inorganic ions, and also water. These are taken up from the soil, by roots, and are transported to all regions of the plant.

The energy requirements of plant cells are, on average, far lower than those of cells in a mammal such as a human. Thus their rate of respiration, and their requirement for oxygen and glucose, is considerably less than that of mammals. They can therefore manage with a much slower transport system than the circulatory system of a mammal.

One of the main requirements of the photosynthetic regions of a plant is sunlight. In order to absorb as much sunlight as possible, plants have thin, flat leaves which present a large surface area to the Sun. In consequence, it is relatively easy for carbon dioxide and oxygen to diffuse into and out of the leaves, reaching and leaving every cell quickly enough so that there is no need for a transport system for these gases.

So, the design of a plant's transport system is quite different from that of a mammal. In fact, plants have *two* transport systems: one for carrying mainly water and inorganic ions from roots to the parts above ground, and one for carrying substances made by photosynthesis from the leaves to other areas. In neither of these systems do fluids move as rapidly as blood does in a mammal, nor is there an obvious pump such as the heart. Neither plant transport system carries oxygen or carbon dioxide, which travel to and from cells and their environment by diffusion alone.

# The transport of water

*Figure 10.1* outlines the pathway taken by water as it is transported through a plant. Water from the soil enters a plant through its root hairs and then moves across the root into the xylem tissue in the centre. Once inside the xylem vessels, the water moves upwards through the root to the stem and from there into the leaves.

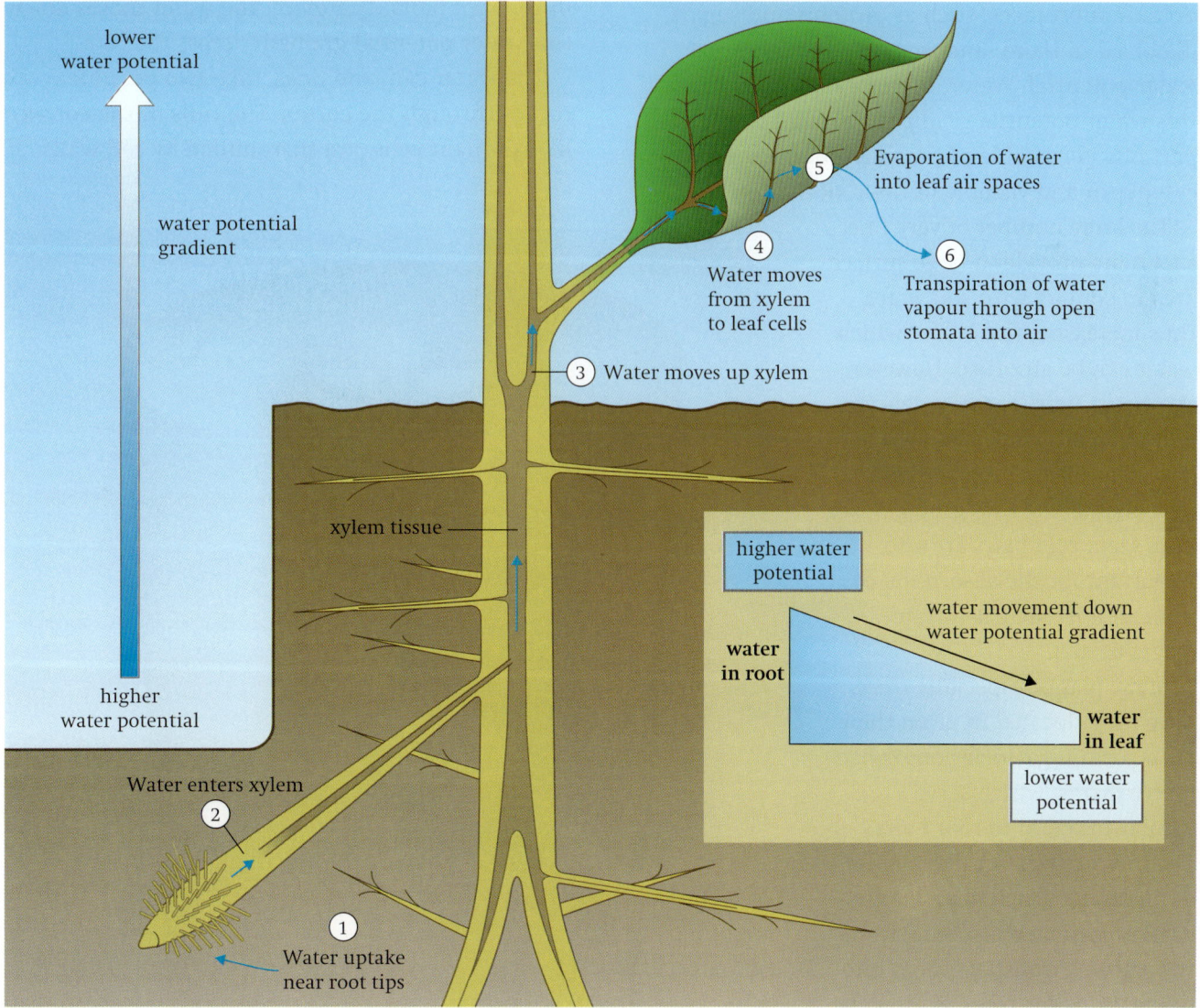

● **Figure 10.1** An overview of the movement of water through a plant. Water moves down a water potential gradient from the soil to the air.

## From soil to root hair

In chapter 4 we looked at the way in which plant roots absorb water and mineral ions. *Figure 4.13a* shows a young root. The tip is covered by a tough, protective root cap and is not permeable to water. However, just behind the tip some of the cells in the outer layer, or **epidermis**, are drawn out into long, thin extensions called **root hairs**. These reach into spaces between the soil particles from where they absorb water.

Water moves into the root hairs down a water potential gradient (*figure 4.13b* and *10.1*). Although soil water contains some inorganic ions in solution, it is a relatively dilute solution and so has a relatively high water potential. However, the cytoplasm and cell sap inside the root hairs have considerable quantities of inorganic ions and organic substances, such as proteins and sugars, dissolved in them, and so have a relatively low water potential. Water, therefore, diffuses down this water potential gradient, through the partially permeable plasma membrane, into the cytoplasm and vacuole of the root hair cell.

The large number of very fine root hairs provides a large surface area in contact with soil water, thus increasing the rate at which water can be absorbed. However, these root hairs are very delicate and often only function for a few days before being replaced by new ones as the root grows. As we have seen, root hairs are also important for the absorption of mineral ions such as nitrate (pages 61–62).

Many plants, especially trees, have fungi located in or on their roots forming associations called **mycorrhizas**, which serve a similar function to root hairs. The mycorrhizas act like a mass of fine roots which absorb nutrients, especially phosphate, from the soil and transport them into the plant. Some trees, if growing on poor soils, are unable to survive without these fungi. In return, the fungi receive organic nutrients from the plant.

### SAQ 10.1

**a** What is the name given to a relationship between two different organisms in which both benefit?

**b** Outline one other example of this type of relationship that also involves plant roots.

## From root hair to xylem

*Figures 10.2* and *10.3* show transverse sections of a young root. Water taken up by root hairs crosses the cortex and enters the xylem in the centre of the root. It does this because the water potential inside the xylem vessels, for reasons to be explained later, is lower than the water potential in the root hairs. Therefore, the water moves down this water potential gradient across the root.

The water can, and does, take two possible routes through the cortex. The cells of the cortex, like all plant cells, are surrounded by cell walls

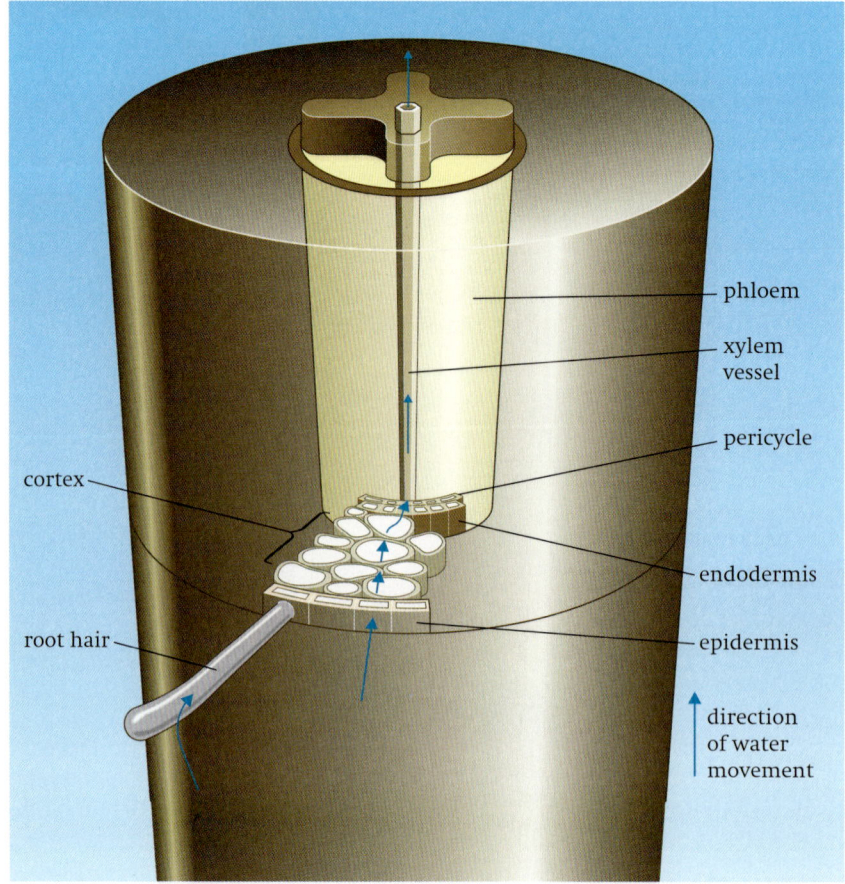

● **Figure 10.2** The pathway of water movement from root hair to xylem.

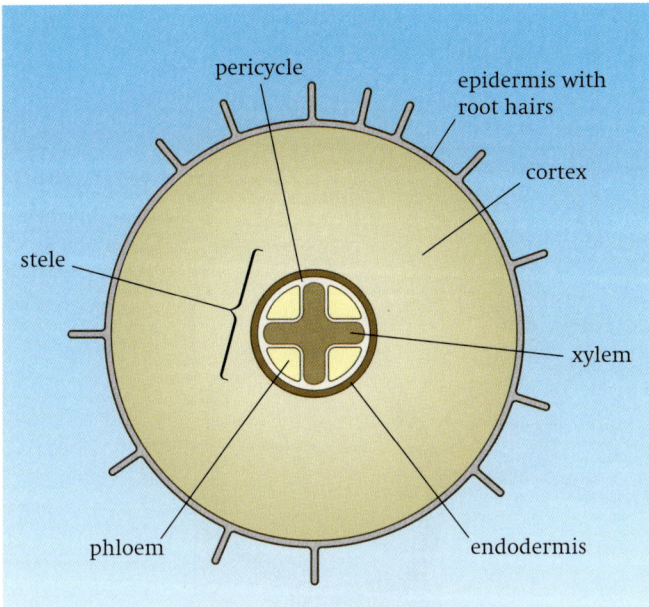

● **Figure 10.3** Transverse section of a young root to show the distribution of tissues.

made of several layers of cellulose fibres criss-crossing one another. Water can soak into these walls rather as it would soak into blotting paper, and can seep across the root from cell wall to cell wall without ever entering the cytoplasm of the cortical cells. This route is called the **apoplast pathway** (*figure 10.4a*). Another possibility is for the water to move into the cytoplasm or vacuole of a cortical cell, and then into adjacent cells

through the interconnecting plasmodesmata (see page 5). This is the **symplast pathway** (*figure 10.4b*). The relative importance of these two pathways varies from plant to plant, and in different conditions. Normally, it is probable that the symplast pathway is more important but, when transpiration rates (see page 131) are especially high, more water travels by the apoplast pathway.

Once the water reaches the stele, the apoplast pathway is abruptly barred. The cells in the outer layer surrounding the stele, the **endodermis**, have a thick, waterproof, waxy band of **suberin** in their cell walls (*figure 10.5*). This band, called the **Casparian strip**, forms an impenetrable barrier to water in the walls of the endodermis cells. The only way for the water to cross the endodermis is through the cytoplasm of these cells. As the endo-dermal cells get older, the suberin deposits become more extensive, except in certain cells called **passage cells**, through which water can continue to pass freely. It is thought that this arrangement gives a plant control over what inorganic ions pass into its xylem vessels, as every-thing has to cross plasma membranes. It may also help with the generation of root pressure (see page 133).

Once across the endodermis, water continues to move down the water potential gradient across the pericycle and towards the xylem vessels.

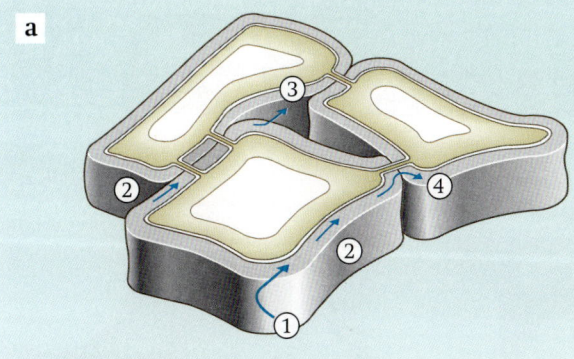

**Apoplast pathway**

① Water enters the cell wall.

② Water moves through the cell wall.

③ Water may move from cell wall to cell wall, through the intercellular spaces.

④ Water may move directly from cell wall to cell wall.

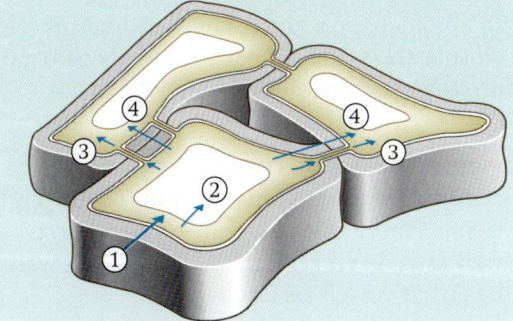

**Symplast pathway**

① Water enters the cytoplasm through the partially permeable plasma membrane.

② Water moves into the sap in the vacuole, through the tonoplast.

③ Water may move from cell to cell through the plasmodesmata.

④ Water may move from cell to cell through adjacent plasma membranes and cell walls.

● **Figure 10.4 a** Apoplast and **b** symplast pathways for movement of water from root hairs to xylem.

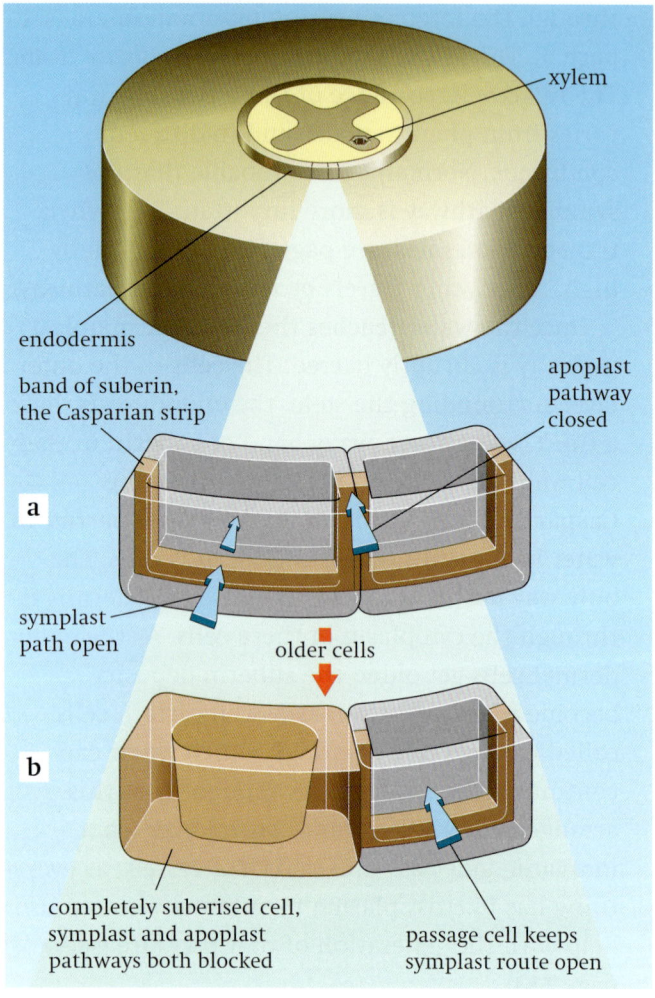

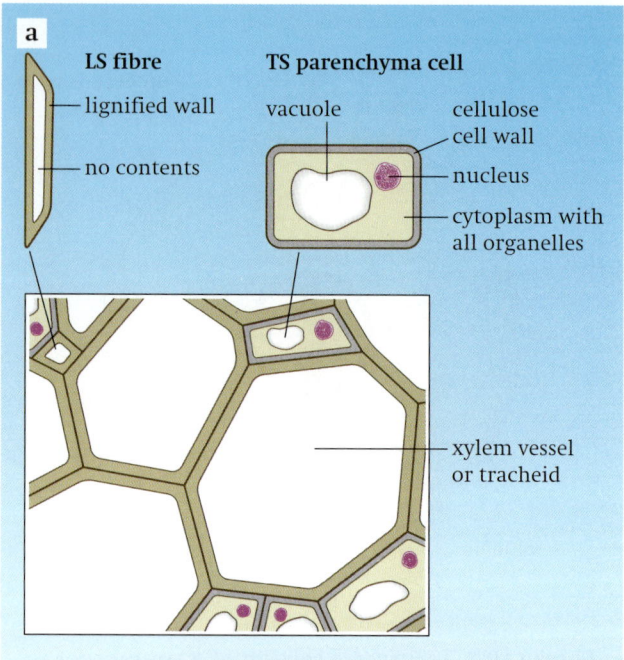

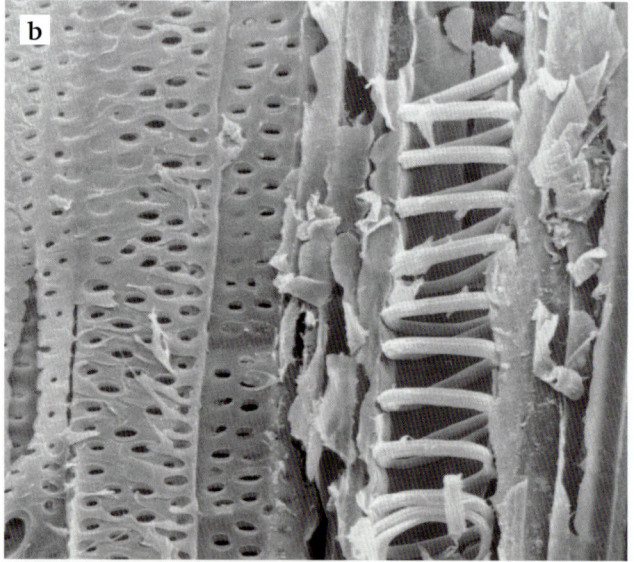

- **Figure 10.5** Suberin deposits close the apoplast pathway for water in the endodermis.
- **a** In a young root the suberin deposits form bands in the cell walls called Casparian strips. The symplast path remains open.
- **b** In an older root, entire cells become suberised, closing the symplast path too. Only passage cells are then permeable to water.

- **Figure 10.6** Xylem tissue.
- **a** Diagram of a transverse section (TS) through xylem tissue. Fibres and parenchyma cells can be seen, as well as wide empty cells with lignified walls that could be either vessels or tracheids.
- **b** Scanning electron micrograph of a longitudinal section through part of a buttercup stem, showing xylem vessels. The young vessel on the right has a spiral band of lignin around it, while those on the left are older and have more extensive coverings of lignin with many pits.
- **c** Light micrograph of a transverse section of xylem vessels. They have been stained so that the lignin appears red. The xylem vessels are the large empty cells. You can also see smaller parenchyma cells between them; these do not have lignified walls, and contain nucleus and cytoplasm.

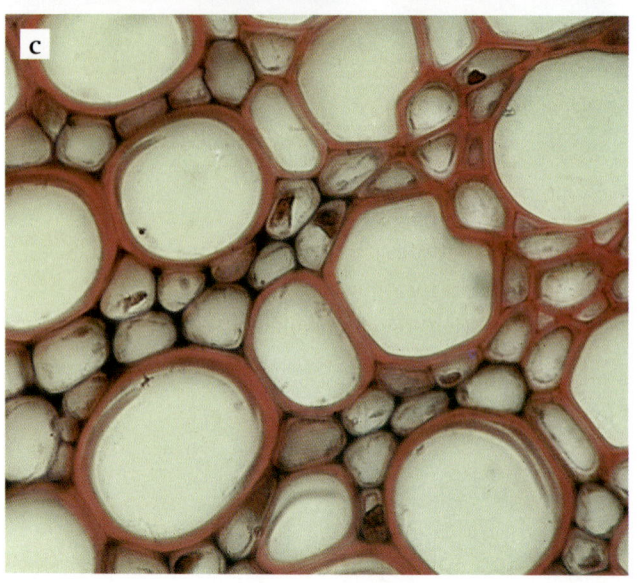

## Xylem tissue

As we saw in chapter 1, a tissue is a group of cells working together to perform a particular function. Xylem tissue (*figure 10.6*) has the dual functions of support and transport. It contains several different types of cell. In angiosperms (that is, all the flowering plants except conifers), xylem tissue contains vessel elements, tracheids, fibres and parenchyma cells.

■ **Vessel elements** and **tracheids** are the cells that are involved with the transport of water, and their structures and functions are described below.

■ **Fibres** are elongated cells with lignified walls (see below) that help to support the plant. They are dead cells; they have no living contents at all.

■ **Parenchyma cells** are 'standard' plant cells. They have unthickened cellulose cell walls and contain all the organelles you would expect a plant cell to contain (see page 16). However, the parenchyma cells in xylem tissue do not usually have chloroplasts as they are not exposed to light. They can be a variety of shapes, but are often isodiametric, that is approximately the same size in all directions.

## Xylem vessels

*Figure 10.7* shows the structure of a typical xylem vessel. Vessels are made up of many elongated **vessel elements** arranged end to end. Each began life as a normal plant cell in whose wall a substance called **lignin** was laid down. Lignin is a very hard, strong substance, which is impermeable to water. As it built up around the cell, the contents of the cell died, leaving a completely empty space, or **lumen**, inside. However, in several parts of the original cell walls, where groups of plasmodesmata were, no lignin was laid down. These non-lignified areas can be seen as 'gaps' in the thick walls of the xylem vessel, and are called **pits**. Pits are not open pores; they are crossed by permeable, unthickened cellulose cell wall.

The end walls of neighbouring vessel elements break down completely, to form a continuous tube rather like a drainpipe running through the plant. This long, non-living tube is a **xylem vessel**.

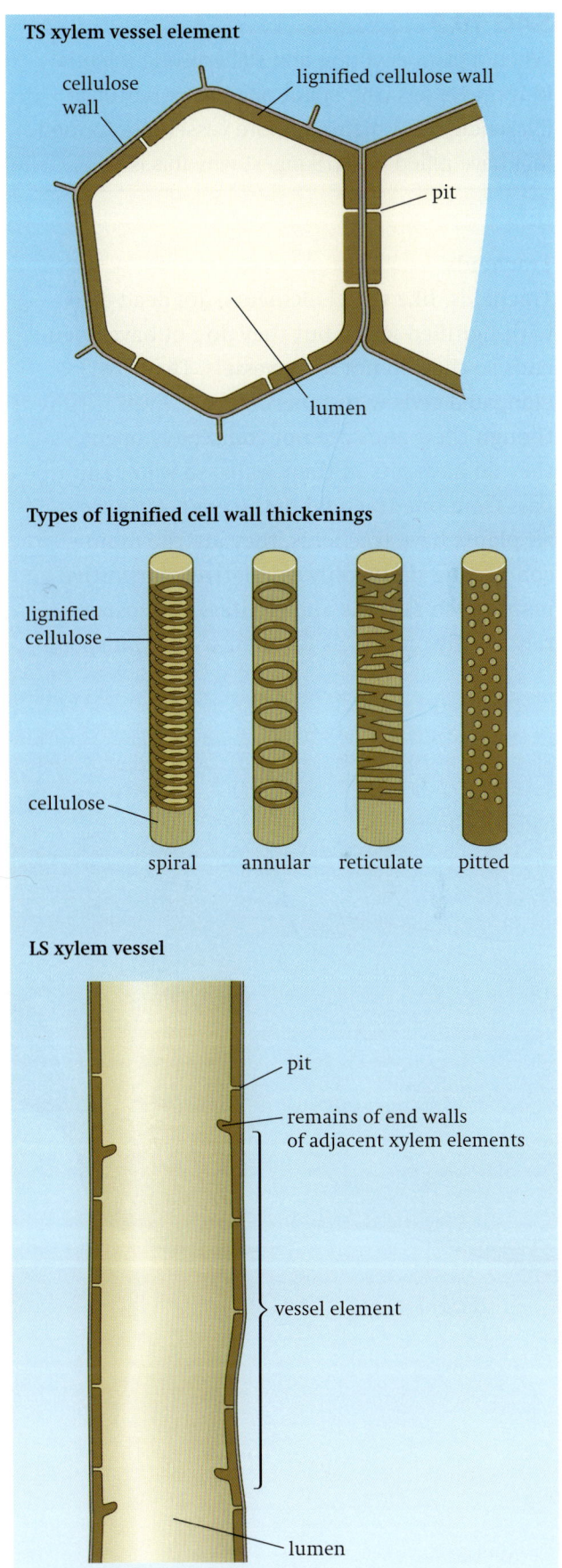

● **Figure 10.7** The structure of xylem vessels.

## SAQ 10.2

When a plant is young and still growing it tends to form vessels with spiral, annular or reticulate thickening. As it matures, more vessels are formed that have pitted walls. Suggest why this is so.

### Tracheids

Tracheids, like vessel elements, are dead cells with lignified walls, but they do not have open ends so they do not form vessels. They are elongated cells with tapering ends. Even though their ends are not completely open, they do have pits in their walls, so water can pass from one tracheid to the next. Although all plants have tracheids, they are the main conducting tissue only in relatively 'primitive' plants such as ferns and conifers. Angiosperms rely mostly on vessels for their water transport.

In the root, water which has crossed the cortex, endodermis and pericycle moves into the xylem vessels through the pits in their walls. It then

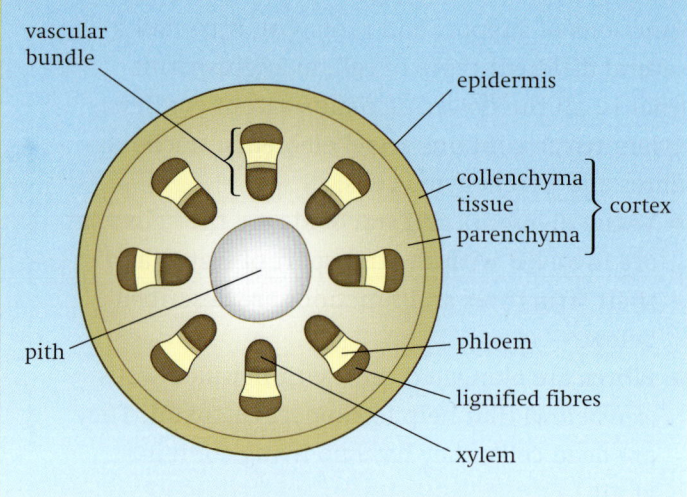

● **Figure 10.8** Transverse section of a young sunflower stem to show the distribution of tissues.

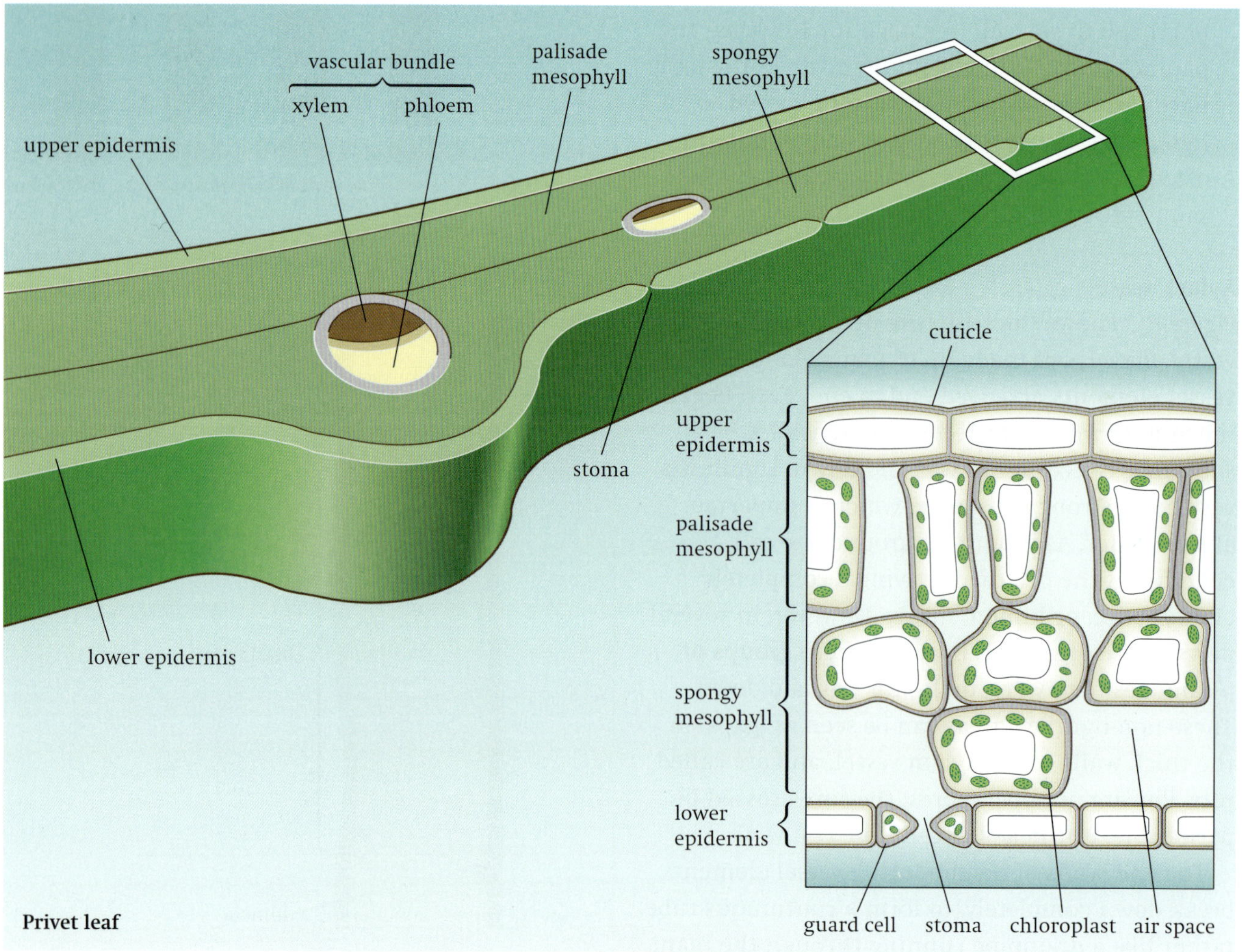

● **Figure 10.9** The structure of a leaf. Water enters the leaf as liquid water in the xylem vessels, and diffuses out as water vapour through the stomata.

Privet leaf

moves up the vessels towards the leaves. Whereas the xylem vessels are in the centre of the root, in the stem they are nearer to the outside (*figure 10.8*).

What causes water to move up xylem vessels? In order to explain this, we must look at what happens to water in a plant's leaves.

## From leaf to atmosphere – transpiration

*Figure 10.9* shows the internal structure of a leaf. The cells in the **mesophyll** ('middle leaf') layers are not tightly packed, and have many spaces around them filled with air. The walls of the mesophyll cells are wet and some of this water evaporates into the air spaces (*figure 10.10*), so that the air inside the leaf is usually saturated with water vapour.

The air in the internal spaces of the leaf has direct contact with the air outside the leaf, through small pores or **stomata**. If there is a water potential gradient between the air inside the leaf and the air outside, then water vapour will diffuse out of the leaf down this gradient. This loss of water vapour from the leaves of a plant is called **transpiration**.

An increase in the water potential gradient between the air spaces in the leaf and the air outside will increase the rate of transpiration. In conditions of low humidity, the gradient is steep, so transpiration takes place more quickly than in high humidity. Transpiration may also be increased by an increase in wind speed or a rise in temperature.

### SAQ 10.3

Suggest how **a** an increase in wind speed, and **b** a rise in temperature, may cause the rate of transpiration to increase.

Another factor which affects the rate of transpiration is the opening or closing of the stomata. The stomata are the means of contact between photosynthesising mesophyll cells and the external air, and must be open to allow carbon dioxide for photosynthesis to diffuse into the leaf. On a bright, sunny day, when the rate of photosynthesis is likely

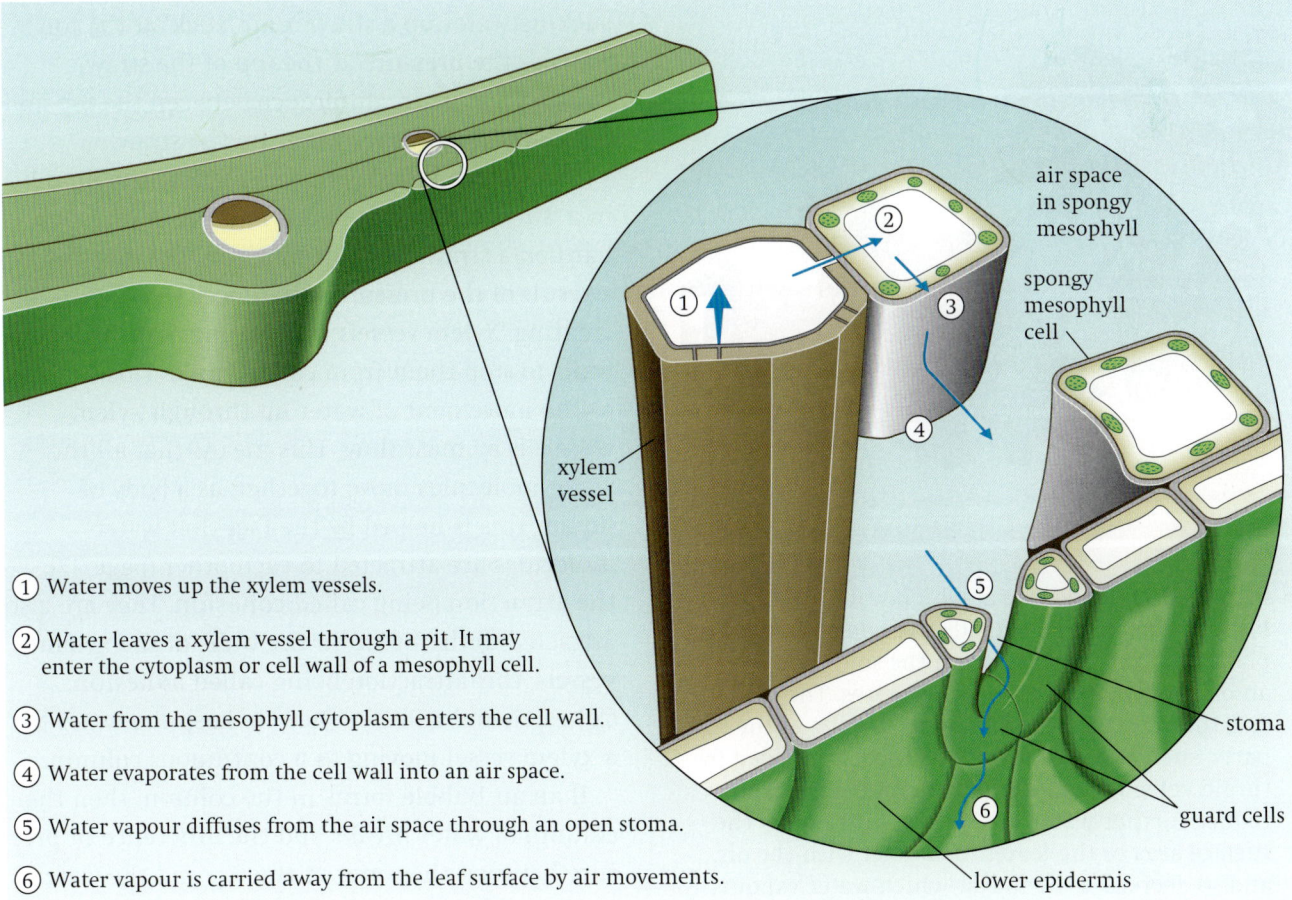

1 Water moves up the xylem vessels.

2 Water leaves a xylem vessel through a pit. It may enter the cytoplasm or cell wall of a mesophyll cell.

3 Water from the mesophyll cytoplasm enters the cell wall.

4 Water evaporates from the cell wall into an air space.

5 Water vapour diffuses from the air space through an open stoma.

6 Water vapour is carried away from the leaf surface by air movements.

● **Figure 10.10** Water movement through a leaf.

to be high, then demand for carbon dioxide by the mesophyll cells means that stomata must be open. This inevitably increases the rate of transpiration. In especially dry conditions, when the water potential gradient between the internal air spaces and the external air is steep, a plant may have to compromise by partially closing its stomata to prevent its leaves drying out, even if this means reducing the rate of photosynthesis (figure 10.11).

In hot conditions, transpiration plays an important role in cooling the leaves. As water evaporates from the cell walls inside the leaf, it absorbs heat energy from these cells, thus reducing their temperature.

## SAQ 10.4

How does this cooling mechanism compare with the main cooling mechanism of mammals?

The amount of water vapour lost by transpiration from the leaves of a plant can be very great. Even

● **Figure 10.11** If the rate at which water vapour is lost by transpiration exceeds the rate at which a plant can take up water from the soil, then the amount of water in its cells decreases. The cells lose turgor (page 57) and the plant wilts as the soft parts, such as leaves, lose the support provided by turgid cells. Wilting actually helps the plant to reduce further water loss, because it reduces the surface area of the leaves in contact with the air, and so decreases the rate at which water vapour diffuses out of the leaves. In this situation the plant will also close its stomata.

in the relatively cool and moist conditions of a temperate country such as Britain, a leaf may lose the volume of water contained in its leaves every 20 minutes. Thus water must move into the leaves equally rapidly to replace this lost water.

## From xylem to leaf

As water evaporates from the cell walls of mesophyll cells, more water is drawn into them to replace it. The source of this water is the xylem vessels in the leaf. Water constantly moves out of these vessels, down a water potential gradient, either into the mesophyll cells or along their cell walls. Some will be used in photosynthesis, but most eventually evaporates and then diffuses out of the leaf.

The removal of water from the top of xylem vessels reduces the hydrostatic pressure. (Hydrostatic pressure is pressure exerted by a liquid.) The hydrostatic pressure at the top of the xylem vessel becomes lower than the pressure at the bottom. This pressure difference causes water to move up the xylem vessels. It is just like sucking water up a straw. Your 'suck' at the top reduces the pressure at the top of the straw, causing a pressure difference between the top and bottom which pushes water up the straw.

The water in the xylem vessels, like the liquid in a 'sucked' straw, is under tension. If you suck hard on a straw, its walls may collapse inwards as a result of the pressure differences you are creating. Xylem vessels have strong, lignified walls to stop them from collapsing in this way.

The movement of water up through xylem vessels is by **mass flow**. This means that all the water molecules move together, as a body of liquid. This is helped by the fact that water molecules are attracted to each other (page 27), the attraction being called **cohesion**. They are also attracted to the lignin in the walls of the xylem vessels, this attraction being called **adhesion**. Cohesion and adhesion help to keep the water in a xylem vessel moving as a continuous column.

If an air bubble forms in the column, then the column of water breaks and the difference in pressure between the water at the top and the water at the bottom cannot be transmitted through the vessel. We say there is an air lock. The water stops moving upwards. The small diameter of xylem vessels helps to prevent such breaks from occurring. The pits in the vessel walls also allow water to move out, which may allow it to move from one vessel to another and so bypass such an air lock. Air bubbles cannot pass through pits. Pits are also important in allowing water to move out of xylem vessels to surrounding living cells.

## Root pressure

You have seen how transpiration *reduces* the water (hydrostatic) pressure at the top of a xylem vessel compared with the pressure at the base, so causing the water to flow up the vessels. Plants may also increase the pressure difference between the top and bottom by *raising* the water pressure at the *base* of the vessels.

The pressure is raised by the active secretion of solutes, for example mineral ions, into the water in the xylem vessels in the root. Cells surrounding the xylem vessels use energy to pump solutes across their membranes and into the xylem by active transport (page 58). The presence of the solutes lowers the water potential of the solution in the xylem, thus drawing in water from the surrounding root cells. This influx of water increases the water pressure at the base of the xylem vessel.

Although root pressure may help in moving water up xylem vessels, it is not essential, and is probably not significant in causing water to move up xylem in most plants. Water can continue to move up through xylem even if the plant is dead. Water transport in plants is largely a **passive** process, fuelled by transpiration from the leaves. The water simply moves down a continuous water potential gradient from the soil to the air.

## SAQ 10.5

Transport of water from the environment to cells occurs in both plants and mammals. For both plants and mammals, state the stages of this transport in which water moves by

**a** osmosis, and    **b** mass flow.

## SAQ 10.6

Explain how each of the following features of xylem vessels adapts them for their function of transporting

water from roots to leaves.
**a** Total lack of cell contents
**b** No end walls in individual xylem elements
**c** A diameter of between 0.01 mm and 0.2 mm
**d** Lignified walls
**e** Pits

## Comparing rates of transpiration

It is not easy to measure the rate at which water vapour is leaving a plant's leaves. This makes it very difficult to investigate directly how different factors, such as temperature, wind speed, light intensity or humidity, affect the rate of transpiration. However, it *is* relatively easy to measure the rate at which a plant stem takes up water. As a very high proportion of the water taken up by a stem is lost in transpiration, and as the rate at which transpiration is happening directly affects the rate of water uptake, this measurement can give a very good approximation of the rate of transpiration.

The apparatus used for this is called a **potometer** (*figure 10.12*). Everything must be completely water-tight and airtight, so that no leakage of water occurs, and so that no air bubbles break the continuous water column. To achieve this, it helps if you can insert the plant stem into the apparatus with everything submerged in water. It also helps to cut the end of the stem with a slanting cut, as air bubbles are less likely to get trapped against it. Potometers can be simpler than this one. You can manage without the reservoir (though this does mean it takes more time and effort to refill the potometer) and the tubing can be straight rather than bent. In other words, you can manage with a straight piece of glass tubing!

As water evaporates from the plant's leaves, it is drawn into the xylem vessels that are exposed at the cut end of the stem. Water is therefore drawn along the capillary tubing. If you record the position of the meniscus at set time intervals, you can plot a graph of distance moved against time. If you expose the plant to different conditions, then you can compare the rates of water uptake (which will have a close relationship to the rates of transpiration) under different conditions.

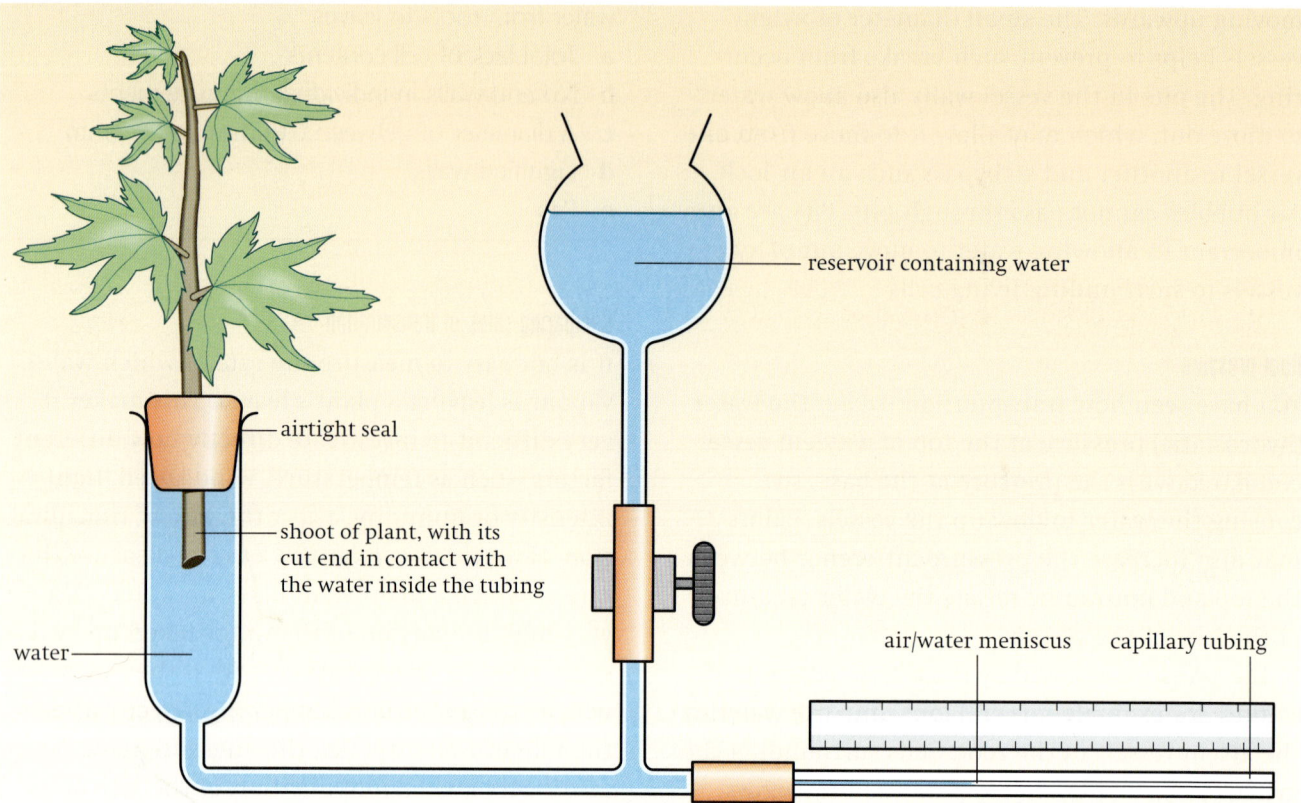

● **Figure 10.12** A potometer.

## Xerophytes

Xerophytes are plants that live in places where water is in short supply. They often have adaptations to reduce the rate of transpiration.

We have already seen that plants are able to close their stomata and allow their leaves to wilt to reduce the rate of transpiration from their leaves (page 132). Many xerophytes, however, have evolved special adaptations of their leaves that keep water loss down to a minimum. Some examples are shown in *Figure 10.14*.

# Translocation

Translocation is the term used to describe the transport of soluble organic substances within a plant. These are substances which the plant itself has made, for example sugars which are made by photosynthesis in the leaves. These substances are sometimes called **assimilates**.

Assimilates are transported in **sieve elements**. Sieve elements are found in **phloem tissue**, along with several other types of cells including **companion cells**, parenchyma and fibres (*figures 10.2, 10.8* and *10.13*). We will see later how the sieve

● **Figure 10.13** Light micrograph of a longitudinal section through phloem tissue. The red triangles are patches of callose that formed at the sieve plates between the sieve tube elements in response to the damage done as the section was being cut. You can see companion cells, with their denser cytoplasm, lying alongside the sieve tube elements. On the far right are some parenchyma cells.

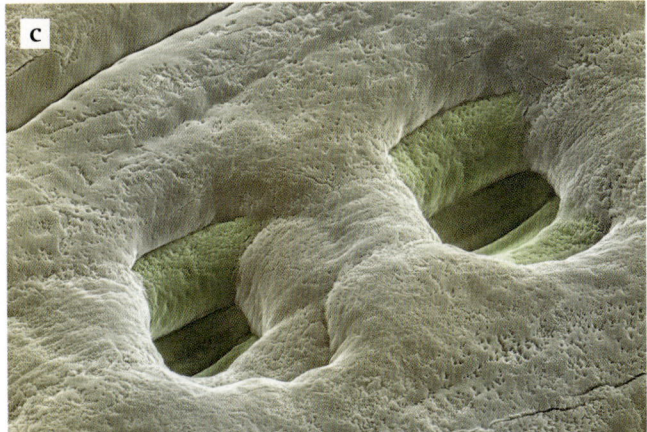

- **Figure 10.14** Some adaptations shown by xerophytes.
a A scanning electronmicrograph of a transverse section through part of a rolled leaf of marram grass, *Ammophila arenaria*. This grass grows on sand dunes, where conditions are very dry. The leaves can roll up, exposing a tough, waterproof cuticle to the air outside the leaf, while the stomata open into the enclosed, humid space in the middle of the 'roll'. Hairs help to trap a layer of moist air close to the leaf surface, reducing the diffusion gradient for water vapour.
b *Opuntia* is a cactus with flattened, photosynthetic stems that store water. The leaves are reduced to spines, which reduces the surface area from which transpiration can take place and protects the plant from being eaten by animals.
c Falsecolour SEM of a needle from a Sitka spruce, (× 1500) a large tree native to Canada and Alaska. Its leaves are in the form of needles, greatly reducing the surface area available for water loss. In addition, they are covered in a layer of waterproof wax and have sunken stomata as shown here.
d TS SEM of *Phlomis italica* leaf showing its 'trichomes'. These are tiny hair-like structures that act as a physical barrier to the loss of water like the marram grass hairs. *Phlomis* is a small shrub that lives in dry habitats in the Mediterranean regions of Europe and North Africa.
e The cardon, *Euphorbia canariensis*, grows in dry areas of Tenerife. It has swollen, succulent stems that store water and photosynthesise. The stems are coated with wax, which cuts down water loss. The leaves are extremely small.

elements and their companion cells work closely together to achieve translocation.

## Sieve elements

*Figure 10.15* shows the structure of a **sieve tube** and its accompanying companion cells. A sieve tube is made up of many elongated sieve elements, joined end to end vertically to form a continuous column. Each sieve element is a living cell. Like a 'normal' plant cell, it has a cellulose cell wall, a plasma membrane and cytoplasm containing endoplasmic reticulum and mitochondria. However, the amount of cytoplasm is very small and only forms a thin layer lining the inside of the wall of the cell. There is no nucleus, nor are there any ribosomes.

Perhaps the most striking feature of sieve elements is their end walls. Where the end walls of two sieve elements meet, a **sieve plate** is formed. This is made up of the walls of both elements, perforated by large pores. These pores are easily visible with a good light microscope. When sieve plates are viewed using an electron microscope, strands of fibrous protein can sometimes be seen passing through these pores from one sieve element to another. However, these strands have been produced by the sieve element in response to the damage caused when the tissue is cut during preparation of the specimen for viewing (see below). In living phloem the protein strands are not present; the pores are open, presenting little barrier to the free flow of liquids through them.

## Companion cells

Each sieve element has at least one companion cell lying close beside it. Companion cells have the structure of a 'normal' plant cell, with a cellulose cell wall, a plasma membrane, cytoplasm, a small vacuole and a nucleus. However, the number of mitochondria and ribosomes is rather larger than normal, and the cells are metabolically very active.

Companion cells are very closely associated with their neighbouring sieve elements. Numerous plasmodesmata pass through their cell walls, making direct contact between the cytoplasms of the companion cell and sieve element.

### The contents of phloem sieve tubes

The liquid inside phloem sieve tubes is called **phloem sap**, or just sap. *Table 10.1* shows the composition of the sap of the castor oil plant, *Ricinus communis*.

### SAQ 10.7

Which of the substances listed in *table 10.1* have been synthesised by the plant?

It is not easy to collect enough phloem sap to analyse its contents. When phloem tissue is cut, the sieve elements respond by rapidly blocking the sieve pores, in a process sometimes called 'clotting'. The pores are blocked first by plugs of phloem protein (as

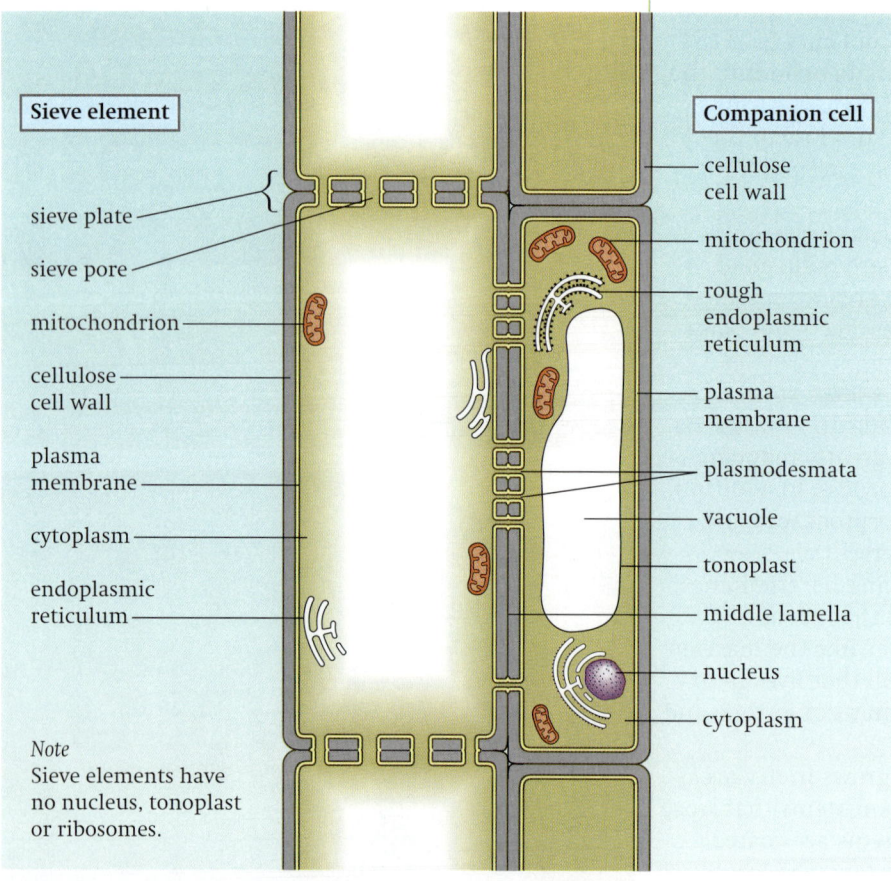

• **Figure 10.15** A phloem sieve tube element and its companion cell.

Labels (left — sieve element): sieve plate, sieve pore, mitochondrion, cellulose cell wall, plasma membrane, cytoplasm, endoplasmic reticulum

*Note*
Sieve elements have no nucleus, tonoplast or ribosomes.

Labels (right — companion cell): cellulose cell wall, mitochondrion, rough endoplasmic reticulum, plasma membrane, plasmodesmata, vacuole, tonoplast, middle lamella, nucleus, cytoplasm

| Solute | Concentration (mol dm$^{-3}$) |
|---|---|
| sucrose | 250 |
| potassium ions | 80 |
| amino acids | 40 |
| chloride ions | 15 |
| phosphate ions | 10 |
| magnesium ions | 5 |
| sodium ions | 2 |
| ATP | 0.5 |
| nitrate ions | 0 |
| plant growth substances (e.g. auxin, cytokinin) | small traces |

● **Table 10.1** Composition of sap.

we saw earlier) and then, within hours, by the carbohydrate **callose**. (Callose has a molecular structure very similar to cellulose. Like cellulose, its molecules are long chains of glucose units, but these are linked by β 1,3 glycosidic bonds, rather than the β 1,4 bonds of cellulose.) However, castor oil plants are unusual in that their phloem sap does continue to flow from a cut for some time, making it relatively easy to collect.

In other plants, aphids may be used to sample sap. Aphids, such as greenfly, feed using tubular mouthparts called stylets. They insert these through the surface of the plant's stem or leaves, into the phloem (*figure 10.16*). Phloem sap flows through the stylet into the aphid. If the stylet is cut near the aphid's head, the sap continues to flow; it seems that the small diameter of the stylet does not allow sap to flow out rapidly enough to switch on the plant's phloem 'clotting' mechanism.

## How translocation occurs

Phloem sap, like the contents of xylem vessels, moves by **mass flow**. However, whereas in xylem vessels differences in pressure are produced by a water potential gradient between soil and air, requiring no energy input from the plant, this is not so in phloem transport. To create the pressure differences needed for mass flow in phloem, the plant has to use energy. Phloem transport can therefore be considered an *active* process, in contrast to the *passive* transport in xylem.

The pressure difference is produced by **active loading** of sucrose into the sieve elements at the place from which sucrose is to be transported. This is usually in a photosynthesising leaf. As sucrose is loaded into the sieve element, this decreases the water potential in the sap inside it. Therefore, water follows the sucrose into the sieve element, moving down a water potential gradient by osmosis.

At another point along the sieve tube, sucrose may be removed by other cells, for example in the root. As sucrose is removed, water again follows by osmosis.

Thus, in the leaf, water moves into the sieve tube. In the root, water moves out of it. This creates a pressure difference; hydrostatic pressure is high in the part of the sieve tube in the leaf, and lower in the part in the root. This pressure difference causes water to flow from the high pressure area to the low pressure area, taking with it any solutes (*figure 10.17*).

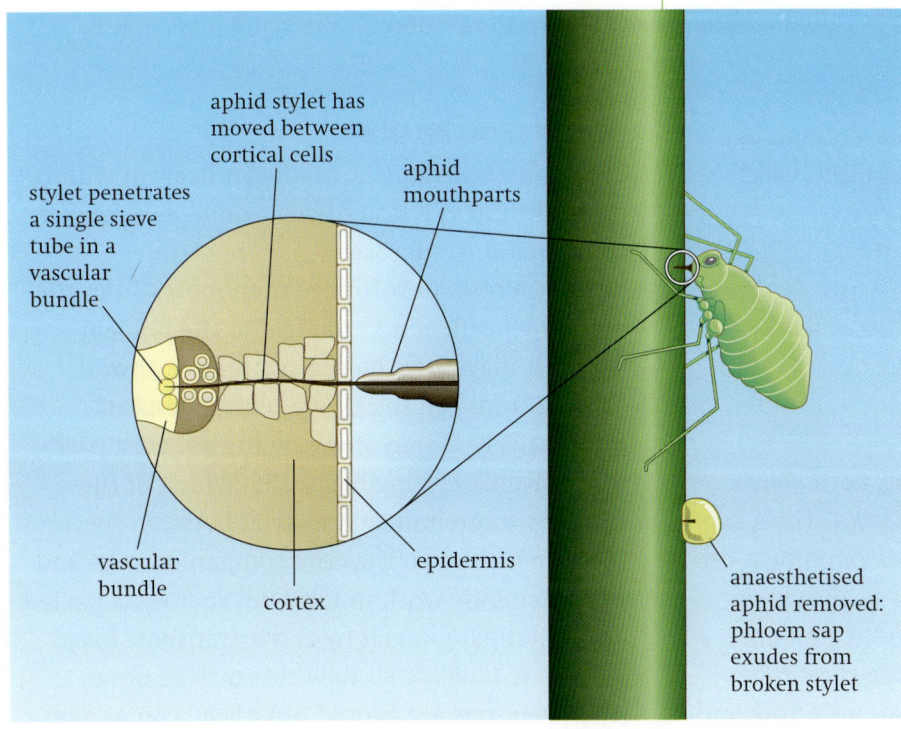

● **Figure 10.16** Using an aphid to collect phloem sap.

● **Figure 10.17** The phloem sap of the sugar maple contains a high concentration of sugar and can be harvested to make maple syrup. Two or three taps are inserted into each tree and the sap runs out under its own pressure through the green and then the black plastic pipelines.

Any area of a plant in which sucrose is loaded into the phloem is called a **source**. Usually, the source is a photosynthesising leaf. Any area where sucrose is taken out of the phloem is called a **sink** (*figure 10.18*).

### SAQ 10.8

Which of the following will be sources, and which will be sinks?

**a** A nectary in a flower.

**b** A developing fruit.

**c** The storage tissue of a potato tuber when the buds are beginning to sprout.

**d** A developing potato tuber.

Sinks can be anywhere in the plant, both above and below the photosynthesising leaves. Thus, sap flows both upwards and downwards in phloem (in contrast with xylem, in which flow is always upwards). Within any vascular bundle, phloem sap may be flowing upwards in some sieve tubes and downwards in others, but it can only flow one way in any particular sieve tube at any one time.

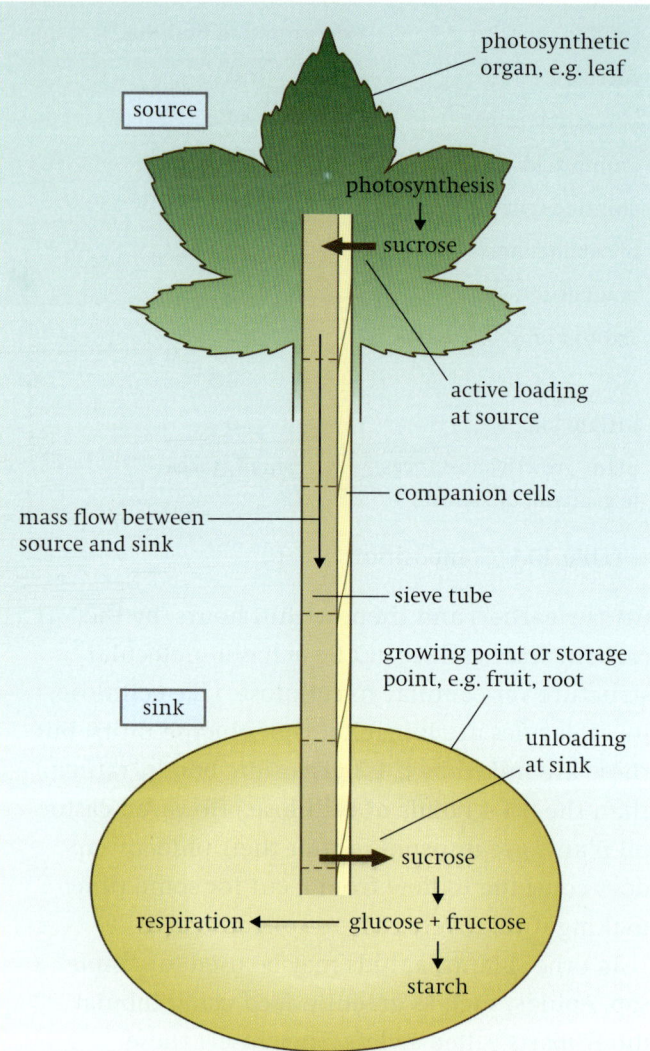

● **Figure 10.18** Sources, sinks and mass flow in phloem.

### Loading of sucrose into phloem

In leaf mesophyll cells, photosynthesis in chloroplasts produces triose sugars, some of which are converted into sucrose.

The sucrose, in solution, then moves from the mesophyll cell, across the leaf to the phloem tissue. It may move by the symplast pathway, moving from cell to cell via plasmodesmata. Alternatively, it may move by the apoplast pathway, travelling along cell walls. Which of these routes is more important varies between species.

It is now known that the companion cells and sieve elements work in tandem. Sucrose is loaded into a companion cell by active transport (page 58). *Figure 10.19* shows how this may be done. Hydrogen ions are moved out of the companion cell, using ATP as an energy source. This creates a

## Evidence for the mechanism of phloem transport

Until the late 1970s and early 1980s, there was considerable argument about whether or not phloem sap did or did not move by mass flow, in the way described above. The stumbling block was the presence of the sieve pores and phloem protein, as it was felt that these must have some important role. Several hypotheses were put forward which tried to provide a role for the phloem protein, and you may come across some of these in various textbooks. It is now known that phloem protein is not present in living, active phloem tissue, and so there is no need to provide it with a role when explaining the mechanism of phloem transport.

The evidence that phloem transport does occur by mass flow is considerable. The rate of transport in phloem is about 10 000 times faster than it would be if substances were moving by diffusion rather than by mass flow. The actual rates of transport measured match closely with those calculated from measured pressure differences at source and sink, assuming that the pores in the sieve plates are open and unobstructed.

There is also considerable evidence for the active loading of sucrose into sieve elements in sources such as leaves. Much experimental work has been done in investigating the sucrose–hydrogen ion co-transporter system in plant cells, although it has so far been very difficult to investigate this in companion cells and phloem sieve elements. Nevertheless, there is much circumstantial evidence that active loading of sucrose into phloem sieve tubes, as described above, does take place. This includes the following observations.

■ Phloem sap always has a relatively high pH, often around 8. This is what would be expected if hydrogen ions were being actively transported out of the cell.

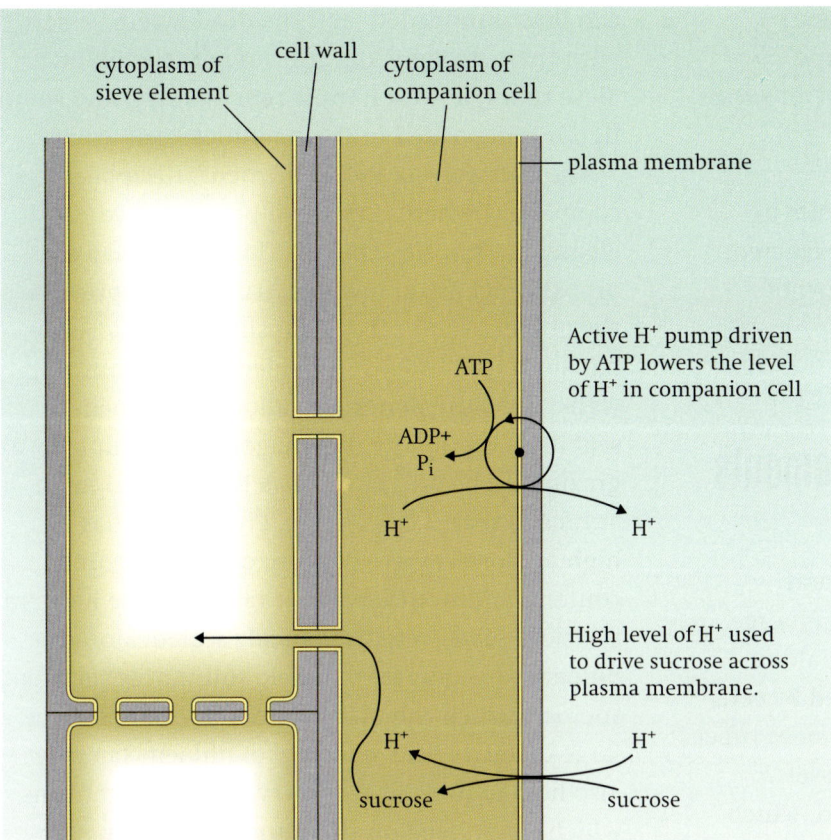

● **Figure 10.19** A possible method by which sucrose is loaded into phloem.

large excess of hydrogen ions outside the companion cell. They can move back into the cell down their concentration gradient, through a protein which acts as a carrier for both hydrogen ions and sucrose at the same time. The sucrose molecules are carried through this **co-transporter** molecule into the companion cell, against the concentration gradient for sucrose. The sucrose molecules can then move from the companion cell into the sieve tube, through the plasmodesmata which connect them.

### Unloading of sucrose from phloem

At the moment, little is known about the way in which sucrose is unloaded from phloem. Unloading occurs into any tissue which requires sucrose. It is probable that sucrose moves out of the phloem into these tissues by diffusion. Once in the tissue, the sucrose is converted into something else by enzymes, so decreasing its concentration and maintaining a concentration gradient. One such enzyme is invertase, which hydrolyses sucrose to glucose and fructose (figure 10.18).

■ There is a difference in electrical potential across the plasma membrane of around −150 mV inside, again consistent with an excess of positive hydrogen ions outside the cell compared with inside.

■ ATP is present in phloem sieve elements in quite large amounts. This would be expected, as it is required for the active transport of hydrogen ions out of the cell.

# Differences between sieve elements and xylem vessels

From this account of translocation, several similarities with the transport of water emerge. In each case, liquid moves by mass flow along a pressure gradient, through tubes formed by cells stacked end to end. So why are phloem sieve tubes so different in structure from xylem vessels?

Unlike water transport through xylem, which occurs through dead xylem vessels, translocation through phloem sieve tubes involves active loading of sucrose at sources, thus requiring living cells.

Xylem vessels have lignified cell walls, whereas phloem tubes do not. The presence of lignin in a cell wall prevents the movement of water and solutes across it, and so kills the cell. This does not matter in xylem, as xylem vessels do not need to be alive; indeed, it is a positive advantage to have an entirely empty tube through which water

can flow unimpeded, and the dead xylem vessels with their strong walls also support the plant. Sieve tubes, however, must remain alive, and so no lignin is deposited in their cellulose cell walls.

The end walls of xylem elements disappear completely, whereas those of phloem sieve elements form sieve plates. These sieve plates probably act as supporting structures to prevent the phloem sieve tube collapsing; xylem already has sufficient support provided by its lignified walls. The sieve plates also allow the phloem to seal itself up rapidly if damaged, for example by a grazing herbivore, rather as a blood vessel in an animal is sealed by clotting. Phloem sap has a high turgor pressure because of its high solute content, and would leak out rapidly if the holes in the sieve plate were not quickly sealed. Moreover, phloem sap contains valuable substances such as sucrose, which the plant cannot afford to lose in large quantity. The 'clotting' of phloem sap may also help to prevent the entry of microorganisms which might feed on the nutritious sap or cause disease.

## SAQ 10.9
Draw up a comparison table between xylem vessels and sieve tubes. Some features which you could include are: cell structure (walls, diameter, cell contents, etc.), substances transported and methods of transport. Include a column giving a brief explanation for the differences in structure.

# SUMMARY

◆ Water is transported through a plant in xylem vessels. This is a passive process, in which water moves down a water potential gradient from soil to air. Water enters root hairs by osmosis, crosses the root either through the cytoplasm of cells or via their cell walls, and enters the dead, empty xylem vessels. Water moves up xylem vessels by mass flow, as a result of pressure differences caused by loss of water from leaves by transpiration. Root pressure can also contribute to this pressure difference.

◆ Transpiration is an inevitable consequence of gaseous exchange in plants. Plants have air spaces within the leaf linked to the external atmosphere through stomata, so that carbon dioxide and oxygen can be exchanged with their environment. Water vapour, formed as water evaporates from wet cell walls, also diffuses through these air spaces and out of the stomata.

◆ The rate of transpiration is affected by several environmental factors, namely temperature, light intensity, wind speed and humidity. It is difficult to measure rate of transpiration directly, but water uptake can be measured using a potometer. Plants that are adapted to live in places where the environmental conditions are likely to cause high rates of transpiration, and where soil water is in short supply, are called xerophytes. They have often evolved adaptations that help to reduce the rate of loss of water vapour from their leaves.

◆ Translocation of organic solutes, such as sucrose, occurs through living phloem sieve tubes. The phloem sap moves by mass flow, as a result of pressure differences produced by active loading of sucrose at sources such as photosynthesising leaves.

## CHAPTER 11

# Gaseous exchange

## By the end of this chapter you should be able to:

1 describe the distribution of alveoli and blood vessels in lung tissue;

2 describe the distribution of cartilage, ciliated epithelium, goblet cells and smooth muscle in the trachea, bronchi and bronchioles;

3 describe the functions of cartilage, cilia, goblet cells, smooth muscle and elastic fibres in the gaseous exchange system;

4 explain the meanings of the terms *tidal volume* and *vital capacity*;

5 describe how to measure a person's pulse rate;

6 understand that pulse rate is a measure of heart rate and explain the significance of resting pulse rate in relation to physical fitness;

7 explain the terms *systolic blood pressure*, *diastolic blood pressure* and *hypertension*.

In this chapter we consider how the structure and functions of our gaseous exchange and cardiovascular systems are linked, how they are affected by physical activity and the contribution this makes to our overall level of health.

## The gaseous exchange system

The gaseous exchange system links the circulatory system (chapter 8) with the atmosphere. It is adapted to:

■ clean and warm the air that enters during breathing;

■ maximise the surface area for diffusion of oxygen and carbon dioxide between the blood and atmosphere;

■ minimise the distance for this diffusion;

■ maintain adequate gradients for this diffusion.

## Lungs

As we saw in chapter 4, the lungs are the site of gaseous exchange between air and blood and they present a huge surface area to the air that flows in and out. They are in the thoracic (chest) cavity surrounded by an airtight space between the pleural membranes. This space contains a small

| Airway | Number | Approximate diameter | Cartilage | Goblet cells | Smooth muscle | Cilia | Site of gas exchange |
|---|---|---|---|---|---|---|---|
| trachea | 1 | 1.8 cm | yes | yes | yes | yes | no |
| bronchus | 2 | 1.2 cm | yes | yes | yes | yes | no |
| terminal bronchiole | 48 000 | 1.0 mm | no | no | yes | yes | no |
| respiratory bronchiole | 300 000 | 0.5 mm | no | no | no | yes | no |
| alveolar duct | $9 \times 10^6$ | 400 μm | no | no | no | no | yes |
| alveoli | $3 \times 10^9$ | 250 μm | no | no | no | no | yes |

● **Table 11.1** The structure of the airways from the trachea to the alveoli. The various airways are shown in *figure 4.12*.

quantity of fluid to allow friction-free movement as the lungs are ventilated by the movement of the diaphragm and ribs. The structure and contents of the thorax are shown in *figure 4.12* on page 60.

## Trachea, bronchi and bronchioles

The lungs are ventilated with air which passes through a branching system of airways (*figure 4.12* and *table 11.1*). Leading from the throat to the lungs is the **trachea**. At the base of the trachea are two **bronchi** (singular **bronchus**), which subdivide and branch extensively forming a bronchial 'tree' in each lung. **Cartilage** in the trachea and bronchi keeps these airways open and air resistance low, and prevents them from collapsing or bursting as the air pressure changes during breathing. In the trachea there is a regular arrangement of C-shaped rings of cartilage; in the bronchi there are irregular blocks of cartilage instead (*figure 11.1*). The small bronchioles are surrounded by smooth muscle which can contract or relax to adjust the diameter of these tiny airways. During exercise they relax to allow a greater flow of air to the alveoli. The absence of cartilage makes these adjustments possible.

### Warming and cleaning the air

As air flows through the nose and the trachea it is warmed to body temperature and moistened by evaporation from the lining, so protecting the delicate surfaces inside the lungs from desiccation (drying out). Protection is also needed against the suspended matter carried in the air, which may include dust, pollen, bacteria, fungal spores, sand and viruses. All are a potential threat to the proper functioning of the lungs. Particles larger than about 5–10 µm are caught on the hairs inside the nose and the **mucus** lining the nasal passages and other airways.

In the trachea and bronchi, the mucus is produced by the **goblet cells** of the ciliated epithelium. The upper part of each goblet cell is swollen with **mucin** droplets which have been secreted by the cell. Mucus is a slimy solution of mucin, which is composed of glycoproteins with many carbohydrate chains that make them sticky and able to trap inhaled particles. The rest of the cell, which contains the nucleus, is quite slender like the stem of a goblet. Mucus is also made by glands beneath the epithelium. Some chemical pollutants, such as sulphur dioxide and nitrogen dioxide, can dissolve in mucus to form an acid solution that irritates the lining of the airways.

Between the goblet cells are the ciliated cells (*figure 11.2*). The continual beating of their cilia carries the carpet of mucus upwards towards the larynx at a speed of about 1 cm per minute. When mucus reaches the top of the trachea it is usually swallowed so that pathogens are destroyed by the acid in the stomach.

Phagocytic white blood cells known as **macrophages** (see chapters 4 and 14) patrol the surfaces of the airways scavenging small particles such as bacteria and fine dust particles. During an infection they are joined by other phagocytic cells which leave the capillaries to help remove pathogens.

## Alveoli

At the end of the pathway between the atmosphere and the bloodstream are the **alveoli** (*figures 11.1c* and *11.3*). These have a very thin epithelial lining and are surrounded by many blood capillaries carrying deoxygenated blood. The short distance between air and blood means that oxygen and carbon dioxide can be exchanged efficiently by diffusion (chapter 4). Alveolar walls contain **elastic fibres** which stretch during breathing and recoil during expiration to help force out air. This elasticity allows alveoli to expand according to the volume of air breathed in. When fully expanded during exercise the surface area available for diffusion increases, and the air is expelled efficiently when the elastic fibres recoil.

*SAQ 11.1*

Describe the pathway taken by a molecule of oxygen as it passes from the atmosphere to the blood in the lungs.

*SAQ 11.2*

Explain how alveoli are adapted for gaseous exchange. (It may help to look back to chapter 4.)

**Figure 11.1**

**a** A light micrograph of part of the trachea in transverse section (×150). The lining is comprised of ciliated epithelium which rests on a basement membrane made of protein fibres. In between the ciliated cells are goblet cells (here stained blue). Beneath the epithelium is an area of loose tissue with blood vessels and glands that secrete mucus. The trachea as a whole is supported by C-shaped rings of cartilage, a portion of which appears as the thick layer running across the bottom of the picture.

**b** A light micrograph of part of a bronchus in transverse section (×1300). Between the ciliated epithelial cells, the goblet cells are stained pink. There are fewer goblet cells per cm² than in the trachea and the epithelial cells are not as tall. Beneath the epithelium there are elastic fibres. Blocks of cartilage, not rings, support the bronchus and part of one can be seen, also stained pink, stretching from top to bottom of the picture.

**c** A light micrograph of a small bronchiole in transverse section (×135). Surrounding the epithelium is smooth muscle. There is no cartilage. Around the bronchiole are some alveloli.

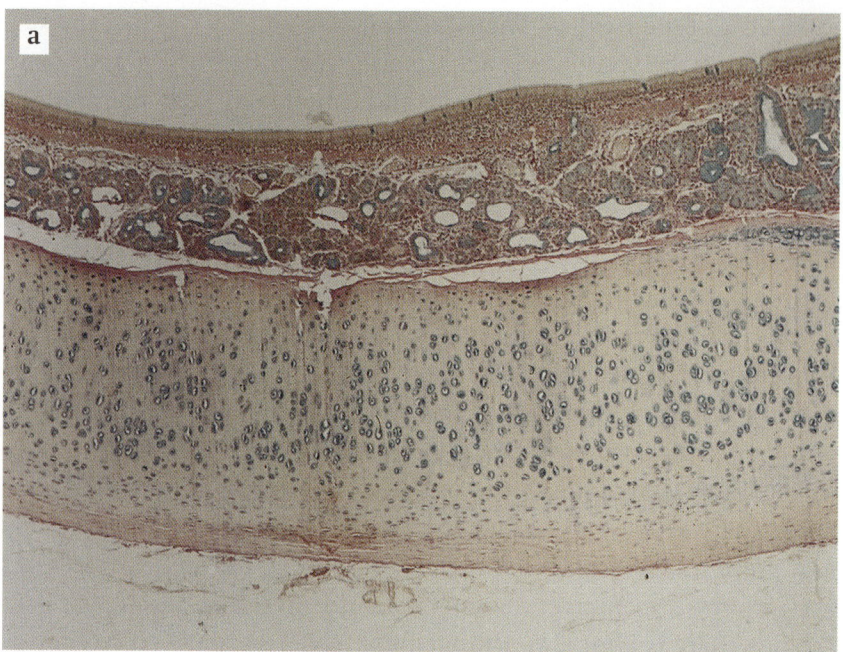

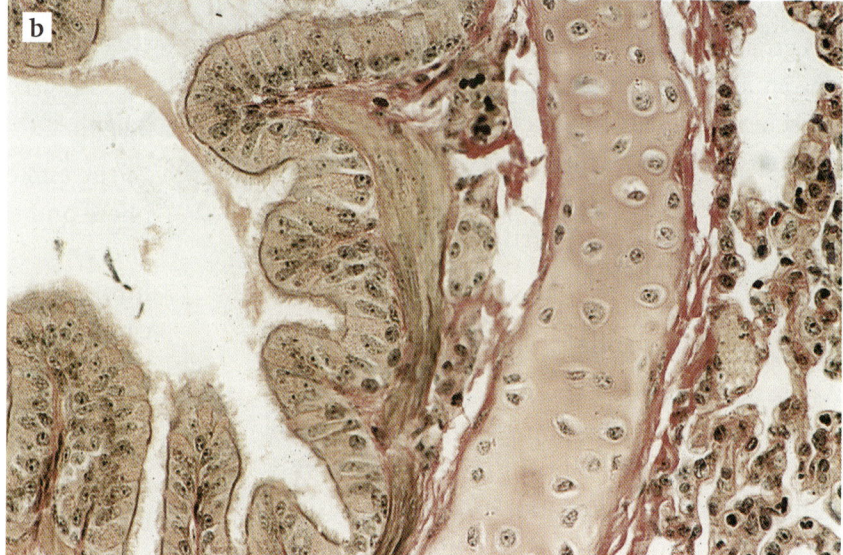

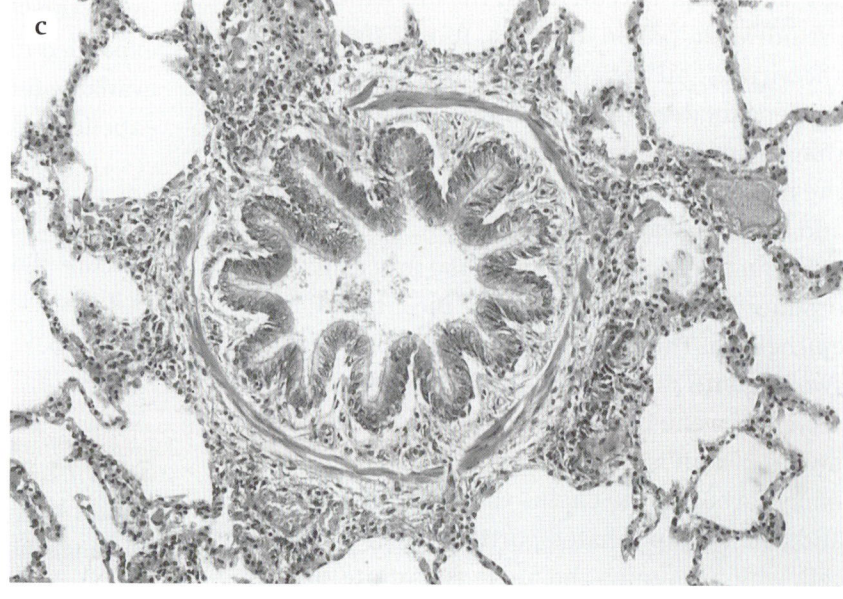

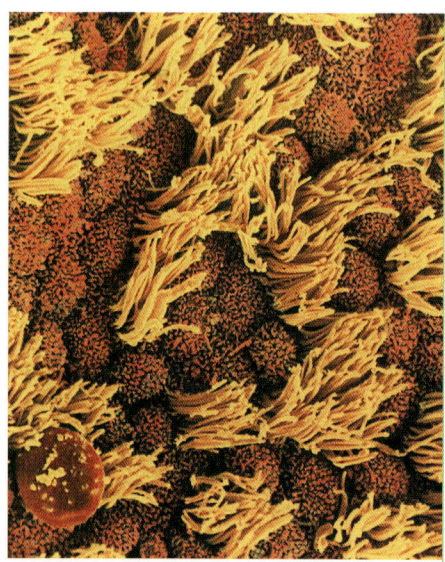

● **Figure 11.2** False-colour SEM of the surface of the trachea, showing large numbers of cilia (yellow) and some mucus-secreting goblet cells (red) (×2300).

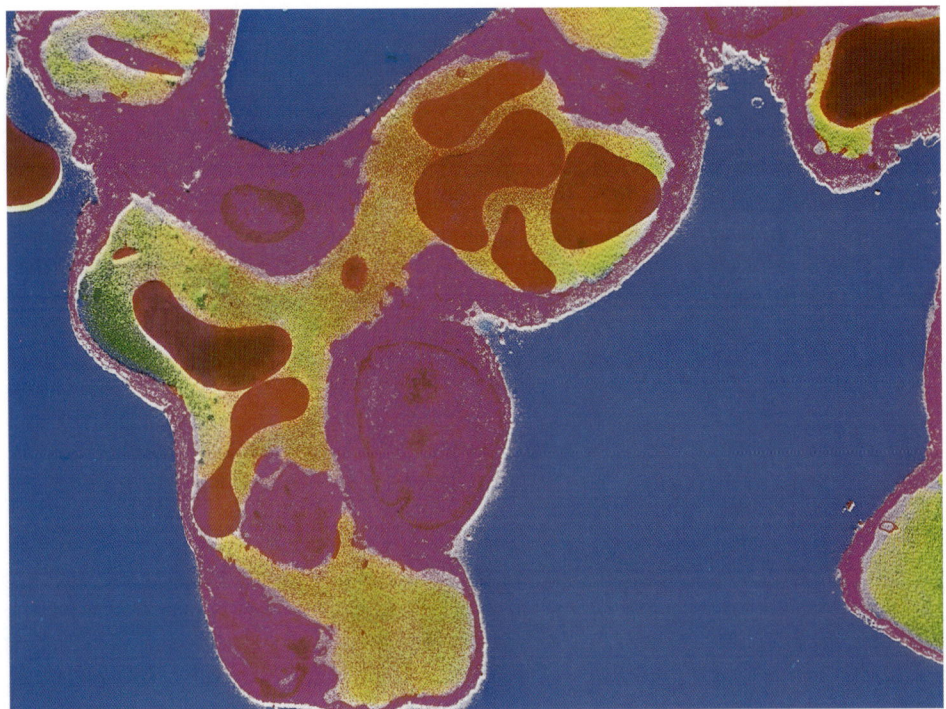

● **Figure 11.3** False-colour TEM of the lining of an alveolus. Red blood cells fill the blood capillaries (yellow) which are separated from the air (blue) by a thin layer of cells (pink).

*SAQ 11.3* _____
Explain the advantage of being able to adjust the diameter of bronchioles.

# Breathing rate and heart rate

As the body varies its level of activity so the rate at which the cells use oxygen also varies. The rate of supply of oxygen to the cells is determined by the rate and depth of breathing and by the rate at which the heart pumps blood around the body.

## Breathing rate and depth

Breathing refreshes the air in the alveoli so that the concentrations of oxygen and carbon dioxide within them remain constant whatever our level of activity. Changing the depth and rate of breathing achieves this. (See *box 11A* for details of how lung volumes and depth of breathing are measured.)

At rest we need to ventilate our lungs with about 6.0 dm$^3$ of air per minute. About 0.35 dm$^3$ of new air enters the alveoli with each breath, representing only about one seventh of the total volume of air in the alveoli. This means that large

changes in the composition of alveolar air never occur. In any case, it is impossible to empty the lungs completely and, even when the chest is compressed during forced exhalation, about 1.0 dm$^3$ of air still remains in the alveoli and the airways. This volume is the **residual volume**. A much larger volume (approximately 2.5 dm$^3$) remains in the lungs after breathing out normally. When breathing deeply the lungs can increase in volume by as much as 3 dm$^3$.

As exercise becomes harder the depth of breathing increases. Often the breathing rate increases too. This gives us the ability to respond to changes in demand for gaseous exchange during exercise. The effect of exercise on breathing is measured by calculating the **ventilation rate**. This is the total volume of air moved into the lungs in one minute. Ventilation rate (expressed as dm$^3$ min$^{-1}$) is calculated as:

tidal volume × breathing rate

A well-trained athlete can achieve adequate ventilation by increasing the tidal volume with only a small increase in the rate of breathing when taking moderate exercise. This is possible because training improves the efficiency of the muscles involved with breathing.

## Pulse rate

When the ventricles of the heart contract, a surge of blood flows into the aorta and the pulmonary arteries under pressure, as we saw in chapter 9. The volume of blood pumped out from each ventricle during each contraction is the **stroke volume**. The total volume pumped out per minute is the **cardiac output**. The surge of blood distends arteries, which contain elastic tissue. The stretch and subsequent recoil of the aorta and the arteries travels as a wave along all the arteries. This is the

## Box 11A Measuring lung volumes

Ventilation brings about changes in lung volume, and these changes can be measured by a **spirometer**. In the spirometer shown in *figure 11.4* a person breathes from a tube connected to an oxygen-containing chamber that floats on a tank of water. The chamber falls during inhalation and rises during exhalation. A canister of soda lime absorbs all the carbon dioxide in the exhaled air. The chamber does not rise to the same height with each breath because oxygen is absorbed in the lungs. The movements of the chamber are recorded on a kymograph trace (*figure 11.5*).

Two measurements can be obtained from the trace.

- **Tidal volume** is the volume of air breathed in and then breathed out during a single breath. The tidal volume at rest is about 0.5 dm$^3$ (500 cm$^3$).
- **Vital capacity** is the maximum volume of air that can be breathed in and then breathed out of the lungs by movement of the diaphragm and ribs. In young men the average is about 4.6 dm$^3$; in young women it is about 3.1 dm$^3$. In trained athletes the figures may be as much as 6.0 dm$^3$ for men and 4.5 dm$^3$ for women.

### SAQ 11.4

a Looking at the trace in *figure 11.5*, measure the tidal volume and vital capacity;

b and, looking at the part labelled A, calculate:
   (i) the rate of breathing in breaths per minute;
   (ii) the ventilation rate (tidal volume × breathing rate);
   (iii) the volume of oxygen absorbed per minute.

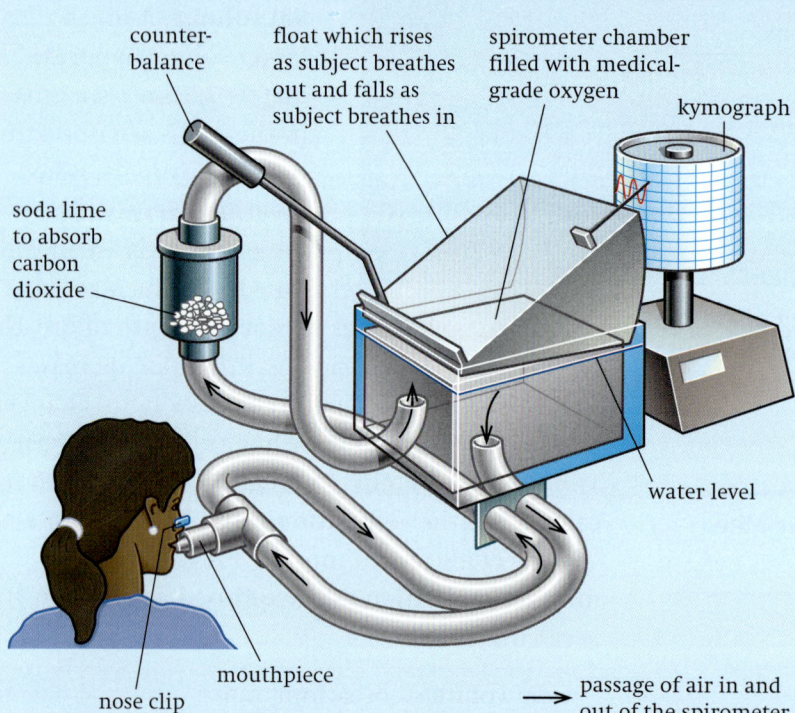

counter-balance

float which rises as subject breathes out and falls as subject breathes in

spirometer chamber filled with medical-grade oxygen

kymograph

soda lime to absorb carbon dioxide

water level

nose clip

mouthpiece

passage of air in and out of the spirometer

● **Figure 11.4** A spirometer.

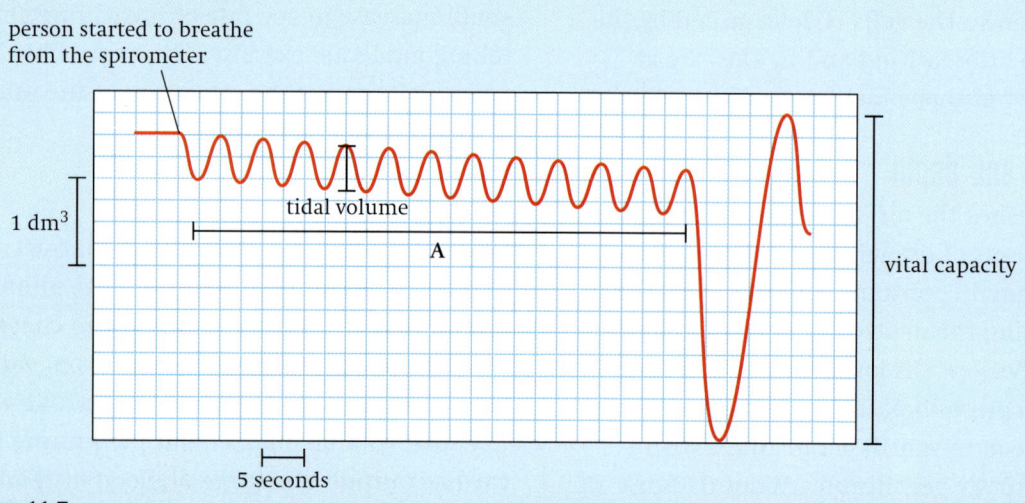

person started to breathe from the spirometer

1 dm$^3$

tidal volume

A

vital capacity

5 seconds

● **Figure 11.5**

A kymograph trace of a 17-year-old male with a mass of 70 kg who breathed normally, took a deep breath and breathed out as much as possible. The kymograph drum revolves at a speed of 2.5 mm s$^{-1}$.

| Pulse rate at rest | Level of fitness |
|---|---|
| less than 50 | outstanding |
| 50–59 | excellent |
| 60–69 | good |
| 70–79 | fair |
| 80 and over | poor |

● **Table 11.2** Resting pulse rates and levels of fitness

**pulse**. The **pulse rate** is identical to the heart rate. It is usually measured at the wrist where the radial artery passes over a bone, or at the carotid artery in the neck. It is counted for 30 seconds with the person sitting still and the result doubled to give the resting pulse rate in beats per minute. The resting pulse rate is an indication of fitness.

At rest, the cardiac output is about 5 dm³ of blood every minute. This can be achieved by having either a large stroke volume with a low pulse rate, or a small stroke volume with a high pulse rate. However it is more efficient to pump slowly as the heart uses less energy than when pumping at a high rate. During exercise, the heart rate increases, providing a faster supply of oxygenated blood to the muscles and of deoxygenated blood to the lungs. If the resting pulse is low and the stroke volume high, only a small increase in pulse is necessary to achieve the required blood supply. People who are physically fit often have a low resting pulse and their pulse rates return to this level quickly after exercise. Endurance athletes in particular usually have large hearts with low pulse rates.

The normal range of resting pulse rates is 60 to 100 beats per minute (*table 11.2*). The average in fit young adults is about 70, and falls with age. The pulse rate is higher during and after exercise, and also after eating or smoking. It is at its lowest when people are asleep.

## Blood pressure

During systole in the cardiac cycle both ventricles contract. Contraction of the left ventricle forces oxygenated blood out of the heart to supply the body (chapter 9). The maximum arterial pressure during this active stroke is the **systolic pressure** and this is the pressure at which blood leaves the heart through the aorta. As the heart relaxes, the pressure in the left ventricle falls, so that the high pressure in the aorta closes the semilunar valve. Elastic recoil of the aorta and the main arteries provides a head of pressure to maintain a steady flow of blood in the arteries towards the capillaries.

The minimum pressure in the arteries is the **diastolic pressure**. The value of the diastolic pressure reflects the resistance of the small arteries and capillaries to blood flow and therefore the load against which the heart must work. If the resistance is high, so is the diastolic pressure. This can be the result of arteries not stretching very well because they have hardened.

Blood pressures are determined using a **sphygmomanometer** (see *box 11B*); it is conventional to give the values in millimetres of mercury (mm Hg) even if the equipment is digital and computerised. Typical blood pressures are:

- systolic – 120 mm Hg (equivalent to 15.8 kPa);
- diastolic – 80 mm Hg (equivalent to 10.5 kPa).

This is often written as 120/80 (120 over 80). Both pressures rise and fall during the day and change in the longer term with age: for a young adult they may be 110/75, but by age 60 years they could be 130/90. At any age, blood pressures may vary slightly from these typical values without causing any health problems.

### SAQ 11.5
Suggest some factors that might affect blood pressure during the day.

### SAQ 11.6
Explain why blood pressure increases with age.

### Hypertension
Blood pressure is a measure of how hard the heart is working to pump blood around the body. Systolic pressure may rise during exercise to 200 mm Hg; diastolic pressure rarely changes very much in healthy people, even during strenuous exercise. If systolic and diastolic blood pressures are high at rest, this indicates that the heart is working too hard at pumping blood. This condition is known as **hypertension**.

## Box 11B Measuring blood pressure

The traditional way to measure blood pressure is with a mercury sphygmomanometer (*figure 11.6*). The rubber cuff of the sphygmomanometer is inflated to give a pressure of 200 mm Hg. This stops the flow of blood into the brachial artery. A stethoscope is placed over the artery and the cuff deflated gradually. The systolic pressure is the pressure when the heart beat is first heard as a soft tapping sound. The cuff is deflated further until the sounds disappear. The diastolic pressure is the pressure when sounds can no longer be heard. Dual blood pressure and pulse rate monitors with digital displays are available for untrained people to use.

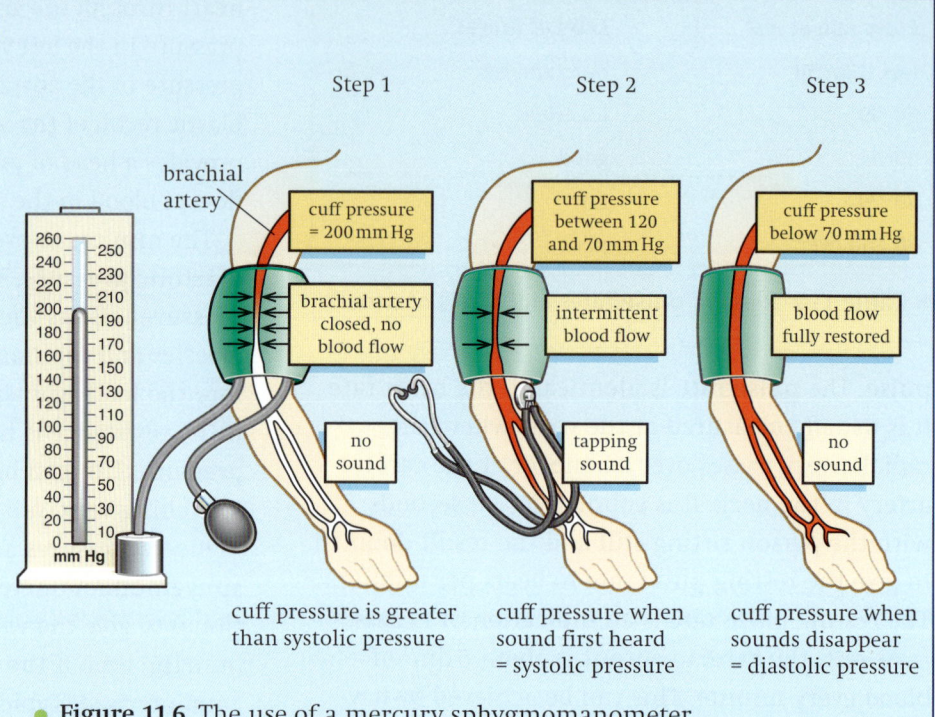

● **Figure 11.6** The use of a mercury sphygmomanometer to measure blood pressure.

Population surveys show that there is a **normal distribution** of blood pressure (a bell-shaped curve). There is no sharp distinction between 'normal' and 'high' blood pressure. However, the risks of cardiovascular diseases such as stroke and coronary heart disease increase considerably with blood pressures in excess of 140/90 (see pages 157–158). The World Health Organisation classifies the resting blood pressures of adults into four groups (*table 11.3*). Hypertension is taken as a blood pressure higher than 140/90. In Britain 15–20% of adults may be hypertensive.

The causes of high blood pressure are generally unknown. In the short term it occurs because of contraction of smooth muscle in the walls of small arteries and arterioles (page 102). This may happen because of an increase in the concentration of the hormone noradrenaline in the blood, which stimulates arterioles to contract. This increases the resistance of the blood vessels and so the heart works harder to force blood through the circulatory system. However, this does not explain long-term hypertension.

Long-term hypertension imposes a strain on the cardiovascular system. If this is not corrected it can lead to heart failure, which occurs when heart muscles weaken and are unable to pump properly. Hypertension is known as the 'silent killer' as there are often no prior symptoms to give a warning of impending heart failure, heart attack or stroke (page 158). In 90% of cases the exact cause of hypertension is unknown, but the condition is known to be closely linked to:

■ excessive alcohol intake;
■ smoking;
■ obesity;
■ too much salt in the diet;
■ genetic factors (people who have close relatives who are hypertensive may also be at risk even if they are not in any of the other high risk categories).

| Category | Blood pressure (mm Hg) | |
| --- | --- | --- |
| | systolic | diastolic |
| optimal | < 120 | < 80 |
| normal | < 130 | < 85 |
| high normal | 130–140 | 85–90 |
| hypertension | > 140 | > 90 |

● **Table 11.3** The World Health Organisation classification of adult blood pressures.

# SUMMARY

◆ Air passes down the trachea and through a branching system of airways in the lungs to reach the alveoli. The airways are lined by a ciliated epithelium with mucus-secreting goblet cells. The epithelium protects the alveoli by moving a carpet of mucus towards the throat where it can be swallowed. The alveoli are the site of gaseous exchange.

◆ A spirometer measures the tidal volume and vital capacity of the lungs. Tidal volume is the volume of air breathed in and then out. At rest it is about $0.5\,dm^3$. Vital capacity is the maximum volume of air that can be breathed out after fully inflating the lungs.

◆ The pulse rate is identical to the heart rate. Resting pulse rate is used as a measurement of fitness since a low rate is associated with a large volume of blood expelled by the heart with each beat.

◆ Blood in the arteries is under pressure which can be measured using a sphygmomanometer. When the heart contracts, this pressure rises. The maximum pressure which corresponds to the emptying of the left ventricle is the systolic pressure. The minimum pressure in the arteries occurs when the left ventricle is relaxed and filling with blood. This is the diastolic pressure.

◆ Hypertension is high blood pressure. The causes of this are unknown, but appear to be related to genetic factors, a diet rich in salt, high consumption of alcohol and obesity. High blood pressure is a contributory factor to coronary heart disease.

## CHAPTER 12

# Smoking and disease

## By the end of this chapter you should be able to:

1 describe the symptoms of chronic bronchitis and emphysema (chronic obstructive pulmonary disease) and lung cancer;

2 describe the effects of tar and carcinogens in tobacco smoke on the gaseous exchange system;

3 evaluate the epidemiological and experimental evidence linking cigarette smoking to disease and early death;

4 describe the effects of nicotine and carbon monoxide in tobacco smoke on the cardiovascular system with reference to atherosclerosis, coronary heart disease and strokes;

5 discuss the possible links between diet and coronary heart disease;

6 discuss the reasons for the global distribution of coronary heart disease;

7 discuss the difficulty in achieving a balance between prevention and cure with reference to coronary heart disease, coronary by-pass surgery and heart transplant surgery.

The World Health Organisation considers smoking to be a disease. Until the end of the nineteenth century, tobacco was smoked almost exclusively by men and in pipes and cigars, involving little inhalation. Then manufacture of cigarettes began. Smoking cigarettes became fashionable for European men during the First World War and in the 1940s women started smoking in large numbers too. In Pakistan, 40% of men and 12% of women smoke. Each day, 1200 young boys take up smoking, as well as many young girls.

## Tobacco smoke

The tobacco companies do not declare the ingredients in their products, but it is known by analysis that there are over 4000 different chemicals in cigarette smoke, many of which are toxic. Tobacco smoke is composed of 'mainstream' smoke (from the filter or mouth end) and 'sidestream' smoke (from the burning tip). When a person smokes, about 85% of the smoke that they release is sidestream smoke. Many of the toxic ingredients are in a higher concentration in sidestream than in mainstream smoke and any other people in the vicinity are also exposed to them. Breathing someone else's cigarette smoke is called **passive smoking**.

Three main components of cigarette smoke pose a threat to human health, damaging in particular either the gaseous exchange or cardiovascular system.

- **Tar** (a mixture of aromatic compounds) settles on the lining of the airways in the lungs and stimulates a series of changes that may lead to obstructive lung diseases and lung cancer (see below). This connection was recognised in the 1950s.

- **Carbon monoxide** diffuses across the walls of the alveoli and into the blood in the lungs. It diffuses into red blood cells where it combines with haemoglobin to form the stable compound carboxyhaemoglobin (see page 113). This means that haemoglobin does not become fully oxygenated. The quantity of oxygen transported in the blood may be 5–10% smaller in a smoker than in a non-smoker. Less oxygen is supplied to the heart muscle, putting a strain on it especially when the heart rate increases during exercise (see chapter 11). Carbon monoxide also damages the lining of the arteries.

- **Nicotine** is the drug in tobacco. It is absorbed very readily by the blood and travels to the brain within a few seconds. It stimulates the nervous system to reduce the diameter of the arterioles and to release the hormone adrenaline from the adrenal glands. As a result, heart rate and blood pressure increase (page 148) and there is a decrease in blood supply to the extremities of the body, such as hands and feet, reducing their supply of oxygen. Nicotine also increases the 'stickiness' of blood platelets, so increasing the risk of blood clotting.

Both carbon monoxide and nicotine increase the risk of developing cardiovascular disease (see pages 156–163), a link that was discovered in the latter half of the twentieth century.

# Lung disease

The gaseous exchange system is naturally efficient and adaptable and training can improve its function and, thereby, our general health. Healthy people breathe with little conscious effort; for people with lung diseases every breath may be a struggle.

The lungs' large surface area of delicate tissue is constantly exposed to moving streams of air that may carry potentially harmful gases and particles. Despite the filtering system in the airways, very small particles (< 2 µm in diameter) can reach the alveoli and stay there. The air that flows down the trachea with each breath fills the huge volume of tiny airways. The small particles settle out easily because the air flow in the depths of the lungs is very slow.

Such deposits make the lungs susceptible to airborne infections such as influenza and pneumonia and, in some people, can cause an allergic reaction. **Allergens** (substances that cause allergies), such as pollen and the faeces of house dust mites, trigger a defence mechanism in the

cells lining the airways. If this is very bad it may cause an asthmatic attack in which the smooth muscles in the airways contract, obstructing the flow of air and making breathing difficult. The body's defence mechanisms may react with the production of more mucus and the collection of white blood cells in the airways. This can block the airways and cause severe bouts of coughing, which can damage the alveoli. Continuous damage can lead to the replacement of the thin alveolar surface by scar tissue, so reducing the surface area for diffusion. This replacement also happens when particles of coal dust or asbestos regularly reach the alveoli.

Chronic (long-term) **obstructive lung diseases**, such as asthma, chronic bronchitis and emphysema, are now prevalent as a result of atmospheric pollution from vehicle and industrial emissions and smoking. After heart disease and strokes, lung diseases are the most common cause of illness and death in the UK. It is estimated that one in seven children in the UK suffers from asthma. There are increasing legal controls in the UK on emissions of pollutants from industrial, domestic and transport fuels and on the condition of the work place, but as yet none regarding tobacco smoke.

## Chronic bronchitis

Tar in cigarette smoke stimulates goblet cells and mucous glands to enlarge and secrete more mucus (page 143). Tar also inhibits the cleaning action of the ciliated epithelium that lines the airways. It destroys many cilia and weakens the sweeping action of those that remain. As a result mucus accumulates in the bronchioles and the smallest of these may become obstructed. As mucus is not moved, or at best only moved slowly, dirt, bacteria and viruses collect and block the bronchioles. This stimulates 'smoker's cough' which is an attempt to move the mucus up the airways. With time the damaged epithelia are replaced by scar tissue and the smooth muscle surrounding the bronchioles and bronchi becomes thicker. This thickening of the airways causes them to narrow and makes it difficult to move air into and out of the lungs.

Infections such as pneumonia easily develop in the accumulated mucus. When there is an infection in the lungs the linings become inflamed and this further narrows the airways. This damage and blocking of the airways is **chronic bronchitis**. Sufferers have a severe cough, producing large quantities of phlegm, which is a mixture of mucus, bacteria and some white blood cells.

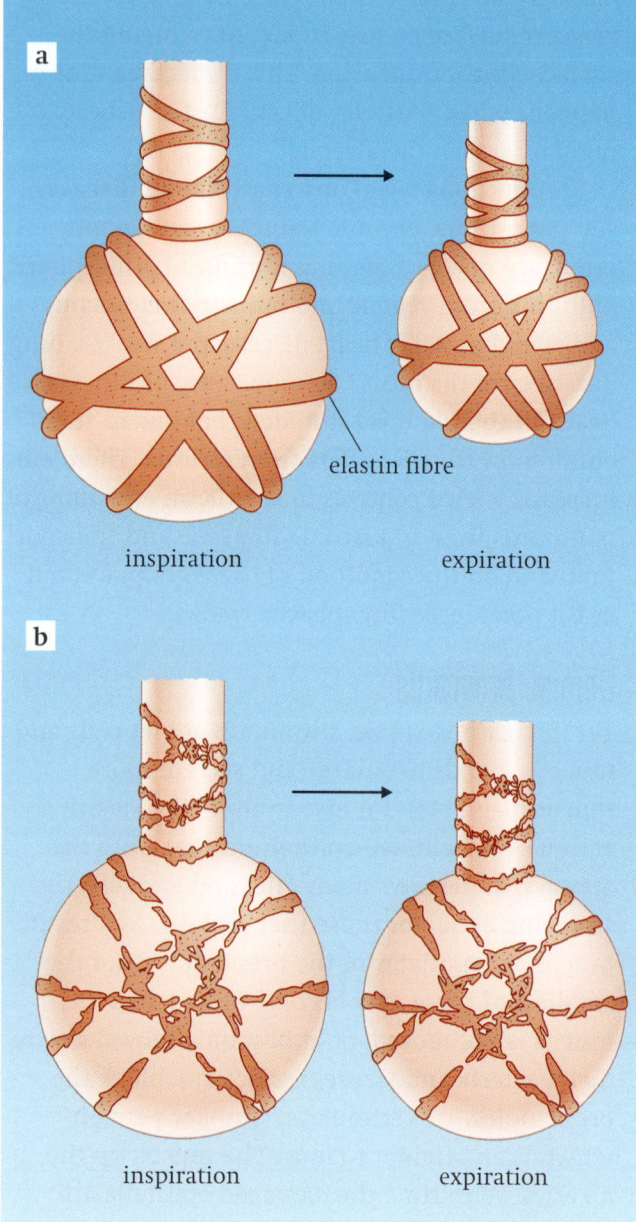

● **Figure 12.1** The development of emphysema.
**a** Healthy alveoli partially deflate when breathing out due to the recoil of elastin fibres.
**b** Phagocytes from the blood make pathways through alveolar walls by digesting elastin and, after many years of this destruction, the alveoli do not deflate very much.

## Emphysema

The inflammation of the constantly infected lungs causes phagocytes to leave the blood and line the airways. Phagocytes are white blood cells that remove bacteria from the body (see chapters 4 and 14). To reach the lining of the lungs from the capillaries, phagocytes release the protein-digesting enzyme elastase. This enzyme destroys elastin in the walls of the alveoli so making a pathway for the phagocytes to reach the surface and remove bacteria. Elastin is responsible for the recoil of the alveoli when we breathe out. With much smaller quantities of elastin in the alveolar walls the alveoli do not stretch and recoil when breathing in and out (*figure 12.1*). As a result, the bronchioles collapse during exhalation trapping air in the alveoli, which often burst. Large spaces appear where they have burst and this reduces the surface area for gaseous exchange. This condition is called **emphysema**.

The loss of elastin makes it difficult to move air out of the lungs. Non-smokers can force out about 4 dm$^3$ of air after taking a deep breath; someone with emphysema may manage to force out only 1.3 dm$^3$ of air. The air remains in the lungs and is not refreshed during ventilation. Together with the reduced surface area for gaseous exchange, this means that many people with emphysema do not oxygenate their blood very well and have a rapid breathing rate.

As the disease progresses the blood vessels in the lungs become more resistant to the flow of blood. To compensate for this increased resistance, the blood pressure in the pulmonary artery increases and, over time, the right side of the heart enlarges.

As lung function deteriorates, wheezing occurs and breathlessness becomes progressively worse. It may become so bad in some people that they cannot get out of bed. People with severe emphysema often need a continuous supply of oxygen through a face mask to stay alive. In *figure 12.2*, you can compare the appearance of diseased and relatively unaffected lung tissue in a computerised tomography (CT) scan.

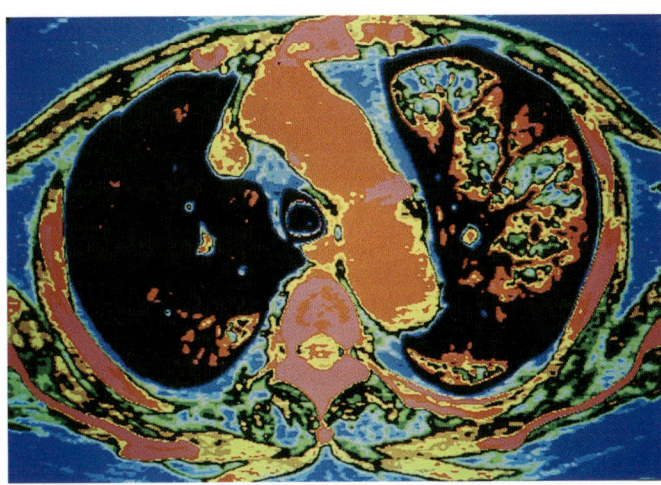

● **Figure 12.2** A computerised tomography scan (CT scan) of a horizontal section through the thorax. The two black regions are the lungs. The right lung is diseased with emphysema, (blue-green areas). The left lung is relatively unaffected at the level of this scan. You can see a cross section of a vertebra at the bottom.

## Chronic obstructive pulmonary disease

Chronic bronchitis and emphysema often occur together and constitute a serious risk to health. The term **chronic obstructive pulmonary disease** refers to the overall condition, which is a progressively disabling disease. The gradual onset of breathlessness only becomes troublesome when about half of the lungs is destroyed. Only in very rare circumstances is it reversible. If smoking is given up when still young, lung function can improve. In older people recovery from chronic obstructive pulmonary disease is not possible.

Chronic obstructive pulmonary disease is responsible for over 30 000 deaths in the UK each year. The UK has the highest death rate for this disease in the world.

### SAQ 12.1
Summarise the changes that occur in the lungs of people with chronic bronchitis and emphysema.

## Lung cancer

Tar in tobacco smoke contains several substances that have been identified as carcinogens (see chapter 6). These react, directly or via breakdown products, with DNA in epithelial cells to produce mutations, which are the first in a series of changes that lead to the development of a malignant tumour (*figures 6.16* and *12.3*).

As the cancer develops, it spreads through the bronchial epithelium and enters the lymphatic tissues (see page 105) in the lung. Cells may break away and spread to other organs (metastasis) so that secondary tumours become established.

Lung cancer takes 20–30 years to develop. Most of the growth of a tumour occurs before there are any symptoms. The most common symptom of lung cancer is coughing up blood, as a result of tissue damage. It is rare for a cancer to be diagnosed before it reaches 1 cm in diameter. By that time it has already doubled in size 30 times from the moment of the malignant mutation.

Tumours in the lungs, such as that shown in *figure 12.3*, are located by one of three methods:

■ bronchoscopy, using an endoscope to allow a direct view of the lining of the bronchi;

■ chest X-ray;

■ CT scan (similar to that shown in *figure 12.2*).

● **Figure 12.3** A scanning electron micrograph of a bronchial carcinoma – a cancer in a bronchus. Cancers often develop at the base of the trachea where it divides into the bronchi as this is where most of the tar is deposited. The disorganised malignant tumour cells at the bottom right are invading the normal tissue of the ciliated epithelium (× 1000).

By the time most lung cancers are discovered they are well advanced and the only hope is to remove them by surgery. If the cancer is small and in one lung, then either a part, or all, of the lung is removed. However, metastasis has usually happened by the time of the diagnosis so, if there are secondary tumours, surgery will not cure the disease. Chemotherapy with anti-cancer drugs or radiotherapy with X-rays (or other form of radiation) will be used.

# Proving the links between smoking and lung disease

As we have seen, cigarette smoking was not widespread until the first half of the twentieth century. In the UK, doctors started noticing cases of lung cancer from the 1930s onwards, and then from the 1950s there was an epidemic (*figure 12.4*). In 1912 there were 374 cases of lung cancer; now there are over 35 000 deaths each year in the UK from the disease.

### Epidemiological evidence

Epidemiologists discovered a correlation between lung cancer and cigarette smoking. *Table 12.1* shows the correlation between the number of cigarettes smoked per day and the risk of early death. Epidemiological data link smoking and lung diseases including cancer in the following ways.

### General
- Up to 50% of smokers may die of smoking-related diseases.
- Smokers are three times more likely to die in middle age than are non-smokers.

### Chronic obstructive pulmonary disease
- Chronic obstructive pulmonary disease is very rare in non-smokers.
- 90% of deaths from chronic obstructive pulmonary disease are attributed to smoking.
- 98% of people with emphysema are smokers.
- 20% of smokers suffer from emphysema.
- Deaths from pneumonia and influenza are twice as high among smokers.

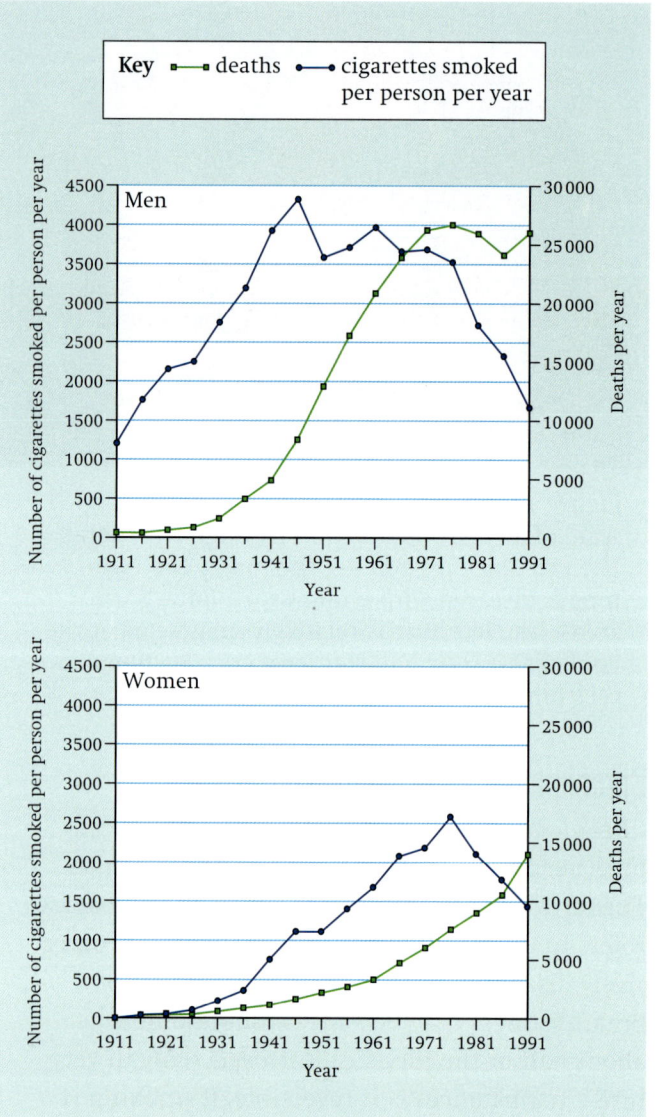

- **Figure 12.4** The smoking epidemic. The correlation between the consumption of cigarettes and deaths from lung cancer in the UK from 1911 to 1991.

| Number of cigarettes smoked per day | Annual death rate per 100 000 men |
|---|---|
| 0 | 10 |
| 1–14 | 78 |
| 15–24 | 127 |
| >25 | 251 |

Note that these are deaths from all causes not just lung cancer.

- **Table 12.1** Results of a study carried out on male doctors in Britain, showing that the risk of early death increases with the number of cigarettes smoked per day.

## Lung cancer

- Smokers are 18 times more likely to develop lung cancer than non-smokers.
- One-third of all cancer deaths are a direct result of cigarette smoking.
- 25% of smokers die of lung cancer.
- The risk of developing lung cancer increases if smokers inhale; start young; increase the number of cigarettes smoked per day; use high tar cigarettes; smoke for a long time (smoking one packet of cigarettes per day for forty years is eight times more hazardous than smoking two packets for twenty years).
- The risk of developing lung cancer starts to decrease as soon as smoking is stopped, but it takes ten or more years to return to the same risk as a non-smoker.

### SAQ 12.2

Summarise the trends shown in *figure 12.4*.

Conclusions drawn from epidemiological data about the risks of developing lung cancer could be criticised because they only show that there is an *association* between the two, not a *causal link* between them. There may be another common factor which is the causative one. For example, exposure to atmospheric pollutants such as sulphur dioxide could be a cause. However, epidemiological studies have ruled out these other factors – comparably close correlations with them cannot be found. With smoking, on the other hand, it is possible to show a direct link with lung cancer because smoking is the common factor in almost all cases. (Reports by the Royal College of Physicians in the UK published in 1962, 1971, 1977, 1983 and 1992 reviewed the epidemiological evidence for the link.)

Cigarette smoking is linked with other cancers. It is a major cause of cancers of the mouth, oesophagus and larynx. It is a cause of bladder cancer and a contributory factor in the development of cancers of the pancreas, kidney and cervix.

### Experimental evidence

Experimental evidence shows a direct causative link between smoking and lung cancer. There are two lines of evidence.

- Tumours similar to those found in humans develop in animals exposed to cigarette smoke.
- Carcinogens have been identified in tar.

In the 1960s experimental animals were used to investigate the effect of cigarette smoke on the lungs (*figure 12.5*). In one study 48 dogs were divided into two groups. One group was made to smoke filter-tipped cigarettes and did not develop cancer. The other group smoked plain (unfiltered) cigarettes and developed abnormalities similar to those in human lung cancer patients. The animals also showed changes similar to those caused by chronic obstructive pulmonary disease. The fact that the group smoking filter-tipped cigarettes remained healthy does not show that the filters remove all the carcinogens from smoke. In fact some of the dogs developed pre-cancerous changes in the cells lining their airways.

Smoking machines, which copy the inhaling pattern of smokers, extract the chemicals contained in smoke. Chemical analysis of the black, oily liquid that accumulates in these machines shows that tar contains a variety of

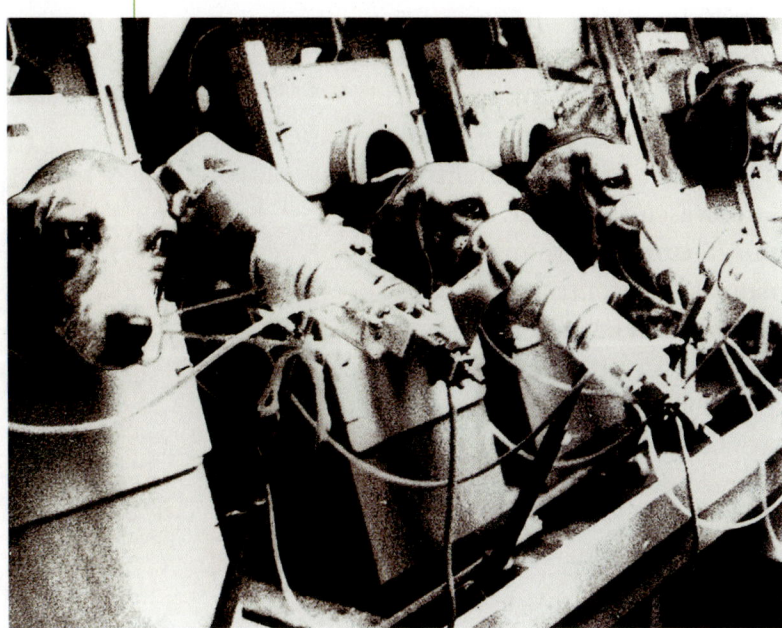

● **Figure 12.5** Beagles used in experiments in the 1960s to investigate the link between smoking and lung cancer. When the results were published they convinced many smokers to change to low tar, filter-tipped brands.

carcinogens and co-carcinogens. The latter are compounds which increase the likelihood that carcinogens will cause mutations in DNA. The most potent carcinogen is benzpyrene. When carcinogens from tar are painted onto the skin of mice, cancerous growths develop. Experiments like this not only confirm the link between smoking and lung cancer, but also help to show how tumours develop in the lungs.

The connection between cigarette smoking and lung cancer is irrefutable. The mechanisms by which carcinogens cause mutations and the factors that influence the growth of cancers are still being investigated. As the popularity of smoking decreases in developed countries, the smoking epidemic is spreading to the developing world, which, it is predicted, will see a rise in smoking-related diseases in the 21st century.

### SAQ 12.3
Summarise the effects of tobacco smoke on the gaseous exchange system.

# Cardiovascular diseases

**Cardiovascular diseases** are degenerative diseases of the heart and circulatory system, such as coronary heart disease and stroke. They are a major cause of death and disability. They are responsible for 20% of all deaths worldwide and up to 50% of deaths in developed countries. Cardiovascular diseases are **multifactorial,** meaning that many factors contribute to the development of these diseases. Smoking is just one among several **risk factors** that increase the chances of developing one of the cardiovascular diseases.

## Atherosclerosis

The main process that leads to cardiovascular diseases is the accumulation of fatty material in artery walls. This reduces the flow of blood to the tissues and may also increase the chance of blood clots forming within the artery, obstructing the flow of blood altogether. If blood cannot flow into capillaries, the surrounding tissue does not receive enough nutrients and oxygen, and may die. This build-up of **atheroma**, which contains cholesterol, fibres, dead muscle cells and platelets, is termed **atherosclerosis**.

The inside of a healthy artery is pale and smooth but yellow fatty streaks can start appearing at any time from childhood. These start with damage to the lining of the arteries. In response to the damage, there is an invasion of phagocytes whose secretions stimulate the growth of smooth muscle cells (to aid repair) and the accumulation of cholesterol.

Cholesterol is a lipid (see page 28) which is needed for the synthesis of vitamin D in the skin, steroid hormones in the ovaries, testes and adrenal glands and, in all cells, plasma membranes (see page 51). It is therefore an essential biochemical. Problems arise, however, if excessive cholesterol collects in tissues.

Cholesterol is insoluble in water and so is transported in the blood plasma in tiny balls of lipid and protein called **lipoproteins** (*figure 12.6*). There are two groups of lipoproteins, **low density** lipoproteins and **high density** lipoproteins. Low density lipoproteins transport cholesterol from the liver to the tissues, including the artery walls.

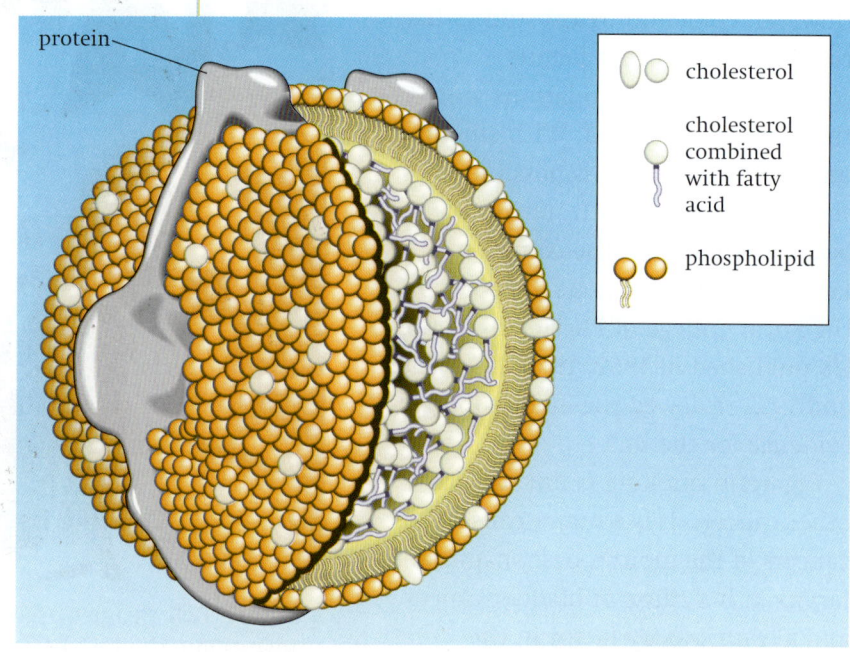

● **Figure 12.6** A low density lipoprotein (LDL). High density lipoproteins (HDLs) have a lower content of cholesterol than LDLs.

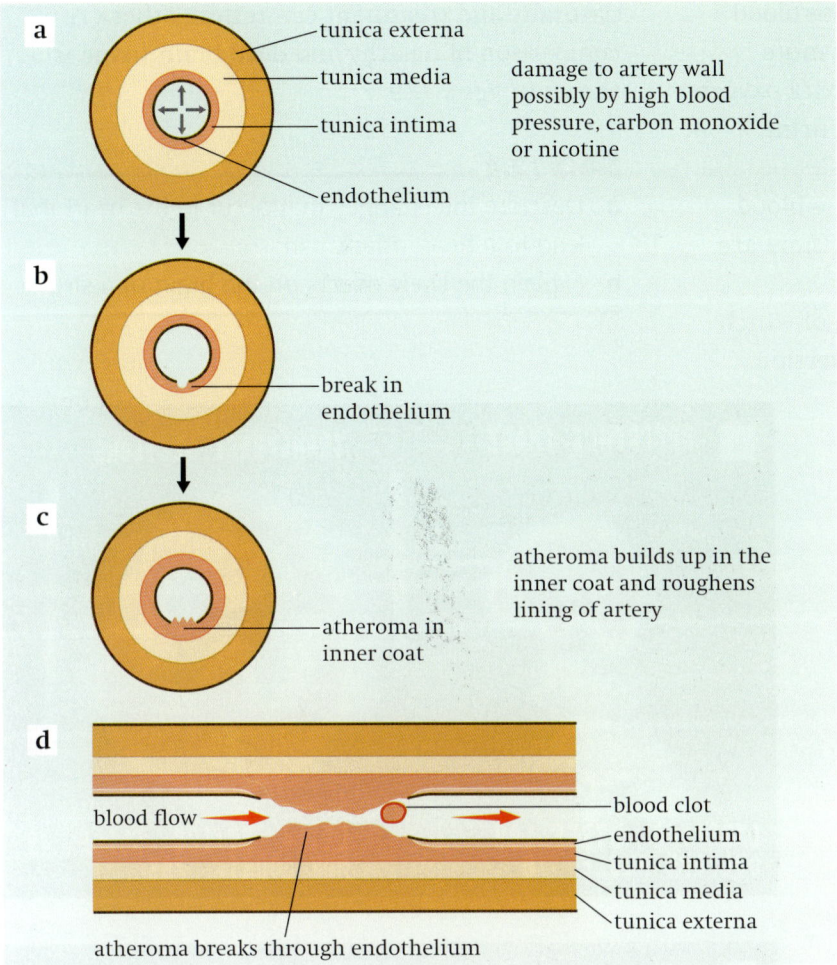

- **Figure 12.7** How damage to artery walls promotes the development of fatty plaques and blood clots. **a–c** Cross sections of an artery showing the development of a plaque; **d** longitudinal section of an artery with a blood clot forming at the site of a plaque.

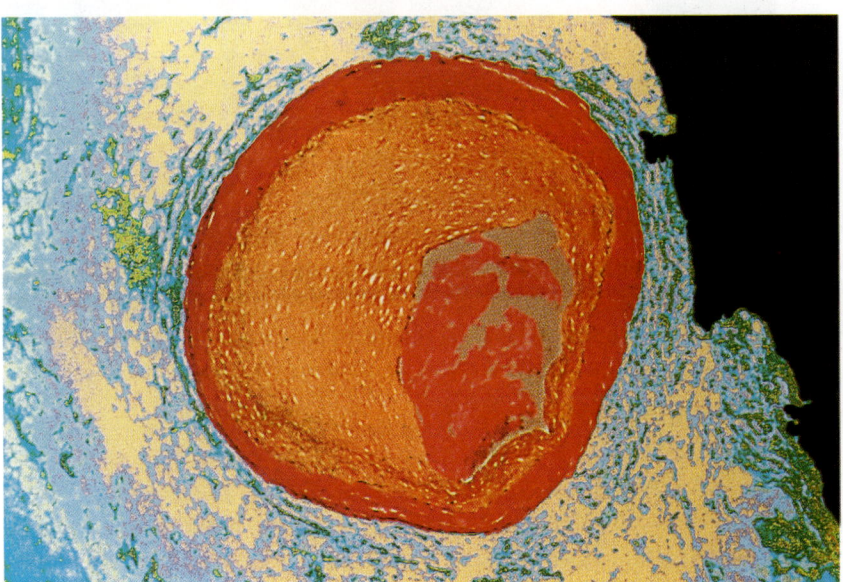

- **Figure 12.8** A cross section of a coronary artery (red) blocked by plaque (orange). Some blood (red) can flow through the lumen of the artery, but is restricted by the plaque.

They tend to deposit their cholesterol at any damaged sites. In contrast, high density lipoproteins remove cholesterol from tissues and transport it to the liver to be excreted. High density lipoproteins therefore help to protect arteries against atherosclerosis.

The cholesterol-rich atheroma forms **plaques** in the lining of the arteries, making them less elastic and restricting the flow of blood (*figures 12.7* and *12.8*). Plaques grow within the artery wall and then push inwards reducing or blocking the passage of blood. The plaque may break through the lining of the artery to give a rough surface. The blood can no longer flow smoothly and tends to clot, forming a **thrombus**. This process of **thrombosis** interrupts blood flow even more, so tissues are starved of oxygen and nutrients. When this happens in a coronary artery, heart muscle may die, causing a heart attack (see chapter 9 and below). If it happens in an artery in the brain the result is a stroke.

## Coronary heart disease

Two coronary arteries branch from the aorta to supply all the muscles of the atria and the ventricles (see *figure 9.2*). **Coronary heart disease** is a disease of these arteries that causes damage to, or malfunction of, the heart.

When atherosclerosis occurs in the lining of the coronary arteries they become narrow, restricting the flow of blood. As a result, the heart has to work harder to force blood through the coronary arteries and this may cause blood pressure to rise. It also means that it is more difficult to supply heart muscle with extra oxygen and nutrients when it has to respond during exercise (chapter 11).

Coronary heart disease develops if the blood supply to the heart muscle is reduced. There are three forms of coronary heart disease.

■ **Angina pectoris**, the main symptom of which is severe chest pain brought on by exertion. The pain starts when exercising, but goes when resting. The pain is caused by a severe shortage of blood to the heart muscle, but there is no death of heart tissue.

■ **Heart attack** which is also known as **myocardial infarction**. When a moderately large branch of a coronary artery is obstructed by a blood clot, part of the heart muscle is starved of oxygen and dies. This causes sudden and severe chest pain. A heart attack may be fatal, but many people survive if they are treated immediately.

■ **Heart failure** due to the blockage of a main coronary artery and the resulting gradual damage of heart muscle. The heart weakens and fails to pump efficiently.

## Stroke

A **stroke** occurs when an artery in the brain bursts so that blood leaks into brain tissue (a brain haemorrhage) or, more commonly, when there is a blockage in a brain artery due to atherosclerosis or a thrombus. The brain tissue in the area supplied by the artery is starved of oxygen and dies (**cerebral infarction**). A stroke may be fatal or cause mild or severe disability. The degree of disability depends on the amount of the brain affected by shortage or lack of oxygen. As a result of a stroke, some people are unable to control part of their body; others may lose their ability to speak or some of their memory. Even personality can change. Some of these faculties can be taken over by other parts of the brain

naturally and treatment can restore others. A comparison of healthy and dead brain tissue is shown in *figure 12.9*.

### SAQ 12.4

**a** Describe the changes that occur in the heart that lead to a heart attack.

**b** Explain the likely effects on the brain of a stroke.

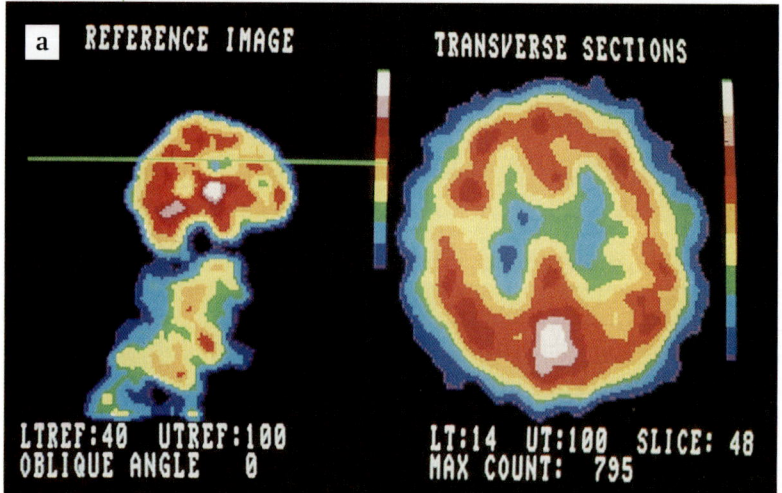

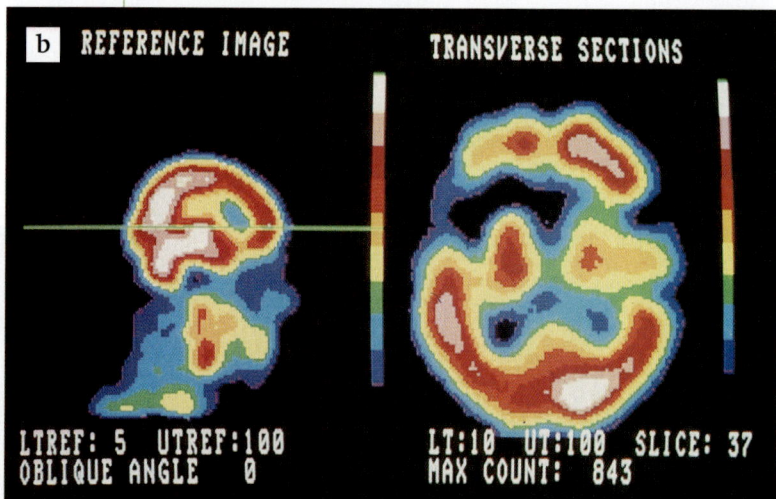

● **Figure 12.9** Scintigram scans of a normal brain and a brain following a stroke. Scintigram scanning shows the pattern of γ rays emitted by a radioactive tracer introduced into the body. The tracer is taken up by different tissues producing a distinct image in organs and tissues which are functioning correctly.  **a** Scan of a normal brain in vertical and horizontal section. The green line on the left shows the position of the horizontal section.  **b** Scan of a brain following a stroke. The black area in the front of the horizontal section shows where tissue has died as a result of an infarction.

# Global distribution of coronary heart disease

Death rates for coronary heart disease are not evenly distributed across the world. The rate is highest in northern Europe and lowest in Japan and France (*figure 12.10*). Coronary heart disease is the major cause of premature death in developed countries and responsible for much ill health and disability.

Coronary heart disease was almost unknown before the twentieth century and until recently it has mainly been confined to developed countries. It is considered to be a disease associated with affluence, but as it is a degenerative disease it is possible that the incidence of coronary heart disease is high simply because people now live longer. High death rates from infectious diseases in earlier centuries obscured the degenerative changes that may have been occurring in people's coronary arteries. In other words, people did not die from coronary heart disease because they died of something else first. As death rates from infectious diseases decrease in the developing world, death rates from cardiovascular diseases, especially coronary heart disease, increase. Coronary heart disease has become the leading cause of death in Argentina, Cuba, Chile, Uruguay and Trinidad and Tobago which have rising standards of living and good public health systems.

In just the same way that death rates for coronary heart disease are not the same over the whole world, so they differ *within* countries. In the UK, the incidence of the disease is highest:

- in Scotland, Northern Ireland and the north and north-west of England;
- among poorer people;
- among certain ethnic groups including south Asians;
- among men.

These epidemiological data suggest that people are not equally at risk of developing coronary heart disease.

## Epidemiological evidence

The evidence for links between smoking and cardiovascular diseases is not as clear cut as it is for smoking and lung cancer. Although many smokers die of coronary heart disease and stroke, so do many non-smokers.

It is known that smoking increases the chances of both the development of atherosclerosis and blood clotting. Both nicotine and carbon monoxide speed up the development of plaques, although it is not known how they do this. They also stimulate the production of one of the blood clotting factors (fibrinogen) and reduce the production of enzymes that remove clots. Nicotine increases blood pressure and heart rate and thus the body's demand for oxygen, but carbon monoxide reduces the blood's ability to carry it.

Smoking also interacts with other risk factors. It increases blood pressure and increases the concentration of cholesterol in the blood; high blood cholesterol is a risk factor for coronary heart disease and hypertension is the major risk factor for stroke.

Smokers not only increase their own risk of developing coronary heart disease or stroke, but also that of other people. Passive smoking is

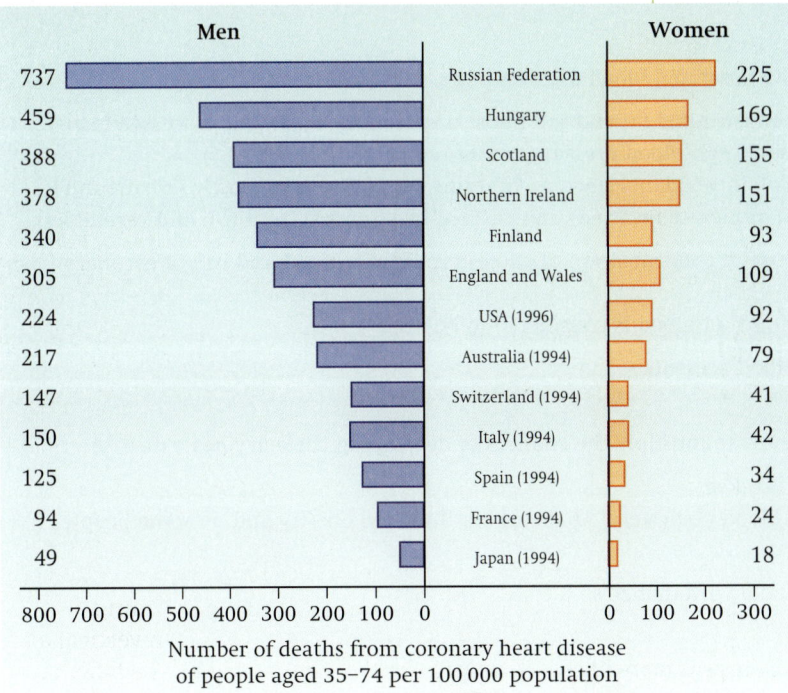

| Men | | Women |
|---|---|---|
| 737 | Russian Federation | 225 |
| 459 | Hungary | 169 |
| 388 | Scotland | 155 |
| 378 | Northern Ireland | 151 |
| 340 | Finland | 93 |
| 305 | England and Wales | 109 |
| 224 | USA (1996) | 92 |
| 217 | Australia (1994) | 79 |
| 147 | Switzerland (1994) | 41 |
| 150 | Italy (1994) | 42 |
| 125 | Spain (1994) | 34 |
| 94 | France (1994) | 24 |
| 49 | Japan (1994) | 18 |

800 700 600 500 400 300 200 100 0     0 100 200 300

Number of deaths from coronary heart disease of people aged 35–74 per 100 000 population

● **Figure 12.10** Death rates from coronary heart disease for men and women aged 35–74 per 100 000 population for selected countries. Unless stated otherwise the figures are for 1997.

believed to be responsible for many deaths from cardiovascular disease and those most at risk are those who live, work or socialise with smokers.

## SAQ 12.5

Describe the effects of tobacco smoke on the cardiovascular system.

*Table 12.2* shows the risk factors that are associated with coronary heart disease. Some of these are fixed, since there is nothing anyone can do about their sex, age or the genes that they have inherited. Others are modifiable. People can take steps to change their way of life to reduce the risks of developing coronary heart disease by giving up smoking, eating less saturated fat, losing weight and taking more exercise.

These factors were identified in long-term epidemiological studies with large groups of people in which information about their way of life, illnesses and causes of death were recorded. A study of the population of Framingham in Massachusetts, USA and a study of civil servants in the UK (the Whitehall study) showed that:

- hypertension, high cholesterol concentrations in the blood, cigarette smoking and diabetes were common factors among inhabitants of Framingham who developed coronary heart disease;
- British civil servants who took regular exercise had half the incidence of coronary heart disease than their less active colleagues.

To find the reasons for the global distribution of coronary heart disease, the World Health Organisation set up a multinational monitoring

| Factor | Effect |
|---|---|
| age | Risk increases with increasing age: about 80% of people who die of coronary heart disease are 65 or older. |
| gender | Men are more at risk than women: protection is given by oestrogen which occurs either naturally or by hormone replacement therapy (HRT). |
| heredity | Children of parents with heart disease are at greater risk than others: genes are involved in developing high blood cholesterol concentrations, high blood pressure and diabetes mellitus. |
| body mass | Being overweight or obese increases risk: excess weight puts a strain on the heart and blood pressure rises. |
| diet | Risk increases with high intake of saturated fat and salt (most prevalent in meat and processed foods): development of atherosclerosis and high blood pressure are more likely. Risk decreases with high intake of antioxidants (such as vitamins E and C, found mostly in fruit and vegetables) and soluble fibre and moderate intake of unsaturated fats (mainly from fish and vegetables). |
| blood cholesterol | There is a direct relationship between concentration of cholesterol in the blood and risk of coronary heart disease: above 250 mg cholesterol per 100 cm$^3$ of blood is considered to be high. |
| high blood pressure | Risk increases with increasing blood pressure. |
| smoking | Combines with all other risk factors to multiply the chances of developing coronary heart disease. |
| exercise | Risk decreases as more exercise is taken: aerobic activity helps to control blood cholesterol, diabetes mellitus and obesity and, in some people, lowers blood pressure. |
| diabetes | There is an increased risk for people with diabetes. |
| alcohol | Risk increases with high intake: blood pressure rises and atherosclerosis is more likely. Risk may decrease with moderate intake. |
| social class | Risk increases with poverty. |

● **Table 12.2** Factors associated with coronary heart disease.

project (known as MONICA) in 1979. Death rates from coronary heart disease were compared with death rates from all causes in a number of different countries. *Table 12.3* shows some results for two European countries, Finland and Spain, with markedly different death rates for coronary heart disease. These rates tend to decline in a north-east to south-west direction across Europe, being particularly high in Finland and low in Spain. Data were collected from two regions within these countries, North Karelia, a region of Finland which used to have the highest death rate from coronary heart disease in the world, and Catalonia in Spain, which has one of the lowest. The MONICA project also collected data on the risk factors associated with coronary heart disease. The results showed that the Finns tended to have much higher blood pressures and blood cholesterol concentrations than the Catalonians. The incidence of obesity was similar in the two populations and more Catalonians smoked. This evidence pointed to the importance of blood pressure and blood cholesterol as key factors in predicting whether someone would suffer from coronary heart disease.

## SAQ 12.6

Looking at *table 12.3*,

a suggest why data were collected for the 35–64 year age group and not for all age groups;

b compare the populations of North Karelia and Catalonia in terms of the proportions of deaths due to (i) all cardiovascular diseases, (ii) coronary heart disease, and (iii) strokes;

c suggest which members of the population of the two regions are most at risk of developing cardiovascular diseases.

It is important to realise that each of the factors listed in *table 12.2* is *associated* with heart disease. None by itself predicts that heart disease will develop. For example, there are many smokers who do not develop heart disease or have a stroke. However, smokers increase their risk of developing heart disease. A smoker who is overweight and eats a diet rich in saturated fat is even more at risk. Recent research suggests that genetic factors are important. Those most at risk have inherited a genetic condition which leads to a very high concentration of blood cholesterol. The disease affects 1 in 500 people in the UK. These people have a very high risk of developing coronary heart disease early in life and they should take special care about controlling other risk factors such as weight, diet and blood pressure. Recent research shows that people can inherit genes that dispose them to high blood pressure and diabetes and increase the risk of heart disease.

### The role of diet in coronary heart disease

Dietary factors have been implicated in heart disease for a long time. People with high levels of saturated fat and cholesterol in their diet tend to have high blood cholesterol levels. The blood cholesterol concentration in the body is not dependent solely on the intake of cholesterol in foods such as meat and eggs; it depends mainly on the saturated fat derived from animal foods. Red meat and dairy products, such as milk and butter, are especially rich in saturated fat. This helps to explain the high incidence of heart disease in countries such as Finland, where the traditional diet is mainly composed of foods rich in animal fat. Some of the lowest rates of heart

| Country or region | Annual mortality (deaths per 100 000 population in age group 35–64 between 1984 and 1986) | | | | | | | |
|---|---|---|---|---|---|---|---|---|
| | total number of deaths (all causes) | | deaths from all cardiovascular diseases | | deaths from coronary heart disease | | deaths from stroke | |
| | men | women | men | women | men | women | men | women |
| Finland | 894 | 314 | 427 | 103 | 317 | 52 | 57 | 32 |
| North Karelia | 1111 | 364 | 600 | 140 | 456 | 75 | 82 | 40 |
| Spain | 634 | 277 | 193 | 77 | 89 | 20 | 43 | 25 |
| Catalonia | 536 | 237 | 138 | 49 | 68 | 16 | 32 | 18 |

● **Table 12.3** Data on death rates collected from Finland and Spain as part of the World Health Organisation MONICA project.

disease are found in countries with high fat intakes, such as Spain and Italy, but the fat is mainly unsaturated. Increasing the intake of unsaturated fats in the diet tends to cause the blood cholesterol concentration to go down so long as saturated fat intake is low.

However, the link between diet and coronary heart disease is not so easily explained. France has one of the lowest rates of heart disease in the world. The consumption of animal fat is as high as it is in the USA where the rate of heart disease is nearly three times greater. This suggests that saturated fat and cholesterol intake alone are not important. The MONICA project found that vitamin E, which is an antioxidant and protects artery walls against atherosclerosis, was much higher in the blood in people from countries with low rates of heart disease.

For a heart attack to occur, blood must clot in the coronary artery – thrombosis must occur. There appears to be no link between fat intake and the risk of thrombosis. Other blood factors, which may or may not be diet-related, are now considered to be important in increasing the risk of thrombosis and therefore deserve as much attention as fat in the diet.

### Prevention and cure of coronary heart disease

The governments of many developed countries have taken steps to reduce the incidence and prevalence of heart disease. They encourage people to reduce the risk of developing the disease by taking more exercise, giving up smoking and decreasing the intake of animal fat in their diet. Death rates in these countries have fallen over the past 25 years, but whether this is as a result of these changes is uncertain. The USA and Australia in particular have seen significant reductions in death rates.

The death rate from coronary heart disease in the UK is one of the highest in the world. Coronary heart disease accounts for a third of all deaths of people between the ages of 45 and 64. Although the death rate is decreasing in the UK, partly as a result of better screening and treatment, it is still higher than in most other countries.

Reducing the incidence of coronary heart disease was the first target listed in the UK Government's long-term health strategy, *Health of the Nation*, published in 1992. The reasons for this are that coronary heart disease is:

- the major cause of premature death in the UK;
- one of the main causes of avoidable ill health;
- a major cost to the National Health Service and to the community.

Treatment for coronary heart disease involves using drugs to lower blood pressure, decrease the risk of blood clotting, prevent abnormal heart rhythms, reduce the retention of fluids and decrease the cholesterol concentration in the blood. If these drug treatments are not successful then a coronary artery **by-pass** operation may be carried out. This involves using a blood vessel from the leg to replace the diseased vessel. The by-pass carries blood from the aorta to a place on the heart beyond the blockage in the coronary artery. Sometimes two or three by-passes are necessary. The number of by-pass operations carried out in the UK increased threefold during the 1980s. In 1997 there were over 24 000 operations, compared with only a few hundred transplants to treat coronary heart disease.

A complete heart transplant is the method of last resort. The costs of the operation are very high and there are difficulties in finding enough donor hearts. Before the operation can be carried out the tissues of the donor and the recipient must be matched so that the new heart is not rejected by the body's immune system. Often heart–lung transplants are more successful than heart-only transplants as the immune system seems to be 'overwhelmed' by the new organs and does not mount such a vigorous defence (chapter 14). Drugs are used to suppress the immune system after the transplant, but these often have unpleasant side-effects and may not prevent rejection. With limited resources, there is the problem of deciding who receives a transplant. In some cases doctors have refused surgery to people who have ignored advice to give up smoking.

Treating coronary heart disease is hugely expensive. There are two ways in which this cost may be reduced. Both involve primary health care, that is advising people to take precautions to avoid developing disease in the first place.

One way is to screen the population to find individuals who are at risk of developing coronary heart disease and target health care to them. The best methods are screening for high blood pressure, high blood cholesterol concentration and monitoring the behaviour of the heart during exercise. Screening can be done by doctors on a regular basis and can be followed with advice to those at risk to give up smoking, adopt a healthier diet and take more exercise. Studies show that methods of lowering blood pressure, such as reducing salt intake and using drugs, are effective. Attempts to change diet are not always so successful, but new drugs that lower blood cholesterol are proving beneficial and are prescribed to those at risk of having a heart attack. It is therefore probably effective to screen the population for the risk factors outlined in *table 12.2* and concentrate on those at highest risk of dying prematurely of heart disease.

The second way is to encourage the population as a whole to adopt a healthy lifestyle to reduce the risks of developing heart disease. Advertising and health education may be able to play a large part in reducing illness and deaths from coronary heart disease by encouraging a healthy way of life from an early age. Encouraging people to engage in different forms of aerobic exercise is seen as an important aspect of the 'population approach' as it often encourages people to change their diet, lose weight, stop smoking and reduce their alcohol intake in the process. Exercise is also one way of decreasing blood pressure. In 1997 the UK Government published its health strategy *Our Healthier Nation*. This includes the target of reducing deaths rates from cardiovascular diseases in people under 75 by at least 40% by 2010. If this reduction is achieved then the health of, and quality of life for, many people should improve. However, in 1996 only £11.6 million was spent on *prevention* of coronary heart disease, less than 1% of the total spending on *treatment* of the disease by the National Health Service. A major problem

with health campaigns is that many people resist the advice they are given until it is too late. As coronary heart disease is a long-term degenerative disease this advice should be provided early in life. Degenerative changes in arteries have been seen in children as young as seven.

The mortality rate from heart disease and strokes in the UK began to decrease in the 1970s. This may have something to do with changing ways of life, for example decreasing the intake of saturated fat and giving up smoking. It may also have much to do with better equipped casualty departments in hospitals and improvements in the treatment of heart attack patients. However, some studies suggest that it might have more to do with better maternal nutrition. These studies show that:

■ higher birth weight is associated with lower blood pressure in middle age;

■ high weight at one year of age is associated with a lower risk of diabetes and a low level in the blood of low density lipoproteins.

It appears from this epidemiological evidence that better maternal nutrition in the early and middle part of the twentieth century meant that later generations were better protected against heart disease. It may also explain why coronary heart disease is more common among poor people than among the affluent.

### SAQ 12.7

It is estimated that smoking is responsible for 90% of all deaths from lung cancer, 76% of deaths from chronic obstructive pulmonary disease and 16% of deaths from stroke and coronary heart disease. Use the data in *table 12.4* to calculate:

**a** the number of deaths in 1993 in (i) men and (ii) women that were attributable to smoking;

**b** the percentage of deaths of (i) men and (ii) women that were attributable to smoking.

| Disease | Men | | Women | | All |
|---|---|---|---|---|---|
| | Number of deaths | Percentage of total deaths | Number of deaths | Percentage of total deaths | |
| lung cancer | 24 963 | 7.87 | 12 757 | 3.75 | 37 720 |
| chronic obstructive pulmonary disease | 18 826 | 5.93 | 11 848 | 3.48 | 30 674 |
| stroke | 26 310 | 8.29 | 44 976 | 13.21 | 71 286 |
| coronary heart disease | 90 774 | 28.62 | 76 831 | 22.57 | 167 605 |
| all diseases | 317 194 | 100.00 | 340 373 | 100.00 | 657 567 |

● **Table 12.4** Deaths from lung cancer, chronic obstructive pulmonary disease, stroke and coronary heart disease for all ages in the UK in 1993.

# SUMMARY

◆ Damage to the bronchioles and alveoli occurs in chronic obstructive pulmonary diseases. In chronic bronchitis the airways are blocked by inflammation and infection; in emphysema the alveoli are destroyed, reducing the surface area for gaseous exchange.

◆ Tobacco smoke contains tar, carbon monoxide and nicotine.

◆ Tar contains carcinogens which cause changes in DNA in bronchial epithelial cells leading to the development of a bronchial carcinoma. This is lung cancer.

◆ Carbon monoxide combines irreversibly with haemoglobin, reducing the oxygen-carrying capacity of the blood.

◆ Nicotine stimulates the nervous system, increasing heart rate and blood pressure.

◆ Epidemiological and experimental evidence show a strong correlation between smoking and lung cancer; smoking damages the cardiovascular system, multiplying the risks of developing coronary heart disease.

◆ Diet is thought to be a contributory factor in the development of coronary heart disease. Saturated fat in the diet is linked with high blood cholesterol concentrations.

◆ Coronary heart disease is most prevalent in Eastern Europe and the UK and USA. France and Japan have a very low prevalence of the disease and death rates are low. The reasons for this global distribution are unclear, although high blood cholesterol concentration and high blood pressure are two risk factors that are implicated in countries where coronary heart disease is a major cause of illness and death.

◆ Coronary heart disease may be treated with drugs to lower blood cholesterol and blood pressure. Coronary artery by-pass surgery involves using a vein from the leg to replace the part or parts of a coronary artery that are damaged. A heart transplant may be necessary, but it is an expensive operation and difficult to find donor hearts so very few are performed.

◆ Primary health care can reduce mortality from coronary heart disease and strokes. Screening people for risk factors of coronary heart disease and stroke allows early intervention. Advertising and education can promote the benefits of exercise, not smoking, avoiding an excessive consumption of alcohol and eating a diet low in saturated fat. These alternatives to treatment and surgery may be more cost effective in the long-term, but they depend on people being willing and able to change their lifestyle.

## CHAPTER 13

# Infectious diseases

**By the end of this chapter you should be able to:**

1 explain what is meant by an infectious disease;

2 describe the causes of cholera, malaria, AIDS and TB;

3 explain how these diseases are transmitted and assess the importance of these diseases worldwide;

4 discuss the roles of social, economic and biological factors in the prevention and control of these diseases;

5 outline the role of antibiotics in the treatment of infectious disease.

Infectious diseases are transmissible, or communicable, diseases. This means that these diseases are caused by pathogens that can spread from infected people to uninfected people. Some diseases can only spread from one person to another by direct contact, as the pathogen cannot survive outside the human body. Others can survive in water, human food, faeces or animals (including insects) and so are transmitted indirectly from person to person. Some people may spread a pathogen even though they do not have the disease themselves. Such people are symptomless **carriers** and it can be very difficult to trace them as the source of an infection.

The way in which a pathogen passes from one host to another is called the **transmission cycle**. Control methods for disease attempt to break transmission cycles by removing the conditions that favour the spread of the pathogen. This is only possible once the cause of the disease and its method of transmission are known and understood.

## Worldwide importance of infectious diseases

The four diseases described in the following pages are of current concern as they have increased in prevalence in recent years. They are all of worldwide importance, posing serious public health problems now and for the foreseeable future.

A new strain of **cholera** appeared in 1992 to begin the eighth pandemic of the disease. **Malaria** has been on the increase since the 1970s and constitutes a serious risk to health in many tropical countries. **AIDS** was officially recognised in 1981, but the infective agent (**HIV**) was in human populations for many years before it was identified. The spread of HIV infection since the early 1980s has been exponential. **Tuberculosis (TB)**, once thought to be nearly eradicated, has shown a resurgence and poses a considerable health risk in both developed and developing countries.

These diseases know no boundaries. As a result of easily accessible international travel it is possible for a person to become infected one day and be half way round the world the next. Tourists, business travellers, migrants, refugees and asylum seekers all represent a challenge to health authorities. There are few effective barriers to disease transmission and the means of treating and curing diseases are rapidly becoming obsolete. For example, since 1940 much of the success in treating bacterial infectious diseases has been due to antibiotics. Now, due to their overuse, some bacteria are resistant to these antibiotics.

# Cholera

The features of cholera are given in *table 13.1*. It is caused by the bacterium *Vibrio cholerae* (*figure 13.1*). As the disease is water-borne, it occurs where people do not have access to proper sanitation, a clean water supply or uncontaminated food. Infected people, three-quarters of whom may be symptomless carriers, pass out large numbers of bacteria in their faeces. If these contaminate the water supply, or if infected people handle food or cooking utensils without washing their hands, then bacteria are transmitted to uninfected people.

To reach their site of action, the small intestine, bacteria have to pass through the stomach. If the contents are sufficiently acidic (< pH4.5) the bacteria are unlikely to survive. If bacteria do reach the small intestine they multiply and secrete a toxin, **choleragen**, which disrupts the functions of the epithelium so that salts and water leave the blood causing severe diarrhoea. This loss of fluid can be fatal if not treated within 24 hours.

Almost all people with cholera who are treated make a quick recovery. A death from cholera is an avoidable death. The disease can be controlled by giving a solution of salts and glucose intra-venously to rehydrate the body (*figure 13.2*). If people can drink, they are given **oral rehydration therapy**. Glucose is effective because it is absorbed into the blood and takes salts (e.g. $Na^+$ and $K^+$) with it. It is important to make sure that a sufferer's fluid intake equals fluid losses in urine and faeces and to maintain the osmotic balance of the blood and tissue fluids (chapter 4).

In developing countries, especially those with large cities which have grown considerably in recent years but as yet have no sewage treatment or clean water, there exist perfect conditions for the spread of the disease. Increasing quantities of

| Pathogen | *Vibrio cholerae* |
|---|---|
| Methods of transmission | food-borne, water-borne |
| Global distribution | Asia, Africa, Latin America |
| Incubation period | 1–5 days |
| Site of action of pathogen | wall of small intestine |
| Clinical features | severe diarrhoea ('rice water'), loss of water and salts, dehydration, weakness |
| Method of diagnosis | microscopical analysis of faeces |
| Annual incidence worldwide | 5.5 million |
| Annual mortality worldwide | 120 000 |

● **Table 13.1** The features of cholera.

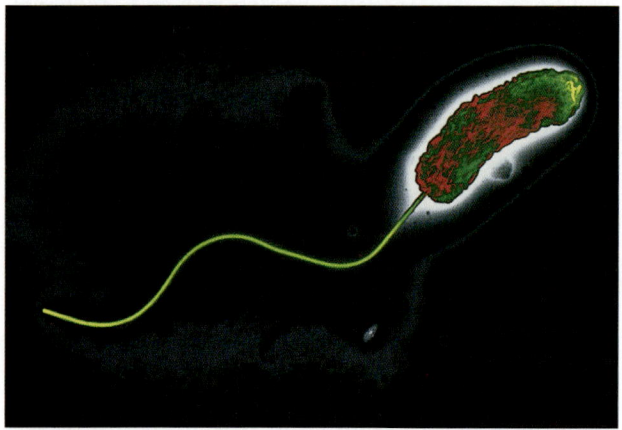

● **Figure 13.1** An electron micrograph of *Vibrio cholerae*. The faeces of a cholera victim are full of these bacteria with their distinctive flagella (× 13 400).

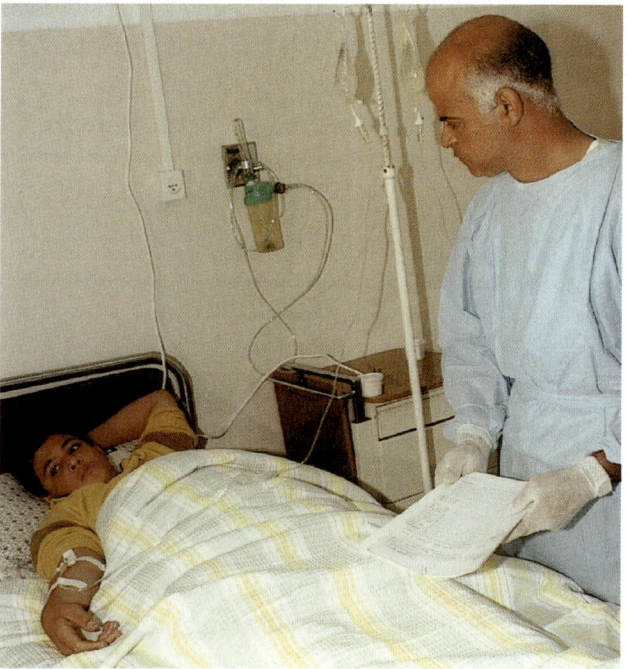

● **Figure 13.2** A Palestinian boy being given intravenous rehydration therapy for cholera in a Gaza Strip hospital, the Middle East, 1994. The drip contains a solution of salts to replace those lost through severe diarrhoea. Cholera causes many deaths when normal life is disrupted by war and other catastrophes.

untreated faeces from a growing population favour cholera's survival. Many countries, saddled with huge debts, do not have the financial resources to tackle large municipal projects such as providing drainage and a clean water supply to large areas of substandard housing. In many countries raw human sewage is used to irrigate vegetables. (This was the source of a cholera outbreak in Santiago, Chile in 1991.) Inadequate cooking or washing in contaminated water are other common causes. Areas of the world where cholera is endemic are West and East Africa and Afghanistan. In Kabul, the capital of Afghanistan, there were 10 000 cases in 1998. A similar situation was seen in large British cities in the nineteenth century until the water-borne nature of the disease was understood. Cholera is now almost unknown in the developed world as a result of sewage treatment and the provision of clean piped water, which is chlorinated to kill bacteria. The transmission cycle has been broken.

Travellers from areas free of cholera to those where cholera is endemic used to be advised to be vaccinated, although the vaccine only provides short-term protection. This recommendation has now largely been dropped. The reasons for this are explained on page 191.

There are 60 different strains of *V. cholerae*. Until the 1990s only the strain known as 01 caused cholera. Between 1817 and 1923 there were six pandemics of cholera. Each originated in what is now Bangladesh and they were caused by the 'classical' strain of cholera 01. A seventh pandemic began in 1961 when a variety of 01, named 'El Tor', originated in Indonesia. El Tor soon spread to India, then to Italy in 1973, reaching South America in January 1991 where it caused an epidemic in Peru. The discharge of a ship's sewage into the sea may have been responsible. Within days of the start of the epidemic the disease had spread 2000 km along the coast and within four weeks had moved inland. In February and March of that year, an average of 2550 cases a day were being reported. People in neighbouring countries were soon infected. In Peru many sewers discharge straight onto shellfish beds. Seafood, especially filter-feeders such as oysters and mussels, become contaminated because they concentrate cholera bacteria when sewage is

pumped into the sea. Fish and shellfish are often eaten raw. As the epidemic developed so rapidly in Peru, the disease probably spread through contaminated seafood.

A new strain, known as *V. cholerae* 0139, originated in Madras in October 1992 and has spread to other parts of India and Bangladesh. This strain threatens to be responsible for an eighth pandemic. It took El Tor 2 years to displace the 'classical' strain in India; 0139 replaced El Tor in less than two months suggesting that it may be more virulent. Many adult cases have been reported, and this may be because previous exposure to El Tor has not given them immunity to 0139.

**SAQ 13.1**
List the ways in which cholera is transmitted from person to person.

**SAQ 13.2**
One person can excrete $10^{13}$ cholera bacteria a day. An infective dose is $10^6$. How many people could one person infect in one day?

**SAQ 13.3**
Explain why there is such a high risk of cholera in refugee camps.

**SAQ 13.4**
Describe the precautions that a visitor to a country where cholera is endemic can take to avoid catching the disease.

# Malaria

The features of this disease are summarised in *table 13.2*. Malaria is caused by one of four species of the protoctist *Plasmodium*, whose life cycle is shown in *figure 13.3*.

Female *Anopheles* mosquitoes feed on human blood to obtain the protein they need to develop their eggs. If the person they bite is infected with *Plasmodium*, they will take up some of the pathogen's gametes with the blood meal. These gametes fuse and develop in the mosquito's gut to

| Pathogen | *Plasmodium falciparum*, *P. vivax*, *P. ovale*, *P. malariae* |
|---|---|
| Method of transmission | insect vector: female *Anopheles* mosquito |
| Global distribution | throughout the tropics and sub-tropics (endemic in 91 countries) |
| Incubation period | from a week to a year |
| Site of action of pathogen | liver, red blood cells, brain |
| Clinical features | fever, anaemia, nausea, headaches, muscle pain, shivering, sweating, enlarged spleen |
| Method of diagnosis | microscopical examination of blood |
| Annual incidence worldwide | 300 million (90% of cases are in Africa) |
| Annual mortality worldwide | 1.5–2.7 million; in tropical Africa malaria kills 1 million children under the age of 5 |

● **Table 13.2** The features of malaria.

form infective stages, which move to the mosquito's salivary glands. When the mosquito feeds again, it injects an anticoagulant that prevents the blood meal from clotting so that it flows out of the host into the mosquito. The infective stages pass out into the blood together with the anticoagulant in the saliva, and the parasites enter the red blood cells, where they multiply (*figure 13.4*). The female *Anopheles* mosquito is therefore a **vector** of malaria and she transmits the disease when she passes the infective stages into an uninfected person. Malaria may also be transmitted during blood transfusion and when unsterile needles are reused. *Plasmodium* can also pass across the placenta from mother to fetus.

*Plasmodium* multiplies in both hosts, the human and the mosquito; at each stage there is a huge increase in the number of parasites and this improves the chances of infecting another mosquito or human host.

If people are continually reinfected they become **immune** to malaria (chapter 14). However, this

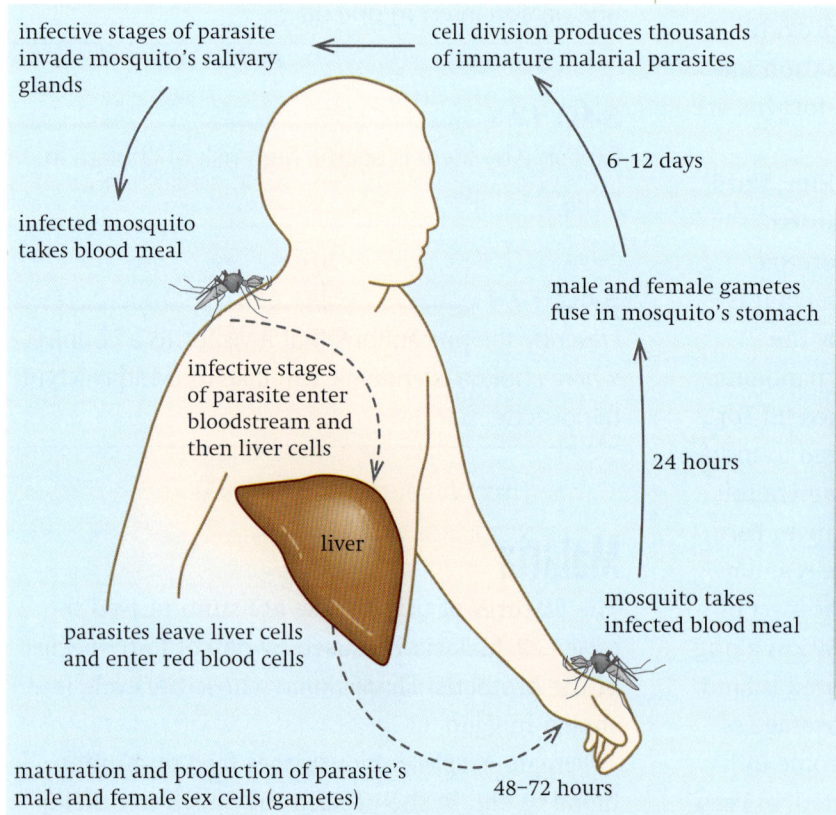

● **Figure 13.3** The life cycle of *Plasmodium*. The parasite has two hosts: the sexual stage occurs in mosquitoes, the asexual stage in humans. The time between infection and appearance of parasites inside red blood cells is 7–30 days in *P. falciparum*; longer in other species.

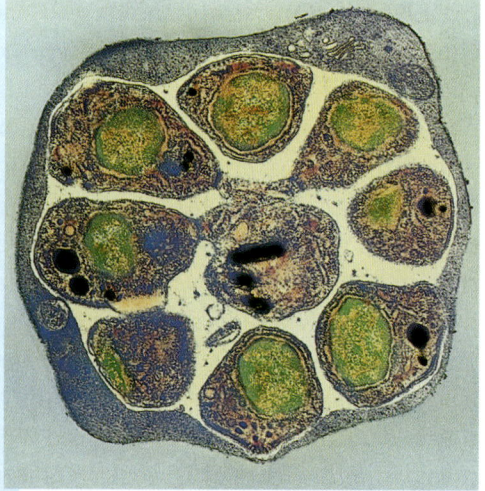

● **Figure 13.4** A transmission electron micrograph of a section through a red blood cell packed tightly with malarial parasites. *Plasmodium* multiplies inside red blood cells; this cell will soon burst, releasing parasites which will infect other red blood cells.

only happens if they survive the first five years of life when mortality from malaria is very high. The immunity only lasts as long as they are in contact with the disease. This explains why epidemics in places where malaria is not endemic can be very serious, and why it is more dangerous in those areas where it only occurs during and after the rainy season. This often coincides with the time of maximum agricultural activity so the disease has a disastrous effect on the economy: people cannot cultivate the land when they are sick.

There are three main ways to control malaria:
- reduce the number of mosquitoes;
- avoid being bitten by mosquitoes;
- use drugs to prevent the parasite infecting people.

It is possible to kill the insect vector and break the transmission cycle. Mosquitoes lay their eggs in water. Larvae hatch and develop in water but breathe air by coming to the surface. Oil can be spread over the surfaces of water to make it impossible for mosquito larvae and pupae to breathe. Marshes can be drained and vegetation cleared. Two biological control measures that can be used are:
- stocking ponds, irrigation and drainage ditches and other permanent bodies of water with fish which feed on mosquito larvae;
- spraying a preparation containing the bacterium *Bacillus thuringiensis*, which kills mosquito larvae, but is not toxic to other forms of life.

Mosquitoes will lay their eggs in small puddles or pools and this makes it impossible to eradicate breeding sites, especially in the rainy season.

The best protection against malaria is to avoid being bitten. People are advised to sleep beneath mosquito nets and use insect repellents. Soaking mosquito nets in insecticide every six months has been shown to reduce mortality from malaria. People should not expose their skin when mosquitoes are active at dusk. Villagers in New Guinea recommend sleeping with a dog or a pig. They say that mosquitoes much prefer animal blood to human blood.

Anti-malarial drugs such as quinine and chloroquine are used to treat infected people. They are also used as **prophylactic** (preventative) drugs, stopping an infection occurring if a person is bitten by an infected mosquito. They are taken before, during and after visiting an area where malaria is endemic. Chloroquine inhibits protein synthesis and prevents the parasite spreading within the body. Another prophylactic, proguanil, has the added advantage of inhibiting the sexual reproduction of *Plasmodium* inside the biting mosquito. Where anti-malarial drugs have been used widely there are strains of drug-resistant *Plasmodium*. Chloroquine resistance is widespread in parts of South America, Africa and New Guinea. Newer drugs, such as mefloquine, are used in these areas. However, mefloquine is expensive and sometimes causes unpleasant side effects such as restlessness, dizziness, vomiting and disturbed sleep.

People from non-malarial countries visiting many parts of the tropics are at great risk of contracting malaria. For example, there were between 1500 and 2300 cases of malaria a year in the UK between 1984 and 1993. Doctors in developed countries, who see very few cases of malaria, often misdiagnose it as influenza since the initial symptoms are similar. Many of these cases are settled immigrants who have been visiting relatives in Africa or India. They do not take prophylactic drugs because they do not realise that they have lost their immunity.

In the 1950s the World Health Organisation coordinated a worldwide eradication programme. Although malaria was cleared from some countries, it was not generally successful. There were two main reasons for this:
- *Plasmodium* became resistant to the drugs used to control it.
- Mosquitoes became resistant to DDT and the other insecticides that were used at the time, such as dieldrin.

This programme was also hugely expensive and often unpopular. People living in areas where malaria was temporarily eradicated by the programme lost their immunity and suffered considerably, even dying, when the disease returned. Some villagers in South-East Asia lost the roofs of their houses because dieldrin killed a parasitic wasp that controlled the numbers of thatch-eating caterpillars. Some spray teams were set upon and killed by angry villagers in New Guinea. The programme could have been more successful if it had been tackled more sensitively, with more involvement of the indigenous people. In the 1970s, war and civil unrest destroyed much of the infrastructure throughout Africa and South-East Asia, making it impossible for mosquito control teams to work effectively.

The reasons for the worldwide concern over the spread of malaria are:

- an increase in drug-resistant forms of *Plasmodium*;
- an increase in the proportion of cases caused by *P. falciparum*, the form that causes severe, often fatal malaria;
- difficulties in developing a vaccine;
- an increase in the number of epidemics because of climatic and environmental changes that favour the spread of mosquitoes;
- the migration of people as a result of civil unrest and war.

Malaria is still one of the world's biggest threats to health. 40% of the world's population live in areas where there is a risk of malaria. In recent years there has been a resurgence of the disease in Africa. Control methods now concentrate on working within the health systems to improve diagnosis, improve the supply of effective drugs and promote appropriate methods to prevent transmission. These methods have proved successful in South

America where the death rate from malaria decreased by 60% between 1994 and 1997 in some countries. Several recent advances give hope that malaria may one day be controlled. The introduction of simple 'dipstick'-type tests for diagnosing malaria means that diagnosis can be done quickly without the need for laboratories. The whole genome of *Plasmodium* has been sequenced and this may lead to the development of effective vaccines. Several vaccines are being trialled, but it is not likely that a successful vaccine will be available for some time. Several drugs are being used in combination to reduce the chances of drug resistance arising.

*SAQ 13.5*

Describe how malaria is transmitted.

*SAQ 13.6*

Describe the factors that make malaria a difficult disease to control.

*SAQ 13.7*

Describe the precautions that people can take to avoid catching malaria.

| Pathogen | Human Immunodeficiency Virus |
|---|---|
| **Methods of transmission** | in semen and vaginal fluids during sexual intercourse, infected blood or blood products, contaminated hypodermic syringes, mother to fetus across placenta, mother to infant in breast milk |
| **Global distribution** | worldwide, especially in sub-Saharan Africa and South-East Asia |
| **Incubation period** | initial incubation a few weeks, but up to ten years or more before symptoms of AIDS may develop |
| **Site of action of pathogen** | T helper lymphocytes, macrophages, brain cells |
| **Clinical features** | HIV infection – flu-like symptoms and then symptomless; AIDS – opportunistic infections including pneumonia, TB, and cancers; weight loss, diarrhoea, fever, sweating, dementia |
| **Method of diagnosis** | blood test for antibodies to HIV |
| **Estimated total number of people infected with HIV worldwide in 2002** | 42 million |
| **Estimated number of new cases of HIV infection worldwide in 2002** | 5 million |
| **Estimated number of deaths from AIDS-related diseases worldwide in 2002** | 3.1 million (one third due to TB) |

● **Table 13.3** The features of HIV/AIDS.

# Acquired Immune Deficiency Syndrome (AIDS)

Features of AIDS and HIV are listed in *table 13.3*. AIDS is caused by the human immunodeficiency virus (HIV) (*figure 13.5*). The virus infects and destroys cells of the body's immune system (*figure 13.6*) so that their numbers gradually decrease. These cells, known as **T helper lymphocytes** (see page 187), control the **immune system**'s response to infection. When the numbers are low, the body is unable to defend itself against infection so allowing a range of pathogens to cause a variety of **opportunistic infections**. AIDS is not a disease. It is a collection of these rare opportunistic diseases associated with immunodeficiency caused by HIV infection. Since HIV is an infective agent, AIDS is called an 'acquired' immunodeficiency to distinguish it from other types, for example an inherited form.

After initial uncertainties in the early 1980s surrounding the emergence of an apparently new disease, it soon became clear that an epidemic

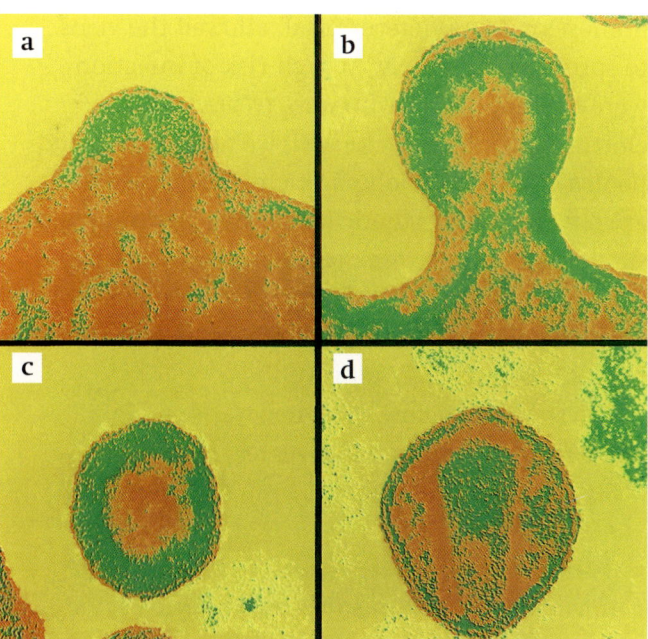

● **Figure 13.6** A series of transmission electron micrographs showing HIV budding from the surface of an infected lymphocyte, becoming surrounded by a membrane derived from the plasma membrane of the host cell. **a** The viral particle first appears as a bump, **b** which then buds out and **c** is eventually cut off. **d** The outer shell of dense material and less dense core are visible in the released virus. (× 176 000)

and then a pandemic was underway. The World Health Organisation estimated that 47 million people had been infected with HIV by 1998 and that 14 million of them had died.

HIV is a virus that is spread by intimate human contact: there is no vector (unlike in malaria) and the virus is unable to survive outside the human body (unlike cholera or malaria pathogens). Transmission is only possible by direct exchange of body fluids. In practice this means that HIV can be spread most easily through sexual intercourse, blood donation, the sharing of intravenous needles and across the placenta from mother to fetus.

The initial epidemic in North America and Europe was amongst those male homosexuals who practised anal intercourse and had many sex partners, two forms of behaviour that put them at risk. The mucous lining of the rectum is not as thick as that of the vagina and there is less natural lubrication. As a result it is easily damaged during intercourse and the virus can pass from semen to blood. Multiple partners, both

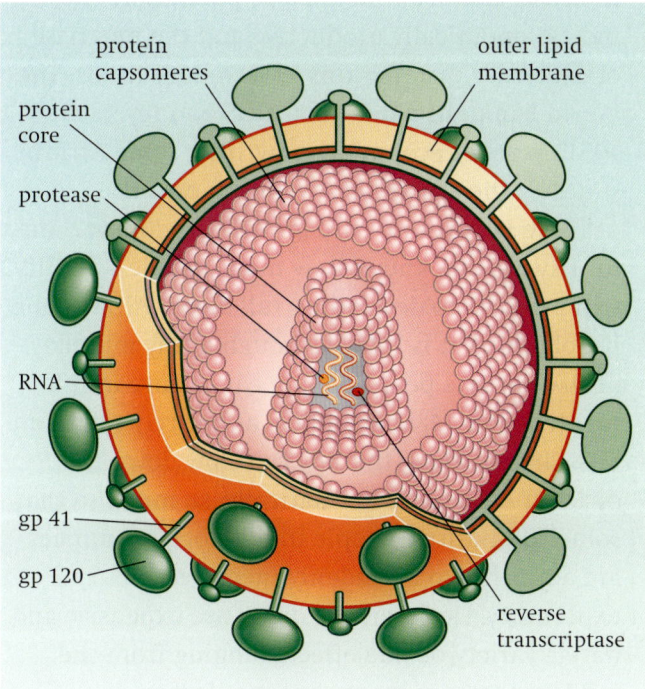

● **Figure 13.5** Human immunodeficiency virus (HIV). The outer envelope contains two glycoproteins gp120 and gp 41. The protein core contains genetic material (RNA) and two enzymes, a protease and reverse transcriptase. Reverse transcriptase uses the RNA as a template to produce DNA (page 237) once the virus is inside a host cell.

homosexual and heterosexual, allowed the virus to spread more widely. At high risk of infection were haemophiliacs who were treated with a clotting substance (factor VIII) isolated from blood pooled from many donors. Such blood products are now largely synthetic (page 239). The transmission of HIV by heterosexual intercourse is rising worldwide. This is particulary rapid in some African states where equal numbers of males and females are now 'HIV positive'.

The statistics below show how serious the pandemic is in sub-Saharan Africa.

- 80% of the world's deaths from AIDS occur in Africa.
- 34 million people are estimated to have been infected with HIV in sub-Saharan Africa since the start of the pandemic and 11.4 million are estimated to have died.
- One-quarter of the population of Zimbabwe is infected with HIV.
- Between 20% and 25% of people aged between 15 and 49 in Botswana and Zimbabwe are infected with HIV.
- 5.9 million children are estimated to have been orphaned by AIDS; in some places this is 25% of the population under 15.
- The prevalence of HIV among women attending antenatal clincs in Zimbabwe was between 20% and 50% in 1997.
- A large proportion of women in Rwanda are HIV positive following the use of rape as a genocidal weapon in the civil war of the early 1990s.
- The average life expectancy in South Africa dropped from 65 to 55 during 1995–1999.

HIV is a slow virus and after infection there may not be any symptoms until years later. Some people who have the virus even appear not to develop any initial symptoms, although there are often flu-like symptoms for several weeks after becoming infected. At this stage a person is HIV positive but does not have AIDS. The infections that can opportunistically develop to create AIDS tend to be characteristic of the condition. Two of these are caused by fungi: oral thrush caused by *Candida albicans*, and a rare form of pneumonia caused by *Pneumocystis carinii*. During the early years of the AIDS epidemic people in developed countries died within twelve hours of contracting this unusual pneumonia. Now this condition is managed much better and drugs are prescribed to prevent the disease developing. As and when the immune system collapses further, it becomes less effective in finding and destroying cancers. A rare form of skin cancer, Kaposi's sarcoma caused by a herpes-like virus, is associated with AIDS. Kaposi's sarcoma and cancers of internal organs are now the most likely causes of death of people with AIDS in developed countries, along with degenerative diseases of the brain, such as dementias.

At about the same time that AIDS was first reported on the west coast of the USA and in Europe, doctors in Central Africa reported seeing people with similar opportunistic infections. We have seen that HIV/AIDS is now widespread throughout sub-Saharan Africa from Uganda to South Africa. It is a serious public health problem here because HIV infection makes people more vulnerable to existing diseases such as malnutrition, TB and malaria. AIDS is having an adverse effect on the economic development of countries in the region as it affects sexually active people in their 20s and 30s who are also potentially the most economically productive and the purchase of expensive drugs drains government funds. The World Bank estimated that AIDS had reversed 10–15 years of economic growth for some African states by the end of the twentieth century.

There is as yet no cure for AIDS and no vaccine for HIV. No-one knows how many people with HIV will progress to developing full-blown AIDS. Some people think it is 100% although a tiny minority of HIV positive people do appear to have immunity (chapter 14) and can live as entirely symptomless carriers. Drug therapy can slow down the onset of AIDS quite dramatically, so much so that some HIV positive people in developed countries are adjusting to a suddenly increased life expectancy. However, the drugs are expensive and have a variety of side effects ranging from the mild and temporary (rashes, headaches, diarrhoea) to the severe and permanent (nerve damage, abnormal fat distribution). If used in combination, two or more drugs which prevent the replication of the virus inside host cells can prolong life, but they do not offer a cure. The drugs are similar to DNA nucleotides (e.g. zidovudine is similar to the nucleotide that contains the base thymine). Zidovudine binds to the viral enzyme reverse transcriptase and blocks its action. This stops the replication of the viral genetic material and leads to an increase in some of the body's lymphocytes. A course of combination therapy can be very complicated to follow. The pattern and timing of medication through the day must be strictly followed. People who are unable to keep to such a regimen can become susceptible to strains of HIV that have developed resistance to the drugs.

The spread of AIDS is difficult to control. The virus's long latent stage means it can be transmitted by people who are HIV positive but who show no symptoms of AIDS and do not know they are infected. The virus changes its surface proteins, which makes it hard for the body's immune system to recognise it (see chapter 14). This also makes the development of a vaccine very difficult.

For the present, public health measures are the only way to stop the spread of HIV. People can be educated about the spread of the infection and encouraged to change their behaviour so as to protect themselves and others. Condoms, femidoms and dental dams are the only effective methods of reducing the risk of infection during intercourse as they form a barrier between body fluids, reducing the chances of transmission of the virus. Uganda and the Philippines are unique in the developing world in their concerted health programmes to promote the use of condoms. As a result, infection rates in these countries have slowed.

## SAQ 13.8

Suggest why the true total of AIDS cases worldwide may be much higher than reported.

## SAQ 13.9

Suggest why condoms are not fully effective at preventing HIV infection.

## SAQ 13.10

Suggest the types of advice which might be offered as part of an AIDS education programme.

**Contact tracing** is an important part of HIV control in the UK. If a person who is diagnosed as HIV positive is willing and able to identify the people who they have put at risk of infection by sexual intercourse or needle sharing, then these people will be offered an HIV test. This test identifies the presence of antibodies to HIV, though these only appear several weeks after the initial infection.

Injecting drug users are advised to give up their habit, stop sharing needles or take their drug in some other way. Needle exchange schemes operate in some places to exchange used needles for sterile ones to reduce the chances of infection with HIV and other blood-borne diseases.

In developed countries, blood collected from blood donors is routinely screened for HIV and heat-treated to kill any viruses. People who think they may have been exposed to the virus are strongly discouraged from donating blood. Both methods are expensive and unlikely to be implemented throughout the developing world for some time. In these countries, people about to have an operation are recommended to donate their own blood before surgery to reduce the risk of infection.

## SAQ 13.11

Children in Africa with sickle cell anaemia or malaria often receive blood transfusions. Explain how this puts them at risk of HIV infection.

Widespread testing of a population to find people who are HIV positive is not expensive, but governments are reluctant to introduce such testing because of the infringement of personal freedom. In the developed world, HIV testing is promoted most strongly to people in high risk groups, such as male homosexuals, prostitutes, injecting drug users and their sexual partners. If tested positive they can be given the medical and psychological support they need. Despite this, people in Britain tend to have an HIV test at a far later stage than any other Europeans. In Africa and South-East Asia the epidemic is not restricted to such easily identifiable groups and widespread testing is not feasible due to the expense of reaching the majority of the population and the difficulty of organising it. People in these regions find out that they are HIV positive when they develop the symptoms of AIDS.

### SAQ 13.12
Explain why the early knowledge of HIV infection is important in transmission control.

HIV positive women in developed countries are advised not to breast feed their children because of the risk of transmitting the virus to their child. Both viral particles and infected lymphocytes are found in breast milk. In developing countries the benefits of breast feeding, such as the protection this gives against other diseases and the lack of clean water to make up supplements, outweigh the risks of transmitting HIV.

# Tuberculosis (TB)

*Table 13.4* gives the main features of this disease. TB is caused by two bacteria, *Mycobacterium tuberculosis* (figure 13.7) and *M. bovis*. These are pathogens that live inside human cells, particularly in the lungs. This is the first site of infection, but the bacteria can spread throughout the whole body and even infect the bone tissue. Some people become infected and develop TB quite quickly, whilst in others the bacteria remain inactive for many years. It is estimated that 30% of the world's population is infected with TB without showing any symptoms of the infection; people with this inactive infection do not spread the disease to others. However, the bacteria can later become active and this is most likely to happen when people are weakened by other diseases, suffer from malnutrition or become infected with HIV. Those who have the active form of TB often suffer from debilitating illness for a long time. They have a persistent cough and, as part of their defence, cells release hormone-like compounds which cause fever and suppress the appetite. As a result sufferers lose weight and often look emaciated (figure 13.8).

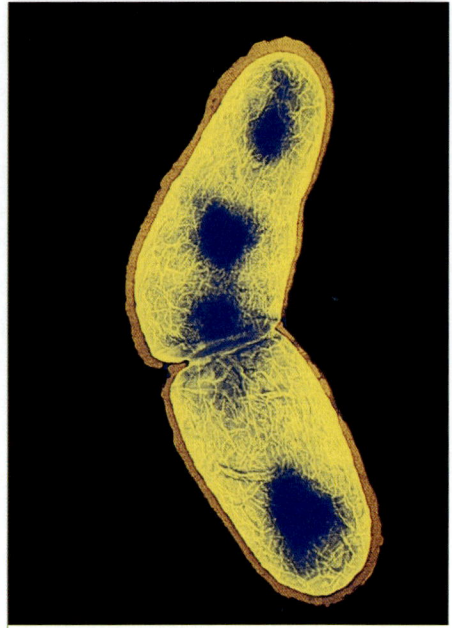

● **Figure 13.7** False-colour transmission electron micrograph of *Mycobacterium tuberculosis* dividing into two. It may multiply like this inside the lungs and then spread throughout the body or lie dormant, becoming active many years later.

TB is often the first opportunistic infection to strike HIV-positive people. HIV infection may reactivate dormant infections of *M. tuberculosis* which may have been present from childhood or, if people are uninfected, make them susceptible to infection. TB is now the

| | |
|---|---|
| **Pathogen** | *Mycobacterium tuberculosis*; *M. bovis* |
| **Methods of transmission** | airborne droplets; via unpasteurised milk |
| **Global distribution** | worldwide |
| **Incubation period** | few weeks or months |
| **Site of action of pathogen** | primary infection in lungs; secondary infections in lymph nodes, bones and gut |
| **Clinical features** | racking cough, coughing blood, chest pain, shortness of breath, fever, sweating, weight loss |
| **Methods of diagnosis** | microscopical examination of sputum for bacteria, chest X-ray |
| **Annual incidence worldwide in 1998** | 8 million (more than 6000 cases in UK) |
| **Annual mortality worldwide in 1998** | 2 million |

● **Table 13.4** The features of tuberculosis.

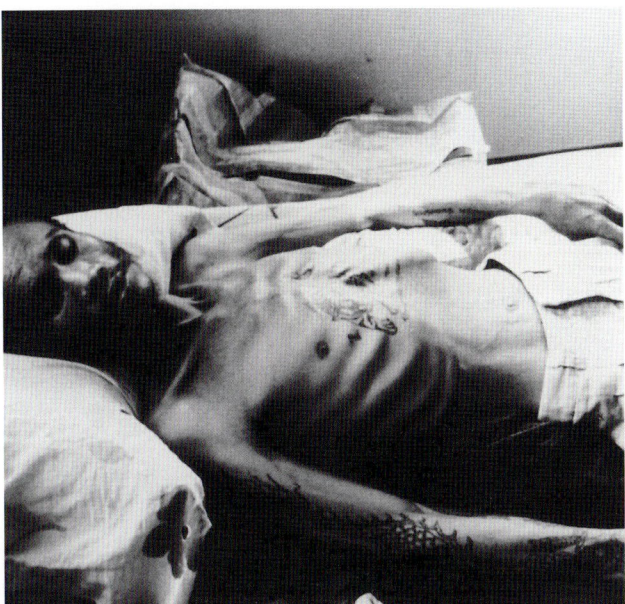

● **Figure 13.8** A TB patient in Thailand. Health workers or members of his family may supervise his drug treatment which can last for up to a year.

leading cause of death of HIV-positive people. The HIV pandemic has been followed very closely by a TB pandemic.

TB is spread when infected people with the active form of the illness cough or sneeze and the bacteria are carried in the air in tiny droplets of liquid. Transmission occurs when people who are uninfected inhale the droplets. TB spreads most rapidly among people living in overcrowded conditions. People who sleep close together in large numbers are particularly at risk. The disease primarily attacks the homeless and people who live in poor, substandard housing; those with low immunity, because of malnutrition or being HIV positive, are also particularly vulnerable.

The form of TB caused by *M. bovis* also occurs in cattle and is spread to humans in meat and milk. It is estimated that there were about 800 000 deaths in the UK between 1850 and 1950 as a result of TB transmitted from cattle. Very few now acquire TB in this way in developed countries for reasons explained later, although this still remains a source of infection in some developing countries.

The incidence of TB in the UK decreased steeply well before the introduction of a vaccine in the 1950s, because of improvements in housing conditions and diet. The antibiotic streptomycin was introduced in the 1940s and this hastened the

decrease in the incidence of TB. This pattern was repeated throughout the developed world.

Once thought to be practically eradicated, TB is now showing a resurgence. There are high rates of incidence all across the developing world and in the countries of the former Soviet Union (*figure 13.9*). Very high rates are found in areas of destitution in inner cities such as New York. The incidence in such areas is as high as in developing countries. The resurgence is due in part to the following factors:

■ some strains of TB bacteria which are resistant to drugs;
■ the AIDS pandemic;
■ poor housing in inner cities in the developed world and rising homelessness;
■ the breakdown of TB control programmes, particularly in the USA; partial treatment for TB increases the chance of drug resistance in *Mycobacterium*;
■ migration from Eastern Europe and developing countries to large cities such as London and New York.

When a doctor first sees a person with the likely symptoms of TB, samples of the sputum (mucus and pus) from their lungs are collected for analysis. The identification of the tuberculosis bacteria can be done very quickly by microscopy. If TB is confirmed, then sufferers should be isolated while they are in the most infectious stage (which is at two to four weeks). This is particularly the case if they are infected with a drug-resistant strain of the bacterium. The treatment involves using several drugs to ensure that all the bacteria are killed. If not, drug-resistant forms remain to continue the infection. The treatment is a long one (nine months to one year) because it takes a long time to kill the bacteria as they are slow growing and are not very sensitive to the drugs used. Unfortunately, many people do not complete their course of drugs as they think that when they feel better they are cured. Anyone who does not complete their treatment may be harbouring drug-resistant bacteria and may spread these to others if the bacteria become active.

Strains of drug-resistant *M. tuberculosis* were identified when treatment with antibiotics began in the 1950s. Antibiotics act as selective agents

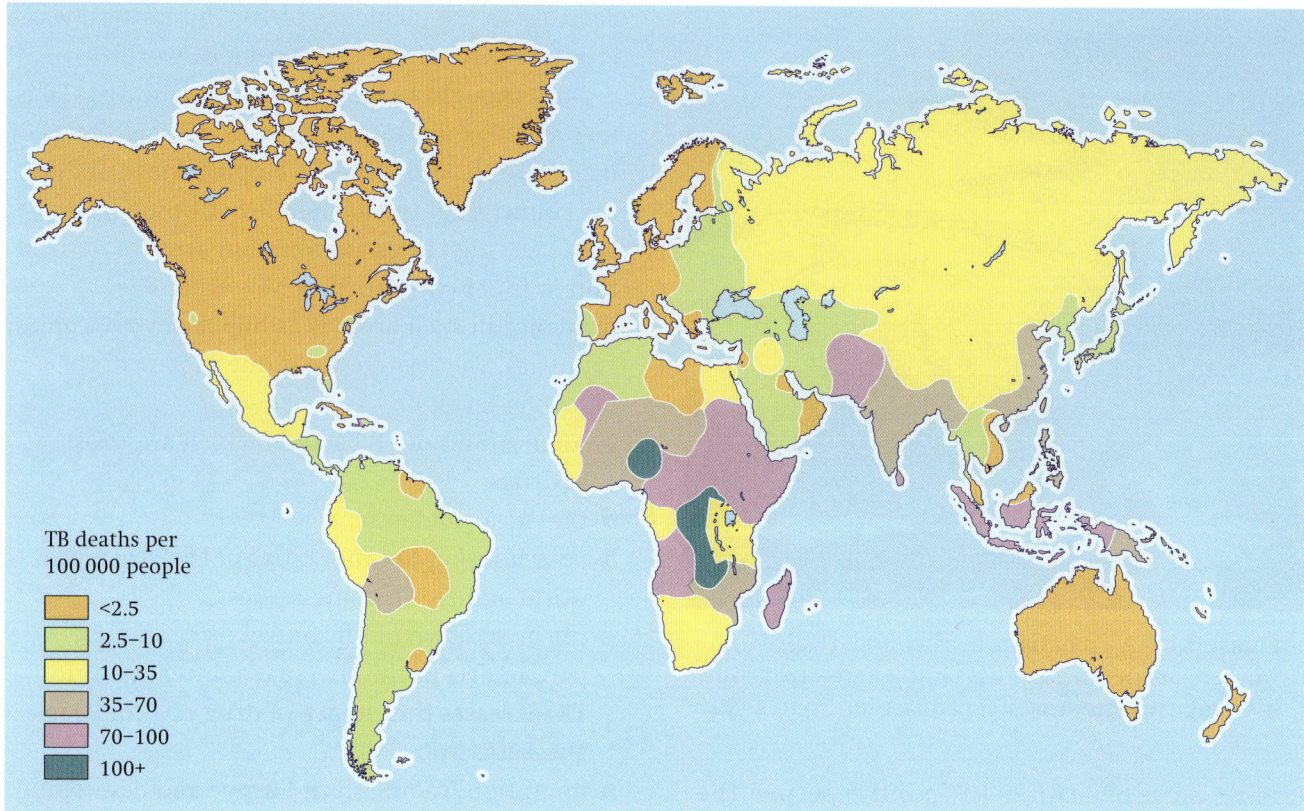

● **Figure 13.9** The global distribution of TB.

TB deaths per
100 000 people

- <2.5
- 2.5–10
- 10–35
- 35–70
- 70–100
- 100+

killing drug-sensitive strains and leaving resistant ones behind. Drug resistance occurs as a result of mutation in the bacterial DNA. Mutation is a random event and occurs with a frequency of about one in every thousand bacteria. If three drugs are used in treatment then the chance of resistance arising by mutation to all three of them is reduced to 1 in a thousand million. If four drugs are used the chance is reduced to 1 in a billion. If TB is not treated or the person stops the treatment before the bacteria are completely eliminated, they spread throughout the body, increasing the likelihood that mutations will arise as the bacteria survive for a long time and multiply. Stopping treatment early can mean that *M. tuberculosis* develops resistance to all the drugs being used. People who do not complete a course of treatment are highly likely to infect others with drug-resistant forms of TB. It is estimated that one person can easily infect ten to fifteen others, especially if they live in overcrowded conditions.

Multiple drug resistant forms of TB (MDR-TB) now exist. In 1995, an HIV unit in London reported an outbreak of MDR-TB with a form of

*M. tuberculosis* that was resistant to five of the major drugs used to treat the disease including isoniazid, which is the most successful drug.

The World Health Organisation now promotes a scheme to ensure that patients complete their course of drugs. DOTS (Direct Observation Treatment, Short Course), involves health workers, or responsible family members, making sure that patients take their medicine regularly for six to eight months. The drugs widely used are isoniazid and rifampicin, often in combination with others. This drug therapy cures 95% of all patients, is twice as effective as other strategies and is helping to reduce the spread of MDR strains.

Contact tracing (see page 173) and their subsequent testing for the bacterium is an essential part of controlling TB. Contacts are screened for symptoms of TB infection, but the diagnosis can take up to two weeks. The spread of the disease among children is prevented, to a large extent, by vaccination. In the UK, teenagers are routinely vaccinated at the age of 13 or 14. The BCG vaccine is derived from *M. bovis* and protects up to 70 to 80% of teenagers in the UK, its effectiveness decreasing with age unless there is exposure to TB.

Studies of the effectiveness of BCG in protecting adults and children give conflicting results. It also appears that the vaccine is effective in some parts of the world (e.g. UK), but less effective in others (e.g. India). Many of the world's victims were not vaccinated.

TB can be transmitted between humans and cattle. To prevent people catching TB in this way, cattle are routinely tested for TB and any found to be infected are destroyed. TB bacteria are killed when milk is pasteurised. These control methods are very effective and have reduced the incidence of human TB caused by *M. bovis* considerably so that it is virtually eliminated in countries where these controls operate. In 1995, there were just eleven cases of TB caused by *M. bovis* in the UK.

### SAQ 13.13

Describe the global distribution of TB and explain the reasons for the distribution you have described.

## Antibiotics

Antibiotics are drugs that are used to treat or cure infections. Effective antibiotics show **selective toxicity**, killing or disabling the pathogen but having no effect on host cells. There is a wide range of antibiotics to treat bacterial and fungal infections, but only very few for viral infections. Antibiotics are derived from living organisms, although other antimicrobial drugs, such as isoniazid used for the treatment of tuberculosis, are synthetic.

Antibiotics interfere with some aspect of growth or metabolism of the target organism such as:

- synthesis of bacterial walls (chapter 1);
- protein synthesis (transcription and translation, chapter 5);
- plasma membrane function (chapter 4);
- enzyme action (chapter 3).

The main sites of action of antibiotics are shown in *figure 13.10*.

Different diseases are treated with different antibiotics. Some kinds of bacteria are completely resistant to particular antibiotics (for example, all strains of *M. tuberculosis* are resistant to penicillins) whilst other bacteria have certain strains that are resistant. **Broad spectrum** antibiotics are effective against a wide range of bacteria, while **narrow spectrum** antibiotics are active only against a few.

One type of antibiotic, penicillins, function by preventing the synthesis of the cross-links between the peptidoglycan polymers in the cell walls of bacteria. This means that they are only active against bacteria and only while they are growing. Many types of bacteria have enzymes for destroying penicillins (penicillinases) and are therefore resistant to these antibiotics.

As a result of all these variables, antibiotics should be chosen carefully. Screening antibiotics against the strain of the bacterium or fungus isolated from sufferers ensures that the most effective antibiotic can be used in treatment. *Figure 13.11* shows the results of an antibiotic sensitivity test carried out on a pathogenic strain of the human gut bacterium *E. coli* (O157). Bacteria are collected from faeces, food or water, and grown on an agar medium. Various antibiotics are absorbed onto discs of filter paper and placed on

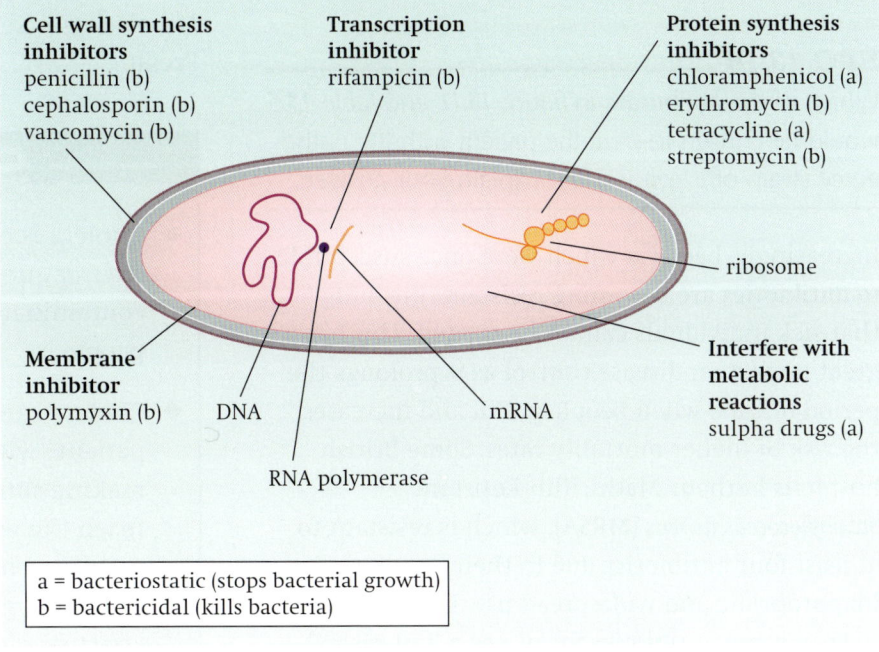

Cell wall synthesis inhibitors
penicillin (b)
cephalosporin (b)
vancomycin (b)

Transcription inhibitor
rifampicin (b)

Protein synthesis inhibitors
chloramphenicol (a)
erythromycin (b)
tetracycline (a)
streptomycin (b)

ribosome

Membrane inhibitor
polymyxin (b)

DNA

mRNA

Interfere with metabolic reactions
sulpha drugs (a)

RNA polymerase

a = bacteriostatic (stops bacterial growth)
b = bactericidal (kills bacteria)

● **Figure 13.10** The sites of action of antibiotics in bacteria.

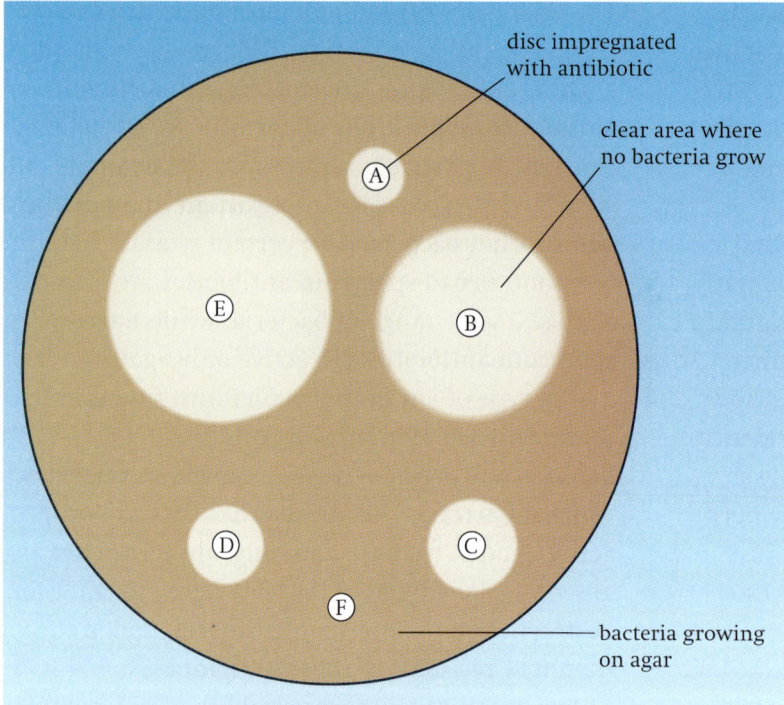

| Antibiotic | Inhibition zone diameter (mm) | |
| --- | --- | --- |
| | Resistant | Sensitive |
| A | ≤11 | ≥14 |
| B | ≤12 | ≥18 |
| C | ≤ 9 | ≥14 |
| D | ≤11 | ≥22 |
| E | ≤12 | ≥15 |
| F | ≤14 | ≥19 |

● **Table 13.5** Inhibition zone diameters for the antibiotics of *figure 13.11*.

If the diameter of the inhibition zone for an antibiotic is equal to or less than the figure given in the first column, the bacteria are resistant to it. If the diameter is equal to or greater than the figure in the right hand column, the bacteria are sensitive and the antibiotic may be chosen for treatment.

● **Figure 13.11** An antibiotic sensitivity test for a pathogenic strain of *Escherichia coli*. *Table 13.5* shows the inhibition zone diameters for the six antibiotics.

the agar plate. The plate is incubated and the diameters of the **inhibition zones** where no bacteria are growing are measured. The diameters are compared with a table similar to *table 13.5* and the most effective antibiotics chosen to treat infected people.

### SAQ 13.14
Which of the antibiotics in *figure 13.11* and *table 13.5* would be chosen to treat the patient with the pathogenic strain of *E. coli* (O157)? Explain your answer.

Increasingly, bacteria which were once susceptible to antibiotics are becoming resistant, meaning that sick individuals cannot be treated. This has a great impact on disease control as it prolongs the period of time when people are ill and increases the risk of higher mortality rates. Some British hospitals harbour Methicillin Resistant *Staphylococcus aureus* (**MRSA**), which is resistant to at least four antibiotics due to their previous inappropriate and widespread use. It is advisable to keep some antibiotics for use as a 'last resort', when everything else has failed, so as to lessen the

chances of more such resistant organisms. Meanwhile, drug companies are continuing to invest in research for new antibiotics, even though they too may quickly become redundant. You can find more information about antibiotics in two of the A Level option texts: *Applications of Genetics* and *Microbiology and Biotechnology* (Cambridge University Press).

## SUMMARY

◆ Cholera is caused by the bacterium *Vibrio cholerae* and is transmitted in water or food contaminated by the faeces of infected people.

◆ Cholera can be controlled by treating patients with oral rehydration therapy and making sure that human faeces do not reach the water supply. The disease is prevented by providing clean, chlorinated, water and good sanitation. There is no effective vaccine.

- Malaria is caused by four species of *Plasmodium*. The most dangerous is *P. falciparum*. The disease is transmitted by female *Anopheles* mosquitoes that transfer *Plasmodium* from infected to uninfected people.

- Malaria is controlled in three main ways: reducing the number of mosquitoes by insecticide spraying or draining breeding sites; using mosquito nets (more effective if soaked in insecticide); using drugs to prevent *Plasmodium* infecting people. There is no effective vaccine.

- AIDS is a set of diseases caused by the destruction of the immune system by infection with human immunodeficiency virus (HIV). HIV is transmitted in certain body fluids: blood, semen, vaginal secretions and breast milk. It also crosses the placenta. It primarily infects economically active members of populations in developing countries and has an extremely adverse effect on social and economic development.

- The transmission of HIV can be controlled by using barrier methods (e.g. condom and femidom) during sexual intercourse. Educating people to practise safer sex is the only control method currently available to health authorities. Contact tracing is used to find people who may have contracted HIV so that they can be tested and counselled. Life expectancy can be greatly extended by the use of combination drug therapy which interferes with the replication of the virus. However, such treatment is expensive, difficult to adhere to and has unpleasant side-effects. There is no vaccine for HIV and no cure for AIDS.

- TB is caused by the bacterium *Mycobacterium tuberculosis* (in developing countries, it may also be caused by *M. bovis*, which also causes a related disease in cattle).

- *M. tuberculosis* is spread when people infected with the active form of the disease release bacteria in droplets of liquid when they cough or sneeze. Transmission occurs when uninfected people inhale the bacteria. This is most likely to happen where people live in overcrowded conditions and especially where many sleep close together. Many people have the inactive form of TB in their lungs, but they do not have the disease and do not spread it. The inactive bacteria may become active in people who are malnourished or who become infected with HIV.

- Drugs are used to treat people with the active form of TB. The treatment may take nine months or more as it is difficult to kill the bacteria. Contact tracing is used to find people who may have caught the disease. These people are tested for TB and treated if found to be infected. The BCG vaccine provides some protection against TB, but its effectiveness varies in different parts of the world.

- Cholera, malaria, AIDS and TB are all increasing in prevalence and pose severe threats to the health of populations in developed and developing countries.

- Public health measures are taken to reduce the transmission of these diseases, but to be effective they must be informed by a knowledge of the life cycle of each pathogen.

- Antibiotics are used to inhibit the growth of pathogenic organisms. Most are only effective against bacteria. The widespread and indiscriminate use of antibiotics has led to the growth of resistant strains of bacteria. This poses a serious challenge to the maintenance of health services in the twenty-first century.

## CHAPTER 14

# Immunity

**By the end of this chapter you should be able to:**

1 describe the structure, origin, maturation and mode of action of phagocytes and lymphocytes;

2 explain the meaning of the term *immune response*;

3 distinguish between the actions of B lymphocytes and T lymphocytes in fighting infection;

4 explain the role of memory cells in long-term immunity;

5 relate the molecular structure of antibodies to their functions;

6 distinguish between active and passive, natural and artificial immunity;

7 explain how vaccination can control disease;

8 discuss the reasons why vaccination has eradicated smallpox but not measles, TB, malaria or cholera.

We now consider in detail something that was mentioned in chapter 13: the body's varying ability to resist infection by pathogens. We have seen that some people experience few or no symptoms when exposed to certain infectious diseases. Even though they may be a carrier of disease to other people, they have **immunity** themselves. How is this possible? The disease measles is used as an example here.

Measles is caused by a virus which enters the body and replicates inside human cells. There are no symptoms for about 8 to 14 days and then a rash appears and a fever develops. Amongst poor people, especially those living in overcrowded conditions, measles can be a serious disease and a major cause of death, especially among infants. Amongst others, after about ten days the disease clears up and there are rarely any complications. Measles used to be a common childhood disease in the UK that most people had only once. In most cases it is very unlikely that anyone surviving the disease will suffer from it again. They are immune. While suffering from the symptoms of the disease the body's defence system has developed a way of recognising the virus and preventing it from doing any harm again. Immunity is the protection against disease provided by the body's defence or **immune system**.

## Defence against disease

We have a variety of mechanisms to protect ourselves against infectious diseases, such as measles and those described in the previous chapter. Many pathogens do not harm us because, if we are healthy, we have physical, chemical and cellular defences that prevent them entering, or if they do enter, from spreading through the body. For example, the epithelia that cover the airways are an effective barrier to the entry of pathogens (see page 143); hydrochloric acid in the stomach kills many bacteria that we ingest with our food and drink; blood clotting is a defence mechanism that stops the loss of blood and prevents the entry of pathogens through wounds in the skin. If pathogens do successfully enter the body, white blood cells (chapter 8) can recognise them as something foreign and destroy them.

White blood cells are part of the immune system and they recognise pathogens by the distinctive, large molecules that cover their surfaces, such as proteins, glycoproteins, lipids and polysaccharides, and the waste materials which some pathogens produce. Any molecule which the body recognises as foreign is an **antigen**.

Lymphocytes are one type of white blood cell. (Their structure and functions are described in detail below.) They play an important role in the **immune response**, as follows. We have many different kinds of lymphocytes each capable of producing a type of protein, termed an **antibody**, that acts against a particular antigen. The antigen and antibody are specific for each other and the lymphocyte's antibodies are only secreted when the appropriate antigen is encountered. The first time this happens in a person's life, the production of antibodies is slow. It may take several weeks to build up enough antibodies to destroy the pathogen. Subsequent encounters, however, see a very rapid response with the result that symptoms can be entirely prevented and the host remains well.

### SAQ 14.1

Explain the terms *antigen*, *antibody* and *immune response*.

# Cells of the immune system

The cells of the immune system originate from the bone marrow. There are two groups of these cells involved in defence:

- phagocytes (neutrophils and macrophages);
- lymphocytes.

All of these cells are visible among red blood cells when a blood smear is stained to show nuclei as shown in *figure 14.1*.

### SAQ 14.2

Looking at *figure 14.1*, **a** describe the differences between the neutrophil and lymphocyte and **b** calculate the actual size of the neutrophil.

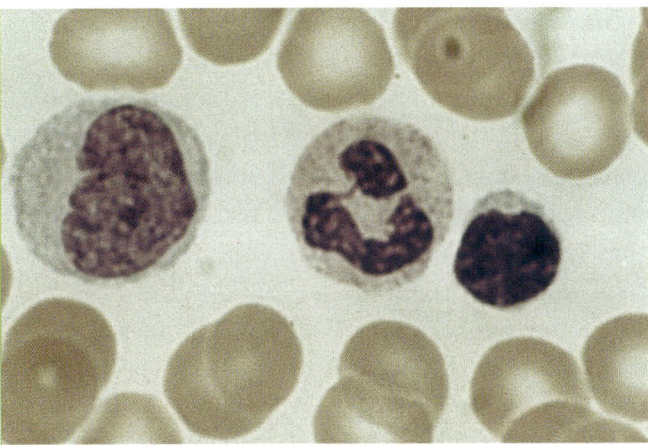

● **Figure 14.1** A monocyte (left), which will develop into a macrophage, a neutrophil (centre) and a lymphocyte (right) together with red blood cells in a blood smear which has been photographed through a light microscope. The cytoplasm of the neutrophil contains vacuoles full of hydrolytic enzymes (× 2200).

## Phagocytes

**Phagocytes** are produced throughout life by the bone marrow. They are stored there before being distributed around the body in the blood. They are scavengers, removing any dead cells as well as invasive microorganisms.

**Neutrophils** are a kind of phagocyte and form about 60% of the white cells in the blood (*figure 14.2*). They travel throughout the body often leaving the blood by squeezing through the walls of capillaries to 'patrol' the tissues. During an infection they are released in large numbers from their stores but they are short-lived cells.

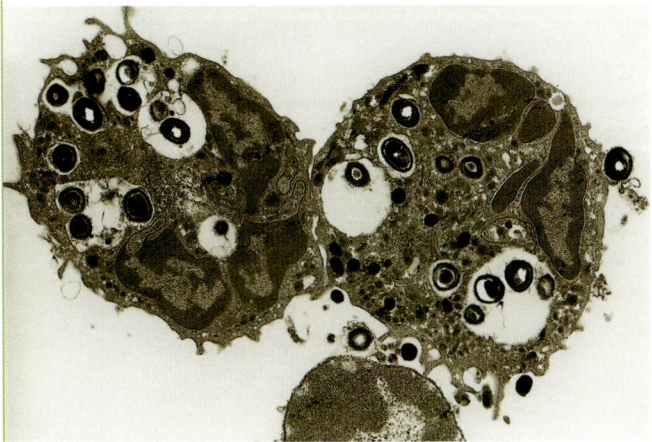

● **Figure 14.2** TEM of two neutrophils that have ingested several *Staphylococcus* bacteria. Notice at the extreme right one bacterium being engulfed. Compare this photograph with *figure 14.3* (× 7500).

**Macrophages** are also phagocytes but are larger then neutrophils and tend to be found in organs, such as the lungs, liver, spleen, kidney and lymph nodes, rather than remaining in the blood. They leave the bone marrow and travel in the blood as **monocytes**, which develop into macrophages once they leave the blood and settle in the organs, removing any foreign matter found there.

Macrophages are long-lived cells and play a crucial role in *initiating* immune responses since they do not destroy pathogens completely, but cut them up to display antigens that can be recognised by lymphocytes.

### Phagocytosis

If pathogens invade the body and cause an infection, some of the cells under attack respond by releasing chemicals such as **histamine**. These, with any chemicals released by the pathogens themselves, attract passing neutrophils to the site. The neutrophils destroy the pathogens by phagocytosis (*figures 4.11a* and *14.3*).

The neutrophils move towards the pathogens, which may be clustered together and covered in antibodies. This further stimulates the neutrophils to attack them. This is because neutrophils have receptor proteins on their surfaces that recognise antibody molecules and attach to them. When this happens the neutrophil's plasma membrane engulfs the pathogen and traps it within a vacuole. Digestive enzymes are secreted into the vacuole so destroying the pathogen.

Neutrophils have a short life – after killing and digesting some pathogens, they die. Dead neutrophils often collect at a site of infection to form pus.

## Lymphocytes

Lymphocytes are smaller than phagocytes. They have a large nucleus that fills most of the cell (*figure 14.1*). There are two types of lymphocyte, both of which are produced before birth in bone marrow.

- **B lymphocytes (B cells)** remain in the bone marrow until they are mature and then spread throughout the body concentrating in lymph nodes and the spleen.
- **T lymphocytes (T cells)** leave the bone marrow and collect in the **thymus** where they mature. The thymus is a gland that lies in the chest just beneath the sternum. It doubles in size between birth and puberty, but after puberty it shrinks. Only mature lymphocytes can carry out immune responses. During the

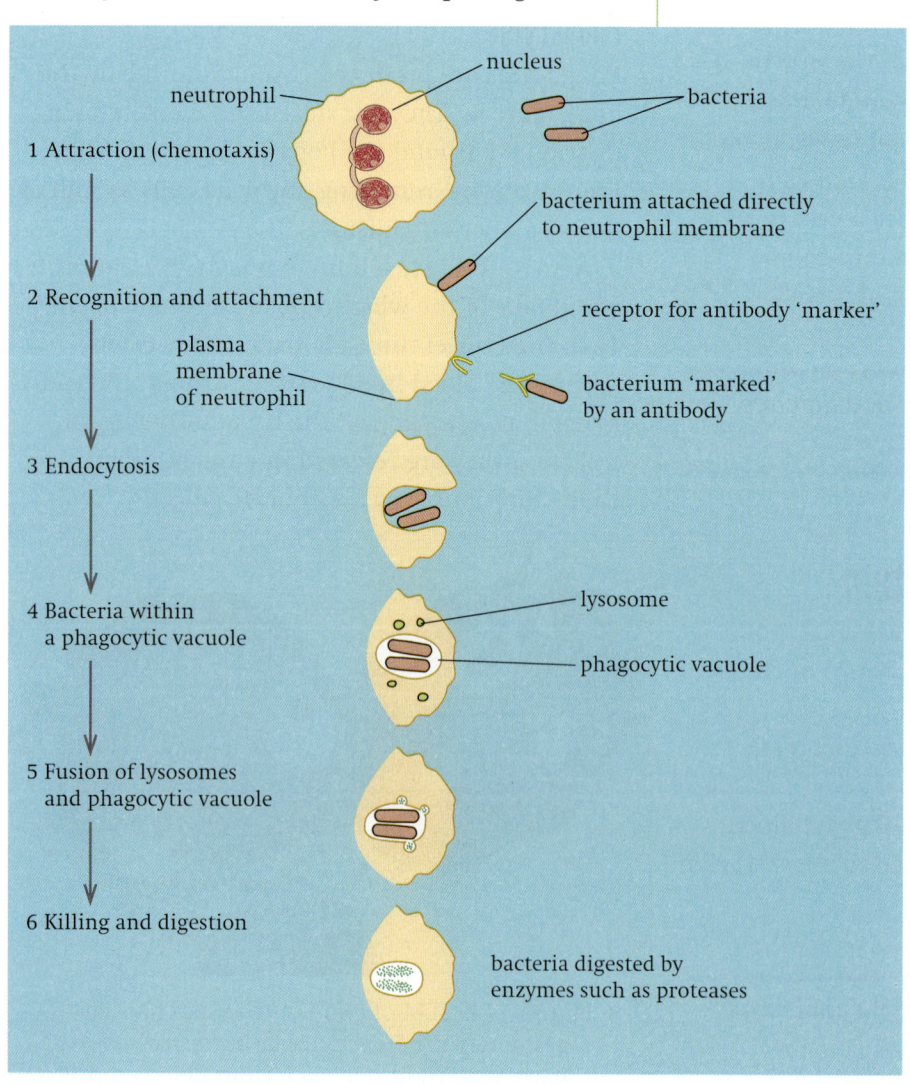

1 Attraction (chemotaxis)

2 Recognition and attachment

3 Endocytosis

4 Bacteria within a phagocytic vacuole

5 Fusion of lysosomes and phagocytic vacuole

6 Killing and digestion

neutrophil

nucleus

bacteria

bacterium attached directly to neutrophil membrane

plasma membrane of neutrophil

receptor for antibody 'marker'

bacterium 'marked' by an antibody

lysosome

phagocytic vacuole

bacteria digested by enzymes such as proteases

● **Figure 14.3** The stages of phagocytosis.

maturation process many different types of B and T lymphocyte develop, perhaps many millions. As we have seen, each type is specialised to respond to one antigen, giving the immune system as a whole the ability to respond to almost any type of pathogen that enters the body. When mature, all these B and T cells circulate between the blood and the lymph (chapter 8). This ensures that they are distributed throughout the body so that they come into contact with any pathogens *and* with each other. Immune responses depend on B and T cells interacting with each other to give an effective defence. We will look in detail at the roles of B and T cells and how they interact in the following section. Briefly, however, some T cells coordinate the immune response, stimulating B cells to divide and then secrete antibodies into the blood; these antibodies destroy the antigenic pathogens. Other T cells seek out and kill any of the body's own cells that are infected with pathogens. To do this they must make direct contact with infected cells.

## SAQ 14.3

State the sites of origin and maturation of B lymphocytes (B cells) and T lymphocytes (T cells).

## SAQ 14.4

Suggest why the thymus gland becomes smaller after puberty.

### B lymphocytes

As each B cell matures it gains the ability to make just one type of antibody molecule. Many different types of B cell develop in each of us, perhaps as many as 10 million. During the maturation process, the genes that code for antibodies are changed in a variety of ways to code for different antibodies. Each cell then divides to give a small number of cells that are able to make the same type of antibody. Each small group of identical cells is called a **clone**. At this stage, the antibody molecules do not leave the B cell but remain in the plasma membrane. Here, part of each antibody forms a protein **receptor**, which can combine specifically with one type of antigen. If that antigen enters the body, there will be some

mature B cells with cell surface receptors that will recognise it (*figure 14.4*).

*Figure 14.5* shows what happens to B cells during the immune response when an antigen enters the body on two separate occasions. When the pathogens first invade the body, some of them are taken up by macrophages in lymph nodes (chapter 8) and elsewhere. The macrophages expose the antigens from the pathogen on their surfaces. Any B lymphocytes whose cell surface receptors fit the antigens respond by dividing repeatedly by mitosis (chapter 6). Huge numbers of identical B cells are produced over a few weeks.

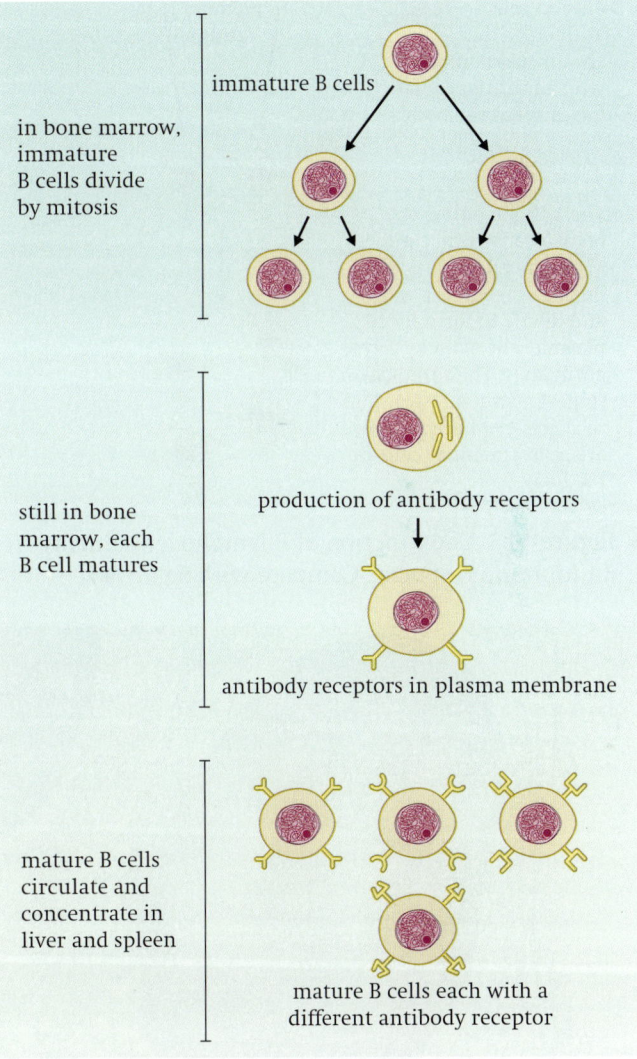

in bone marrow, immature B cells divide by mitosis

immature B cells

still in bone marrow, each B cell matures

production of antibody receptors

antibody receptors in plasma membrane

mature B cells circulate and concentrate in liver and spleen

mature B cells each with a different antibody receptor

● **Figure 14.4** Origin and maturation of B lymphocytes. As they mature in bone marrow, the cells become capable of secreting one type of antibody molecule with a specific shape. Some of these molecules become receptor proteins in the plasma membrane and act like markers. By the time of birth, there are millions of different B cells, each with a specific antibody receptor.

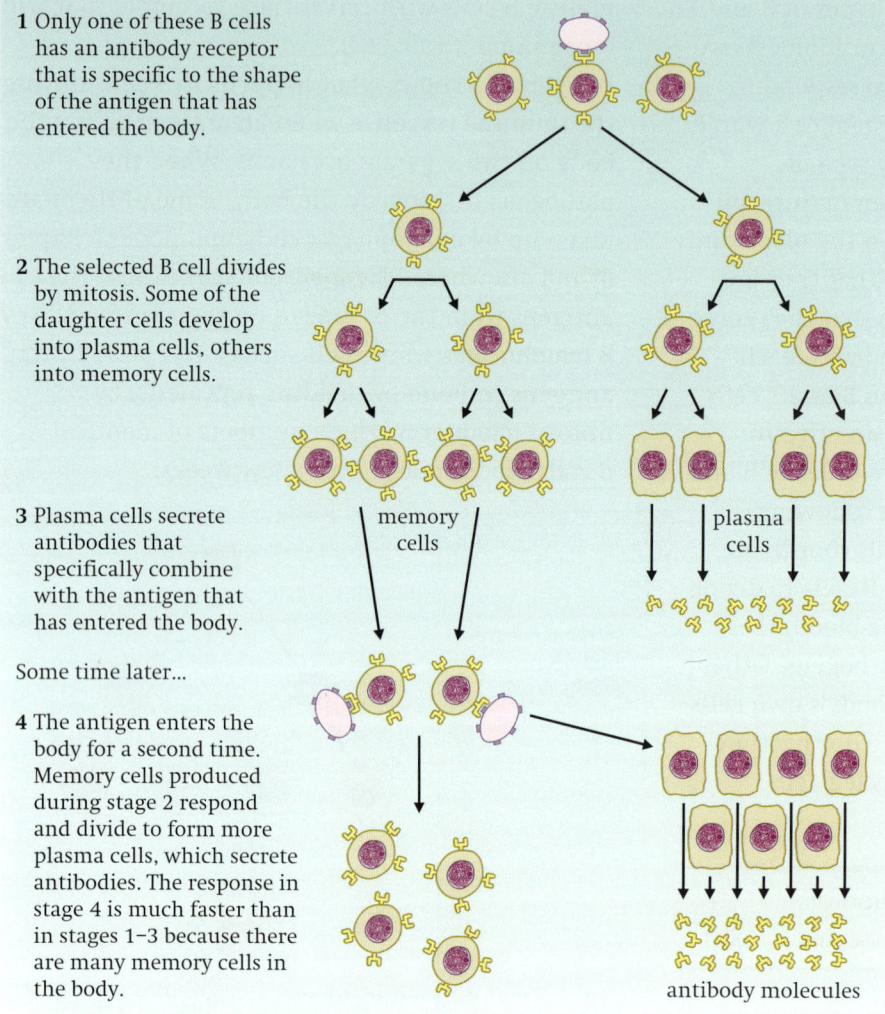

1 Only one of these B cells has an antibody receptor that is specific to the shape of the antigen that has entered the body.

2 The selected B cell divides by mitosis. Some of the daughter cells develop into plasma cells, others into memory cells.

3 Plasma cells secrete antibodies that specifically combine with the antigen that has entered the body.

Some time later...

4 The antigen enters the body for a second time. Memory cells produced during stage 2 respond and divide to form more plasma cells, which secrete antibodies. The response in stage 4 is much faster than in stages 1–3 because there are many memory cells in the body.

memory cells

plasma cells

antibody molecules

● **Figure 14.5** The function of B lymphocytes during an immune response. Compare with *figure 14.7*.

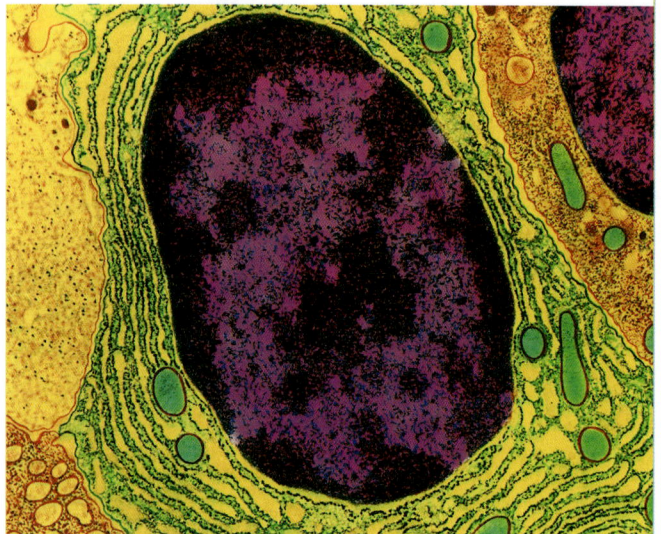

● **Figure 14.6** TEM of the contents of a plasma cell. There is an extensive network of RER (green) in the cytoplasm for the production of antibody molecules, which plasma cells secrete into blood or lymph by exocytosis (chapter 4).

Some of these activated B cells become **plasma cells** that produce antibody molecules very quickly – up to several thousand a second. Plasma cells secrete antibodies into the blood, lymph or onto the linings of the lungs and the gut (*figure 14.6*). These cells do not live long: after several weeks their numbers decrease. The antibody molecules they have secreted stay in the blood for longer, however, until they too eventually decrease in concentration.

Other B cells become **memory cells**. These cells remain circulating in the body for a long time. If the same antigen is reintroduced a few weeks or months after the first infection, memory cells divide rapidly and develop into plasma cells and more memory cells. This is repeated on every subsequent invasion by the same antigen, meaning that the infection can be destroyed and removed before any symptoms of disease develop.

*Figure 14.7* shows the changes in the concentration of antibody molecules in the blood when the body encounters an antigen. The first or **primary response** is slow because, at this stage, there are very few B cells that are specific to the antigen. The **secondary response** is faster because there are now many memory cells, which quickly divide and differentiate into plasma cells. Many more antibodies are produced in the secondary response.

Memory cells are the basis of **immunological memory**; they last for many years, often a lifetime. This explains why someone is very unlikely to catch measles twice. There is only one strain of the virus that causes measles and each time it infects the body there is a fast secondary response. However, we do suffer repeated infections of the common cold and influenza because

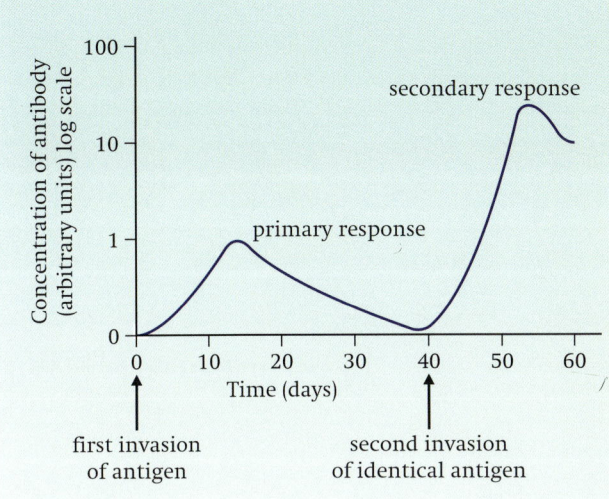

● **Figure 14.7** The changes in antibody concentration in the blood during a primary and secondary response to the same antigen.

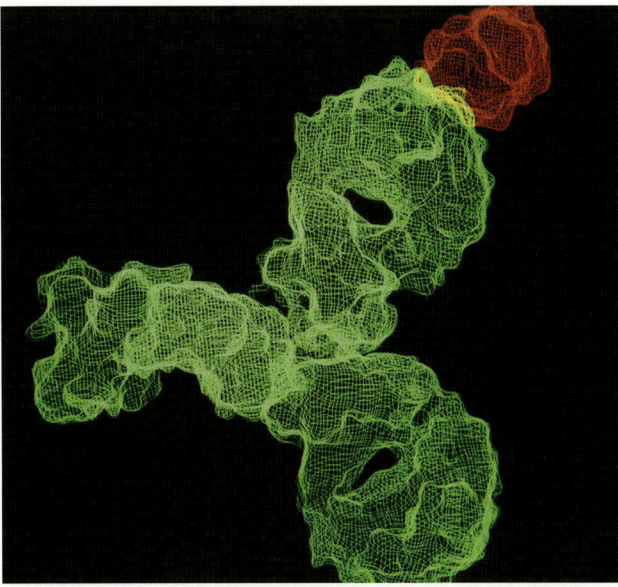

● **Figure 14.8** A model of an antibody made using computer graphics. The main part is the antibody molecule and the small part in the top right-hand corner (red) is an antigen at one of the two antigen binding sites. Compare this with *figure 14.9*.

there are many different and new strains of the viruses that cause these diseases, each one having different antigens. Each time a pathogen with different antigens infects us, the primary response must occur before we become immune and during that time we often become ill.

### Antibodies

Antibodies are all globular glycoproteins (chapter 2) and form the group of plasma proteins called **immunoglobulins**. The basic molecule common to all antibodies consists of four polypeptide chains: two 'long' or 'heavy' chains and two 'short' or 'light' chains (*figures 14.8* and *14.9*). Disulphide bridges hold the chains together. Each molecule has two identical antigen binding sites which are formed by both light and heavy chains. The sequences of amino acids in these regions make the specific three-dimensional shape which binds to just one type of antigen. This is the **variable region** which is different on each type of antibody molecule produced. The 'hinge' region gives the flexibility for the antibody molecule to bind around the antigen.

*Figure 14.10* shows the different ways in which antibodies work to protect the body from pathogens. As we saw earlier, some antibodies act as labels to identify antigens as appropriate targets for phagocytes to destroy. A special group of antibodies are **antitoxins** which block the

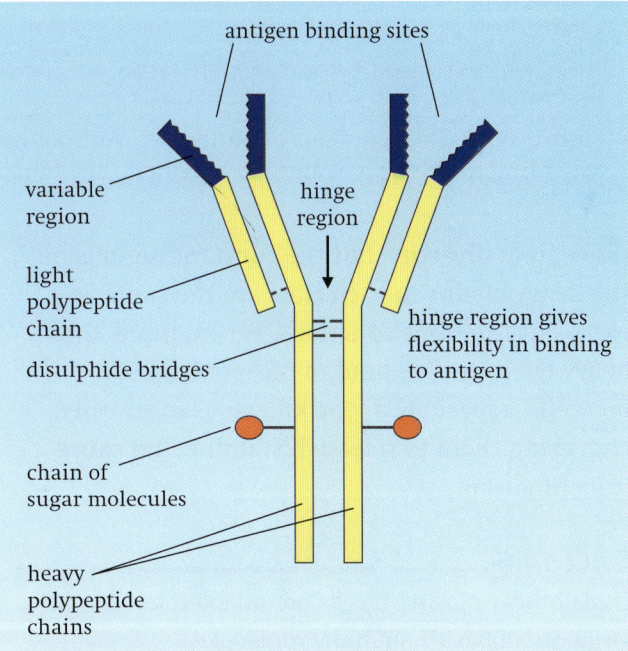

● **Figure 14.9** A diagram of an antibody molecule. Antigen–antibody binding occurs at the variable regions. An antigen fits into the binding site like a substrate fitting into the active site of an enzyme.

toxins released by bacteria such as those that cause diphtheria and tetanus.

*Table 14.1* shows the four classes of antibody. Even though they have different structures, these different classes of antibody share common sub-units in their structure. The variable regions

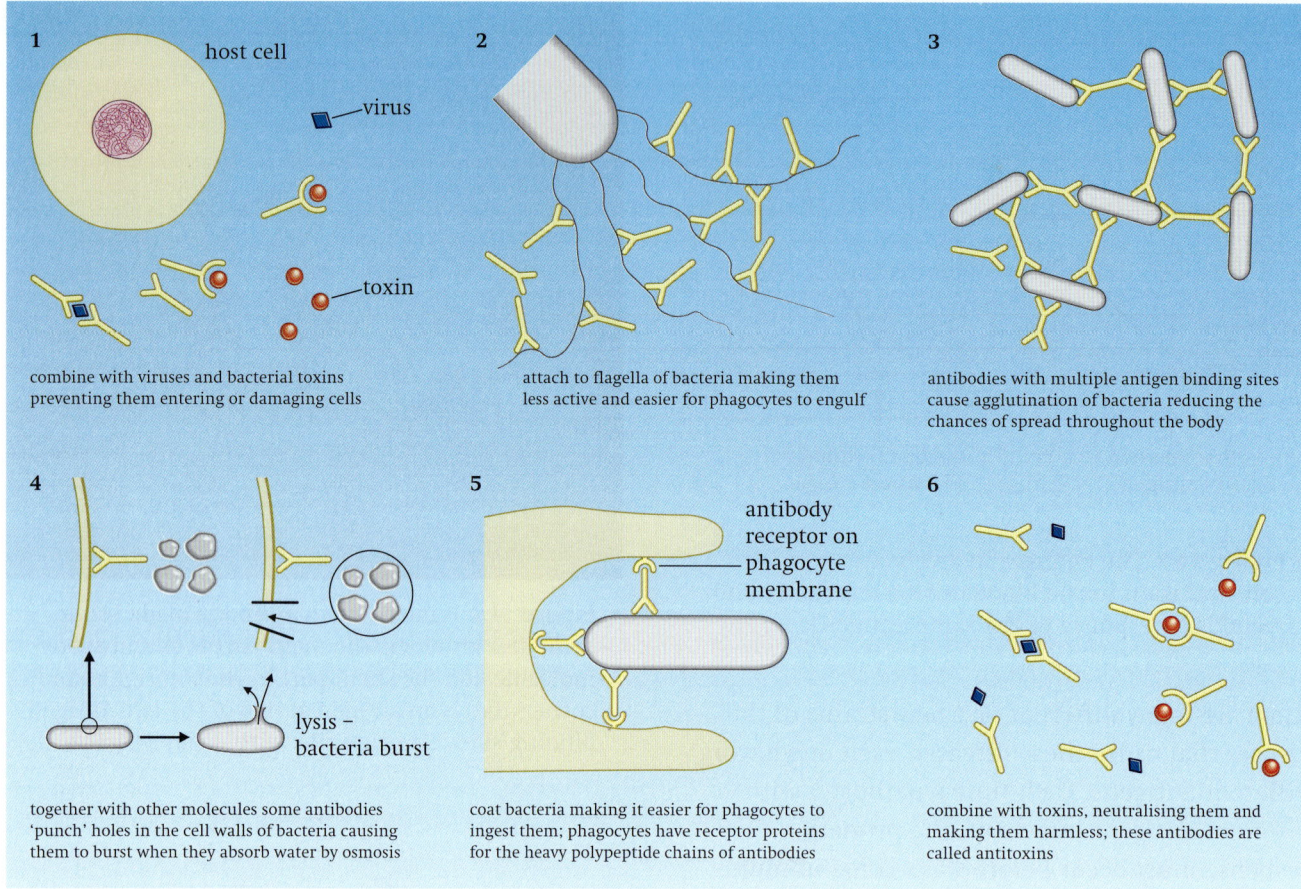

1 — host cell — virus — toxin

combine with viruses and bacterial toxins preventing them entering or damaging cells

2 — attach to flagella of bacteria making them less active and easier for phagocytes to engulf

3 — antibodies with multiple antigen binding sites cause agglutination of bacteria reducing the chances of spread throughout the body

4 — lysis – bacteria burst

together with other molecules some antibodies 'punch' holes in the cell walls of bacteria causing them to burst when they absorb water by osmosis

5 — antibody receptor on phagocyte membrane

coat bacteria making it easier for phagocytes to ingest them; phagocytes have receptor proteins for the heavy polypeptide chains of antibodies

6 — combine with toxins, neutralising them and making them harmless; these antibodies are called antitoxins

● **Figure 14.10** The functions of antibodies. Antibodies have different functions according to the type of antigen to which they bind.

show great diversity, but the constant regions of the heavy chains in each class are the same and carry out the same functions. For example, the heavy chains of IgE bind to receptors on some of our cells, especially a type known as **mast cells**, triggering them to release histamine and cause inflammation.

### SAQ 14.5
Explain how plasma B cells are adapted to secrete large quantities of antibody molecules.

### SAQ 14.6
Explain why B cells divide by mitosis during an immune response.

### SAQ 14.7
Explain why polysaccharides would not be suitable for making antibody molecules.

### SAQ 14.8
Explain why only some B cells respond during an immune response to a pathogen.

### SAQ 14.9
There are many different strains of the rhinovirus, which causes the common cold. Explain why people can catch several different colds in the space of a few months.

### T lymphocytes
Mature T cells have specific cell surface receptors called T cell receptors (*figure 14.11*). These have a structure similar to antibodies and they are each specific to one antigen. T cells are activated when they encounter this antigen in contact with another host cell. Sometimes this is a macrophage that has engulfed a pathogen and cut it up to expose the pathogen's surface molecules or it may be a body cell that has been invaded by a

| Antibody class | Relative molecular mass | Number of antigen binding sites | Sites of action | Functions |
|---|---|---|---|---|
| Immunoglobulin G (IgG) | 150 000 | 2 | blood<br>tissue fluid<br>(can cross placenta) | • enhances activity of macrophages<br>• act as antitoxins<br>• causes agglutination |
| Immunoglobulin M (IgM) | 970 000 | 10 | blood<br>tissue fluid<br>(cannot cross placenta) | • causes agglutination |
| Immunoglobulin A (IgA) | 160 000 or 320 000 | 2 or 4 | saliva, tears<br>bronchial secretions<br>mucus secretions of<br>  small intestine<br>prostate and vaginal<br>  secretions<br>nasal fluid<br>colostrum/breast milk | • inhibits bacteria adhering to<br>  host cells<br>• prevents bacteria forming colonies<br>  on mucous membranes |
| Immunoglobulin E (IgE) | 180 000 | 2 | tissues | • heavy chains of IgE activate mast<br>  cells to release histamine<br>• involved in response to infections<br>  by worms, and allergic responses<br>  to harmless substances,<br>  e.g. pollen (hay fever) |

● **Table 14.1** The four different classes of antibody and their functions. Whatever the class of antibody, each antibody molecule produced by a single clone of plasma cells possesses just one type of variable region and binds to one antigen.

pathogen and is similarly displaying the antigen on its plasma membrane as a kind of 'help' signal. Those T cells that have matching receptors respond to the antigen by dividing.

There are two main types of T cell:
■ **T helper cells**;
■ **killer T cells** (or T cytotoxic cells).

When T helper cells are activated they release hormone-like **cytokines** that stimulate appropriate B cells to divide, develop into plasma cells and secrete antibodies. Some T helper cells secrete cytokines that stimulate macrophages to carry out phagocytosis more vigorously. Killer T cells search the body for cells that have become invaded by pathogens and are displaying foreign antigens from the pathogens on their plasma membranes. Killer T cells recognise the antigens, attach themselves to the surface of infected cells and secrete toxic

substances, such as hydrogen peroxide, killing the body cells and the pathogens inside (*figure 14.12*).

In addition to the helper cells and killer cells, **memory** T cells are produced which remain in the body and become active very quickly during the secondary response to antigens.

### SAQ 14.10
Explain why people are often ill for several weeks after they catch a disease, even though they can make antibodies against the disease.

### SAQ 14.11
Outline the functions of B lymphocytes and T lymphocytes and describe how they interact during an immune response.

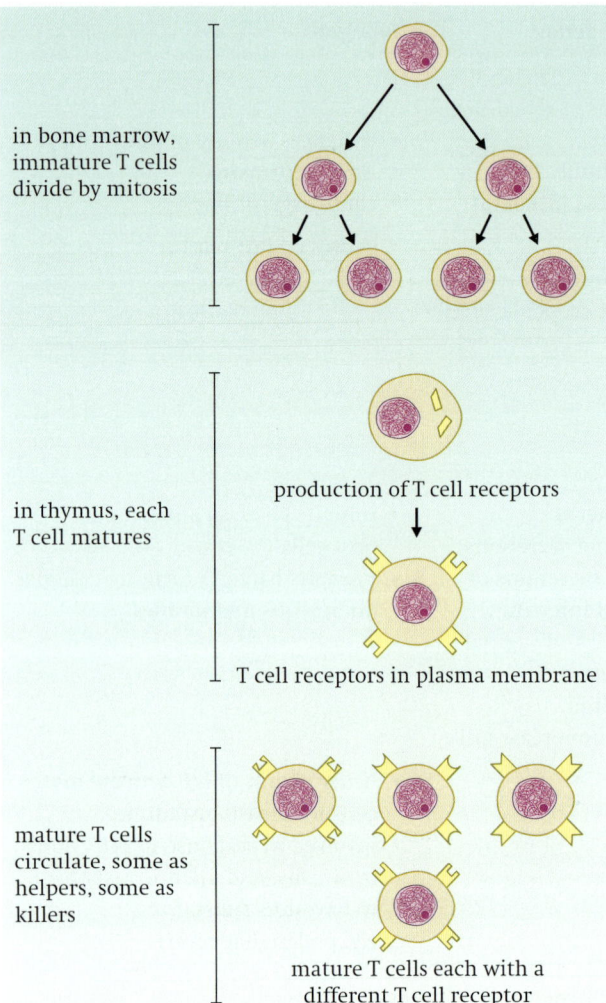

● **Figure 14.11** Origin and maturation of T lymphocytes. As T cells mature in the thymus gland they produce T cell receptor proteins. Each cell has a specific receptor. Some cells become T helper cells, others become killer T cells.

# Active and passive immunity

The type of immunity described so far occurs during the course of an infection. This form of immunity is **active** because lymphocytes are activated by antigens on the surface of pathogens that have invaded the body. As this activation occurs naturally during an infection it is called **natural active immunity**. The immune response can also be activated artificially either by injecting antigens into the body or taking them by mouth. This is the basis of **artificial active immunity**, more commonly known as **vaccination**. The immune response is similar to that following an infection, and the effect is the same – long-term immunity. In both natural and artificial active immunity anti-

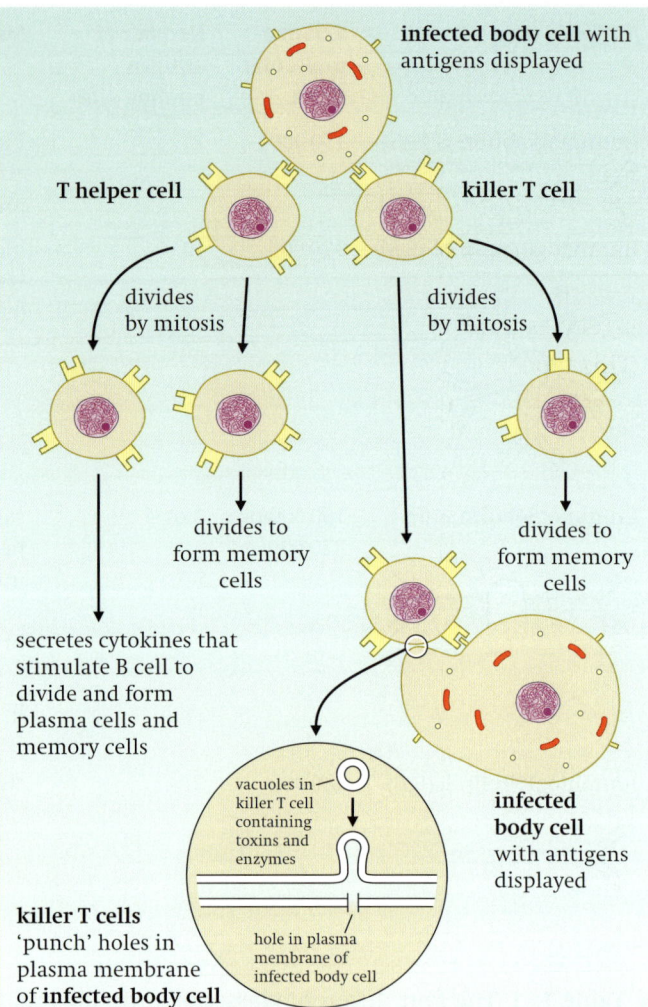

● **Figure 14.12** The functions of T lymphocytes during an immune response. T helper cells and killer T cells with T cell receptor proteins specific to the antigen respond and divide by mitosis. Activated T helper cells stimulate B cells to divide and develop into plasma cells (*figure 14.5*). Killer T cells attach themselves to infected cells and kill them.

body concentrations in the blood follow patterns similar to those shown in *figure 14.7*.

In both forms of active immunity, it takes time for sufficient active B and T cells to be produced to give an effective defence. If a person becomes infected with a potentially fatal disease such as tetanus, a more immediate defence is needed for survival. Tetanus kills quickly, before the body's natural primary response can take place. So people who have a wound that may be infected with the bacterium that causes tetanus are given an injection of **antitoxin**. This is a preparation of human antibodies against the tetanus toxin. The antibodies are collected from blood donors who have recently been vaccinated against tetanus. Antitoxin provides immediate protection but this is only temporary as the antibodies are not produced by the body's own B cells and are therefore regarded as foreign themselves. They are removed from the circulation by phagocytes in the liver and spleen.

This type of immunity to tetanus is **passive immunity** because the B and T cells have not been activated and plasma cells have not produced any antibodies. More specifically, it is **artificial passive immunity**: the antibodies have come from another person who has encountered the antigen.

The immune system of a newborn infant is not as effective as that of a child or an adult. However, babies are not entirely unprotected against pathogens because antibodies from their mothers cross the placenta during pregnancy and remain in the infant for several months (*figure 14.13*). For example, antibodies against measles may last for four months or more in the infant's blood. **Colostrum**, the thick yellowish fluid produced by a mother's breasts for the first four or five days after birth, is rich in IgA. Some of these antibodies remain on the surface of the infant's gut wall while others pass into the blood undigested. IgA acts in the gut to prevent the growth of bacteria and viruses and also circulates in the blood. This is **natural passive immunity**. The features of active and passive immunity are compared in *table 14.2*.

**SAQ 14.12**
Explain the difference between artificial active immunisation (vaccination) and artificial passive immunisation.

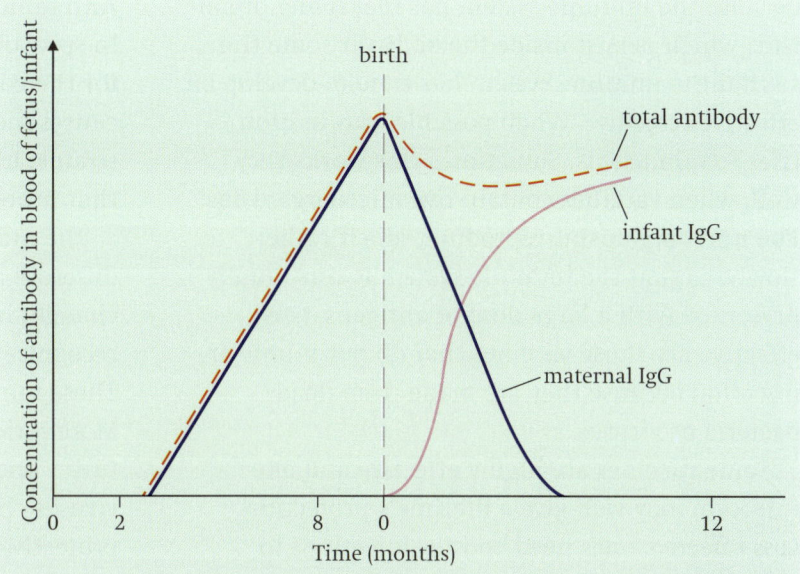

● **Figure 14.13** The concentrations of antibody in the blood of a fetus and an infant.

**SAQ 14.13**
Explain the pattern of maternal and infant IgG shown in *figure 14.13*.

**SAQ 14.14**
Explain the advantages of natural passive immunity for newborn infants.

## Vaccination

A **vaccine** is a preparation containing antigenic material, which may be a whole live micro-organism, a dead one, a harmless version (known as an attenuated organism), a harmless form of a toxin (toxoid) or a preparation of surface antigens. Vaccines are either given by injection into a vein or muscle, or are taken orally.

Immunity derived from a natural infection is often extremely good at providing protection

| Immunity | | | Features | | |
|----------|------------------------|-----------------|--------------------------------------------|------------------------------|-------------|
| | antigen encountered | immune response | time before antibodies appear in blood | production of memory cells | protection |
| **Active** | yes | yes | several weeks during primary response | yes | permanent |
| **Passive** | no | no | immediate | no | temporary |

● **Table 14.2** Features of active and passive immunity.

because the immune system has met living organisms which persist inside the body for some time, so that the immune system has time to develop an effective response. When possible, vaccination tries to mimic this. Sometimes this works very well, when vaccines contain live microorganisms. The microorganisms reproduce, albeit rather slowly, so that the immune system is continually presented with a large dose of antigens. Less effective are those vaccines that do not mimic an infection because they are made from dead bacteria or viruses.

Some vaccines are highly effective and one injection may well give a lifetime's protection. Less effective ones need booster injections to stimulate secondary responses that give enhanced protection (*figure 14.7*). It is often a good idea to receive booster injections if you are likely to be exposed to a disease, even though you may have been vaccinated as a child.

## Problems with vaccines

### Poor response

Some people do not respond at all, or not very well, to vaccinations. This may be because they have a defective immune system and as a result do not develop the necessary B and T cell clones. It may also be because they suffer from malnutrition, particularly protein energy malnutrition, and do not have enough protein to make antibodies or clones of lymphocytes. These people are at a high risk of developing infectious diseases and transmitting them to people who have no immunity.

People vaccinated with a live virus may pass it out in their faeces during the primary response and may infect others. This is why it is better to vaccinate a large number of people at the same time to give **herd immunity**, or to ensure that all children are vaccinated within a few months of birth. Herd immunity interrupts transmission in a population, so that those who are susceptible never encounter the infectious agents concerned.

## SAQ 14.15
Explain why malnourished children give very weak responses to vaccines.

## Antigenic variation

In spite of years of research, there are no vaccines for the common cold. The type of rhinovirus that causes most colds has at least 113 different strains. It may be impossible to develop a vaccine that protects against all of these.

The influenza virus mutates regularly to give different antigens. When there are only minor changes in the viral antigen, memory cells will still recognise them and start a secondary response. These minor changes are called **antigenic drift**. More serious are major changes in antigen structure – known as **antigenic shift** – when influenza viruses change their antigens considerably and the protective immunity given by vaccination against a previous strain is ineffective against the new one. The World Health Organisation (WHO) recommends the type of vaccine to use according to the antigens that are common at the time. The vaccine is changed almost every year.

There are, as yet, no effective vaccines against the diseases which are caused by protoctists such as malaria and sleeping sickness. This is because these pathogens are eukaryotes with many more genes than bacteria and viruses. They can have many hundreds, or even thousands, of antigens on their cell surfaces. *Plasmodium* passes through three stages in its life cycle while it is in the human host. Each stage has its own specific antigens. This means that effective vaccines would have to contain antigens to all three stages or be specific to the infective stage. The latter would only work if the immune system can give an effective response in the short period of time (a few hours) between the mosquito bite and the infection of liver cells (*figure 13.3*). *Trypanosoma*, the causative agent of sleeping sickness, has a total of about a thousand different antigens and changes them every four or five days. This makes it impossible for the immune system to respond effectively. After several weeks the body is completely overwhelmed by the parasite, with fatal consequences.

## SAQ 14.16
Explain why humans cannot produce an effective immune response to an infection by *Trypanosoma*.

### Antigenic concealment

Some pathogens evade attack by the immune system by living inside cells. For example when *Plasmodium* enters liver cells or red blood cells, it is protected against antibodies in the plasma. Some parasitic worms conceal themselves by covering their bodies in host proteins, so they remain invisible to the immune system. Other pathogens suppress the immune system by parasitising cells such as macrophages and T cells. It is very difficult to develop effective vaccines against these pathogens because there is such a short period of time for an immune response to occur before the pathogen 'hides'.

Another example is *Vibrio cholerae* (the causative agent of cholera), which remains in the intestine where it is beyond the reach of many antibodies. The cholera vaccine is injected rather than taken orally and so it is unable to stimulate antibody production in the intestine, nor can it stimulate the production of an antitoxin against choleragen. An oral vaccine against cholera has been developed.

### SAQ. 14.17

Name one pathogen that parasitises
**a** macrophages and **b** T helper cells.

## The eradication of smallpox

Smallpox was an acute, highly infectious disease caused by the variola virus and transmitted by direct contact. It was a terrible disease. Red spots containing a transparent fluid would appear all over the body (*figure 14.14*). These then filled with thick pus. Eyelids became swollen and could become 'glued' together. Sufferers often had to be prevented from tearing at their flesh. Many people who recovered were permanently blind and disfigured by scabs left when the pustules dried out. Smallpox killed between 12 and 30% of its victims.

WHO started an eradication programme in 1956; in 1967 it stated its intention to rid the world of the disease within ten years. There were two main aspects of the programme: vaccination and surveillance. Successful attempts were made across the world to vaccinate in excess of 80% of populations at risk of the disease. When a case of smallpox was reported, everyone in the household

● **Figure 14.14** A parent and child of the Kampa people of the Amazon region, South America. They clearly show the characteristic pustules of smallpox.

and the 30 surrounding households, as well as other relatives and possible contacts in the area, were vaccinated. This **ring vaccination** protected everyone who could possibly have come into contact with a person with the disease, reduced the chances of transmission and contained the disease. The last strongholds of smallpox were in East Africa, Afghanistan and the Indian subcontinent. Eradication was most difficult in Ethiopia and Somalia, where many people lived in remote districts well away from main roads which were no more than dirt tracks. In the late 1970s the two countries went to war and, even though large parts of Ethiopia were overrun by the Somalis, the eradication programme continued. The last case of smallpox was reported in Somalia in 1977. WHO finally declared the world free of smallpox in 1980.

The eradication programme was successful for a number of reasons.

■ The variola virus was stable; it did not mutate and change its surface antigens. This meant that the same vaccine could be used everywhere in the world throughout the campaign.

It was therefore cheap to produce.

- The vaccine was made from a harmless strain of a similar virus (vaccinia) and was effective because it was a 'live' vaccine.
- The vaccine was freeze-dried and could be kept at high temperatures for as long as 6 months. This made it suitable for use in the tropics.
- Infected people were easy to identify.
- The vaccine was easy to administer and was even more effective after the development of a stainless steel, reusable needle for its delivery. This 'bifurcated needle' had two prongs, which were used to push the vaccine into the skin.
- The smallpox virus did not linger in the body after an infection to become active later and form a reservoir of infection.
- The virus did not infect animals, which made it easier to break the transmission cycle.
- Many 16- to 17-year-olds became enthusiastic vaccinators and suppliers of information about cases; this was especially valuable in remote areas.

The eradication of smallpox is a medical success story. It has been more difficult to repeat this success with other infectious diseases. This is partly because of the more unstable political situation in the late 1970s and 1980s particularly in Africa, Latin America and parts of Asia such as Afghanistan. Public health facilities are difficult to organise in developing countries with poor infrastructure, few trained personnel and limited financial resources. They are almost impossible to maintain during periods of civil unrest or during a war. Nevertheless, WHO declared the Americas to be free of polio in 1991 and vaccination programmes have been organised in Asia to

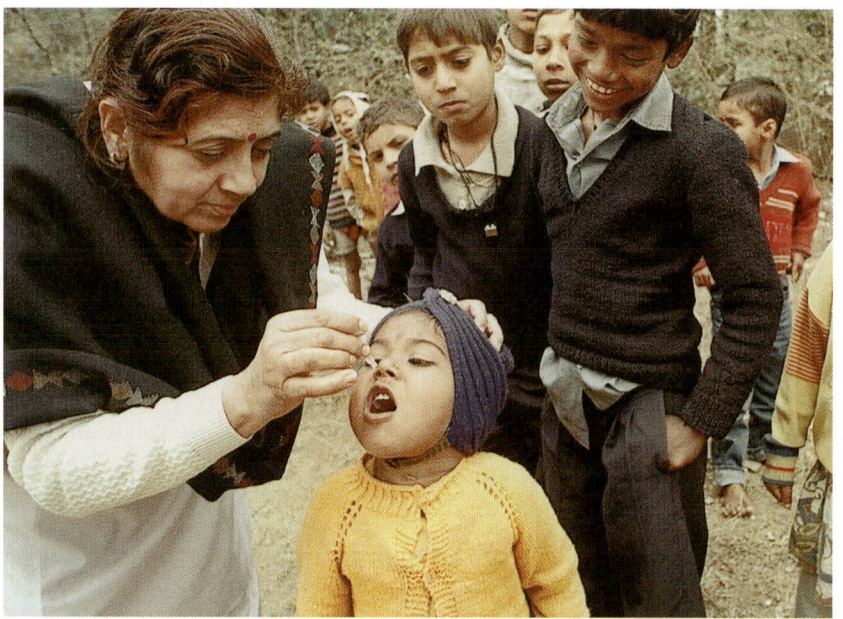

● **Figure 14.15** Indian children queuing to be given an oral vaccine against polio on 17th January 1999. The Indian government organised a programme to vaccinate 130 million children during the day in an attempt to eradicate the disease.

try to eradicate the disease from the world (*figure 14.15*).

# Measles

Measles is caused by a virus which is spread by airborne droplets. As we saw at the beginning of this chapter, it causes a rash and fever. There can be fatal complications. The disease rarely affects infants under eight months of age as they have passive immunity in the form of antibodies that have crossed the placenta from their mother. Measles used to be a common childhood disease in the UK and other developed countries, but is now quite rare because most children are vaccinated. However, epidemics do occur in the developed world, for example in the USA in 1989–90 when there were 55 000 cases and 132 deaths.

Measles is a major disease in developing countries, particularly in cities where people live in overcrowded, insanitary conditions and where there is a high birth rate. The measles virus is transmitted easily in these conditions and it infects mainly malnourished infants suffering from vitamin A deficiency. Measles is responsible for many cases of childhood blindness and it also causes severe brain damage, which can be fatal. In 1993 it was estimated that there were over 45 million cases of measles and 1.16 million deaths making it the ninth leading cause of death worldwide. Most of those who die from measles are young malnourished children who do not have the resistance to fight it.

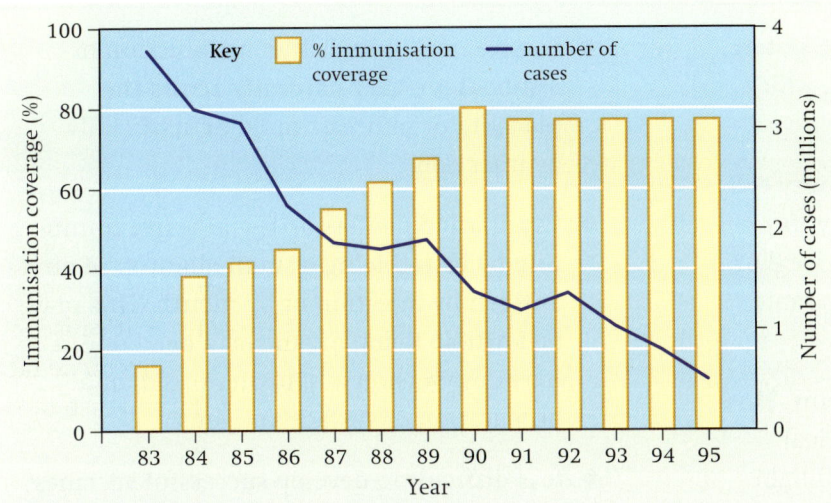

● **Figure 14.16** The global measles immunisation programme coordinated by WHO has increased the immunisation coverage to about 80% of young children. As a result, the number of cases reported each year has decreased significantly.

Measles is a preventable disease and one that can be eradicated by a worldwide surveillance and vaccination programme. However, a programme of one-dose-vaccination has not eliminated the disease in any country despite high coverage of the population. This is explained by the poor response to the vaccine shown by some children who need several boosters to develop full

immunity. In large cities with high birth rates and shifting populations, it can be difficult to give boosters, follow up cases of measles and trace contacts. Migrants and refugees can form reservoirs of infection, experiencing epidemics within their communities and spreading the disease to surrounding populations. This makes measles a very difficult disease to eliminate even with high vaccination coverage.

Measles is highly infectious and it is estimated that herd immunity of 93–95% is required to prevent transmission in a population. As the currently available vaccine has a success rate of 95%, this means that the whole population needs to be vaccinated and infants must be vaccinated within about eight months of birth. Many countries achieve up to 80% or more coverage with measles vaccination (*figure 14.16*), and it is hoped to declare the Americas free of the disease early in the twenty-first century. With coverage of under 50% in Africa, it is likely that the disease will still persist there for many years to come.

# SUMMARY

◆ Phagocytes and lymphocytes are the cells of the immune system.

◆ Phagocytes originate in the bone marrow and are produced there throughout life. There are two types: neutrophils circulate in the blood and enter infected tissues; macrophages are more stationary inside tissues. They destroy bacteria and viruses by phagocytosis.

◆ Lymphocytes also originate in bone marrow, but migrate just before and after birth to other sites in the body. There are two types: B lymphocytes (B cells) and T lymphocytes (T cells).

◆ Antigens are 'foreign' macromolecules that stimulate the immune system.

◆ During an immune response, those B and T cells that have receptors specific to the antigen are activated.

◆ When B cells are activated they form plasma cells which secrete antibodies.

◆ T lymphocytes do not secrete antibodies; their surface receptors are similar to antibodies and identify antigens. They mature in the thymus and develop into either T helper cells or killer T cells (cytotoxic T cells). T helper cells secrete

◆ cytokines that control the immune system, activating B cells and killer T cells, which kill infected host cells.

◆ During an immune response, memory cells are formed which retain the ability to divide rapidly and develop into active B or T cells on a second exposure to the same antigen (immunological memory).

◆ Antibodies are globular glycoproteins. They all have one or more pairs of identical heavy polypeptides and of identical light polypeptides. Each type of antibody interacts with one antigen via the specific shape of its variable region. Each molecule of the simplest antibody (IgG) can bind to two antigen molecules. Larger antibodies (IgM and IgA) have more than two antigen binding sites.

◆ Antibodies agglutinate bacteria; prevent viruses infecting cells; coat bacteria and viruses to aid phagocytosis; act with plasma proteins to burst bacteria; neutralise toxins.

◆ Active immunity is the production of antibodies and active T cells during a primary immune response to an antigen acquired either naturally by infection or artificially by vaccination. This gives permanent immunity.

◆ Passive immunity is the introduction of antibodies either naturally across the placenta or in breast milk, or artificially by injection.

◆ Vaccination confers artificial active immunity by introducing a small quantity of an antigen by injection or by mouth. This may be a whole living organism, a dead one, a harmless version of a toxin (toxoid) or a preparation of surface antigens.

◆ It is difficult to develop successful vaccines against diseases caused by organisms that have many different strains, or express different antigens during their life cycle within humans (antigenic variation), or infect parts of the body beyond the reach of antibodies (antigenic concealment).

◆ Smallpox was eradicated by a programme of surveillance, contact tracing and 'ring' vaccination, using a 'live' vaccine against the only strain of the smallpox virus.

◆ Measles is a common cause of death amongst infants in poor communities. It is difficult to eradicate because a wide coverage of vaccination has not been achieved and malnourished children do not respond well to just one dose of the vaccine.

# Part 2
# A Level

# Energy and respiration

## The need for energy in living organisms

All living organisms require a continuous supply of energy to stay alive, either from the absorption of light energy or from chemical potential energy. The process of photosynthesis transfers light energy to chemical potential energy and so almost all life on Earth depends on photosynthesis, either directly or indirectly. Photosynthesis supplies living organisms with two essential requirements: an energy supply and usable carbon compounds.

All biological macromolecules, such as carbohydrates, lipids, proteins and nucleic acids, contain carbon. All living organisms therefore need a source of carbon. Organisms which can use an inorganic carbon source in the form of carbon dioxide are called **autotrophs**. Those needing a ready-made organic supply of carbon are **heterotrophs**. (An *organic* molecule is a compound including carbon and hydrogen. The term originally meant a molecule derived from an organism, but now includes all compounds of carbon and hydrogen even if they do not occur naturally.)

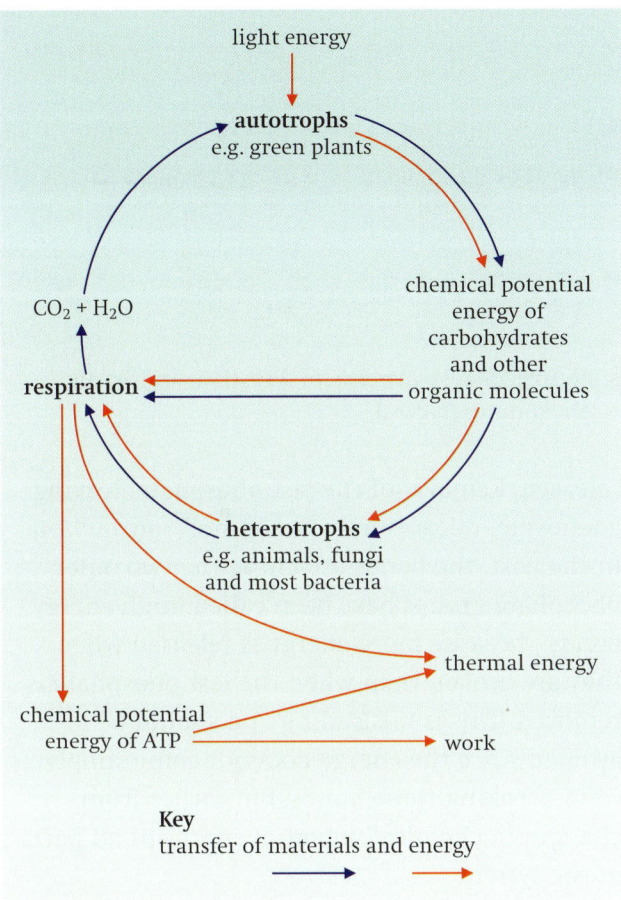

● **Figure 15.1** Transfer of materials and energy in an ecosystem.

Organic molecules can be used by living organisms in two ways. They can serve as 'building bricks' for making other organic molecules that are essential to the organism, and they can represent chemical potential energy which can be released by breaking down the molecules in respiration (page 200). This energy can then be used for all forms of work. Heterotrophs depend on autotrophs for both materials and energy (*figure 15.1*).

# Work

Work in a living organism includes:

■ the synthesis of complex substances from simpler ones (anabolic reactions) such as the synthesis of polypeptides from amino acids;

■ the active transport of substances against a diffusion gradient such as the activity of the sodium-potassium pump;

■ mechanical work such as muscle contraction and other cellular movements, for example the movement of cilia and flagella, amoeboid

movement and the movement of vesicles through cytoplasm;

■ in a few organisms, bioluminescence and electrical discharge.

Mammals and birds use thermal energy from metabolic reactions to maintain a constant body temperature.

Two of these forms of work, active transport and muscle contraction, will be looked at in more detail later (*boxes 15A* and *15B*).

For a living organism to do work, energy-requiring reactions must be linked to those that yield energy. In the complete oxidation of glucose ($C_6H_{12}O_6$) in aerobic conditions a large quantity of energy is made available:

$$C_6H_{12}O_6 + 6O_2 \rightarrow 6CO_2 + 6H_2O + 2870 \, kJ$$

Reactions such as this take place in a series of small steps, each releasing a small quantity of the total available energy. You may remember that multi-step reactions allow precise control via feed-back mechanisms (see chapter 3) but this and other such advantages are in addition to the basic fact that the cell could not usefully harness the total available energy if all of it were made available at one instant.

Although the complete oxidation of glucose to carbon dioxide and water has a very high energy yield, the reaction does not happen easily. Glucose is actually quite stable, because of the **activation energy** that has to be overcome before any reaction takes place (*figure 15.2*). In living

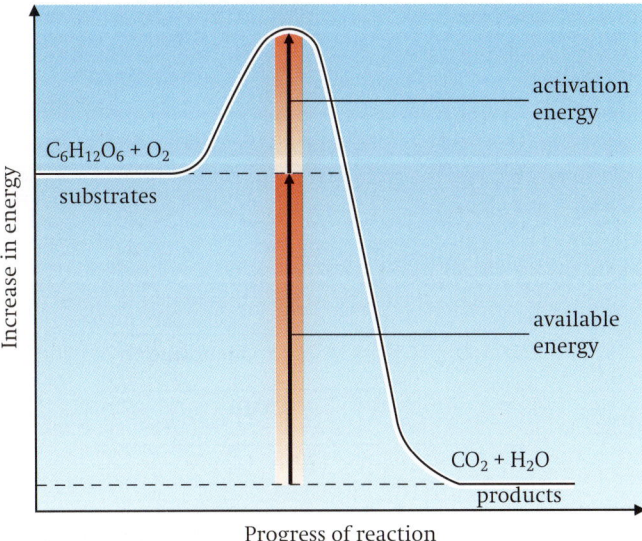

● **Figure 15.2** Oxidation of glucose.

organisms this is overcome by lowering the activation energy using enzymes (see page 42) and also by raising the energy level of the glucose by phosphorylation (page 202).

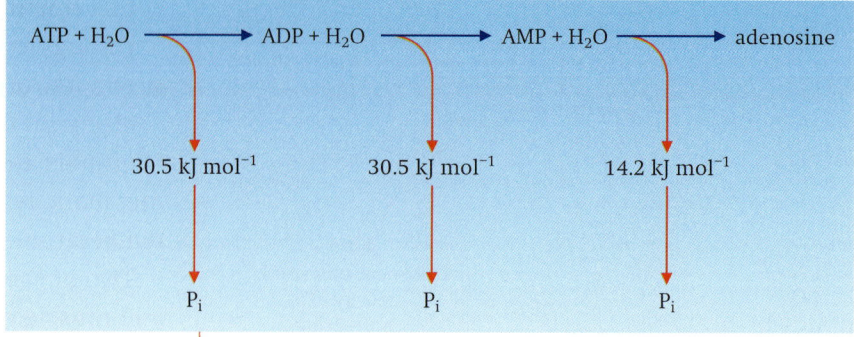

using enzymes (see page 42)

● **Figure 15.4** Hydrolysis of ATP. ($P_i$ is inorganic phosphate, $H_3PO_4$.)

Theoretically, the energy released from each step of respiration could be harnessed directly to some form of work in the cell. However, a much more flexible system actually occurs in which energy-yielding reactions in *all* organisms are linked to the production of an intermediary molecule, **ATP** (adenosine triphosphate).

# ATP

## ATP as energy 'currency'

The structure of adenosine triphosphate (ATP) is shown in *figure 15.3*. It consists of adenine (an organic base) and ribose (a pentose sugar), which together make adenosine (a nucleoside). This is combined with three phosphate groups to make ATP. ATP is therefore a nucleotide (see page 65). ATP is a small, water-soluble molecule. This allows it to be easily transported around the cell.

When a phosphate group is removed from ATP, adenosine diphosphate (ADP) is formed and $30.5 \, kJ \, mol^{-1}$ of energy is released. Removal of a second phosphate produces adenosine monophosphate (AMP) and $30.5 \, kJ \, mol^{-1}$ of energy is again

released. Removal of the last phosphate, leaving adenosine, releases only $14.2 \, kJ \, mol^{-1}$ (*figure 15.4*). In the past, the bonds attaching the two outer phosphate groups have been called 'high-energy bonds', because more energy is released when they are broken than when the last phosphate is removed. This is misleading and should be avoided since the energy does not come simply from breaking those bonds, but rather from changes in chemical potential energy of all parts of the system.

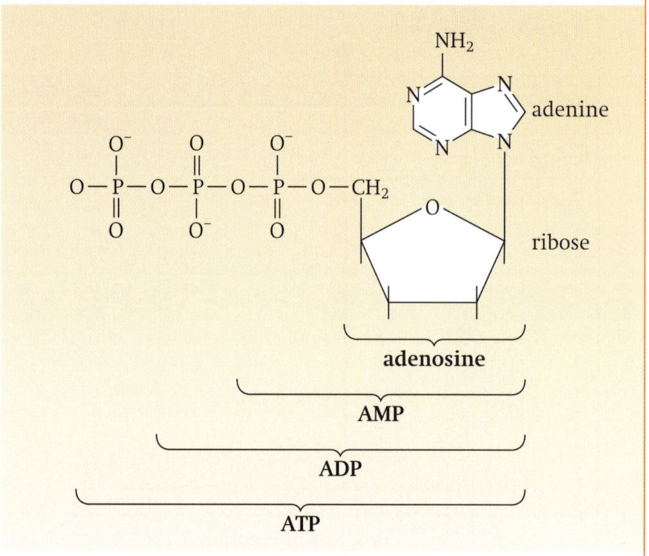

These reactions are all reversible and it is the interconversion of ATP and ADP that is all-important in providing energy for the cell:

$$ATP + H_2O \rightleftharpoons ADP + H_3PO_4 \pm 30.5 \, kJ$$

The rate of interconversion, or turnover, is enormous. It is estimated that a resting human uses about 40 kg of ATP in 24 hours, but at any one time contains only about 5 g of ATP. During strenuous exercise, ATP breakdown may be as much as 0.5 kg per minute.

The cell's energy-yielding reactions are linked to ATP synthesis. The ATP is then used by the cell in all forms of work. ATP is the universal intermediary molecule between energy-yielding and energy-requiring reactions used in a cell, whatever its type. In other words, ATP is the 'energy currency' of the cell. The cell 'trades' in ATP rather than making use of a number of different intermediates.

Energy transfers are inefficient. Some energy is converted to thermal energy whenever energy is transferred. At the different stages in a multi-step reaction, such as respiration, the energy made available may not perfectly correspond with the energy needed to synthesise ATP. Any 'excess'

● **Figure 15.3** Structure of ATP.

energy is converted to thermal energy. Also, many energy-requiring reactions in cells use less energy than that released by hydrolysis of ATP to ADP. Again, any extra energy will be released as thermal energy.

Be careful to distinguish between molecules used as energy *currency* and as energy *storage*. An energy currency molecule acts as the immediate donor of energy to the cell's energy-requiring reactions. An energy storage molecule is a short-term (glucose or sucrose) or long-term (glycogen, starch or triglyceride) store of chemical potential energy.

## Synthesis of ATP

Energy for ATP synthesis can become available in two ways. In respiration, energy released by reorganising chemical bonds (chemical potential energy) during glycolysis and the Krebs cycle (pages 201–204) is used to make some ATP. However, most ATP in cells is generated using electrical potential energy. This energy is from the transfer of electrons by electron carriers in mitochondria and chloroplasts (page 204). It is stored as a difference in hydrogen ion concentration across some phospholipid membranes in mitochondria and chloroplasts

which are essentially impermeable to hydrogen ions. Hydrogen ions are then allowed to flow down their concentration gradient through a protein which spans the phospholipid bilayer. Part of this protein acts as an enzyme which synthesises ATP, and is called **ATP synthase**. The transfer of three hydrogen ions allows the production of one ATP molecule provided that ADP and an inorganic phosphate group ($P_i$) are available inside the organelle. This process occurs in both mitochondria (page 206) and chloroplasts (page 218) and is summarised in *figure 15.5*. The process was first proposed by Peter Mitchell in 1961 and is called **chemiosmosis**.

ATP synthase has three binding sites (*figure 15.6*) and a part of the molecule ($\gamma$) that rotates as hydrogen ions pass. This produces structural changes in the binding sites and allows them to pass sequentially through three phases:

- binding ADP and $P_i$;
- forming tightly bound ATP;
- releasing ATP.

### SAQ 15.1
Write the equation for the reaction catalysed by ATP synthase.

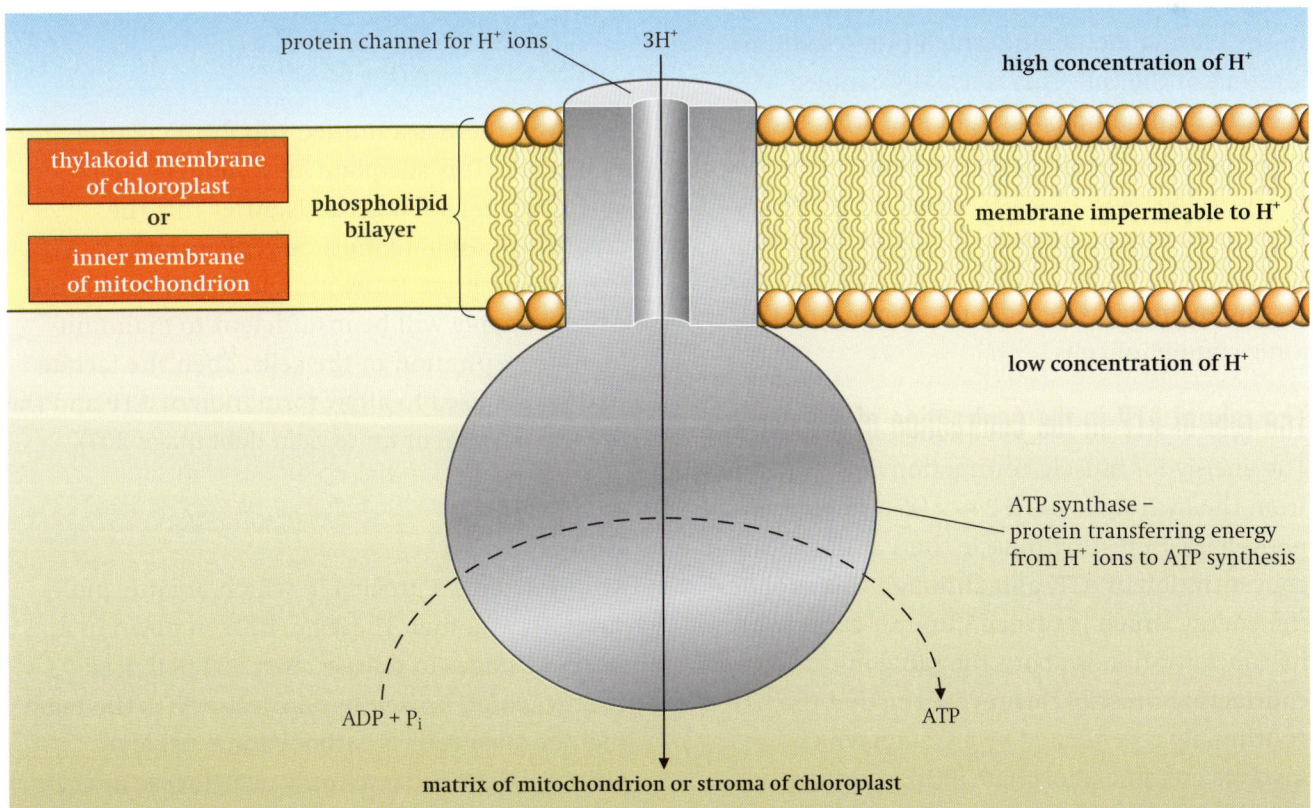

● **Figure 15.5** ATP synthesis.

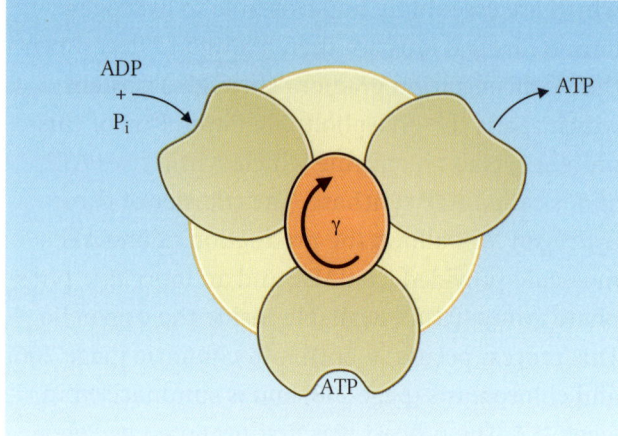

● **Figure 15.6** Transverse section (TS) of ATP synthase showing its activity.

## The role of ATP in active transport

**Active transport** is the movement of molecules or ions across a differentially permeable membrane against a concentration gradient. Energy is needed, in the form of ATP, to counteract the tendency of these particles to move by diffusion down the gradient.

All cells show differences in concentration of ions, in particular sodium and potassium ions, inside the cell with respect to the surrounding solution. Most cells seem to have sodium pumps in the plasma membrane which pump sodium ions out of the cell. This is usually coupled with the ability to pump potassium ions from the surrounding solution into the cell (*box 15A*).

The importance of active transport in ion movement into and out of cells should not be underestimated. About 50% of the ATP used by a resting mammal is devoted to maintaining the ionic content of cells.

## The role of ATP in the contraction of muscle

The energy for muscle contraction (*box 15B*) comes from the hydrolysis of ATP to ADP and inorganic phosphate. In resting muscle there is only a small concentration of ATP, and although this supplies the energy which is turned into muscular work, its concentration is about the same in resting and contracting muscle. During contraction the ATP is continually regenerated by a system which involves creatine phosphate (PCr). A resting muscle may contain around 20 mmol kg$^{-1}$ of PCr compared with 6 mmol kg$^{-1}$ of ATP.

### Box 15A The sodium–potassium pump

The sodium–potassium pump is a protein which spans the plasma membrane. It has binding sites for sodium ions (Na$^+$) and for ATP on the inner side, and for potassium ions (K$^+$) on the outer side. The protein acts as an ATPase, and catalyses the hydrolysis of ATP to ADP and inorganic phosphate, releasing energy to drive the pump. Changes in the shape of the protein move sodium and potassium ions across the membrane in opposite directions. For each ATP used, two potassium ions move into the cell and three sodium ions move out of the cell. Since only two potassium ions are added to the cell contents for every three sodium ions removed, a potential difference is created across the membrane which is negative inside with respect to the outside. Both sodium and potassium ions leak back across the membrane, down their diffusion gradients. However plasma membranes are much less permeable to sodium ions than potassium ions, so this diffusion actually increases the potential difference across the membrane.

This potential difference is most clearly seen as the resting potential of a nerve cell (see page 278). One of the specialisations of a nerve cell is an exaggeration of the potential difference across the plasma membrane as a result of the activity of the sodium–potassium pump.

The ADP produced during muscle contraction is reconverted to ATP by transferring a phosphate group from creatine phosphate, leaving creatine (Cr).

$$ADP + PCr \rightarrow ATP + Cr$$

However, there is a limited supply of creatine phosphate. It is adequate for a sudden, short sprint lasting a few seconds. After this the creatine phosphate must be replenished via ATP from respiration. If the muscle is very active, the oxygen supply will be insufficient to maintain aerobic respiration in the cells. Then the lactate pathway is used to allow formation of ATP and the muscle cells incur an oxygen debt (page 207).

## Respiration

**Respiration** is a process in which organic molecules act as a fuel. These are broken down in a series of stages to release chemical potential energy which is used to synthesise ATP. The main fuel for most cells is carbohydrate, usually glucose. Many cells can only use glucose as their respiratory substrate, but others break down fatty acids, glycerol and amino acids in respiration.

## Box 15B  Muscle contraction

A sarcomere contracts by sliding the thin **actin** filaments over the thick **myosin** filaments. Myosin filaments are made up of many myosin molecules, each with a flexible 'head'. This head is an ATPase molecule, which can hydrolyse ATP to ADP and $P_i$. In resting muscle, the ADP and $P_i$ are bound to the head.

When the muscle is activated by a nerve impulse, calcium ions are released from the **sarcoplasmic reticulum** (specialised endoplasmic reticulum). They allow the myosin head to bind to the portion of actin filament next to it. The myosin head then tilts about 45°, moving the attached actin filament about 10 nm in relation to the myosin, towards the centre of the sarcomere. This is the 'power stroke'. The combined effect of millions of such power strokes makes the muscle contract. At the same time, the ADP and $P_i$ are released from the head.

Then another ATP binds to the head and is hydrolysed to ADP and $P_i$, releasing energy that allows the actin and myosin to separate. The head tilts back to its original position, ready for the cycle to repeat – which can happen about five times per second. Note that the hydrolysis of ATP and the power stroke do not occur at the same time.

When excitation ceases, ATP is again needed to pump calcium ions back into the sarcoplasmic reticulum.

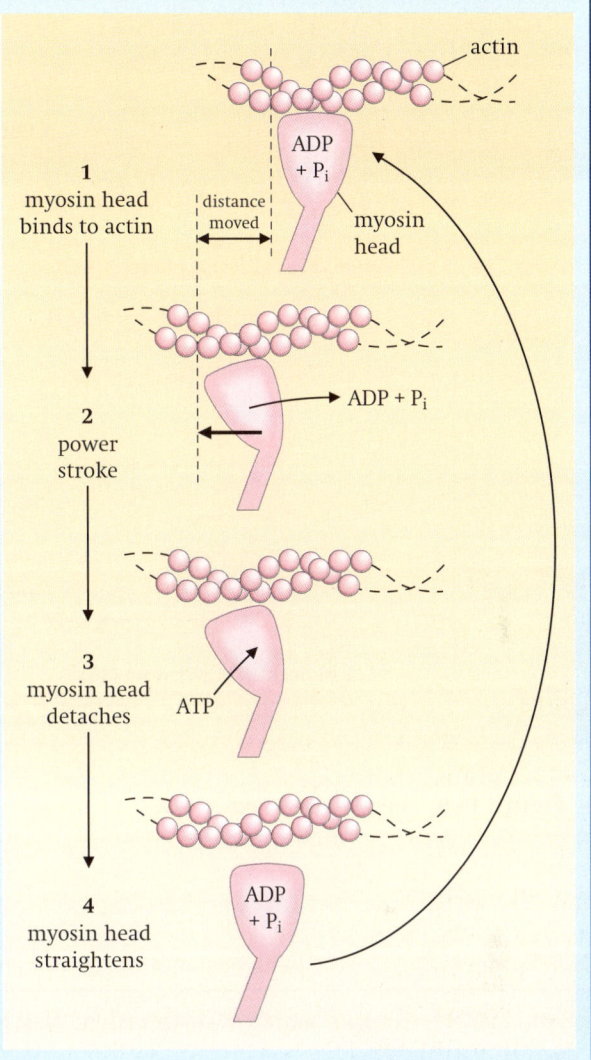

● **Figure 15.7** The action of ATP in muscle contraction.

Glucose breakdown can be divided into four stages: **glycolysis**, the **link reaction**, the **Krebs cycle** and **oxidative phosphorylation** (*figure 15.8*).

## The glycolytic pathway

Glycolysis is the splitting, or **lysis** of glucose. It is a multi-step process in which a glucose molecule with six carbon atoms is eventually split into two molecules of pyruvate, each with three carbon atoms. Energy from ATP is needed in the first steps, but energy is *released* in later steps, when it can be used to make ATP. There is a net gain of two ATP

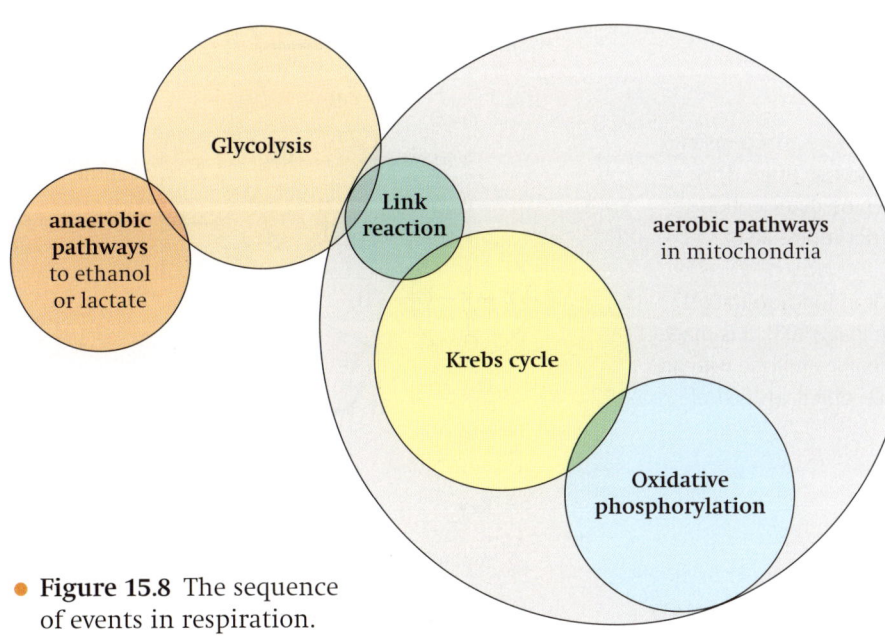

● **Figure 15.8** The sequence of events in respiration.

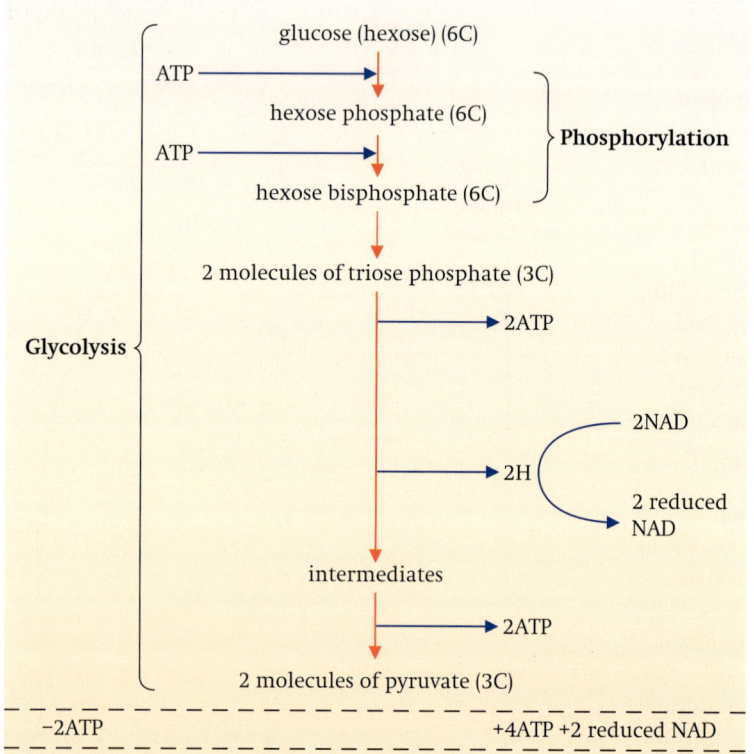

● **Figure 15.9** The glycolytic pathway.

molecules per molecule of glucose broken down. Glycolysis takes place in the cytoplasm of a cell. A simplified flow diagram of the pathway is shown in *figure 15.9*.

In the first stage, **phosphorylation**, glucose is phosphorylated using ATP. As we saw on page 197, glucose is energy-rich, but does not react easily. To tap the bond energy of glucose, energy must first be used to make the reaction easier (*figure 15.2*). Two ATP molecules are used for each molecule of glucose to make hexose bisphosphate, which breaks down to produce two molecules of triose phosphate.

Hydrogen is then removed from triose phosphate and transferred to the carrier molecule NAD (nicotinamide adenine dinucleotide). The structure of NAD is shown in *box 15C, figure 15.10*. Two molecules of reduced NAD are produced for each molecule of glucose entering glycolysis. The hydrogens carried by reduced NAD can easily be transferred to other molecules and are used in oxidative phosphorylation to generate ATP (page 204).

## Box 15C Hydrogen carrier molecules: NAD, NADP and FAD

NAD (nicotinamide adenine dinucleotide) is made of two linked nucleotides. Both nucleotides contain ribose. One nucleotide contains the nitrogenous base adenine. The other has a nicotinamide ring, which can accept a hydrogen ion and two electrons, thereby becoming reduced.

$$NAD + 2H \rightleftharpoons reduced\ NAD$$
$$NAD^+ + 2H \rightleftharpoons NADH^+ + H^+$$

A slightly different form of NAD has a phosphate group instead of the hydrogen on carbon 1 in one of the ribose rings. This molecule is called NADP (nicotinamide adenine dinucleotide phosphate) and is used as a hydrogen carrier molecule in photosynthesis.

FAD (flavin adenine dinucleotide) is similar in function to NAD and is used in respiration in the Krebs cycle (page 203). It is made of one nucleotide containing ribose and adenine and one with an unusual structure involving a linear molecule, ribitol, instead of ribose.

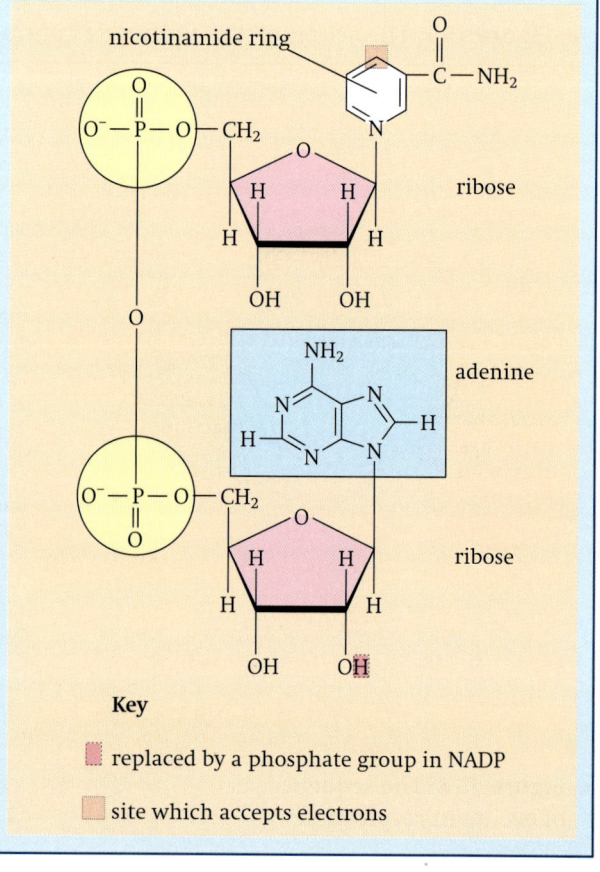

● **Figure 15.10** NAD.

**Key**

▇ replaced by a phosphate group in NADP

▇ site which accepts electrons

The end-product of glycolysis, pyruvate, still contains a great deal of chemical potential energy. When free oxygen is available, some of this energy can be released via the Krebs cycle and oxidative phosphorylation. However, the pyruvate first enters the link reaction, which takes place in the mitochondria (page 206).

### SAQ 15.2
How does the linkage between the nucleotides in NAD differ from that in a polynucleotide? (You may need to refer back to page 67 to answer this question.)

## The link reaction
Pyruvate passes by active transport from the cytoplasm, through the outer and inner membranes of a mitochondrion and into the mitochondrial matrix. Here it is decarboxylated (that is carbon dioxide is removed), dehydrogenated and combined with coenzyme A (CoA) to give acetyl coenzyme A. This is known as the **link reaction** (figure 15.11). Coenzyme A is a complex molecule of a nucleoside (adenine + ribose) with a vitamin (pantothenic acid), and acts as a carrier of acetyl groups to the Krebs cycle. The hydrogen removed from pyruvate is transferred to NAD.

pyruvate + CoA + NAD
$\rightleftharpoons$ acetyl CoA + $CO_2$ + reduced NAD

Fatty acids from fat metabolism may also be used to produce acetyl coenzyme A. Fatty acids are broken down in the mitochondrion in a cycle of reactions in which each turn of the cycle shortens the fatty acid chain by a two-carbon acetyl unit. Each of these can react with coenzyme A to produce acetyl coenzyme A, which, like that produced from pyruvate, now enters the Krebs cycle.

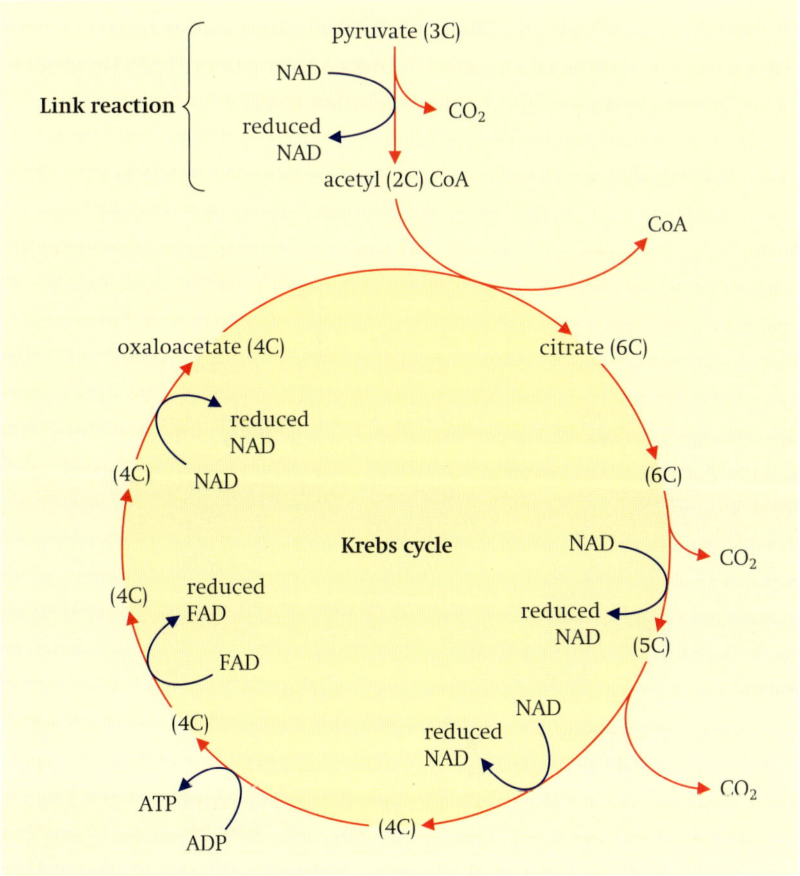

## The Krebs cycle
The Krebs cycle (also known as the citric acid cycle or tricarboxylic acid cycle) was discovered in 1937 by Hans Krebs. It is shown in figure 15.11.

The Krebs cycle is a closed pathway of enzyme-controlled reactions:

- acetyl coenzyme A combines with a four-carbon compound (oxaloacetate) to form a six-carbon compound (citrate);
- the citrate is decarboxylated and dehydrogenated in a series of steps, to yield carbon dioxide, which is given off as a waste gas, and hydrogens which are accepted by the carriers NAD and FAD (flavin adenine dinucleotide) (box 15C);
- oxaloacetate is regenerated to combine with another acetyl coenzyme A.

For each turn of the cycle, two carbon dioxide molecules are produced, one FAD and three NAD molecules are reduced, and one ATP molecule is generated via an intermediate compound.

Although part of aerobic respiration, the reactions of the Krebs cycle make no use of molecular oxygen. However, oxygen is necessary for the final stage which is called oxidative phosphorylation.

● **Figure 15.11** The link reaction and the Krebs cycle.

The most important contribution of the Krebs cycle to the cell's energetics is the release of hydrogens, which can be used in oxidative phosphorylation to provide energy to make ATP.

## SAQ 15.3
Explain how the events of the Krebs cycle can be cyclical.

## Oxidative phosphorylation and the electron transport chain
In the final stage of aerobic respiration, the energy for the phosphorylation of ADP to ATP comes from the activity of the electron transport chain. This takes place in the mitochondrial membranes.

Reduced NAD and reduced FAD are passed to the electron transport chain. Here, hydrogens are removed from the two hydrogen carriers and each is split into its constituent hydrogen ion ($H^+$) and electron. The electron is transferred to the first of a series of electron carriers (box 15D), whilst the hydrogen ion remains in solution in the mitochondrial matrix. Once the electron is transferred to oxygen (also in solution in the matrix), a hydrogen ion will be drawn from solution to reduce the oxygen to water (figure 1.12).

### Box 15D Electron carrier molecules
Electron carriers are almost all proteins and mostly **cytochromes**. These proteins have a haem prosthetic group (see page 36) in which the iron atom oscillates between the $Fe^{2+}$ and $Fe^{3+}$ states as it accepts an electron from the previous carrier, and passes it on to the next carrier, in the electron transport chain.

The transfer of electrons along the series of electron carriers makes energy available which is used to convert ADP + $P_i$ to ATP. As an electron passes from a carrier at a higher energy level to one that is lower, energy is released. This is usually lost as heat, but at particular points in the chain the energy released is sufficient to produce ATP.

Potentially, three molecules of ATP can be produced from each reduced NAD molecule and two ATP from each reduced FAD molecule (figure 15.12). However, this yield cannot be realised unless ADP and $P_i$ are available inside the mitochondrion. About 25% of the total energy yield of electron transfer is used to transport ADP into the mitochondrion, and ATP into the cytoplasm. Hence, each reduced NAD molecule entering the chain produces on average two and a half molecules of ATP and each reduced FAD produces one and a half molecules of ATP.

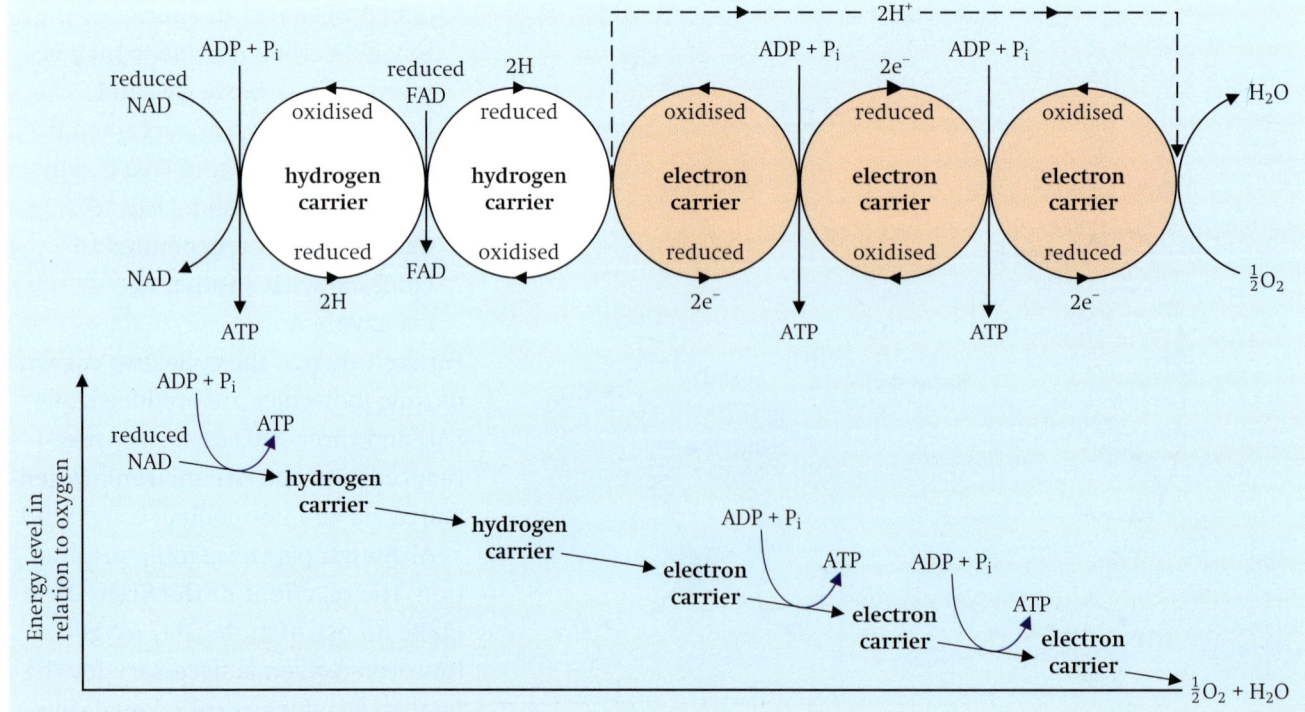

● **Figure 15.12** Oxidative phosphorylation: the electron transport chain.

The most widely accepted explanation for the synthesis of ATP in oxidative phosphorylation is that of chemiosmosis (page 199). The energy released by the electron transport chain is used to pump hydrogen ions from the mitochondrial matrix into the space between the two membranes of the mitochondrial envelope. The concentration of hydrogen ions in the intermembrane space therefore becomes higher than that in the matrix, so a concentration gradient is set up. Hydrogen ions pass back into the mitochondrial matrix through protein channels in the inner membrane. Associated with each channel is the enzyme ATP synthase. As the ions pass through the channel, their electrical potential energy is used to synthesise ATP (*figure 15.5*).

The sequence of events in respiration and their sites are shown in *figure 15.13*. The balance sheet of ATP use and synthesis for each molecule of glucose entering the respiration pathway is shown in *table 15.1*.

|  | ATP used | ATP made | Net gain in ATP |
|---|---|---|---|
| Glycolysis | −2 | 4 | +2 |
| Link reaction | 0 | 0 | 0 |
| Krebs cycle | 0 | 2 | +2 |
| Oxidative phosphorylation | 0 | 28 | +28 |
| Total | −2 | 34 | +32 |

● **Table 15.1** Balance sheet of ATP use and synthesis for each molecule of glucose entering respiration.

### SAQ 15.4

Calculate the number of reduced NAD and reduced FAD molecules produced for each molecule of glucose entering the respiration pathway when oxygen is available.

### SAQ 15.5

Using your answer to SAQ 15.4, calculate the number of ATP molecules produced for each molecule of glucose in oxidative phosphorylation.

### SAQ 15.6

Explain why the important contribution of the Krebs cycle to cellular energetics is the release of hydrogens and not the direct production of ATP.

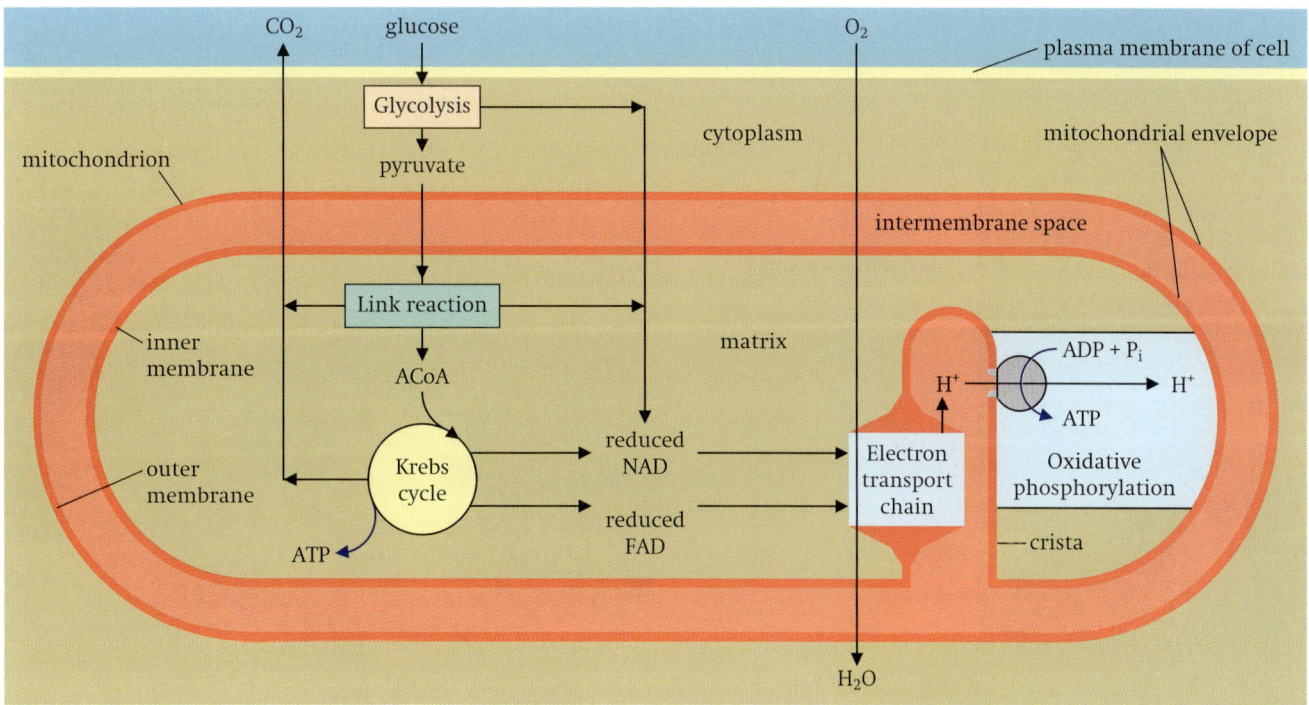

● **Figure 15.13** The sites of the events of respiration in a cell.

# Mitochondrial structure and function

In eukaryotic organisms, the mitochondrion is the site of the Krebs cycle and the electron transport chain. Mitochondria are rod-shaped or filamentous organelles about 0.5–1.0 µm in diameter. Time-lapse photography shows that they are not rigid, but can change their shape. The number of mitochondria in a cell depends on its activity. Mammalian liver cells contain between 1000 and 2000 mitochondria, occupying 20% of the cell volume.

The structure of a mitochondrion is shown in *figures 1.19* and *15.4*. Like a chloroplast, each mitochondrion is surrounded by an **envelope** of two phospholipid membranes (see page 51). The outer membrane is smooth, but the inner is much folded inwards to form **cristae** (singular **crista**). These give the inner membrane a large total surface area. Cristae in mitochondria from different types of cells show considerable variation, but, in general, mitochondria from active cells have longer, more densely packed cristae than those from less active cells. The two membranes have different compositions and properties. The outer membrane is relatively permeable to small molecules, whilst the inner membrane is less permeable. The inner membrane is studded with tiny spheres, about 9 nm in diameter, which are attached to the inner membrane by stalks (*figure 15.15*). The spheres are the enzyme **ATP synthase**.

The inner membrane is the site of the electron transport chain and contains the proteins necessary for this. The space between the two membranes of the envelope usually has a lower pH than the matrix of the mitochondrion as a result of the hydrogen ions that are released into the intermembrane space by the activity of the electron transport chain.

The **matrix** of the mitochondrion is the site of the link reaction and the Krebs cycle, and contains the enzymes needed for these reactions. It also contains small (70 S) **ribosomes** and several identical copies of looped mitochondrial **DNA**.

ATP is formed in the matrix by the activity of ATP synthase on the cristae. The energy for the production of ATP comes from the hydrogen ion gradient between the intermembrane space and the matrix. The ATP can be used for all the energy-requiring reactions of the cell, both inside and outside the mitochondrion.

### SAQ 15.7

Explain how the structure of a mitochondrion is adapted for its functions in aerobic respiration.

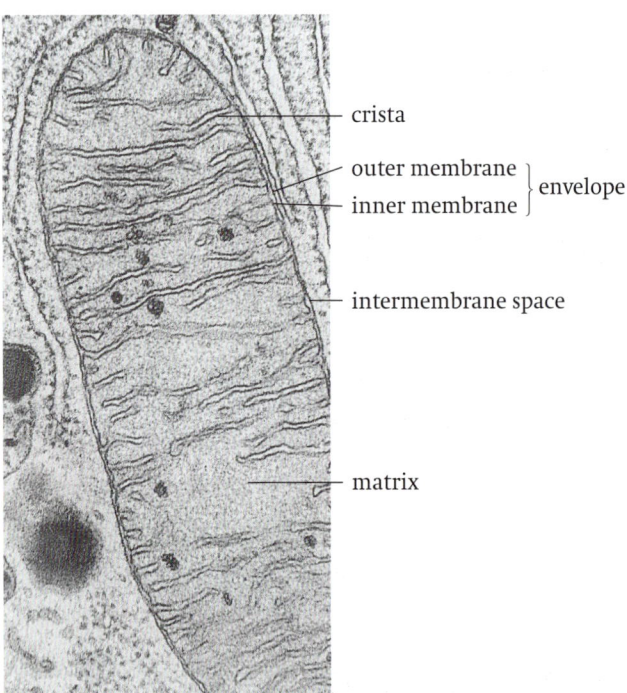

crista

outer membrane
inner membrane } envelope

intermembrane space

matrix

● **Figure 15.14** Transmission electron micrograph of a mitochondrion from a pancreas (× 15 000).

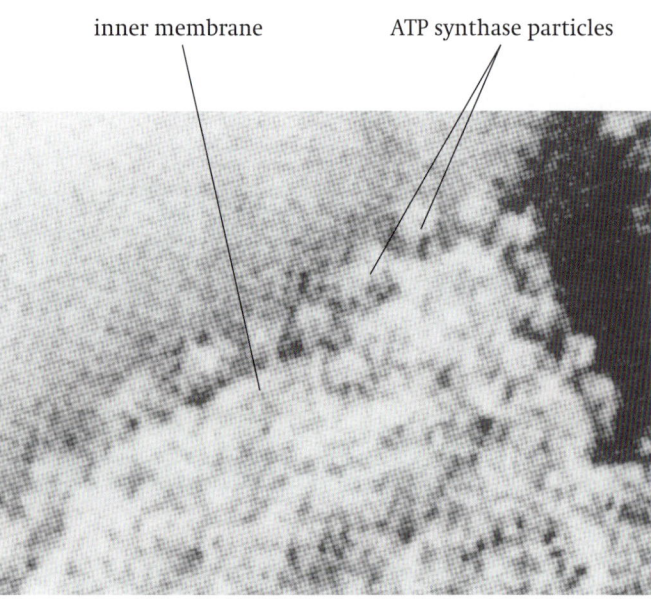

inner membrane          ATP synthase particles

● **Figure 15.15** TEM of ATP synthase particles on the inner membrane of a mitochondrion.

# Anaerobic respiration

When free oxygen is not present, hydrogen cannot be disposed of by combination with oxygen. The electron transfer chain therefore stops working and no further ATP is formed by oxidative phosphorylation. If a cell is to gain even the two ATP molecules for each glucose yielded by glycolysis, it is essential to pass on the hydrogens from the reduced NAD that are also made in glycolysis. There are two different anaerobic pathways which solve the problem of 'dumping' hydrogen. Both pathways take place in the cytoplasm of the cell.

In various microorganisms such as yeast, and in some plant tissues, the hydrogen from reduced NAD is passed to ethanal ($CH_3CHO$). This releases the NAD and allows glycolysis to continue. The pathway is shown in *figure 15.16*. First, pyruvate is decarboxylated to ethanal; then the ethanal is reduced to ethanol ($C_2H_5OH$) by the enzyme alcohol dehydrogenase. The conversion of glucose to ethanol is referred to as **alcoholic fermentation**.

In other microorganisms, and in mammalian muscles when deprived of oxygen, pyruvate acts as the hydrogen acceptor and is converted to lactate by the enzyme lactate dehydrogenase (named after the reverse reaction, which it also catalyses). Again, the NAD is released and allows glycolysis to continue in anaerobic conditions. This pathway is shown in *figure 15.17*.

These reactions 'buy time'. They allow the continued production of at least some ATP even

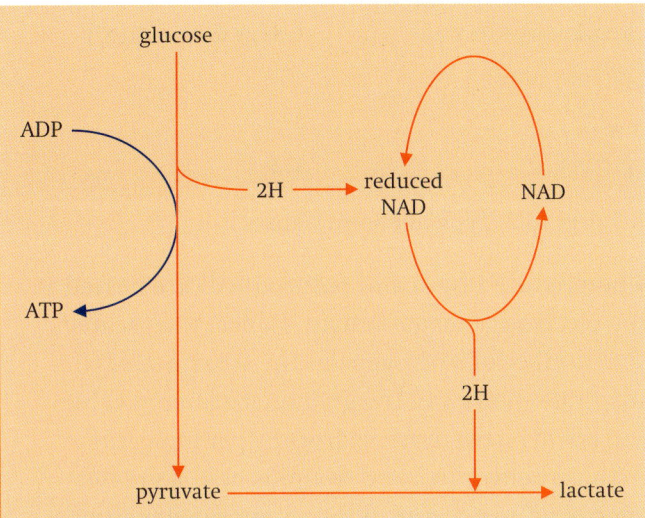

● **Figure 15.17** Anaerobic respiration: the lactate pathway.

though oxygen is not available as the hydrogen acceptor. However, since the products of anaerobic reaction, ethanol or lactate, are toxic, the reactions cannot continue indefinitely. The pathway leading to ethanol cannot be reversed and the remaining chemical potential energy of ethanol is wasted. The lactate pathway can be reversed in mammals. Lactate is carried by the blood plasma to the liver and converted back to pyruvate. The liver oxidises some (20%) of the incoming lactate to carbon dioxide and water via aerobic respiration when oxygen is available again. The remainder of the lactate is converted by the liver to glycogen. The oxygen needed to allow this removal of lactate is called the **oxygen debt**.

# Respiratory substrates

Although glucose is the essential respiratory substrate for some cells, such as neurones in the brain, red blood cells and lymphocytes, other cells can oxidise lipids and amino acids. When lipids are respired, carbon atoms are removed in pairs, as acetyl CoA, from the fatty acid chains and fed into the Krebs cycle. The carbon–hydrogen skeletons of amino acids are converted into pyruvate or into acetyl CoA.

## Energy values of respiratory substrates

Most of the energy liberated in aerobic respiration comes from the oxidation of hydrogen to water

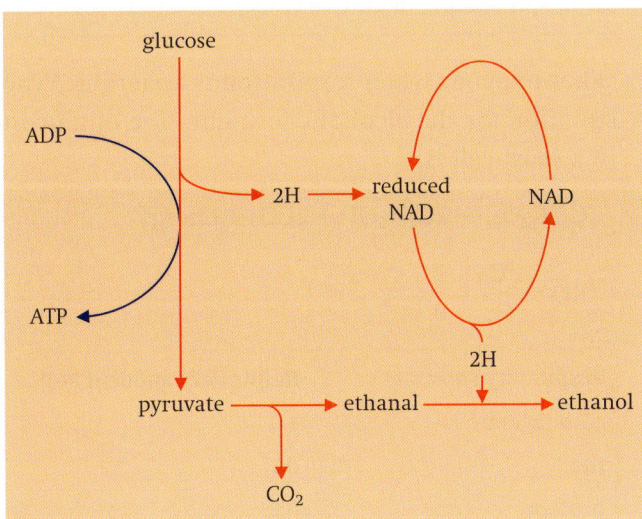

● **Figure 15.16** Anaerobic respiration: the ethanol pathway.

| Respiratory substrate | Energy density (kJ g⁻¹) |
|---|---|
| carbohydrate | 15.8 |
| lipid | 39.4 |
| protein | 17.0 |

● **Table 15.2** Typical energy values.

when reduced NAD and reduced FAD are passed to the electron transport chain. Hence, the greater the number of hydrogens in the structure of the substrate molecule, the greater the energy value. Fatty acids have more hydrogens per molecule than carbohydrates and so lipids have a greater energy value per unit mass, or **energy density**, than carbohydrates or proteins.

The energy value of a substrate is determined by burning a known mass of the substance in oxygen in a **calorimeter** (*figure 15.18*).

The energy liberated by oxidising the substrate can be determined from the rise in temperature of a known mass of water in the calorimeter. Typical energy values are shown in *table 15.2*.

## Respiratory quotient (RQ)

The overall equation for the aerobic respiration of glucose shows that the number of molecules, and hence the volumes, of oxygen used and carbon dioxide produced are the same:

$$C_6H_{12}O_6 + 6O_2 \rightarrow 6CO_2 + 6H_2O + energy$$

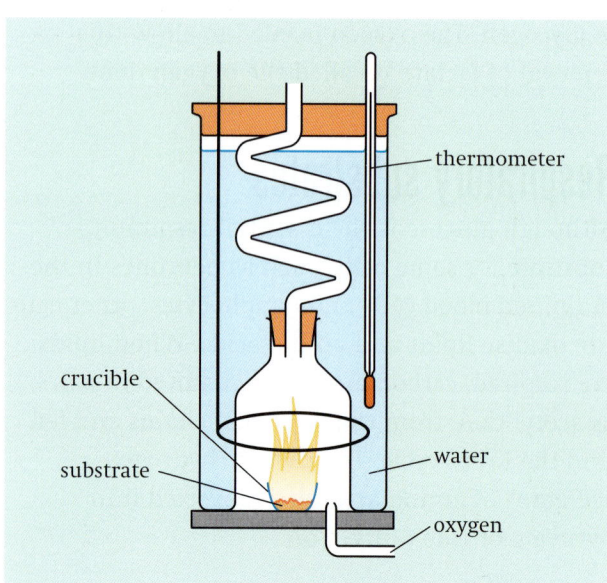

● **Figure 15.18** A simple calorimeter in which the energy value of a respiratory substrate can be measured.

So the **ratio** of $O_2$ taken in and $CO_2$ released is $1 : 1$. However, when other substrates are respired, the ratio of the volumes of oxygen used and carbon dioxide given off differ. It follows that measuring this ratio, called the **respiratory quotient** (**RQ**), shows what substrate is being used in respiration. It can also show whether or not anaerobic respiration is occurring.

$$RQ = \frac{volume\ of\ carbon\ dioxide\ given\ out\ in\ unit\ time}{volume\ of\ oxygen\ taken\ in\ in\ unit\ time}$$

Or, from an equation,

$$RQ = \frac{moles\ or\ molecules\ of\ carbon\ dioxide\ given\ out}{moles\ or\ molecules\ of\ oxygen\ taken\ in}$$

For the aerobic respiration of glucose,

$$RQ = \frac{CO_2}{O_2} = \frac{6}{6} = 1.0$$

When the fatty acid oleic acid (from olive oil) is respired aerobically the equation is:

$$C_{18}H_{34}O_2 + 25.5\ O_2 \rightarrow 18CO_2 + 17H_2O + energy$$

For the aerobic respiration of oleic acid,

$$RQ = \frac{CO_2}{O_2} = \frac{18}{25.5} = 0.7$$

Typical RQs for the aerobic respiration of different substrates are shown in *table 15.3*.

### SAQ 15.8

Calculate the RQ for the aerobic respiration of the fatty acid, stearic acid ($C_{18}H_{36}O_2$).

What happens when respiration is anaerobic? The equation for the alcoholic fermentation of glucose in a yeast cell is:

$$C_6H_{12}O_6 \rightarrow 2C_2H_5OH + 2CO_2 + energy$$

$$RQ = \frac{CO_2}{O_2} = \frac{2}{0} = \infty$$

| Respiratory substrate | Respiratory quotient (RQ) |
|---|---|
| carbohydrate | 1.0 |
| lipid | 0.7 |
| protein | 0.9 |

● **Table 15.3** Respiratory quotients of different substrates.

In reality, some respiration in the yeast cell will be aerobic and so a small volume of oxygen will be taken up and the RQ will be <2. High values of RQ indicate that anaerobic respiration is occurring: note that no RQ can be calculated for muscle cells using the lactate pathway since no carbon dioxide is produced:

glucose ($C_6H_{12}O_6$)
→ 2 lactic acid ($C_3H_6O_3$) + energy

Oxygen uptake during respiration can be measured using a **respirometer**. A respirometer suitable for measuring the rate of oxygen consumption of small terrestrial invertebrates at different temperatures is shown in *figure 15.19*.

Carbon dioxide produced in respiration is absorbed by a suitable chemical, such as soda-lime or a concentrated solution of potassium hydroxide or sodium hydroxide. Any decrease in the volume of air surrounding the organisms results from their oxygen consumption. Oxygen consumption in unit time can be measured by reading the level of the manometer fluid against the scale. Changes in temperature and pressure alter the volume of air in the apparatus and so the temperature of the surroundings must be kept constant whilst readings are taken, for example by using a thermostatically controlled water bath. The presence of a control tube containing an equal volume of inert material to the volume of the organisms used helps to compensate for changes in atmospheric pressure. Once measurements have been taken at a series of temperatures, a graph can be plotted of oxygen consumption against temperature.

The same apparatus can be used to measure the RQ of an organism. First, oxygen consumption at a particular temperature is found ($x$ cm$^3$ min$^{-1}$). Then the respirometer is set up with the same organism at the same temperature, but with no chemical to absorb carbon dioxide. The manometer scale will show whether the volumes of oxygen absorbed and carbon dioxide produced are the same. When the volumes *are* the same, the level of the manometer fluid will not change and the RQ = 1. When more carbon dioxide is produced than oxygen absorbed, the scale will show an increase in the volume of air in the respirometer (by $y$ cm$^3$ min$^{-1}$). The RQ can then be calculated:

$$RQ = \frac{CO_2}{O_2} = \frac{x + y}{x}$$

Conversely, when less carbon dioxide is produced than oxygen absorbed, the volume of air in the respirometer will decrease (by $z$ cm$^3$ min$^{-1}$) and the calculation will be:

$$RQ = \frac{CO_2}{O_2} = \frac{x - z}{x}$$

**SAQ 15.9**
Outline the steps you would take to investigate the effect of temperature on respiration rate.

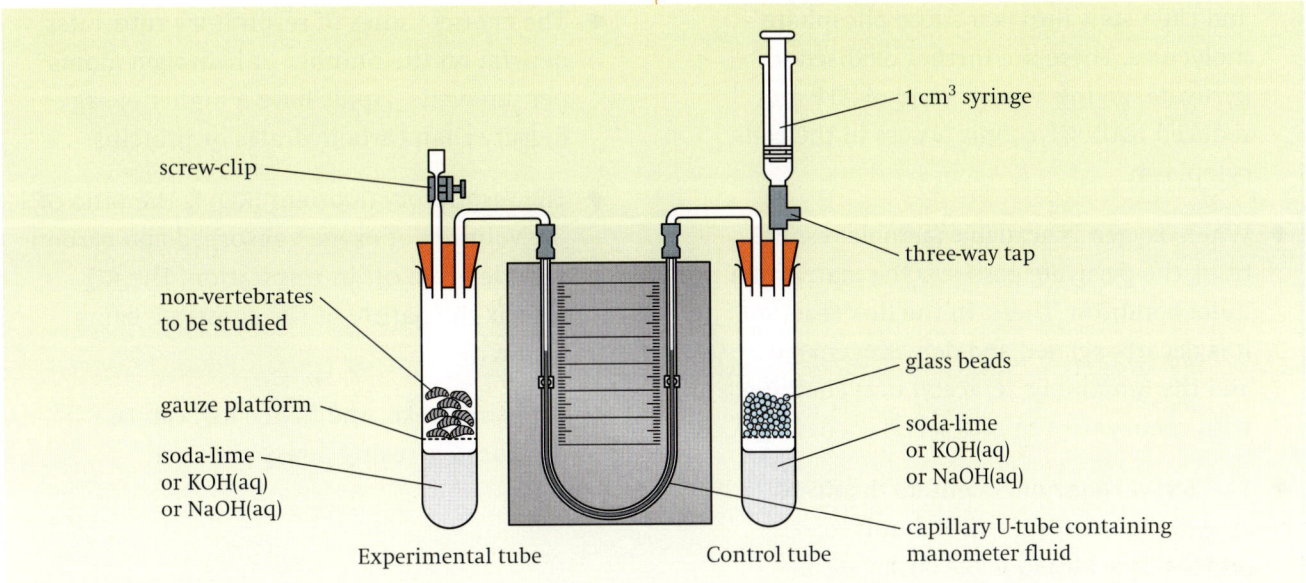

**Figure 15.19** A respirometer.

# SUMMARY

- Organisms must do work to stay alive. The energy input necessary for this work is either light, for photosynthesis, or the chemical potential energy of organic molecules. Photosynthesis traps light energy as chemical bond energy which can later be released and used by cells.

- Work includes anabolic reactions, active transport and mechanical work.

- Reactions which release energy must be harnessed to energy-requiring reactions. This 'harnessing' involves an intermediary molecule, ATP. This can be synthesised from ADP and phosphate using energy, and hydrolysed to ADP and phosphate to release energy. ATP therefore acts as an energy currency.

- Respiration is the sequence of enzyme-controlled steps by which an organic molecule, usually glucose, is broken down so that its chemical potential energy can be used to make the energy currency, ATP.

- In aerobic respiration, the sequence involves four main stages: glycolysis, the link reaction, the Krebs cycle and oxidative phosphorylation.

- In glycolysis, glucose is first phosphorylated and then split into two triose phosphate molecules. These are further oxidised to pyruvate, giving a small yield of ATP and reduced NAD. Glycolysis occurs in the cell cytoplasm.

- When oxygen is available (aerobic respiration), the pyruvate passes to the matrix of a mitochondrion. There, in the link reaction, it is decarboxylated and dehydrogenated and the remaining 2C acetyl unit combined with coenzyme A to give acetyl coenzyme A.

- The acetyl coenzyme A enters the Krebs cycle and donates the acetyl unit to oxaloacetate (4C) to make citrate (6C).

- The Krebs cycle decarboxylates and dehydrogenates citrate to oxaloacetate in a series of small steps. The oxaloacetate can then react with another acetyl coenzyme A from the link reaction.

- Dehydrogenation provides hydrogen atoms which are accepted by the carriers NAD and FAD. These pass to the inner membrane of the mitochondrial envelope where the hydrogens are split into hydrogen ions and electrons.

- The electrons are passed along a series of carriers. Some of the energy released in this process is used to phosphorylate ADP to ATP. The phosphorylation depends on a gradient of hydrogen ions set up across the inner membrane of the mitochondrial envelope.

- At the end of the carrier chain, electrons and protons are recombined and reduce oxygen to water.

- In the absence of oxygen as a hydrogen acceptor (anaerobic respiration), a small yield of ATP is made by dumping hydrogen into other pathways in the cytoplasm which produce ethanol or lactate.

- The energy values of respiratory substrates depend on the number of hydrogen atoms per molecule. Lipids have a higher energy density than carbohydrates or proteins.

- The respiratory quotient (RQ) is the ratio of the volumes of oxygen absorbed and carbon dioxide given off in respiration. The RQ reveals the nature of the substrate being respired.

- Oxygen uptake, and hence RQ, can be measured by using a respirometer.

## CHAPTER 16

# Photosynthesis

**By the end of this chapter you should be able to:**

1 explain that photosynthesis traps light energy as chemical energy in organic molecules, and that respiration releases this energy in a form which can be used by living organisms;

2 describe the photoactivation of chlorophyll that results in the conversion of light energy into chemical energy of ATP and reduction of NADP;

3 describe in outline the Calvin cycle involving the light-independent fixation of carbon dioxide by combination with a five-carbon compound (RuBP) to yield a three-carbon compound, GP (PGA), and the subsequent conversion of this compound into carbohydrates, amino acids and lipids;

4 describe the structure of a dicotyledonous leaf, a palisade cell and a chloroplast and relate these structures to their roles in photosynthesis;

5 discuss limiting factors in photosynthesis.

## An energy transfer process

As you have seen at the beginning of chapter 15, the process of photosynthesis transfers light energy into chemical potential energy of organic molecules. This energy can then be released for work in respiration (*figure 15.1*). Almost all the energy transferred to all the ATP molecules in all living organisms is derived from light energy used in photosynthesis by autotrophs. Such **photoautotrophs** are green plants, the photo-synthetic prokaryotes and both single-celled and many-celled protoctists (including the green, red and brown algae). A few autotrophs do *not* depend on light energy, but use chemical energy sources. These **chemoautotrophs** include the nitrifying bacteria which are so important in the nitrogen cycle. These bacteria obtain their energy from oxidising ammonia to nitrite, or nitrite to nitrate.

## An outline of the process

Photosynthesis is the trapping (fixation) of carbon dioxide and its subsequent reduction to carbohydrate, using hydrogen from water.

An overall equation for photosynthesis in green plants is:

$$nCO_2 + nH_2O \xrightarrow[\text{in the presence of chlorophyll}]{\text{light energy}} (CH_2O)n + nO_2$$

carbon dioxide + water → carbohydrate + oxygen

Hexose sugars and starch are commonly formed, so the following equation is often used:

$$6CO_2 + 6H_2O \xrightarrow[\text{in the presence of chlorophyll}]{\text{light energy}} C_6H_{12}O_6 + 6O_2$$

carbon dioxide + water → carbohydrate + oxygen

Two sets of reactions are involved. These are the **light-dependent reactions**, for which light energy is necessary, and the **light-independent reactions**, for which light energy is not needed. The light-dependent reactions only take place in the presence of suitable pigments which absorb certain wavelengths of light. Light energy is necessary for the splitting of water into hydrogen and oxygen; oxygen is a waste product. Light energy is also needed to provide chemical energy (ATP) for the reduction of carbon dioxide to carbohydrate in the light-independent reactions.

# Trapping light energy

Light energy is trapped by photosynthetic pigments. Different pigments absorb different wavelengths of light. The photosynthetic pigments of higher plants form two groups: the **chlorophylls** and the **carotenoids** (table 16.1).

Chlorophylls absorb mainly in the red and blue-violet regions of the light spectrum. They reflect

| Pigment | | Colour |
|---|---|---|
| Chlorophylls: | chlorophyll *a* | yellow-green |
| | chlorophyll *b* | blue-green |
| Carotenoids: | β carotene | orange |
| | xanthophyll | yellow |

● **Table 16.1** The colours of the commonly occurring photosynthetic pigments.

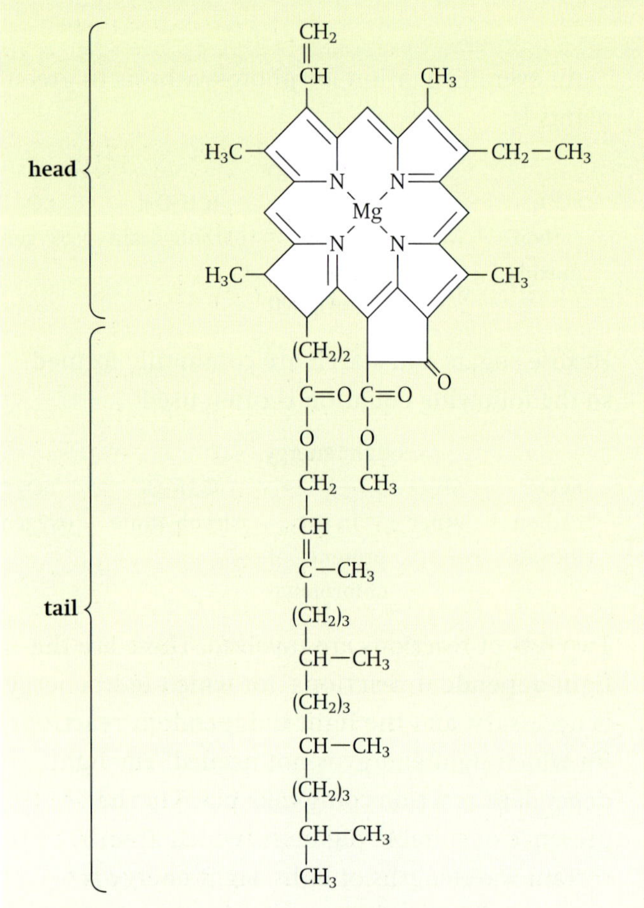

● **Figure 16.1** Structure of chlorophyll *a*.

green light, which is why plants look green. The structure of chlorophyll *a* is shown in *figure 16.1*. The carotenoids absorb mainly in the blue-violet region of the spectrum.

An **absorption spectrum** is a graph of the absorbance of different wavelengths of light by a pigment. The absorption spectra of chlorophyll *a* and *b*, and of the carotenoids can be seen in *figure 16.2a*.

An **action spectrum** is a graph of the rate of photosynthesis at different wavelengths of light (*figure 16.2b*). This shows the effectiveness of the different wavelengths, which is, of course, related to their absorption and to their energy content. The shorter the wavelength, the greater the energy it contains.

In the process of photosynthesis, the light energy absorbed by the photosynthetic pigments is converted to chemical energy. The absorbed light energy excites electrons in the pigment molecules. If you illuminate a solution of chlorophyll *a* or *b* with ultraviolet light, you will see a red fluorescence. (In the absence of a safe ultraviolet light, you can illuminate the pigment with a standard fluorescent tube.) The ultraviolet light is absorbed and electrons are excited but, in a solution which only contains extracted pigment, the absorbed energy cannot usefully be passed on to do work. The electrons return to their unexcited state and the absorbed energy is transferred to the surroundings as thermal energy and as light at a longer (less energetic) wavelength than that which was absorbed, and is seen as the red fluorescence. In the functioning photosynthetic system it is this energy that drives the process of photosynthesis.

The photosynthetic pigments fall into two categories: **primary pigments** and **accessory pigments**. The primary pigments are two forms of chlorophyll *a* with slightly different absorption peaks. The accessory pigments include other forms of chlorophyll *a*, chlorophyll *b* and the carotenoids. The pigments are arranged in light-harvesting clusters called **photosystems**. In a photosystem, several hundred accessory pigment molecules surround a primary pigment molecule and the energy of the light absorbed by the different pigments is passed to the primary pigment (*figure 16.3*). The primary pigments are

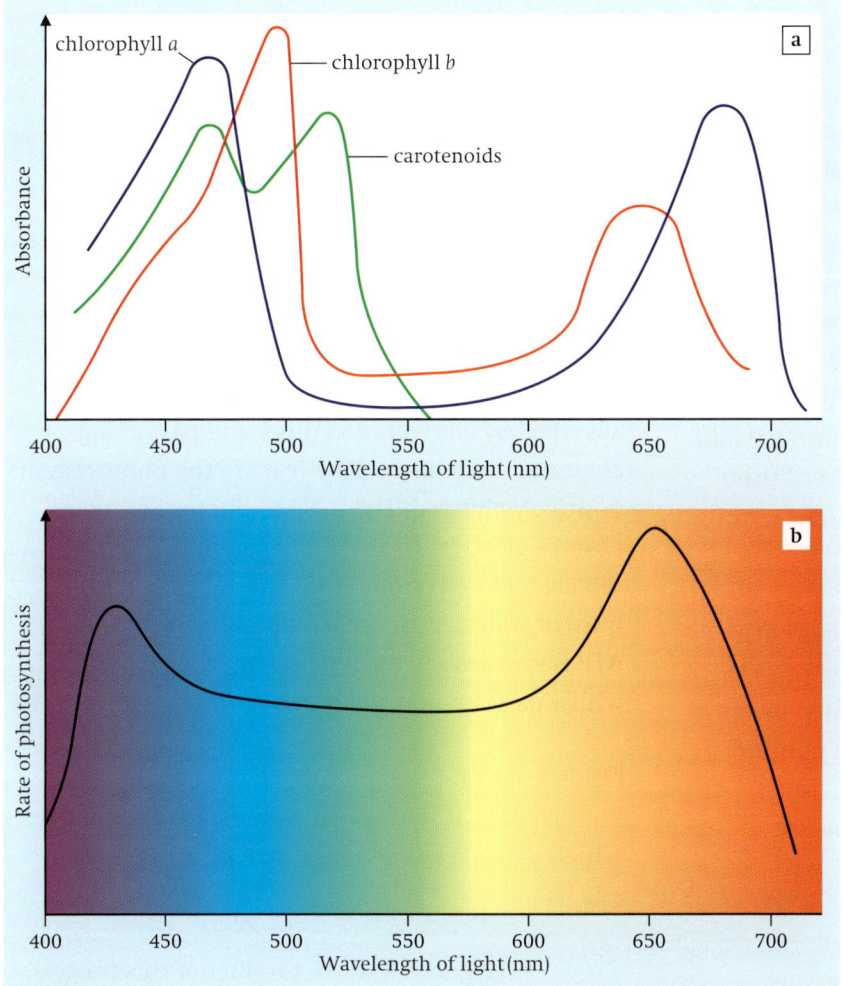

● **Figure 16.2**
**a** Absorption spectra of chlorophyll *a*, *b* and carotenoid pigments.
**b** Photosynthetic action spectrum.

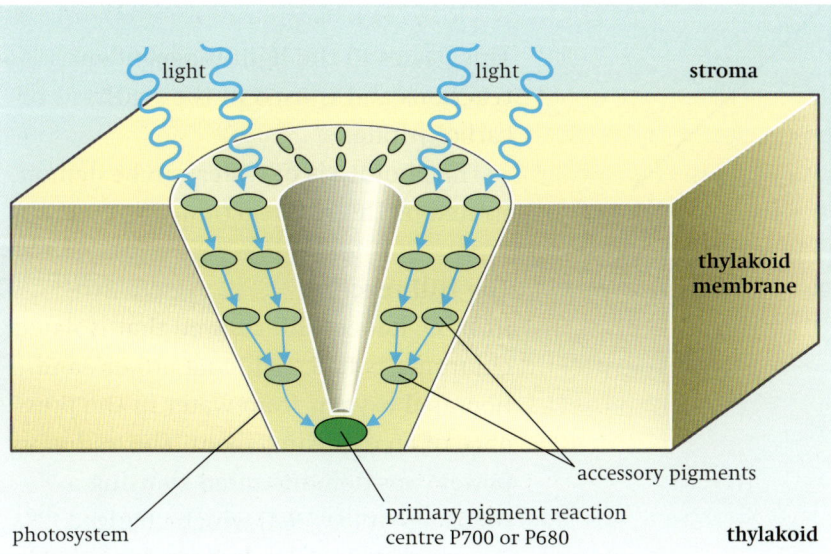

● **Figure 16.3** A photosystem: a light-harvesting cluster of photosynthetic pigments in a chloroplast thylakoid membrane. Only a few of the pigment molecules are shown.

said to act as **reaction centres**. **Photosystem I** is arranged around a molecule of chlorophyll *a* with a peak absorption at 700 nm. The reaction centre of photosystem I is therefore known as **P700**. **Photosystem II** is based on a molecule of chlorophyll *a* with a peak absorption of 680 nm. The reaction centre of photosystem II is therefore known as **P680**.

**SAQ 16.1**
Compare the absorption spectra shown in *figure 16.2a* with the action spectrum shown in *figure 16.2b*.
**a** Identify and explain any similarities in the absorption and action spectra.
**b** Identify and explain any differences between the absorption and action spectra.

# The light-dependent reactions of photosynthesis

These reactions include the synthesis of ATP in photophosphorylation and the splitting of water by photolysis to give hydrogen ions. The hydrogen ions combine with a carrier molecule NADP (*box 15C* on page 202) to make reduced NADP. ATP and reduced NADP are passed from the light-dependent to the light-independent reactions.

Photophosphorylation of ADP to ATP can be cyclic or non-cyclic depending on the pattern of electron flow in one or both photosystems.

## Cyclic photophosphorylation

Cyclic photophosphorylation involves only photosystem I. Light is absorbed by photosystem I and is passed to chlorophyll *a* (P700). An electron in the chlorophyll *a* molecule is excited to a higher energy level and is emitted from the chlorophyll molecule. Instead of falling back into the photosystem and losing its energy as fluorescence, it is captured by an electron acceptor and passed back to a chlorophyll *a* (P700) molecule via a chain of electron carriers. During this process enough energy is released to synthesise ATP from ADP and an inorganic phosphate group ($P_i$). The ATP then passes to the light-independent reactions.

## Non-cyclic photophosphorylation

Non-cyclic photophosphorylation involves both photosystems in the so-called 'Z scheme' of electron flow (*figure 16.4*). Light is absorbed by both photosystems and excited electrons are emitted from the primary pigments of both reaction centres (P680 and P700). These electrons are

### Box 16A Redox reactions

These are oxidation–reduction reactions and involve the transfer of electrons from an electron donor (reducing agent) to an electron acceptor (oxidising agent). Sometimes hydrogen atoms are transferred, so that dehydrogenation is equivalent to oxidation. Chains of electron carriers involve electrons passing via redox reactions from one carrier to the next. Such chains occur in both chloroplasts and mitochondria. During their passage, electrons fall from higher to lower energy states.

absorbed by electron acceptors and pass along chains of electron carriers leaving the photosystems positively charged. The P700 of photosystem I absorbs electrons from photosystem II. P680 receives replacement electrons from the splitting (photolysis) of water. As in cyclic photophosphorylation, ATP is synthesised as the electrons lose energy whilst passing along the carrier chain.

### Photolysis of water

Photosystem II includes a water-splitting enzyme which catalyses the breakdown of water:

$$H_2O \rightarrow 2H^+ + 2e^- + \tfrac{1}{2}O_2$$

Oxygen is a waste product of this process. The hydrogen ions combine with electrons from photosystem I and the carrier molecule NADP to give reduced NADP.

$$2H^+ + 2e^- + NADP \rightarrow \text{reduced NADP}$$

This passes to the light-independent reactions and is used in the synthesis of carbohydrate.

The photolysis of water can be demonstrated by the Hill reaction.

### The Hill reaction

In 1939, Robert Hill showed that isolated chloroplasts had 'reducing power', and liberated oxygen from water in the presence of an oxidising agent. The 'reducing power' was demonstrated by using a redox agent (*box 16A*) which changed colour on reduction. Hill used $Fe^{3+}$ ions as his acceptor, but various redox agents, such as the blue dye DCPIP (dichlorophenolindophenol), can substitute for the

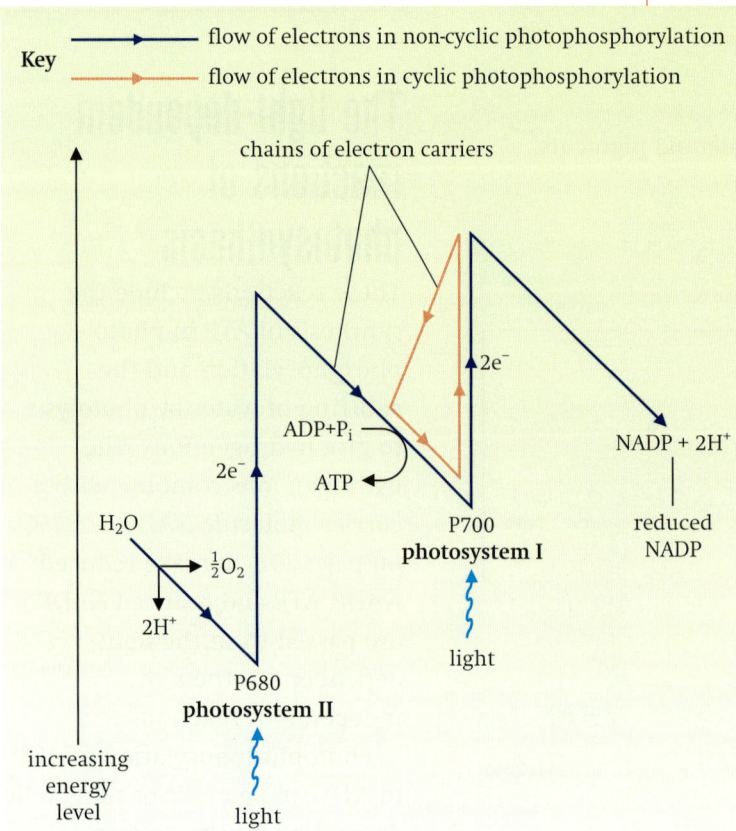

● **Figure 16.4** The 'Z scheme' of electron flow in photophosphorylation.

plant's NADP in this system. DCPIP becomes colourless when reduced:

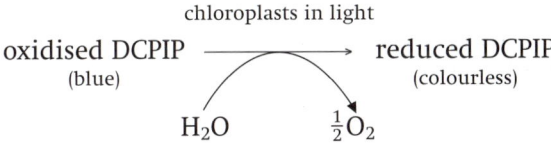

$$\text{oxidised DCPIP} \xrightarrow{\text{chloroplasts in light}} \text{reduced DCPIP}$$
$$\text{(blue)} \qquad\qquad\qquad\qquad \text{(colourless)}$$
$$H_2O \qquad \tfrac{1}{2}O_2$$

*Figure 16.5* shows classroom results of this reaction.

## SAQ 16.2

Examine the two curves shown in *figure 16.5* and explain:

**a** the downward trend of the two curves;

**b** the differences between the two curves.

## SAQ 16.3

Explain what contribution the discovery of the Hill reaction made to an understanding of the process of photosynthesis.

# The light-independent reactions of photosynthesis

The fixation of carbon dioxide is a light-independent process in which carbon dioxide combines with a five-carbon sugar, ribulose bisphosphate (RuBP), to give two molecules of a three-carbon compound, glycerate 3-phosphate (GP). (This compound is also sometimes known as PGA.)

GP, in the presence of ATP and reduced NADP from the light-dependent stages, is reduced to triose phosphate (three-carbon sugar).

This is the point at which carbohydrate is produced in photosynthesis. Some of these triose phosphates condense to form hexose phosphates, sucrose, starch and cellulose or are converted to acetylcoenzyme A (page 203) to make amino acids and lipids. Others regenerate RuBP. This cycle of events was worked out by Calvin, Benson and Bassham between 1946 and 1953, and is usually called the Calvin cycle (*figure 16.6*). The enzyme ribulose

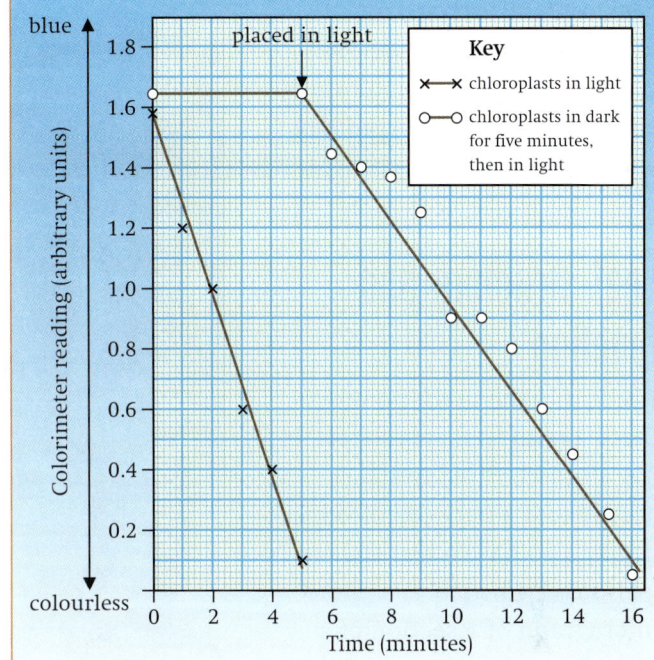

● **Figure 16.5** The Hill reaction. Chloroplasts were extracted from lettuce and placed in buffer solution with DCPIP. The colorimeter reading is proportional to the amount of DCPIP remaining unreduced.

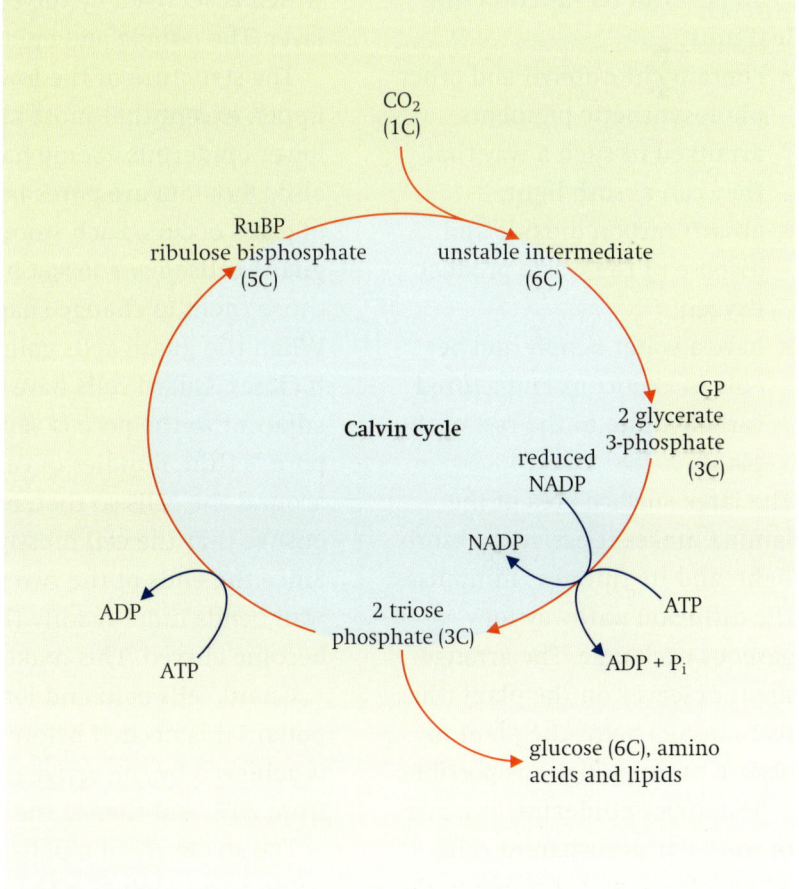

● **Figure 16.6** The Calvin cycle.

bisphosphate carboxylase (rubisco), which catalyses the combination of carbon dioxide and RuBP, is the most common enzyme in the world.

# Leaf structure and function

The leaf is the main photosynthetic organ in dicotyledons. It has a broad, thin lamina, a midrib and a network of veins. It may also have a leaf stalk (petiole). *Figure 16.7* is a photomicrograph of a section of a typical leaf from a mesophyte, that is a plant adapted for 'middling' terrestrial conditions (it is adapted neither for living in water nor for withstanding excessive drought).

To perform its function the leaf must:

- contain chlorophyll and other photosynthetic pigments arranged in such a way that they can absorb light;
- absorb carbon dioxide and dispose of the waste product oxygen;
- have a water supply and be able to export manufactured carbohydrate to the rest of the plant.

The large surface area of the lamina makes it easier to absorb light, and its thinness minimises the diffusion pathway for gaseous exchange. The arrangement of leaves on the plant (the leaf mosaic) helps the plant to absorb as much light as possible.

The upper **epidermis** is made of thin, flat, transparent cells which allow light through to the cells of the mesophyll below,

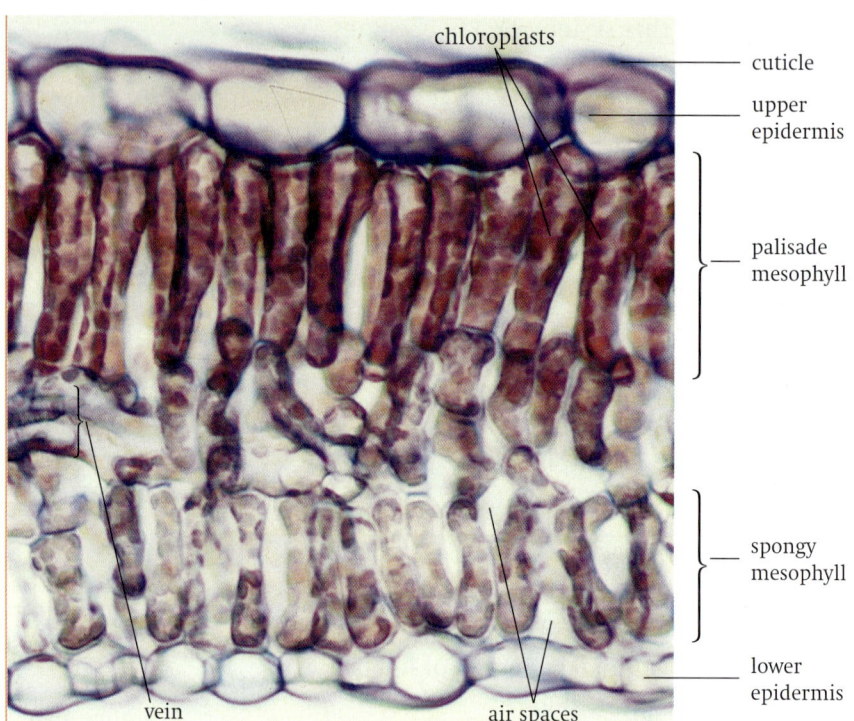

● **Figure 16.7** Photomicrograph of a TS of *Hypericum* leaf (× 1600). See also *figure 1.25*.

where photosynthesis takes place. A waxy transparent **cuticle**, which is secreted by the epidermal cells, provides a watertight layer. The cuticle and epidermis together form a protective layer.

The structure of the lower epidermis is similar to that of the upper, except that most mesophytes have many **stomata** in the lower epidermis. (Some have a few stomata in the upper epidermis also.) Stomata are pores in the epidermis through which diffusion of gases occurs. Each stoma is bounded by two sausage-shaped **guard cells** (*figure 16.8*). Changes in the turgidity of these guard cells cause them to change shape so that they open and close the pore. When the guard cells gain water, the pore opens; as they lose water it closes. Guard cells have unevenly thickened cell walls. The wall adjacent to the pore is very thick, whilst the wall furthest from the pore is thin. Bundles of cellulose microfibrils are arranged as hoops around the cells so that, as the cell becomes turgid, these hoops ensure that the cell mostly increases in length and not diameter. Since the ends of the two guard cells are joined and the thin outer wall bends more readily than the thick inner one, the guard cells become curved. This makes the pore between the cells open.

Guard cells gain and lose water by osmosis. A decrease in water potential is needed before water can enter the cells by osmosis. This is achieved by the active removal of hydrogen ions, using energy from ATP, and thence the intake of potassium ions (*figure 16.9*).

The structure of a **palisade cell** is shown in *figure 16.10*. The palisade mesophyll is the main site of photosynthesis, as there are more chloroplasts per cell than in the spongy mesophyll. The cells

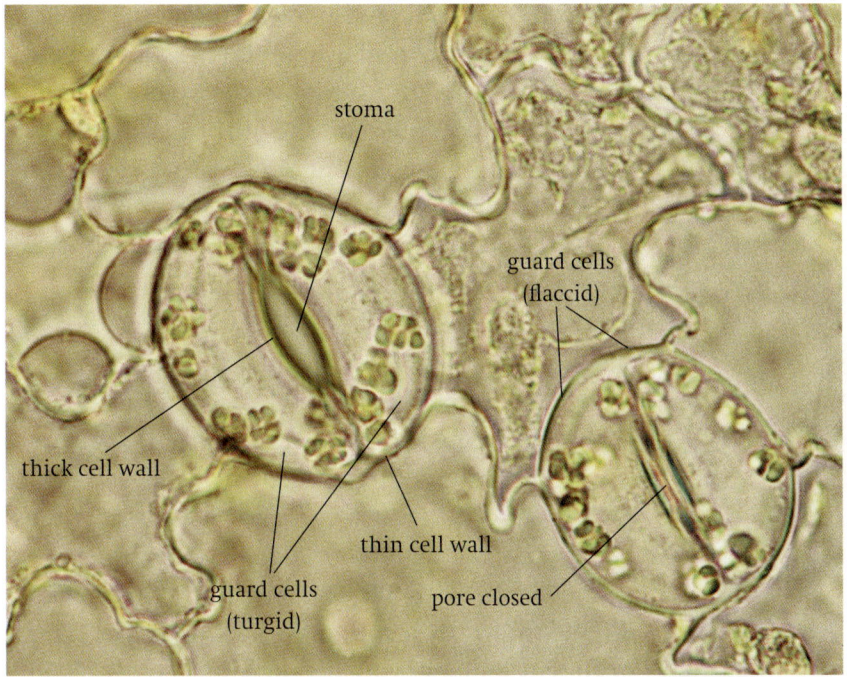

● **Figure 16.8** Photomicrograph of stomata and guard cells in *Tradescantia* leaf epidermis (× 4200).

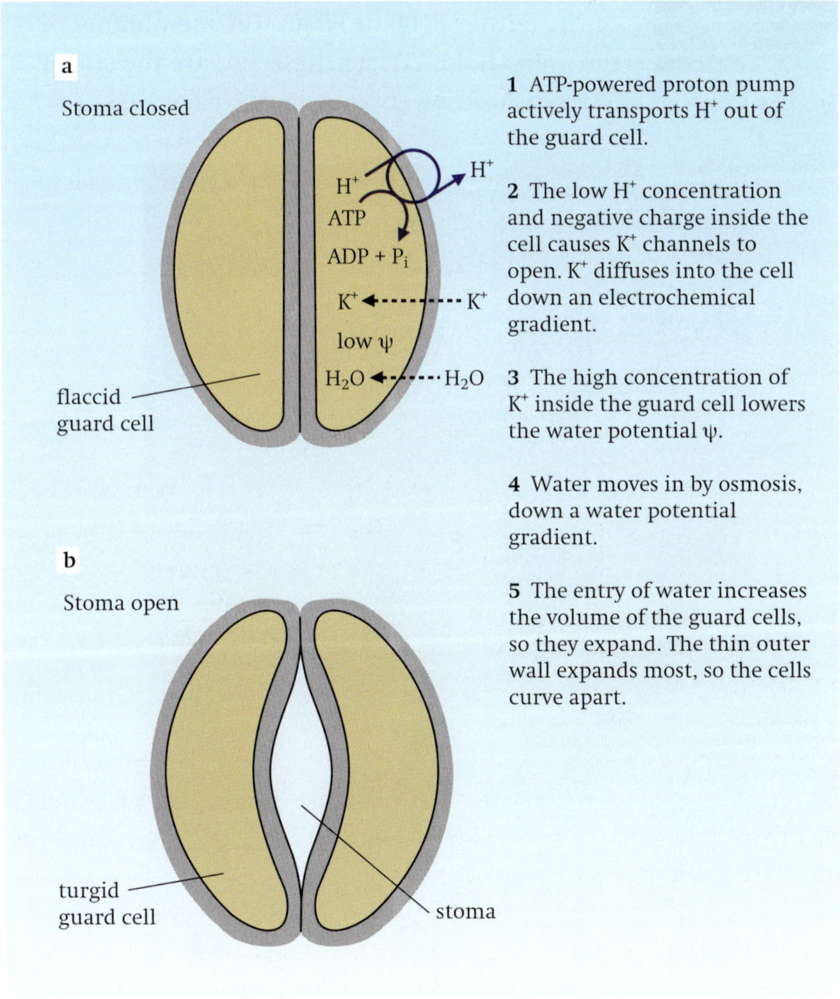

**a**

Stoma closed

flaccid guard cell

1 ATP-powered proton pump actively transports $H^+$ out of the guard cell.

2 The low $H^+$ concentration and negative charge inside the cell causes $K^+$ channels to open. $K^+$ diffuses into the cell down an electrochemical gradient.

3 The high concentration of $K^+$ inside the guard cell lowers the water potential $\psi$.

4 Water moves in by osmosis, down a water potential gradient.

5 The entry of water increases the volume of the guard cells, so they expand. The thin outer wall expands most, so the cells curve apart.

**b**

Stoma open

turgid guard cell

stoma

● **Figure 16.9** How a stoma is opened.

show several adaptations for light absorption.

- They are long cylinders arranged at right-angles to the upper epidermis. This reduces the number of light-absorbing cross walls in the upper part of the leaf so that as much light as possible can reach the chloroplasts.
- The cells have a large vacuole with a thin peripheral layer of cytoplasm. This restricts the chloroplasts to a layer near the outside of the cell where light can reach them most easily.
- The chloroplasts can be moved (by proteins in the cytoplasm – they cannot move themselves) within the cells, to absorb the most light or to protect the chloroplasts from excessive light intensities.

The palisade cells also show adaptations for gaseous exchange.

- The cylindrical cells pack together with long, narrow air spaces between them. This gives a large surface area of contact between cell and air.
- The cell walls are thin, so that gases can diffuse through them more easily.

**Spongy mesophyll** is mainly adapted as a surface for the exchange of carbon dioxide and oxygen. The cells contain chloroplasts, but in smaller numbers than in palisade cells. Photosynthesis occurs in the spongy mesophyll only at high light intensities. The irregular packing of the cells and the large air spaces thus produced provide a large surface area of moist cell wall for gaseous exchange.

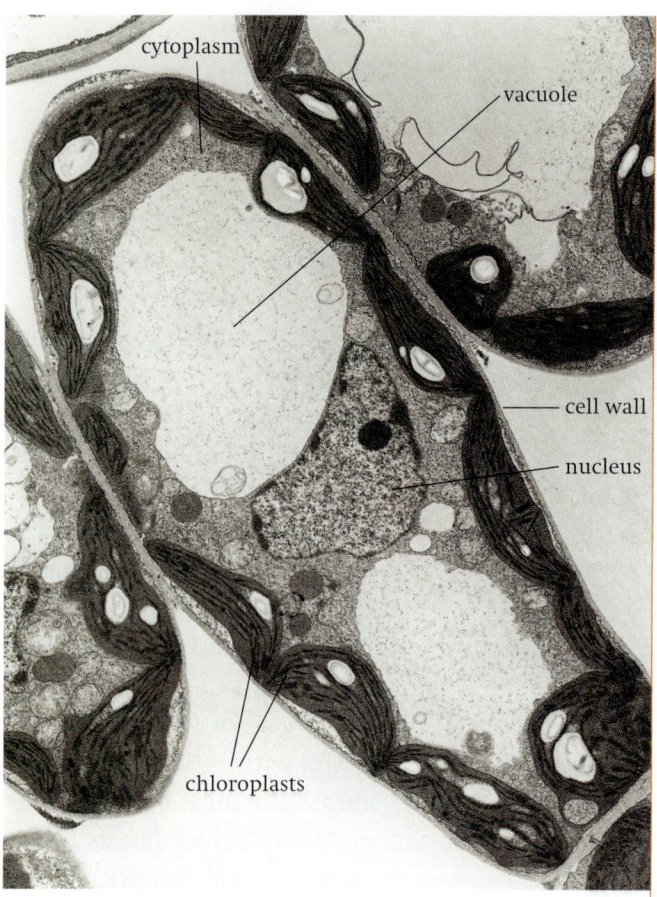

● **Figure 16.10** TEM of a palisade cell from soya bean leaf (× 4200).

surrounded by an envelope of two phospholipid membranes. A system of membranes also runs through the ground substance, or **stroma**. The membrane system is the site of the light-dependent reactions of photosynthesis. It consists of a series of flattened fluid-filled sacs, or **thylakoids**, which in places form stacks, called **grana**, that are joined to one another by membranes. The membranes of the grana provide a large surface area which holds the pigments, enzymes and electron carriers needed for the light-dependent reactions. They make it possible for a large number of pigment molecules to be arranged so that they can absorb as much light as necessary. The pigment molecules are also arranged in particular light-harvesting clusters for efficient light absorption. In each photosystem the different pigments are arranged in the thylakoid in funnel-like structures (*figure 16.3*). Each pigment passes energy to the next member of the cluster, finally 'feeding' it to the chlorophyll *a* reaction centre (either P700 or P680). The membranes of the grana hold ATP synthase and are the site of ATP synthesis by chemiosmosis (page 199).

The veins in the leaf help to support the large surface area of the leaf. They contain xylem, which brings in the water necessary for photosynthesis and for cell turgor, and phloem, which takes the products of photosynthesis to other parts of the plant.

# Chloroplast structure and function

In eukaryotic organisms, the photosynthetic organelle is the **chloroplast**. In dicotyledons, chloroplasts can be seen with a light microscope and appear as biconvex discs about 3–10 μm in diameter. There may be only a few chloroplasts in a cell or as many as 100 in some palisade mesophyll cells.

The structure of a chloroplast is shown in *figure 16.11*. Each chloroplast is

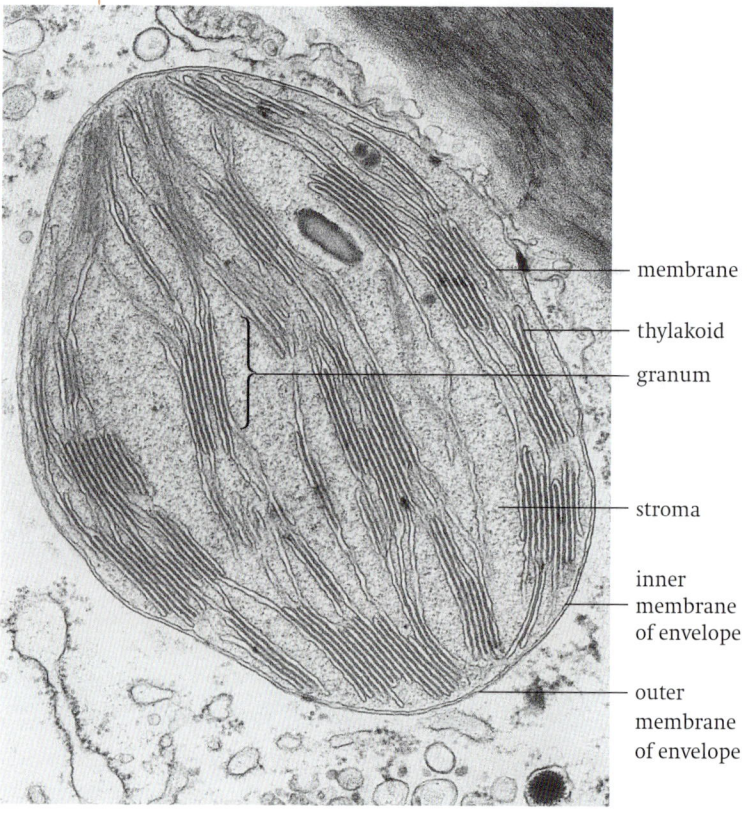

● **Figure 16.11** TEM of a chloroplast from *Potamogeton* leaf (× 27 000). See also *figure 1.23*.

The stroma is the site of the light-independent reactions. It contains the enzymes of the Calvin cycle, sugars and organic acids. It bathes the membranes of the grana and so can receive the products of the light-dependent reactions. Also within the stroma are small (70 S) ribosomes, a loop of DNA, lipid droplets and starch grains. The loop of DNA codes for some of the chloroplast proteins, which are made by the chloroplast's ribosomes. However, other chloroplast proteins are coded for by the DNA in the plant cell nucleus.

## SAQ 16.4 _____

List the features of a chloroplast that aid photosynthesis.

# Factors necessary for photosynthesis

You can see from the equation on page 211 that certain factors are necessary for photosynthesis to occur, namely the presence of a suitable photosynthetic pigment, a supply of carbon dioxide, water and light energy.

## Factors affecting the rate of photosynthesis

The main external factors affecting the rate of photosynthesis are light intensity, temperature and carbon dioxide concentration.

In the early 1900s F. F. Blackman investigated the effects of light intensity and temperature on the rate of photosynthesis. At constant temperature, the rate of photosynthesis varies with the light intensity, initially increasing as the light intensity increases (*figure 16.12*). However, at higher light intensities this relationship no longer holds and the rate of photosynthesis reaches a plateau.

The effect on the rate of photosynthesis of varying the temperature at constant light intensities can be seen in *figure 16.13*. At high light intensities the rate of

photosynthesis increases as the temperature is increased over a limited range. At low light intensities, increasing the temperature has little effect on the rate of photosynthesis.

These two experiments illustrate two important points. Firstly, from other research we know that photochemical reactions are not generally affected by temperature. However, these experiments clearly show that temperature affects the rate of photosynthesis, so there must be two sets of reactions in the full process of photosynthesis. These are a light-dependent photochemical stage and a light-independent, temperature-dependent stage. Secondly, Blackman's experiments illustrate the concept of 'limiting factors'.

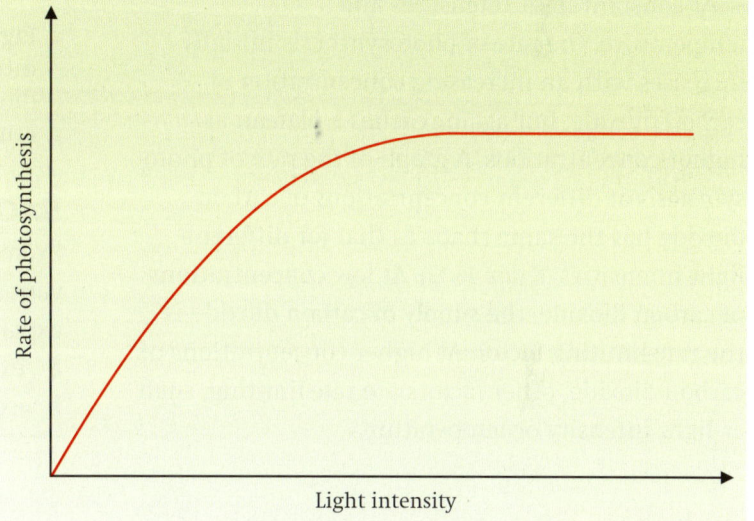

● **Figure 16.12** The rate of photosynthesis at different light intensities and constant temperature.

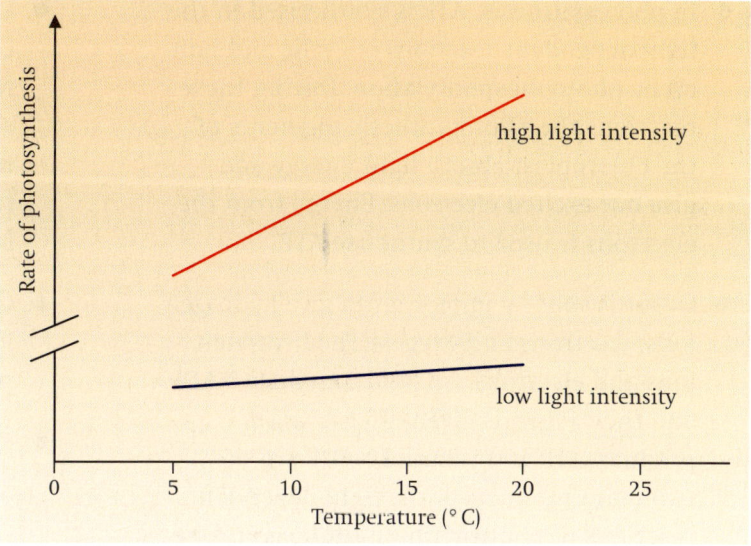

● **Figure 16.13** The rate of photosynthesis at different temperatures and constant light intensities.

## Limiting factors

The rate of any process which depends on a series of reactions is limited by the slowest reaction in the series. In biochemistry, if a process is affected by more than one factor, the rate will be limited by the factor which is nearest its lowest value.

Look again at *figure 16.12*. At low light intensities, the limiting factor governing the rate of photosynthesis is the light intensity; as the intensities increase so does the rate. But at high light intensities one or more other factors must be limiting, such as temperature or carbon dioxide supply.

At constant light intensities and temperature, the rate of photosynthesis initially increases with an increasing concentration of carbon dioxide, but again reaches a plateau at higher concentrations. A graph of the rate of photosynthesis at different concentrations of carbon dioxide has the same shape as that for different light intensities (*figure 16.12*). At low concentrations of carbon dioxide, the supply of carbon dioxide is the rate-limiting factor. At higher concentrations of carbon dioxide, other factors are rate-limiting, such as light intensity or temperature.

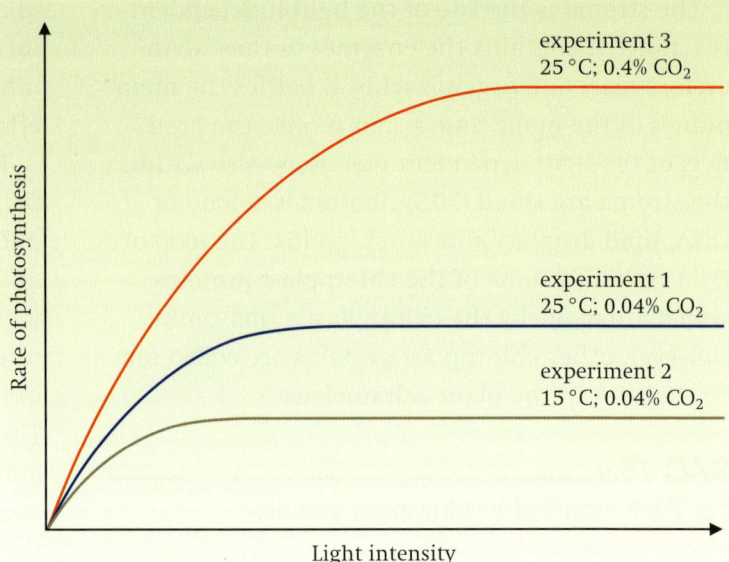

● **Figure 16.14** The rate of photosynthesis at different temperatures and different carbon dioxide concentrations. (0.04% $CO_2$ is about atmospheric concentration.)

### SAQ 16.5

Examine *figure 16.14* which shows the effect of various factors on the rate of photosynthesis and explain the differences in the results of:

**a** experiments 1 and 2;

**b** experiments 1 and 3.

# SUMMARY

♦ In photosynthesis, ATP is synthesised in the light-dependent reactions of cyclic and non-cyclic photophosphorylation. During these reactions the photosynthetic pigments of the chloroplast absorb light energy and give out excited electrons. Energy from the electrons is used to synthesise ATP.

♦ Water is split by photolysis to give hydrogen ions, electrons and oxygen. The hydrogen ions and electrons are used to reduce NADP and the oxygen is given off as a waste product. ATP and reduced NADP are the two main products of the light-dependent reactions of photosynthesis and pass to the light-independent reactions.

♦ Carbon dioxide is trapped and reduced to carbohydrate in the light-independent reactions of photosynthesis, using ATP and reduced NADP from the light-dependent reactions. This fixation of carbon dioxide requires an acceptor molecule, ribulose bisphosphate, and involves the Calvin cycle.

♦ Chloroplasts, mesophyll cells and whole leaves are all adapted for the process of photosynthesis.

♦ The rate of photosynthesis is subject to various limiting factors.

## CHAPTER 17

# Meiosis, genetics and gene control

## By the end of this chapter you should be able to:

1 understand the roles of meiosis and fertilisation in sexual reproduction;

2 describe, with the aid of diagrams, the behaviour of chromosomes during meiosis, and the associated behaviour of the nuclear envelope, plasma (cell surface) membrane and centrioles;

3 explain how meiosis and fertilisation can lead to variation;

4 explain the terms *gene*, *allele*, *locus*, *phenotype*, *genotype*, *dominant*, *recessive* and *codominant*;

5 use genetic diagrams to solve problems involving monohybrid and dihybrid crosses, including those involving sex linkage, codominance and multiple alleles;

6 understand the use of the test cross, and use genetic diagrams to solve problems involving such crosses;

7 use the chi-squared ($\chi^2$) test to test the significance of the difference between observed and expected results in genetic crosses;

8 explain, using sickle cell anaemia and other examples, how mutation may affect phenotype;

9 explain, with examples, how environment may affect phenotype;

10 outline the principles of gene manipulation by biotechnology (genetic engineering);

11 describe how bacteria have been genetically modified to synthesise human insulin and outline the production of human factor VIII from genetically modified animal cells;

12 describe the benefits and hazards of genetic engineering and discuss its social and ethical implications.

All species of living organisms are able to reproduce. Reproduction may be **asexual** or **sexual**. In asexual reproduction, a single organism produces offspring that are genetically identical to itself (see *figure 6.13*). The cells of the new organisms are formed as a result of mitosis in eukaryotes or binary fission in prokaryotes. However, in sexual reproduction, the offspring that are produced are genetically different from each other and from their parent or parents. Each parent produces specialised reproductive cells, known as

**gametes**, that fuse together in **fertilisation** to produce the first cell of the new organism – a **zygote**.

You saw in AS (page 80) that if a life cycle involves sexual reproduction, then it is necessary for the number of chromosomes to be halved at some point (*figure 17.1*). This is done by a special type of cell division called **meiosis**. In animals such as humans, for example, meiosis occurs as gametes are formed inside the testes and ovaries. The cells from which the gametes will be produced

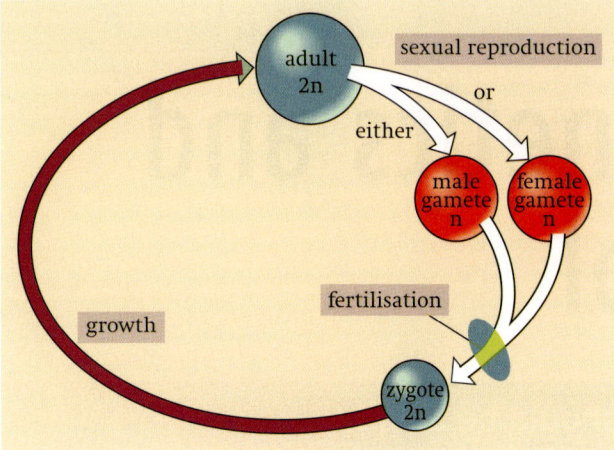

● **Figure 17.1** Outline of the life cycle of an animal.

are normal **diploid** (2n) cells, each containing two complete sets of chromosomes. As a result of meiosis, the gametes contain only half the normal number of chromosomes, and they are said to be **haploid** (n) cells. Thus, when two gametes fuse together at fertilisation, the zygote that is formed obtains two complete sets of chromosomes, returning to the diploid condition.

As you will see in this chapter, meiosis does more than halve the number of chromosomes in a cell. Meiosis also introduces **genetic variation** into the gametes and therefore the zygotes that are produced. Genetic variation may also arise as a result of **mutation**, which can occur at any stage in a life cycle. Such variation is the raw material on which natural selection has worked to produce the huge range of species that live on Earth, and we will look at this in chapter 18.

## Meiosis

The process of meiosis is best described by means of annotated diagrams (*figure 17.2*). An animal cell is shown where 2n = 4, and different colours represent maternal and paternal chromosomes.

Unlike mitosis (see page 81), meiosis involves two divisions, called meiosis I and meiosis II. **Meiosis I** is a reduction division, resulting in two daughter nuclei with *half* the number of chromosomes of the parent nucleus. In **meiosis II**, the chromosomes behave as in mitosis, so that each of the two haploid daughter nuclei divides again. Meiosis therefore results in a total of four haploid nuclei. Note that it is the behaviour of the

chromosomes in *meiosis I* that is particularly important and contrasts with mitosis.

*Figure 17.2* summarises the process of meiosis diagrammatically. *Figure 17.3* shows photographs of the process as seen with a light microscope.

### SAQ 17.1
Name the stage of meiosis at which each of the following occurs. Remember to state whether the stage you name is during division I or division II.
a  Homologous chromosomes pair to form bivalents.
b  Crossing over between chromatids of homologous chromosomes takes place.
c  Homologous chromosomes separate.
d  Centromeres split and chromatids separate.
e  Haploid nuclei are first formed.

### SAQ 17.2
A cell with 3 sets of chromosomes is said to be triploid, 3n. A cell with 4 sets of chromosomes is said to be tetraploid, 4n. Could meiosis take place in a 3n or a 4n cell? Explain your answer.

Two of the events that take place during meiosis help to produce genetic variation between the daughter cells that are produced. These are **independent assortment** of the homologous chromosomes, and **crossing over**, which happens between the chromatids of homologous chromosomes. When these genetically different gametes fuse, randomly, at fertilisation, yet more variation is produced amongst the offspring. In order to understand how these events produce variation, we first need to consider the **genes** that are carried on the chromosomes, and the way in which these are passed on from parents to offspring. This branch of biology is known as **genetics**.

## Genetics

You will remember that a **gene** is a length of DNA that codes for the production of a polypeptide molecule. The code is held in the sequence of bases in the DNA. A triplet of three bases 'stands for' one amino acid in the protein that will be constructed on the ribosomes in the cell (see chapter 5). One chromosome contains enough DNA to code for many polypeptides.

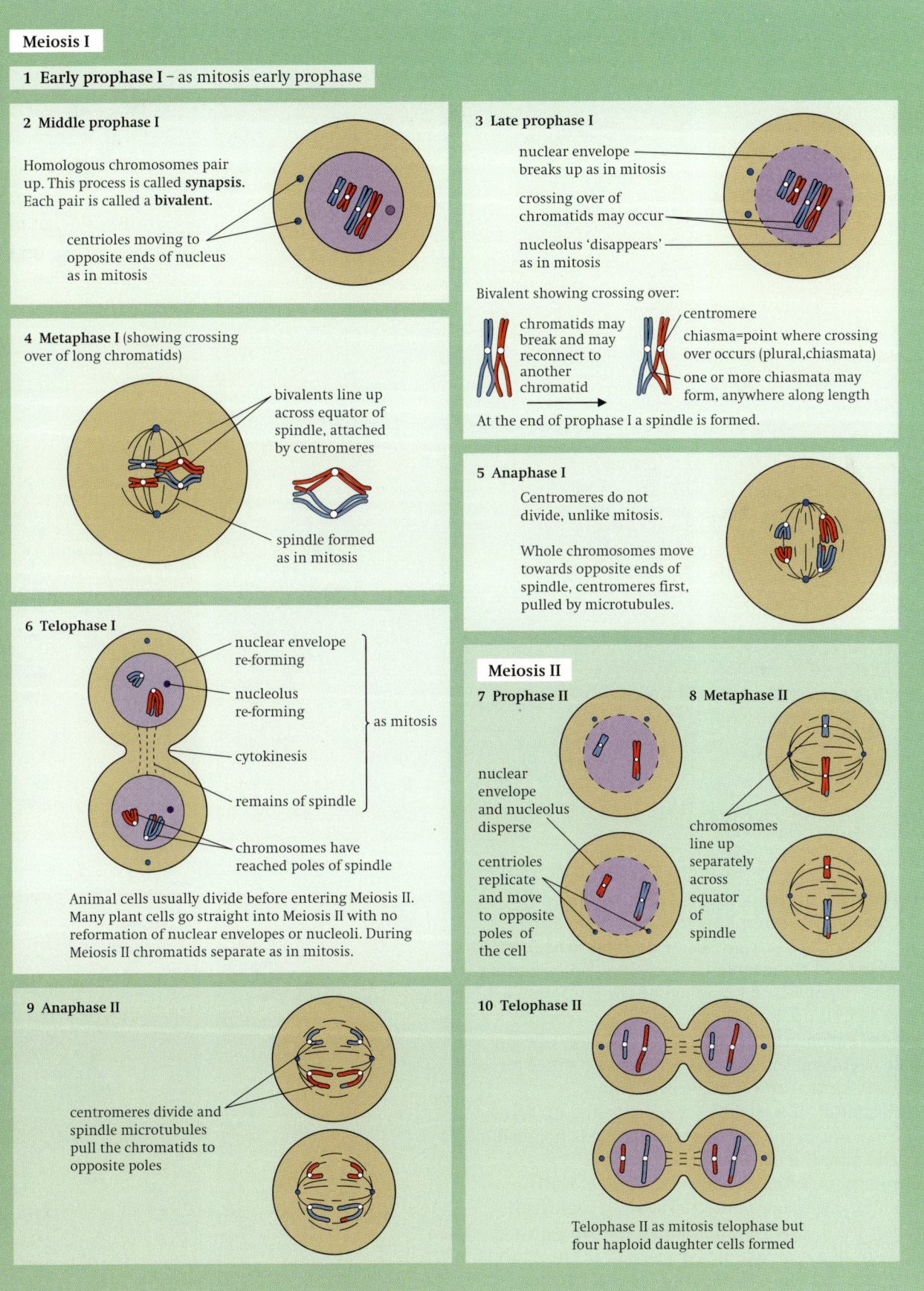

**Meiosis I**

**1 Early prophase I** – as mitosis early prophase

**2 Middle prophase I**

Homologous chromosomes pair up. This process is called **synapsis**. Each pair is called a **bivalent**.

centrioles moving to opposite ends of nucleus as in mitosis

**3 Late prophase I**

nuclear envelope breaks up as in mitosis

crossing over of chromatids may occur

nucleolus 'disappears' as in mitosis

Bivalent showing crossing over:

chromatids may break and may reconnect to another chromatid

centromere

chiasma=point where crossing over occurs (plural,chiasmata)

one or more chiasmata may form, anywhere along length

At the end of prophase I a spindle is formed.

**4 Metaphase I** (showing crossing over of long chromatids)

bivalents line up across equator of spindle, attached by centromeres

spindle formed as in mitosis

**5 Anaphase I**

Centromeres do not divide, unlike mitosis.

Whole chromosomes move towards opposite ends of spindle, centromeres first, pulled by microtubules.

**6 Telophase I**

nuclear envelope re-forming

nucleolus re-forming

cytokinesis

remains of spindle

chromosomes have reached poles of spindle

as mitosis

Animal cells usually divide before entering Meiosis II. Many plant cells go straight into Meiosis II with no reformation of nuclear envelopes or nucleoli. During Meiosis II chromatids separate as in mitosis.

**Meiosis II**

**7 Prophase II**

nuclear envelope and nucleolus disperse

centrioles replicate and move to opposite poles of the cell

**8 Metaphase II**

chromosomes line up separately across equator of spindle

**9 Anaphase II**

centromeres divide and spindle microtubules pull the chromatids to opposite poles

**10 Telophase II**

Telophase II as mitosis telophase but four haploid daughter cells formed

● **Figure 17.2** Meiosis and cytokinesis in an animal cell. Compare this process with nuclear division by mitosis, shown in *figure 6.10*.

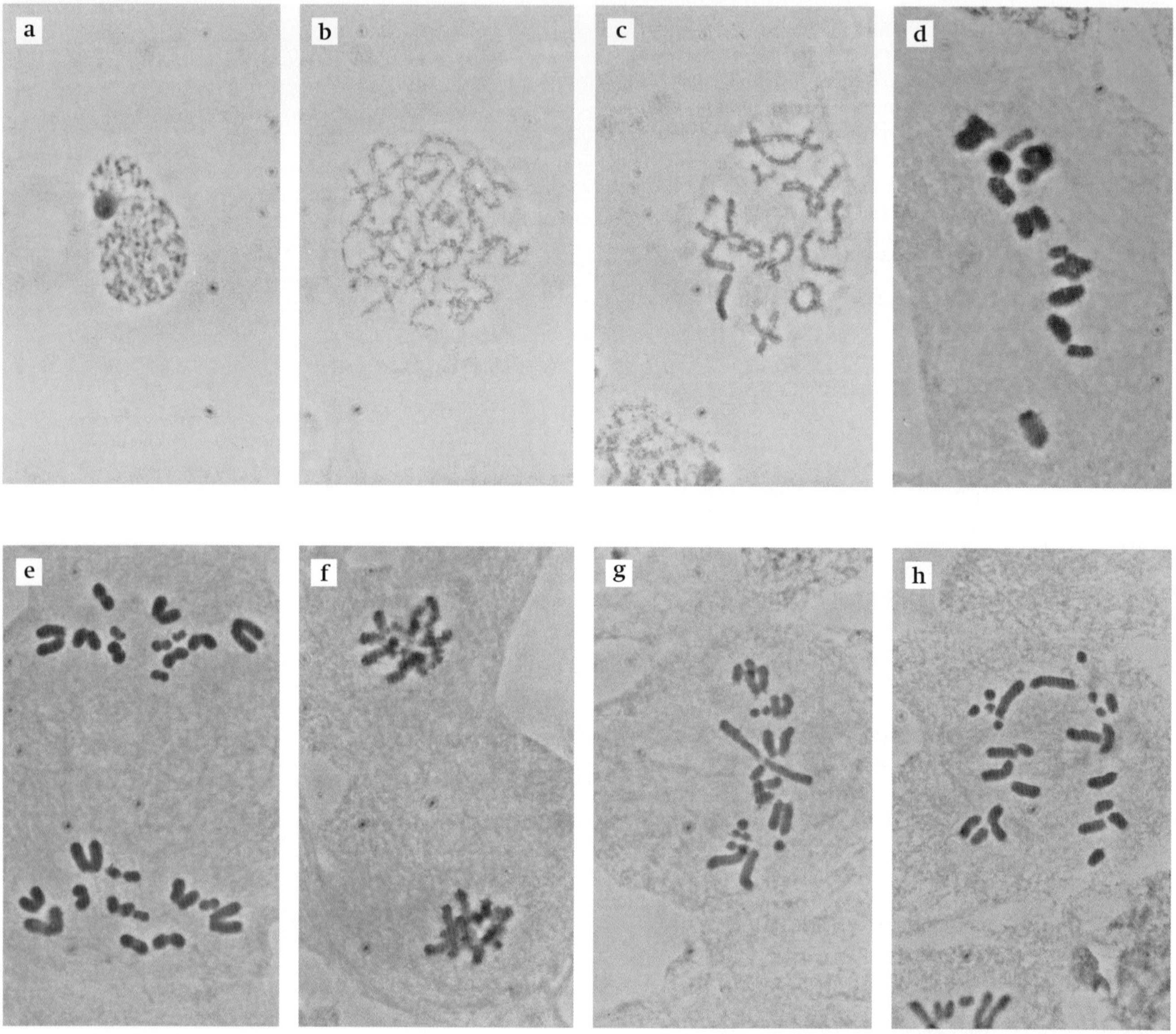

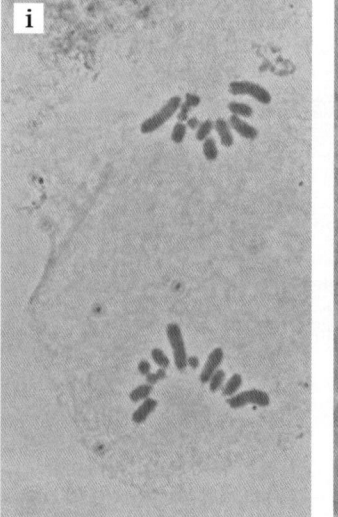
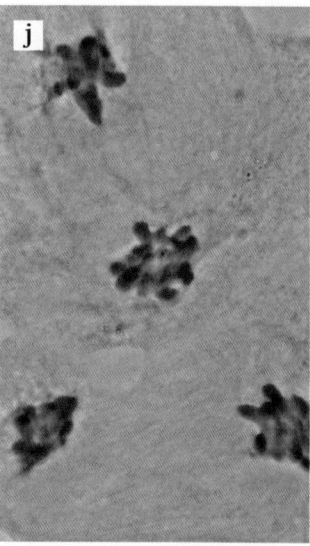

● **Figure 17.3** Stages of meiosis in an animal cell (locust) (× 950). Interphase (not part of meiosis) is also shown. **a** *interphase* nucleus; **b meiosis I**, *early prophase I*: chromosomes condensing and becoming visible; **c** *prophase I*: homologous chromosomes have paired up, forming bivalents, and crossing over of chromatids is occurring; members of each pair of chromosomes are repelling each other but are still held at the crossing-over points (**chiasmata**); **d** *metaphase I*: bivalents line up across the equator of the spindle; the spindle is not visible in the photo; **e** *anaphase I*; homologous chromosomes move to opposite poles of the spindle; **f** *telophase I* and *cytokinesis*; **g meiosis II**, *metaphase II*; single chromosomes line up across the equator of a new spindle; **h** *anaphase II*: chromatids separate and move to opposite poles of the new spindle; **i** *late anaphase II*; **j** *telophase II*.

Chromosomes that contain DNA for making the same polypeptides are said to be **homologous**. A diploid cell contains two of each type of chromosome, one homologue from the mother and one homologue from the father. So, in a diploid cell, there are two copies of each gene. The two copies lie in the same position, or **locus**, on the two homologous chromosomes (*figure 17.4*).

## Alleles

The number of chromosomes per cell is characteristic for each species. Human cells contain 46 chromosomes, two each of 23 types. Each type is numbered and has its own particular genes. (They are shown in *figure 6.3*.) For example, the gene which codes for the production of the β polypeptide of the haemoglobin molecule (see pages 34–35) is on chromosome 11. Each cell contains two copies of this gene, one maternal in origin (from the mother) and one paternal (from the father).

There are several forms or varieties of this gene. One variety contains the base sequence CCTGAGGAG, and codes for the normal β polypeptide. Another variety contains the base sequence CCTGTGGAG, and codes for a different

sequence of amino acids which forms a variant of the β polypeptide know as the **sickle cell** β polypeptide. These different varieties of the same gene are called **alleles**.

## Genotype

Most genes, including the β polypeptide gene, have several different alleles. For the moment, we will consider only the above two alleles of the β polypeptide gene.

For simplicity, the different alleles of a gene can be represented by symbols. In this case, they can be represented as follows:

$H^N$ = the allele for the normal β polypeptide;
$H^S$ = the allele for the sickle cell β polypeptide.

The letter H stands for the locus of the haemoglobin gene, while the superscripts N and S stand for particular alleles of the gene.

In a human cell, which is diploid, there are two copies of the β polypeptide gene. The two copies might be:

$$H^N H^N \quad \text{or} \quad H^S H^S \quad \text{or} \quad H^N H^S.$$

The alleles that an organism has form its **genotype**. In this case, where we are considering just two different alleles, there are three possible genotypes.

### SAQ 17.3

If there were three different alleles, how many possible genotypes would there be?

A genotype in which the two alleles of a gene are the same, for example $H^N H^N$, is said to be **homozygous** for that particular gene. A genotype in which the two alleles of a gene are different, for example $H^N H^S$, is said to be **heterozygous** for that gene. The organism can also be described as homozygous or heterozygous for that characteristic.

### SAQ 17.4

How many of the genotypes in your answer to SAQ 17.3 are homozygous, and how many are heterozygous?

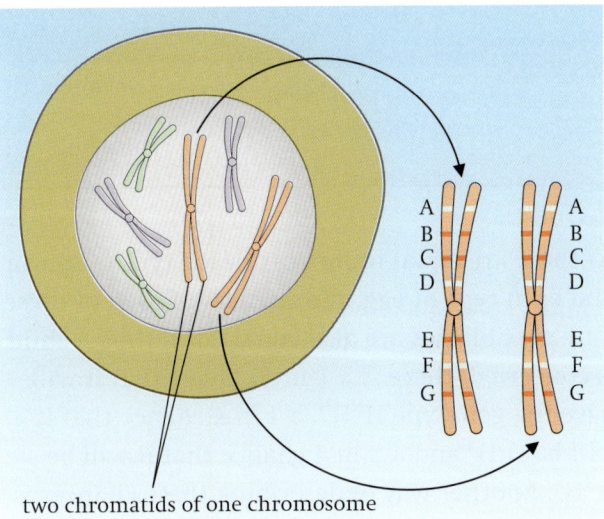

two chromatids of one chromosome

● **Figure 17.4** Homologous chromosomes carry the same genes at the same loci. Just seven genes, labelled A–G, are shown on these chromosomes, but in reality there are often hundreds or thousands of genes on each chromosome. This is a diploid cell as there are two complete sets of chromosomes (2n = 6).

# Genotype affects phenotype

A person with the genotype $H^N H^N$ has two copies of the gene in each cell coding for the production of the normal β polypeptide. All of their haemoglobin will be normal.

A person with the genotype $H^S H^S$ has two copies of the gene in each cell coding for the production of the sickle cell β polypeptide. All of their haemoglobin will be sickle cell haemoglobin, which is inefficient at transporting oxygen. The person will have **sickle cell anaemia**. This is a very dangerous disease, in which great care has to be taken not to allow the blood to become short of oxygen, or death may occur (page 251). A person with the genotype $H^N H^S$ has one allele of the haemoglobin gene in each cell coding for the production of the normal β polypeptide, and one coding for the production of the sickle cell β polypeptide. Half of their haemoglobin will be normal, and half will be sickle cell haemoglobin. They will have **sickle cell trait**, and are sometimes referred to as **carriers**. They will probably be completely unaware of this, because they have enough normal haemoglobin to carry enough oxygen, and so will have no problems at all. They will appear to be perfectly healthy. Difficulties arise only very occasionally, for example if a person with sickle cell trait does strenuous exercise at high altitudes, when oxygen concentrations in the blood might become very low (page 113).

The observable characteristics of an individual are called their **phenotype**. We will normally use the word 'phenotype' to describe just the one or two particular characteristics that we are interested in. In this case, we are considering the characteristic of having, or not having, sickle cell anaemia (*table 17.1*).

| Genotype | Phenotype |
|----------|-----------|
| $H^N H^N$ | normal |
| $H^N H^S$ | normal, but with sickle cell trait |
| $H^S H^S$ | sickle cell anaemia |

● **Table 17.1** Genotypes and phenotypes for sickle cell anaemia.

# Inheriting genes

In sexual reproduction, haploid gametes are made, following meiosis, from diploid body cells. Each gamete contains one of each pair of chromosomes. Therefore, each gamete contains only one copy of each gene.

Think about what will happen when sperm are made in the testes of a man who has the genotype $H^N H^S$. Each time a cell divides during meiosis, four gametes will be made, two of them with the $H^N$ allele and two with the $H^S$ allele. Of all the millions of sperm that are made in his lifetime, half will have the genotype $H^N$ and half will have the genotype $H^S$ (*figure 17.5*).

Similarly, a heterozygous woman will produce eggs of which half have the genotype $H^N$ and half have the genotype $H^S$.

This information can be used to predict the possible genotypes of children born to a couple who are both heterozygous. Each time fertilisation occurs, either a $H^N$ sperm or a $H^S$ sperm may fertilise either a $H^N$ egg or a $H^S$ egg. The possible results can be shown like this:

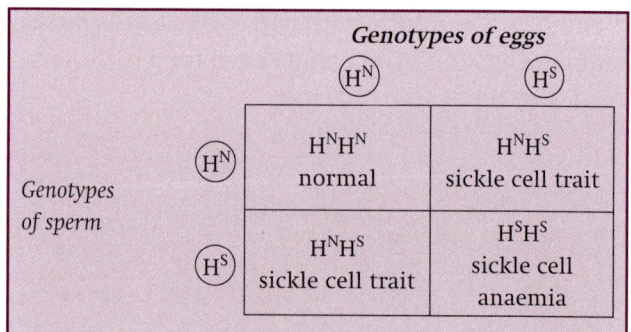

As there are equal numbers of each type of sperm and each type of egg, the chances of each of these four possibilities are also equal. Each time a child is conceived, there is a 1 in 4 chance that it will have the genotype $H^N H^N$, a 1 in 4 chance that it will be $H^S H^S$ and a 2 in 4 chance that it will be $H^N H^S$. Another way of describing these chances is to say that the probability of a child being $H^S H^S$ is 0.25, the probability of being $H^N H^N$ is 0.25, and the probability of being $H^N H^S$ is 0.5. It is important to realise that these are only *probabilities*. It would not be surprising if this couple had two children, both of whom had the genotype $H^S H^S$ and so suffered from sickle cell anaemia.

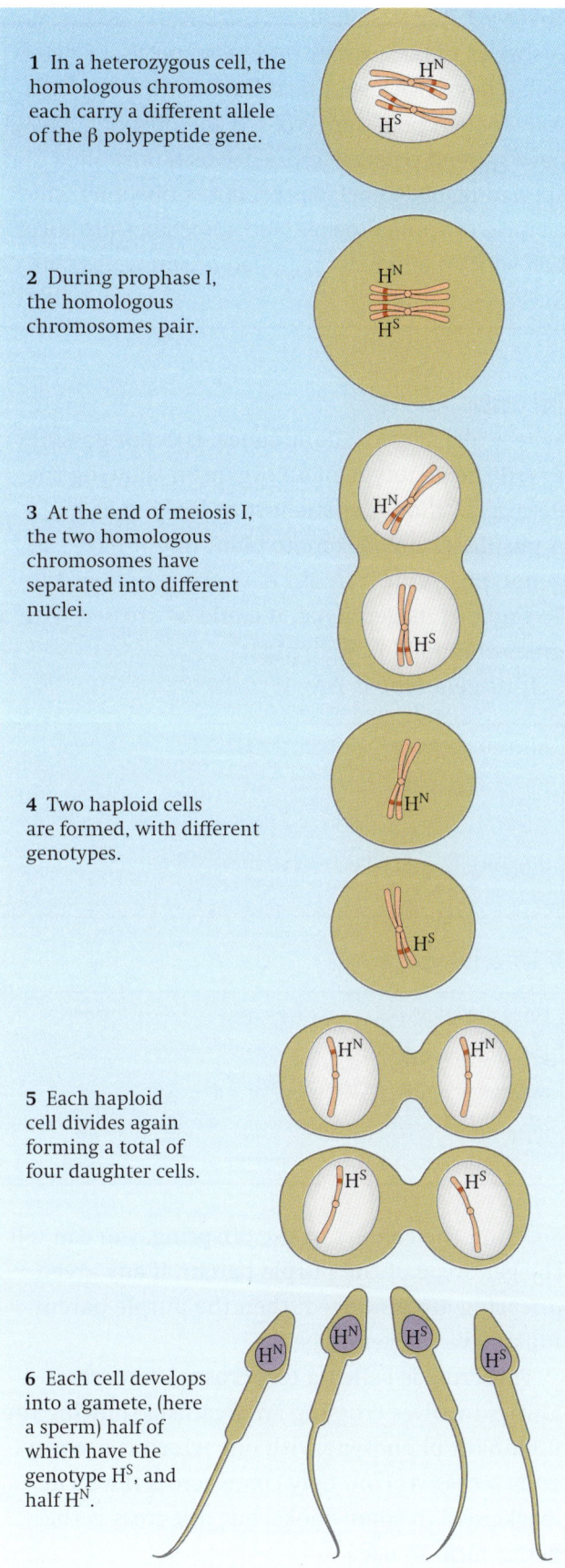

1 In a heterozygous cell, the homologous chromosomes each carry a different allele of the β polypeptide gene.

2 During prophase I, the homologous chromosomes pair.

3 At the end of meiosis I, the two homologous chromosomes have separated into different nuclei.

4 Two haploid cells are formed, with different genotypes.

5 Each haploid cell divides again forming a total of four daughter cells.

6 Each cell develops into a gamete, (here a sperm) half of which have the genotype H$^S$, and half H$^N$.

- Figure 17.5 Meiosis of a heterozygous cell produces gametes of two different genotypes. Only one pair of homologous chromosomes is shown.

## Genetic diagrams

A genetic diagram is the standard way of showing the genotypes of offspring that might be expected from two parents. To illustrate this, let us consider a different example: flower colour in snapdragons (*Antirrhinum*).

One of the genes for flower colour has two alleles, namely C$^R$ which gives red flowers, and C$^W$ which gives white flowers. The phenotypes produced by each genotype are:

| Genotype | Phenotype |
|----------|-----------|
| C$^R$C$^R$ | red |
| C$^R$C$^W$ | pink |
| C$^W$C$^W$ | white |

What colour flowers would be expected in the offspring from a red and a pink snapdragon?

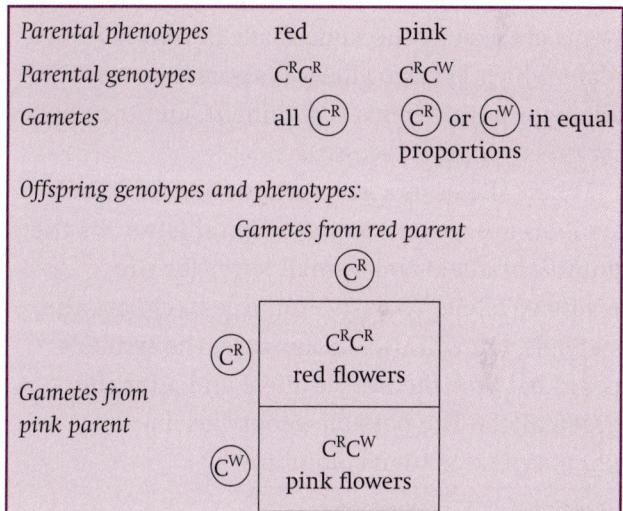

| Parental phenotypes | red | pink |
|---|---|---|
| Parental genotypes | C$^R$C$^R$ | C$^R$C$^W$ |
| Gametes | all C$^R$ | C$^R$ or C$^W$ in equal proportions |

*Offspring genotypes and phenotypes:*

Gametes from red parent: C$^R$

Gametes from pink parent:

| | C$^R$ |
|---|---|
| C$^R$ | C$^R$C$^R$ red flowers |
| C$^W$ | C$^R$C$^W$ pink flowers |

Thus, you would expect about half of the offspring to have red flowers and half to have pink flowers.

### SAQ 17.5

Red Poll cattle are homozygous for an allele which gives red coat colour. White Shorthorn are homozygous for an allele which gives white coat colour. When crossed, the offspring all have a mixture of red and white hairs in their coats, producing a colour called roan.

a Suggest suitable symbols for the two alleles of the coat colour gene.

b List the three possible genotypes for the coat colour gene and their phenotypes.

c Draw genetic diagrams to show the offspring expected from the following matings:
(i) a Red Poll with a roan; (ii) two roans.

# Dominance

In the examples used so far, both of the alleles in a heterozygous organism have an effect on the phenotype. A person with the genotype $H^N H^S$ has some normal haemoglobin and some sickle haemoglobin. A snapdragon with the genotype $C^R C^W$ has some red colour and some white colour, so that the flowers appear pink. Alleles which behave like this are said to be **codominant** alleles.

Frequently, however, only one allele has an effect in a heterozygous organism. This allele is said to be the **dominant** allele, while the one which has no effect is **recessive**. An example is stem colour in tomatoes. There are two alleles for stem colour, one of which produces green stems, and the other purple stems. In a tomato plant which has one allele for purple stems and one allele for green stems, the stems are exactly the same shade of purple as in a plant which has two alleles for purple stems. The allele for purple stems is dominant, and the allele for green stems is recessive.

When alleles of a gene behave like this, their symbols are written using a capital letter for the dominant allele and a small letter for the recessive allele. You are often free to choose the symbols you will use. In this case, the symbols could be A for the purple allele and **a** for the green allele. The possible genotypes and phenotypes for stem colour are:

| Genotype | Phenotype |
|----------|-----------|
| AA | purple stem |
| Aa | purple stem |
| aa | green stem |

It is a good idea, when choosing symbols to use for alleles, to use letters where the capital looks very different from the small one. If you use symbols such as S and s or P and p, it can become difficult to tell them apart if they are written down quickly.

## SAQ 17.6

In mice, the gene for eye colour has two alleles. The allele for black eyes is dominant, while the allele for red eyes is recessive.

Choose suitable symbols for these alleles, and then draw a genetic diagram to show the probable results of a cross between a heterozygous black-eyed mouse and a red-eyed mouse.

## SAQ 17.7

A species of poppy may have plain petals, or petals with a large black spot near the base. If two plants with spotted petals are crossed, the offspring always have spotted petals. A cross between unspotted and spotted plants sometimes produces offspring which all have unspotted petals, and sometimes produces half spotted and half unspotted offspring. Explain these results.

## Test crosses

Where alleles show dominance, it is not possible to tell the genotype of an organism showing the dominant characteristic just by looking at it. A purple-stemmed tomato plant might have the genotype AA, or it might have the genotype Aa. To find out its genotype, it could be crossed with a green-stemmed tomato plant.

If its genotype is AA:

| | | |
|---|---|---|
| *Parental phenotypes* | purple | green |
| *Parental genotypes* | AA | aa |
| *Gametes* | (A) | (a) |
| *Offspring* | all Aa purple | |

If its genotype is Aa:

| | | |
|---|---|---|
| *Parental phenotypes* | purple | green |
| *Parental genotypes* | Aa | aa |
| *Gametes* | (A) or (a) | (a) |
| *Offspring* | Aa  or  aa | |
| | purple     green | |

So, from the colours of the offspring, you can tell the genotype of the purple parent. If any green offspring are produced, then the purple parent must have the genotype Aa.

This cross is called a **test cross**. A test cross always involves crossing an organism showing the dominant phenotype with one which is homozygous recessive. (You may come across the term 'backcross' in some books, but test cross is the better term to use.)

## SAQ 17.8

In dalmatian dogs, the colour of the spots is determined by a gene which has two alleles. The allele for black spots is dominant, and the allele for brown spots is recessive.

A breeder wanted to know the genotype of a black-spotted bitch. She crossed her with a brown-spotted dog, and a litter of three puppies was produced, all of which were black. The breeder concluded that her bitch was homozygous for the allele for black spots. Was she right? Explain your answer.

## Multiple alleles

So far, we have considered just two alleles, or varieties, of any one gene. Most genes, however, have more than two alleles. An example of this situation, known as **multiple alleles**, is the gene for human blood groups.

The four blood groups A, B, AB and O are all determined by a single gene. Three alleles of this gene exist, $I^A$, $I^B$, and $I^o$. Of these, $I^A$ and $I^B$ are codominant, while $I^o$ is recessive to both $I^A$ and $I^B$. As a diploid cell can carry only two alleles, the possible genotypes and phenotypes are as shown in *table 17.2*.

## SAQ 17.9

A man of blood group B and a woman of blood group A have three children. One is group A, one group B and one group O. What are the genotypes of these five people?

| Genotype | Blood group |
|---|---|
| $I^A I^A$ | A |
| $I^A I^B$ | AB |
| $I^A I^o$ | A |
| $I^B I^B$ | B |
| $I^B I^o$ | B |
| $I^o I^o$ | O |

● **Table 17.2** Genotypes and phenotypes for blood groups.

● **Figure 17.6** Colour variations in rabbits, caused by multiple alleles of a single gene: **a** agouti; **b** albino; **c** chinchilla; **d** Himalayan.

## SAQ 17.10

Coat colour in rabbits is determined by a gene with four alleles. The allele for agouti (normal) coat is dominant to all of the other three alleles. The allele for albino coat is recessive to the other three alleles. The allele for chinchilla (grey) coat is dominant to the allele for Himalayan (white with black ears, nose, feet and tail) (*figure 17.6*).

**a** Write down the ten possible genotypes for coat colour, and their phenotypes.

**b** Draw genetic diagrams to explain each of the following.

  (i)   An albino rabbit is crossed with a chinchilla rabbit, producing offspring which are all chinchilla. Two of these chinchilla offspring are then crossed, producing 4 chinchilla offspring and 2 albino.

  (ii)  An agouti rabbit is crossed with a Himalayan rabbit, producing 3 agouti offspring and 3 Himalayan.

  (iii) Two agouti rabbits produce a litter of 5 young, three of whom are agouti and two chinchilla. The two chinchilla young are then crossed, producing 4 chinchilla offspring and 1 Himalayan.

## Sex inheritance

In humans, sex is determined by one of the 23 pairs of chromosomes. These chromosomes are called the **sex chromosomes**. The other 22 pairs are called **autosomes**.

The sex chromosomes differ from the autosomes in that the two sex chromosomes in a cell are not always alike. They do not always have the same genes in the same position, and so they are not always homologous. This is because there are two types of sex chromosome, known as the X and Y chromosomes, because of their shapes. The Y chromosome is much shorter than the X, and carries fewer genes. A person with two X chromosomes is female, while a person with one X and one Y chromosome is male.

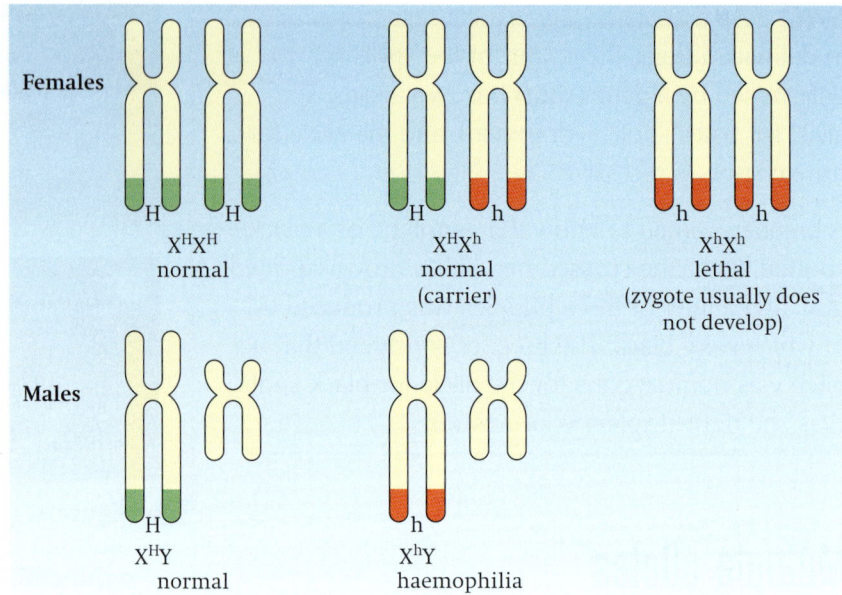

● **Figure 17.7** The possible genotypes and phenotypes for haemophilia.

### SAQ 17.11

Draw a genetic diagram to explain why there is always an equal chance that a child will be male or female. (You can do this in just the same way as the other genetic diagrams you have drawn, but using symbols to represent whole chromosomes, not genes.)

## Sex linkage

The X chromosome contains many different genes. (You can see some of these in *figure 6.5*.) One of them is a gene that codes for the production of a protein needed for blood clotting, called **factor VIII**. There are two alleles of this gene, the dominant one **H** producing normal factor VIII, and the recessive one **h** resulting in lack of it. People who are homozygous for the recessive allele suffer from the disease haemophilia, in which the blood fails to clot properly.

The fact that this gene is on the X chromosome, and not on an autosome, affects the way that it is inherited. Females, who have two X chromosomes, have two copies of the gene. Males, however, who have only one X chromosome, have only one copy of the gene. Therefore, the possible genotypes for men and women are different. They are shown in *figure 17.7*.

The factor VIII gene is said to be **sex-linked**. A sex-linked gene is one which is found on a part of the X chromosome not matched by the Y, and therefore not found on the Y chromosome.

Genotypes including sex-linked genes are always represented by symbols which show that they are on an X chromosome. Thus the genotype of a woman who has the allele H on one of her X chromosomes and the allele h on the other is written as $X^HX^h$.

You can draw genetic diagrams to show how sex-linked genes are inherited in exactly the same way as for other genes. For example, the following diagram shows the children that could be born to a couple where the man does not have haemophilia, while the woman is a carrier for the disease.

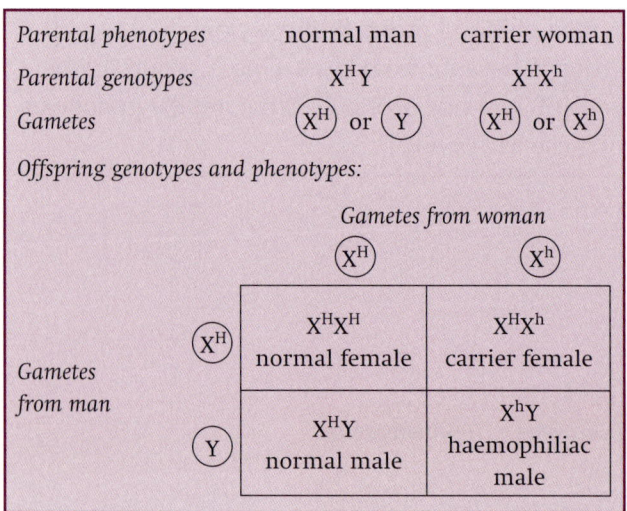

Each time this couple have a child, therefore, there is a 0.25 probability that it will be a normal girl, a 0.25 probability that it will be a normal boy, a 0.25 probability that it will be a carrier girl and a 0.25 probability that it will be a boy with haemophilia.

## SAQ 17.12

Can a man with haemophilia pass on the disease to:

**a** his son?

**b** his grandson?

Draw genetic diagrams to explain your answers.

## SAQ 17.13

One of the genes for colour vision in humans is found on the X chromosome, but not on the Y chromosome. The dominant allele of this gene gives normal colour vision, while a recessive allele produces red–green colour blindness.

**a** Choose suitable symbols for these alleles, and then write down all of the possible genotypes for a man and for a woman.

**b** A couple who both have normal colour vision have a child with colour blindness. Explain how this may happen, and state what the sex of the colour blind child must be.

**c** Is it possible for a colour blind girl to be born? Explain your answer.

## SAQ 17.14

One of the genes for coat colour in cats is sex-linked. The allele $C^O$ gives orange fur, while $C^B$ gives black fur. The two alleles are codominant, and when both are present the cat has patches of orange and black, which is known as tortoiseshell.

**a** Explain why male cats cannot be tortoiseshell.

**b** Draw a genetic diagram to show the expected genotypes and phenotypes of the offspring from a cross between an orange male and a tortoise-shell female cat. (Remember to show the X and Y chromosomes, as well as the symbols for the alleles.)

# Dihybrid crosses

So far, we have considered the inheritance of just one gene. Such examples are called **monohybrid crosses**. **Dihybrid crosses** look at the inheritance of two genes at once.

You have already seen that, in tomato plants, there is a gene which codes for stem colour. This gene has two alleles:

| stem colour gene | **A** = allele for purple stem and **a** = allele for green stem |
|---|---|
| where A is dominant and a is recessive. | |

A different gene, at a different locus on a different chromosome, codes for leaf shape. Again, there are two alleles:

| leaf shape gene | **D** = allele for cut leaves (jagged edges) **d** = allele for potato leaves (smooth edges) |
|---|---|
| where D is dominant and d is recessive. | |

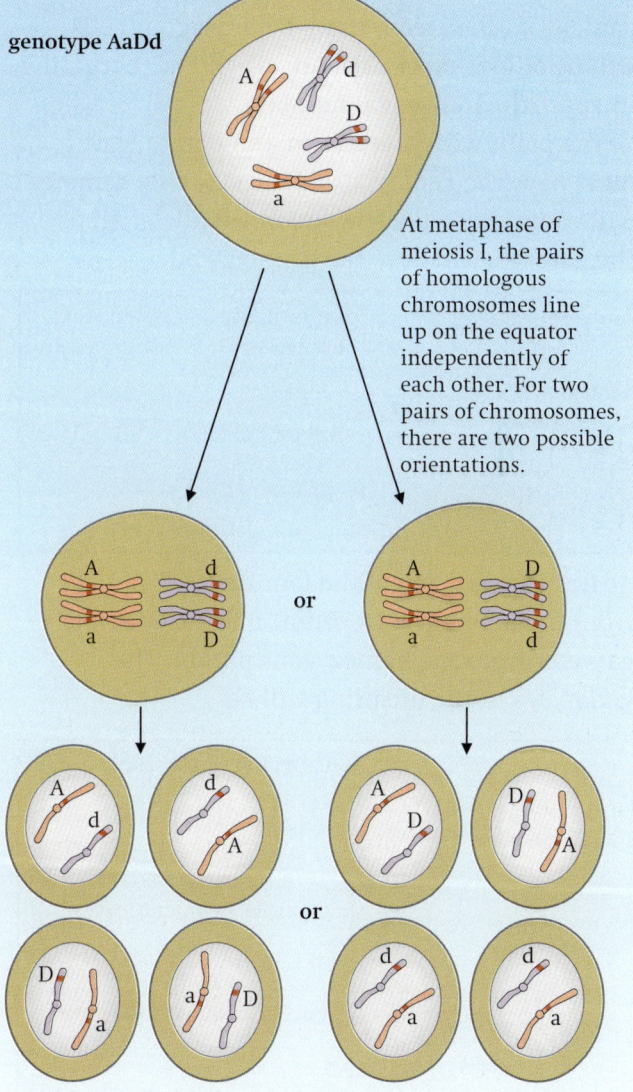

At the end of meiosis II each orientation gives two types of gamete. There are therefore four types of gamete altogether.

● **Figure 17.8** Independent assortment of homologous chromosomes during meiosis I results in a variety of genotypes in the gametes formed.

What will happen if a plant which is heterozygous for both of these genes is crossed with a plant with green stem and potato leaves?

*Figure 17.8* shows the alleles in a cell of the plant which is heterozygous for both genes. When this cell undergoes meiosis to produce gametes, the pairs of homologous chromosomes line up independently of each other on the equator during metaphase I. There are two ways in which the two pairs of chromosomes can do this. If there are many such cells undergoing meiosis, then the chromosomes in roughly half of them will probably line up one way, and the other half will line up the other way. This is **independent assortment**. We can therefore predict that the gametes formed from these heterozygous cells will be of four types, **AD**, **Ad**, **aD** and **ad**, occurring in approximately equal numbers.

The plant with green stem and potato leaves must have the genotype **aadd**. Each of its gametes will contain one **a** allele and one **d** allele. All of the gametes will have the genotype **ad**.

| Parental phenotypes | purple stem, cut leaves | green stem, potato leaves |
|---|---|---|
| Parental genotypes | AaDd | aadd |
| Gametes | (AD) or (Ad) | all (ad) |
| | or (aD) or (ad) | |
| | in equal proportions | |

At fertilisation, any of the four types of gamete from the heterozygous parent may fuse with the gametes from the homozygous parent. The genotypes of the offspring will be:

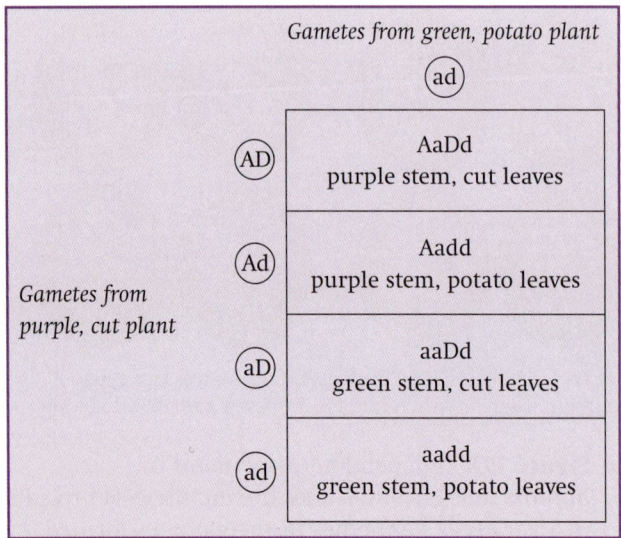

| | Gametes from green, potato plant |
|---|---|
| | (ad) |
| Gametes from purple, cut plant | |
| (AD) | AaDd purple stem, cut leaves |
| (Ad) | Aadd purple stem, potato leaves |
| (aD) | aaDd green stem, cut leaves |
| (ad) | aadd green stem, potato leaves |

From this cross, therefore, we would expect approximately equal numbers of the four possible phenotypes. This 1 : 1 : 1 : 1 ratio is typical of a dihybrid cross between a heterozygous organism and a homozygous recessive organism, where the alleles show complete dominance.

If *both* parents are heterozygous, then things become a little more complicated, because both of them will produce four kinds of gametes.

| Parental phenotypes | purple stem, cut leaves | purple stem, cut leaves |
|---|---|---|
| Parental genotypes | AaDd | AaDd |
| Gametes | (AD) or (Ad) | (AD) or (Ad) |
| | or (aD) or (ad) | or (aD) or (ad) |
| | in equal proportions | in equal proportions |

*Offspring genotypes and phenotypes:*

| | | Gametes from one parent | | | |
|---|---|---|---|---|---|
| | | (AD) | (Ad) | (aD) | (ad) |
| Gametes from other parent | (AD) | AADD purple, cut | AADd purple, cut | AaDD purple, cut | AaDd purple, cut |
| | (Ad) | AADd purple, cut | AAdd purple, potato | AaDd purple, cut | Aadd purple, potato |
| | (aD) | AaDD purple, cut | AaDd purple, cut | aaDD green, cut | aaDd green, cut |
| | (ad) | AaDd purple, cut | Aadd purple, potato | aaDd green, cut | aadd green, potato |

If you sort out the numbers of each phenotype amongst these sixteen possibilities, you will find that the offspring would be expected to occur in the following ratio:

9 purple, cut : 3 purple, potato : 3 green, cut : 1 green, potato.

This 9 : 3 : 3 : 1 ratio is typical of a dihybrid cross between two heterozygous organisms, where the two alleles show complete dominance and where the genes are on different chromosomes.

### SAQ 17.15

Explain the contribution made to the variation amongst the offspring of this cross by:
**a** independent assortment;
**b** random fertilisation.

## SAQ 17.16

Draw genetic diagrams to show the genotypes of the offspring from each of the following crosses.

**a** AABb × aabb

**b** GgHh × gghh

**c** TTyy × ttYY

**d** eeFf × Eeff

## SAQ 17.17

The allele for grey fur in a species of animal is dominant to white, and the allele for long tail is dominant to short.

**a** Using the symbols G and g for coat colour, and T and t for tail length, draw a genetic diagram to show the genotypes and phenotypes of the offspring you would expect from a cross between a pure-breeding grey animal with a long tail and a pure-breeding white animal with a short tail.

**b** If this first generation of offspring were bred together, what would be the expected phenotypes in the second generation of offspring, and in what ratios would they occur?

## SAQ 17.18

In a species of plant, the allele for tall stem is dominant to short. The two alleles for leaf colour, giving green or white in the homozygous condition, are codominant, producing variegated leaves in the heterozygote.

A plant with tall stems and green leaves was crossed with a plant with short stems and variegated leaves. The offspring from this cross consisted of plants with tall stems and green leaves and plants with tall stems and variegated leaves in the ratio of 1 : 1. Construct a genetic diagram to explain this cross.

## SAQ 17.19

In a species of animal, it is known that the allele for black eyes is dominant to the allele for red eyes, and that the allele for long fur is dominant to the allele for short fur.

**a** What are the possible genotypes for an animal with black eyes and long fur?

**b** How could you find out which genotype this animal had?

# The $\chi^2$ (chi-squared) test

If you look back at the cross between the two heterozygous tomato plants, on page 232, you will see that we would expect to see a 9 : 3 : 3 : 1 ratio of phenotypes in the offspring. It is important to remember that this ratio represents the *probability* of getting these phenotypes, and we would probably be rather surprised if the numbers came out absolutely precisely to this ratio.

But just how much difference might we be happy with, before we began to worry that perhaps the situation was not quite what we had thought? For example, let us imagine that the two plants produced a total of 144 offspring. If the parents really were both heterozygous, and if the purple stem and cut leaf alleles really are dominant, and if the alleles really do assort independently, then we would expect the following numbers of each genotype to be present in the offspring:

$$\text{purple, cut} = \tfrac{9}{16} \times 144 = 81$$

$$\text{purple, potato} = \tfrac{3}{16} \times 144 = 27$$

$$\text{green, cut} = \tfrac{3}{16} \times 144 = 27$$

$$\text{green, potato} = \tfrac{1}{16} \times 144 = 9$$

But imagine that, amongst these 144 offspring, the results we actually observed were as follows:

| | | | |
|---|---|---|---|
| purple, cut | 86 | green, cut | 24 |
| purple, potato | 26 | green, potato | 8 |

We might ask: are these results sufficiently close to the ones we expected that the differences between them have probably just arisen by chance, or are they so different that something unexpected must be going on?

To answer this question, we can use a statistical test called the $\chi^2$ (**chi-squared**) **test**. This test allows us to compare our observed results with the expected results, and decide whether or not there is a significant difference between them.

The first stage in carrying out this test is to work out the expected results, as we have already done. These, and the observed results, are then recorded in a table like the one overleaf. We then calculate the difference between each set of results, and square each difference. (Squaring it gets rid of any minus signs – it is irrelevant whether the differences are negative or positive.) Then we divide each squared difference by the expected value, and add up all of these answers:

$$\chi^2 = \Sigma \frac{(O - E)^2}{E}$$

where   $\Sigma$ = the sum of
   O = the observed value
   E = the expected value

| Phenotypes of plants | purple stems, cut leaves | | purple stems, potato leaves | | green stems, cut leaves | | green stems, potato leaves |
|---|---|---|---|---|---|---|---|
| Observed number (O) | 86 | | 26 | | 24 | | 8 |
| Expected ratio | 9 | : | 3 | : | 3 | : | 1 |
| Expected number (E) | 81 | | 27 | | 27 | | 9 |
| O – E | +5 | | –1 | | –3 | | –1 |
| (O –E)$^2$ | 25 | | 1 | | 9 | | 1 |
| (O – E)$^2$/E | 0.31 | | 0.04 | | 0.33 | | 0.11 |
| $\Sigma(O - E)^2/E = 0.79$ | | | | | | | |
| $\chi^2 = 0.79$ | | | | | | | |

So now we have our value of $\chi^2$. Next we have to work out what it means.

To do this, we look in a table that relates $\chi^2$ values to probabilities (*table 17.3*). The probabilities given in the table are

the probability that the differences between our expected and observed results are due to chance.

For example, a probability of 0.05 means that we would expect these differences to occur in 5 out of every hundred experiments, or 1 in 20, just by chance. A probability of 0.01 means that we would expect them to occur in 1 out of every hundred experiments, just by chance. In biological experiments, we usually take a probability of 0.05 as being the critical one. If our $\chi^2$ value represents a probability of 0.05 or larger, then we can be fairly certain that the differences between our observed and expected results are due to chance – the differences between them are not **significant**. However, if the probability is smaller than this, then it is likely that the difference *is* significant, and we must reconsider our assumptions about what was going on in this cross.

There is one more aspect of our results to consider, before we can look up our value of $\chi^2$ in the table. This is the number of **degrees of freedom** in our results. This takes into account the number of comparisons made. (Remember that to get our value of $\chi^2$, we added up all our calculated values, so obviously the larger the number of observed and expected values we have, the larger $\chi^2$ is likely to be. We need to compensate for this.) To work out the number of degrees of freedom, simply calculate the (number of classes of data – 1). Here we have four classes of data (the four possible sets of phenotypes), so the degrees of freedom = (4 – 1) = 3.

Now, at last, we can look at the table to determine whether our results show a significant deviation from what we expected. The numbers in the body of the table are $\chi^2$ values. We look at the third row in the table (because that is the one relevant to 3 degrees of freedom), and find the $\chi^2$ value that represents a probability of 0.05. You can see that this is 7.82. Our calculated value of $\chi^2$ was 0.79. So our value is a much, much smaller value than the one we have read from the table. In fact, we cannot find anything like this number in the table – it would be way off the left hand side, representing a probability of much more than 0.1 (1 in 10) that the difference in our results is just due to chance. So we can say that the difference between our observed and expected results is almost certainly due to chance, and there is **no significant difference** between what we expected, and what we actually got.

| Degrees of freedom | Probability greater than | | | |
|---|---|---|---|---|
| | 0.1 | 0.05 | 0.01 | 0.001 |
| 1 | 2.71 | 3.84 | 6.64 | 10.83 |
| 2 | 4.60 | 5.99 | 9.21 | 13.82 |
| 3 | 6.25 | 7.82 | 11.34 | 16.27 |
| 4 | 7.78 | 9.49 | 13.28 | 18.46 |

● **Table 17.3** Table of $\chi^2$ values.

**SAQ 17.20**
Look back at your answer to SAQ 17.17b. In the actual crosses between the animals in this generation, the numbers of each phenotype obtained in the offspring were:
grey, long  54
grey, short  4
white, long  4
white, short  18

Use a $\chi^2$ test to determine whether or not the difference between these observed results and the expected results is significant.

# Mutations

You have seen that most genes have several different variants, called alleles. A gene is a made up of a sequence of nucleotides, each with its own base. The different alleles of a gene contain slightly different sequences of bases.

These different alleles originally arose by a process called **mutation**. Mutation is an unpredictable change in the genetic material of an organism. A change in the structure of a DNA molecule, producing a different allele of a gene, is a **gene mutation**. Mutations may also cause changes in the structure or number of whole chromosomes in a cell, in which case they are known as **chromosome mutations**.

Mutations may occur completely randomly, with no obvious cause. However, there are several environmental factors that significantly increase the chances of a mutation occurring. All types of **ionising radiation** (alpha, beta and gamma radiation) can damage DNA molecules, altering the structure of the bases within them. **Ultraviolet radiation** has a similar effect, as do many chemicals, for example mustard gas. A substance that increases the chances of mutation occurring is said to be a **mutagen**.

In gene mutations, there are three different ways in which the sequence of bases in a gene may be altered. These are:
- **base substitution**, where one base simply takes the place of another. For example:
  CCT GAG GAG may change to CCT GTG GAG;
- **base addition**, where one or more extra bases are added to the sequence. For example:
  CCT GAG GAG may change to CCA TGA GGA G;
- **base deletion**, where one or more bases are lost from the sequence. For example:
  CCT GAG GAG may change to CCG AGG AG.

Base additions or deletions usually have a very significant effect on the structure, and therefore the function, of the polypeptide that the allele codes for. If you look up the amino acids that are coded for by the 'normal' sequence shown above in Appendix 2, you will see that it is Gly Leu Leu. But the new sequence resulting from the base addition codes for Gly Thr Pro, and that resulting from the base deletion is Gly Ser. Base additions or deletions always have large effects, because they alter every set of three bases that 'follows' them in the DNA molecule. They are said to cause **frame shifts** in the code. Often, the effects are so large that the protein that is made is totally useless. Or they may introduce a 'stop' triplet part way through a gene, so that a complete protein is never made at all.

Base substitutions, on the other hand, often have no effect at all. A mutation that has no apparent effect on an organism is said to be a **silent mutation**. Base substitutions are often silent mutations because many amino acids have more than one triplet code (see Appendix 2 again), so even if one base is changed the same amino acid is still coded for. You have seen above that a change from CCT to CCA or CCG makes no difference – the amino acid that will be slotted into the chain at that point will still be Gly.

However, base substitutions *can* have very large effects. If, for example, the base sequence ATG (coding for Tyr) mutated to ATT, this has produced a 'stop' triplet, so the synthesis of the protein would stop at this point.

## Sickle cell anaemia

One example of a base substitution that has a significant effect on the phenotype is the one involved in the inherited blood disorder sickle cell anaemia. (We have already looked at the inheritance of this disease, on pages 226–227.)

Haemoglobin is the red pigment in red blood cells which carries oxygen around the body.

A haemoglobin molecule is made up of four polypeptide chains, each with one iron-containing haem group in the centre. Two of these polypeptide chains are called α chains, and the other two β chains. (The structure of haemoglobin is described on page 34.)

The gene which codes for the amino acid sequence in the β chains is not the same in everyone. In most people, the β chains begin with the amino acid sequence:

Val-His-Leu-Thr-Pro-Glu-Glu-Lys-

But in some people, the base sequence CTT is replaced by CAT, and the amino acid sequence becomes:

Val-His-Leu-Thr-Pro-Val-Glu-Lys-

This small difference in the amino acid sequence makes little difference to the haemoglobin molecule when it is combined with oxygen. But when it is not combined with oxygen, the 'unusual' β chains make the haemoglobin molecule much less soluble. The molecules tend to stick to each other, forming long fibres inside the red blood cells. The red cells are pulled out of shape, into a half-moon or sickle shape. When this happens, the distorted cells become quite useless at transporting oxygen. They also get stuck in small capillaries, stopping any unaffected cells from getting through (*figure 17.9*).

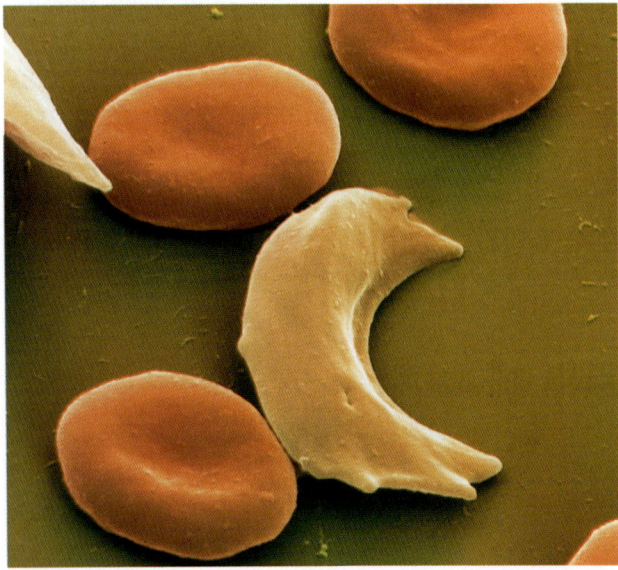

● **Figure 17.9** A scanning electron micrograph of red blood cells from a person with sickle cell anaemia. You can see both normal and sickled cells. (× 12 200)

A person with these unusual β chains can suffer severe anaemia (lack of oxygen transported to the cells) and may die. It is especially common in some parts of Africa, and in India. You can read about the reasons for this distribution on pages 251–252.

## Phenylketonuria

**Phenylketonuria**, or **PKU**, is another disease caused by an abnormal base sequence in part of a DNA molecule. However, in PKU the affected gene codes for an enzyme, not for an oxygen-carrying pigment.

The enzyme affected in PKU is phenylalanine hydroxylase. People with the disease lack this enzyme because their DNA does not carry the correct code for making it. Phenylalanine hydroxylase helps to catalyse the conversion of the amino acid phenylalanine to tyrosine, which can then be converted into melanin.

$$\text{phenylalanine} \xrightarrow{\text{phenylalanine hydroxylase}} \text{tyrosine} \longrightarrow \text{melanin}$$

Phenylalanine is found in many different kinds of foods. Melanin is the brown pigment in skin and hair. If phenylalanine cannot be converted to tyrosine, then little melanin is formed, so people with PKU frequently have a lighter skin and hair colour than normal.

However, a far more important problem which arises is that phenylalanine accumulates in the blood and tissue fluid. This causes severe brain damage in young children. Children with untreated PKU become severely mentally retarded.

All babies born in the UK are tested for PKU at birth, simply by testing the phenylalanine levels in their blood. This testing is very important because brain damage can be completely prevented if a child with PKU is, at birth, put on to a diet which does not contain phenylalanine.

The Human Genome Project is providing us with considerable amounts of information about the genes involved in diseases such as sickle cell anaemia and PKU.

# Environment and phenotype

In all the examples in this chapter, we have so far assumed that the genotype of the organism will always affect its phenotype in the same way. This is not always true.

Consider human height. If you have inherited a number of alleles for tallness from your parents, you have the *potential* to grow tall. However, if your diet is poor while you are growing, your cells might not be supplied with sufficient nutrients to allow you to develop this potential. You will not grow as tall as you could. Part of your environment, your diet, has also affected your height. Many characteristics of organisms are affected in this way by both genes and environment.

Another example is the development of the dark tips to ears, nose, paws and tail in the Himalayan colouring of rabbits (*figure 17.6*). This colouring is caused by an allele which allows the formation of the dark pigment only at low temperature. The parts of the rabbit which grow dark fur are the coldest parts. If an area somewhere else on its body is plucked of fur and kept cold, the new fur growing in this region will be dark.

# Gene technology

The structure of DNA, and the way in which it codes for protein synthesis, was worked out during the 1950s and 1960s. Since then, this knowledge has developed to the level at which we can change the DNA in a cell, and so change the proteins which that cell synthesises. This is called genetic engineering or **gene technology**.

## Insulin production

To explain the principles of gene technology, we will look at one example, that is the use of genetically modified bacteria to mass-produce human insulin.

One form of diabetes mellitus is caused by the inability of the pancreas to produce insulin. People with this disease need regular injections of insulin which, until recently, was extracted from the pancreases of pigs or cattle. This extraction was expensive, and many people did not like the idea of using insulin from an animal. Moreover, insulin from pigs or cattle is not identical to human insulin and so can have side-effects.

In the 1970s, biotechnology companies began to work on the idea of inserting the gene for human insulin into a bacterium, and then using this bacterium to make insulin. They tried several different approaches, finally succeeding in the early 1980s.

The procedure had several stages as described below and shown in *figure 17.10*.

### Isolating the insulin gene

Insulin is a small protein. The first task was to isolate the gene coding for human insulin from all the rest of the DNA in a human cell. In this instance, there were problems in doing this directly. Instead, mRNA carrying the code for making insulin was extracted from the cells in a human pancreas that synthesise insulin, called β cells.

The mRNA was then incubated with an enzyme called **reverse transcriptase** which comes from a special group of viruses called **retroviruses**. As the name suggests, this enzyme does something which does not normally happen in human cells – it reverses transcription, causing DNA to be made from RNA. Complementary DNA (cDNA) molecules were formed from the mRNA from the pancreas cells. First, single-stranded molecules were formed, which were then converted to double-stranded DNA. These DNA molecules carried the code for making insulin; that is, they were insulin genes.

In order to enable these insulin genes to stick onto other DNA at a later stage in the procedure, they were given 'sticky ends'. This was done by adding lengths of single-stranded DNA made up of guanine nucleotides to each end, using enzymes.

### Inserting the gene into a vector

In order to get the human insulin gene into a bacterium a go-between, called a **vector**, has to be used. In this instance the vector was a **plasmid**. A plasmid is a small, circular piece of DNA which can be found in many bacteria (page 17). Plasmids are able to insert themselves into bacteria so, if you can put your piece of human DNA into a plasmid, the plasmid can take it into a bacterium. (Viruses can act in a similar way to plasmids.)

To get the plasmids, the bacteria containing them were treated with enzymes to dissolve their

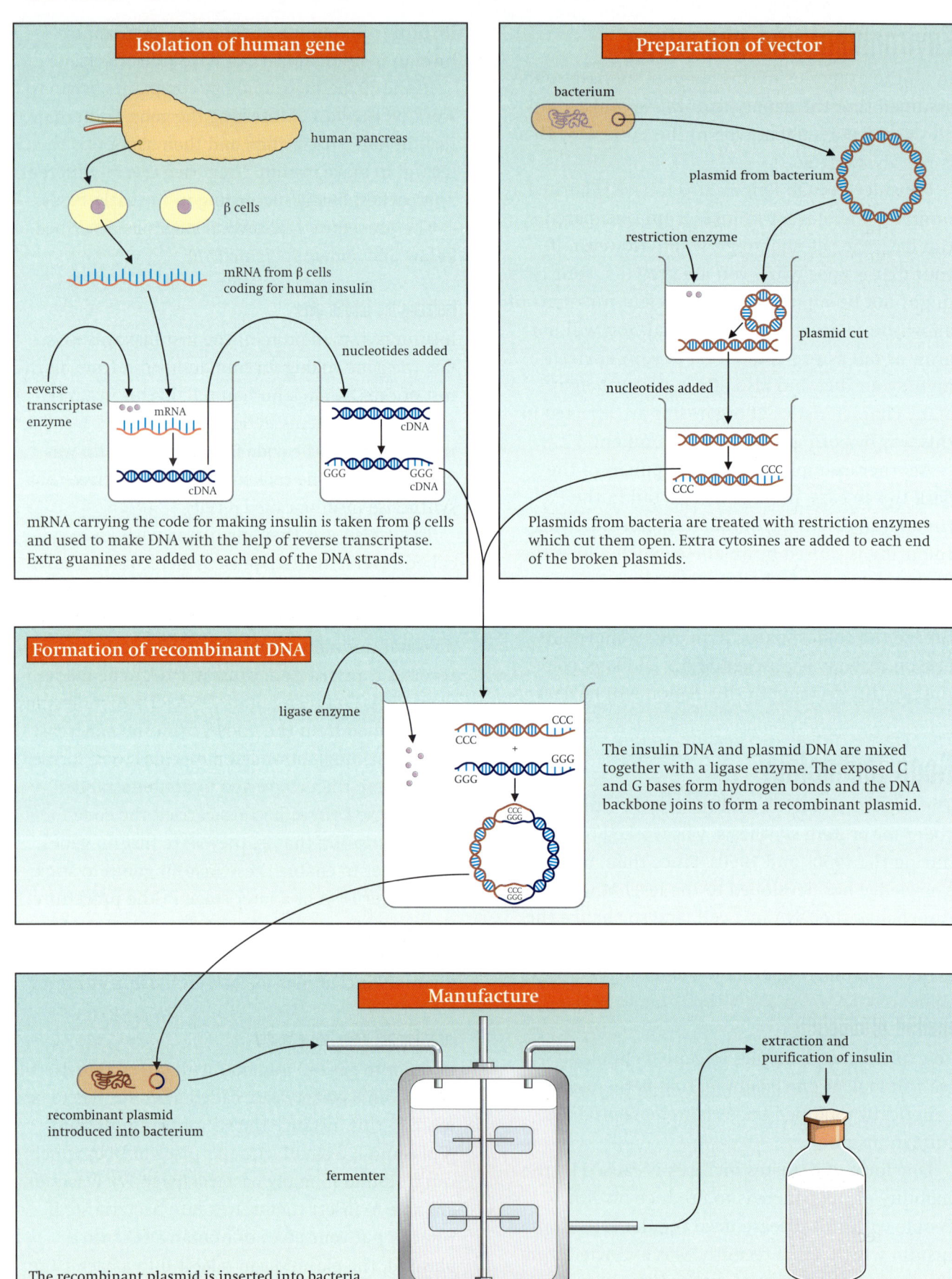

**Isolation of human gene**

human pancreas

β cells

mRNA from β cells coding for human insulin

reverse transcriptase enzyme

mRNA

cDNA

nucleotides added

cDNA

GGG    GGG
cDNA

mRNA carrying the code for making insulin is taken from β cells and used to make DNA with the help of reverse transcriptase. Extra guanines are added to each end of the DNA strands.

**Preparation of vector**

bacterium

plasmid from bacterium

restriction enzyme

plasmid cut

nucleotides added

CCC
CCC

Plasmids from bacteria are treated with restriction enzymes which cut them open. Extra cytosines are added to each end of the broken plasmids.

**Formation of recombinant DNA**

ligase enzyme

CCC        CCC
+
GGG
GGG

CCC
GGG

CCC
GGG

The insulin DNA and plasmid DNA are mixed together with a ligase enzyme. The exposed C and G bases form hydrogen bonds and the DNA backbone joins to form a recombinant plasmid.

**Manufacture**

extraction and purification of insulin

recombinant plasmid introduced into bacterium

fermenter

The recombinant plasmid is inserted into bacteria. These bacteria can now produce insulin.

pure human insulin

● **Figure 17.10** Producing insulin from genetically modified bacteria.

cell walls. They were then centrifuged, so that the relatively large bacterial chromosomes were separated from the much smaller plasmids and the cell debris. The circular DNA molecule making up the plasmid was then cut open using a **restriction enzyme**. Once again, sticky ends were added, but this time the nucleotides used to make these single strands contained cytosine.

The cut plasmid and the cut human DNA were mixed together, and the C and G bases on their sticky ends paired up. The nucleotide backbones were linked using an enzyme called **DNA ligase**, so that the human insulin gene became part of the plasmid. This created **recombinant** DNA.

### Inserting the gene into the bacteria

The plasmids were now mixed with bacteria. In the case of insulin, the bacterium was *E. coli*. A small proportion, perhaps 1%, of the bacteria took up the plasmids containing the insulin gene. These bacteria were separated from the others using antibiotic resistance provided by another gene which was introduced at the same time as the human insulin gene. When the bacteria were treated with antibiotic, only the ones containing the resistance gene (and therefore the insulin gene) survived.

The genetically modified bacteria are now cultured on a large scale. They secrete insulin, which is extracted, purified and sold for use by people with diabetes. It is called recombinant insulin because it is produced by organisms containing a combination of their own and human DNA.

## Other uses of gene technology

Insulin production was one of the earliest success stories for gene technology. Since then, there have been many others. Other human protein hormones have been synthesised, for example human growth hormone. Enzymes are made for use in the food industry, for example, or in biological washing powders.

Gene technology can introduce genes into any organism, not just bacteria. Recent developments give hope for the success of **gene therapy** in humans, in which 'good' copies of genes are inserted into cells of people with 'defective' ones.

This could be used to treat genetic diseases. However, at the moment there are problems in getting the genes into enough cells for there to be any useful effect.

Genetically modified hamster cells are used by several companies to produce a protein called **human factor VIII**. This protein is essential for blood clotting, and people who cannot make it suffer from haemophilia. The human gene for making factor VIII has been inserted into hamster kidney and ovary cells that are then cultured in fermenters. The cells constantly produce factor VIII, which is extracted and purified before being used to treat people with haemophilia. These people need regular injections of factor VIII, which, before the availability of the recombinant factor VIII, came from donated blood. This carried risks of infection, such as with HIV (see chapter 13). Recombinant factor VIII avoids such problems.

Genes can be inserted into plants, too. Genes conferring resistance to pests can be extracted from a wild plant and inserted into a crop plant, for example. However, people who are the possible future consumers of foods derived from such crops are not at all satisfied that this new technology is entirely safe (either for other organisms in the environment or for themselves) or desirable. There has therefore been considerable opposition to field trials of genetically modified crops.

## Benefits and hazards of genetic engineering

We can now create genetically modified organisms for specific uses. In the past, such organisms were derived from selective breeding or arose by chance mutation. In contrast to organisms produced by selective breeding, there is a tendency to see genetically engineered organisms as unnatural and intrinsically unsafe.

As a result of genetic engineering, micro-organisms now produce many substances that they would not normally produce. Most of these modified microorganisms are kept in industrial fermenters. Provided that proper containment precautions are used, they cannot affect the general environment. In many cases the strain of organism used, for instance of *E. coli*, survives very poorly in the general environment. The obvious danger is a breakdown of containment, and regulations exist to help prevent contact between such microorganisms and the outside environment.

### SAQ 17.21

Explain the potential dangers of contact between genetically modified microorganisms and the outside environment.

A totally different set of problems emerges when genetically engineered organisms, such as crop plants and organisms for the biological control of pests, are intended for use in the general environment. Can such organisms be used safely?

The United Kingdom has a good reputation for rigorous consideration of the risks and benefits of releasing genetically modified organisms. Anyone wishing to conduct a field trial must assess the risk to the environment. The Department of the Environment then decides, after taking advice from the Advisory Committee on Releases to the Environment (ACRE), whether to issue consent. The approval process is meant to take no longer than 90 days. This procedure has sometimes been seen as a disadvantage for researchers in the UK in comparison with Japan and the USA where approval is given in a shorter time. However, a so-called 'fast track' procedure has been allowed in the UK since 1993 for experiments thought to be less hazardous. Much less information needs to be given about these experiments and permission to perform the trial can be given in 30 days. A low-hazard release might be a species that has no natural relatives in the UK and has been modified with a well-known gene, whilst a high-risk release might be a native weed with a gene that had not been transferred before. 'Fast-tracking' has raised many concerns, not least because 30 days is a short time for any oppo-nents of the release to express their fears and have them acted upon.

Consider a field trial that sparked a blaze of publicity in 1994. David Bishop and his colleagues at Oxford sought permission to release genetically

modified viruses onto cabbages to test their effect on the caterpillars of the cabbage looper moth. The virus, a baculovirus, had been discovered in a moth that is not native to the UK, the alfalfa looper. The virus had been modified by adding a gene coding for one of the proteins of scorpion venom, which is lethal to insects but harmless to other animals. An earlier field trial had shown that unmodified virus killed the caterpillars, but that modified virus killed them more quickly, reducing damage to cabbages (*table 17.4*). The chance of the virus spreading was also reduced, since the modified virus killed caterpillars so rapidly that they produced very few viruses. The new field trial was to measure the effect of the virus on caterpillars of six other moths, and to see how long the virus survived in the soil and on the cabbages. The cabbages were grown in fine netting enclosures to keep out birds and other animals that might spread the virus. Traps prevented insects from getting in.

The opponents of the trial feared that the virus might escape and harm other species of moth, particularly since the field experiment was sited close to the University of Oxford's nature reserve, Wytham Wood. Another potential hazard is that the virus might swap (recombine) genes with other viruses. This is theoretically possible if two different viruses infect the same insect at the same time.

In an attempt to allay fears over the safety of the 1994 trials, the preliminary results were, most unusually, given to a public meeting in November 1994. These results showed that non-target species were much less susceptible to the virus than were

| Treatment of cabbage plants | Mean leaf area per cabbage plant eaten by caterpillars (cm$^2$) |
|---|---|
| untreated (control) | 107 |
| sprayed with unmodified virus: | |
|    low dose ($10^6$ virus particles per m$^2$) | 100 |
|    high dose ($10^7$–$10^8$ virus particles per m$^2$) | 63 |
| sprayed with genetically modified virus: | |
|    low dose | 70 |
|    high dose | 50 |

● **Table 17.4** Effect on damage to cabbage plants by looper caterpillars treated with genetically modified and unmodified virus.

the cabbage looper caterpillars. Of the target caterpillars, 80% died after 8 days, in comparison with 1% of non-target species. The results also showed that the genetically engineered virus is some way from being marketed as a biopesticide, since although it kills caterpillars faster than the unmodified virus, the caterpillars still have time to eat holes in the cabbage leaves. Tests will take place for a further five years on the same site, but it will not be possible to test all species of British moths and butterflies.

The aim of this work is to produce a viral insecticide that kills only a specific pest, replacing chemical insecticides that kill a range of organisms and may also damage other organisms. Unmodified viruses have been successfully used in other countries to control particular insects for some time.

## SAQ 17.22

Examine *table 17.4* and compare the effects, on damage to cabbage plants by looper caterpillars, of treatment with genetically modified and unmodified viruses.

Consider another experiment, relevant to herbicide-resistant plants. The concerns about such genetically engineered crops are that:

- the modified crop plants will become agricultural weeds or invade natural habitats;
- the introduced gene(s) will be transferred by pollen to wild relatives whose hybrid offspring will become more invasive;
- the introduced gene(s) will be transferred by pollen to unmodified plants growing on a farm with 'organic' certification;
- the modified plants will be a direct hazard to humans, domestic animals or other beneficial animals, by being toxic or producing allergies;
- the herbicide that can now be used on the crop will itself leave toxic residues in the crop.

The results of an investigation to compare invasiveness of normal and genetically modified oilseed rape plants was published by M. Crawley and colleagues at Silwood Park (Imperial College, UK) in 1993. Three genetic lines were compared: non-engineered oilseed rape and two different genetically engineered versions of the same

cultivar. The rates of population increase were compared in plants grown in a total of 12 different environments. In each of these, various treatments were applied, including cultivated and uncultivated background vegetation, and presence and absence of various herbivores and pathogens. There was no evidence that genetic engineering increased the invasiveness of these plants. Where differences between them existed, the genetically engineered plants were slightly less invasive than the unmodified plants.

The risk of pollen transfer, by wind or by insect, is real. Although 'safe' planting distances are specified for trials of genetically modified plants (for example 200 m for oilseed rape) pollen from various plants has been found between 1000 and 1500 m away. Bees visiting some flowers have been found to forage for distances of more than 4000 m. 'Safe' planting distances should be increased to allow the organic farming industry to maintain its 'GM-free' certification.

In an attempt to assess the risk of introduced genes being spread from genetically engineered bacteria to wild strains, the Agricultural and Food Research Council in the UK investigated the interaction of different strains of *Rhizobium* in field conditions. A strain of *R. leguminosum* whose cells contain a plasmid that can be transferred to other bacteria was used. The plasmid was 'marked' by inserting a gene for antibiotic resistance into it. Large numbers of samples of bacteria over a two-year period revealed that the genetically engineered strain remained in the field soil, but there was no evidence of transfer of the marked plasmid to naturally occurring *Rhizobium*.

Crop plants which contain the toxin genes from *Bacillus thuringiensis* produce their own insecticides. However, a small number of insect populations have evolved resistance to these toxins. The danger is that large numbers of crop plants containing the genes may simply accelerate the evolution of resistance to the toxins.

The pollen of maize engineered with *Bacillus thuringiensis* toxin (Bt maize) expresses the gene and has been found to disperse at least 60 m by wind. In the USA, milkweed frequently grows around the edge of maize fields and is fed upon by the caterpillars of the Monarch butterfly. Half of the summer population of Monarch butterflies is found in the maize-growing areas of the USA. An experiment was set up in which caterpillars were fed milkweed leaves dusted with pollen from Bt maize, pollen from unmodified maize or no pollen at all. Caterpillar survival after 4 days of feeding on leaves dusted with pollen from Bt maize was 56%, whereas no caterpillars died after eating leaves dusted with pollen from unmodified maize or leaves with no pollen.

Safety concerns about the Flavr Savr tomato come not from its transgene, but because it also contains a bacterial gene coding for a protein that gives resistance to two antibiotics, kanamycin and neomycin. This gene was inserted so that researchers could identify whether the transgene had been taken up. During testing, exposure to kanamycin kills any plant cells that are not transgenic. The United States Food and Drug Administration found no evidence that the protein giving antibiotic resistance would poison consumers or trigger allergies, nor that it would interfere when people took antibiotics. It declared the tomatoes "as safe as tomatoes bred by conventional means". However, there remains a risk that the gene might be transferred to other bacteria, adding to the problem of antibiotic-resistant disease.

## Ethical implications of genetic engineering

Ethics are sets of standards by which a particular group of people agree to regulate their behaviour, distinguishing an acceptable from an unacceptable activity. Ethics change with time, because people alter their views according to their knowledge and experience.

In 1974, genetic engineers worldwide accepted a self-imposed ban on some recombinant DNA experiments on the basis that they were too risky. Four years later, their general view was that the risks had been greatly over-estimated. Nevertheless, genetic engineering is a relatively new development, experience of it is limited and a large number of people know virtually nothing about it. Also, development of the techniques has been

rapid. The public was introduced to its first genetically engineered animal in 1982 (*figure 17.11*) and now transgenic animals are standard tools in research and in the production of pharmaceuticals.

An EU committee set up to investigate ethical aspects of biotechnology has as one of its aims that of improving public understanding and acceptance. A survey published in 1993 by the Open University found that four out of five people do not trust industry to tell the truth about genetic engineering and two out of three felt that industry takes shortcuts with safety.

Discuss with your friends whether:

- genetic engineering is in principle acceptable, and if so, in what circumstances;
- it is acceptable to patent a genetically engineered organism or to patent a gene sequence;
- it is acceptable to engineer any organism to produce a product useful to humans;
- it is acceptable to engineer animals to show human diseases for research into those diseases;
- genetically modified food is acceptable;
- products on sale are adequately labelled to indicate that genetic engineering was involved in their production.

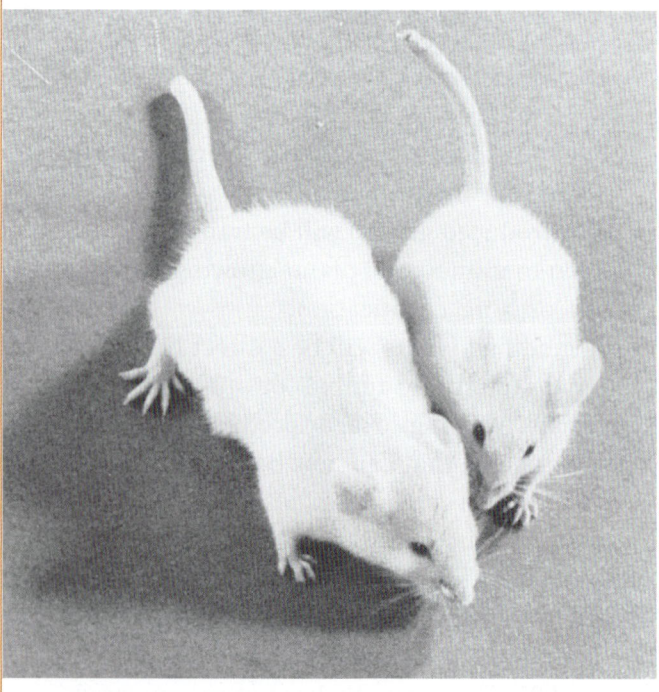

● **Figure 17.11** 'Supermouse' (left) is a transgenic mouse with a rat growth hormone gene. It is almost twice the mass of its normal brother (right).

# SUMMARY

◆ Diploid organisms contain two copies of each gene in each of their cells. In sexual reproduction, gametes are formed containing one copy of each gene. Each offspring receives two copies of each gene, one from each of its parents.

◆ Gametes are haploid cells, and they are formed from diploid cells by meiosis. Meiosis consists of two divisions. The first division, meiosis I, separates the homologous chromosomes, so that each cell now has only one of each pair. The second division, meiosis II, separates the chromatids of each chromosome. Meiotic division therefore produces four cells, each with one complete set of chromosomes.

◆ The cells produced by meiosis are genetically different from each other and from their parent cell. This results from independent assortment of the chromosomes as the bivalents line up on the equator during metaphase I, and also from crossing over between the chromatids of homologous chromosomes during prophase I.

◆ Genetic variation also results from random fertilisation, as gametes containing different varieties of genes fuse together to form a zygote.

◆ Different varieties of a gene are called alleles. Alleles may show dominance or codominance. An organism possessing two identical alleles of a gene is homozygous; an organism possessing two different alleles of a gene is heterozygous. If a gene has several different alleles, such as the gene for human blood groups, these are known as multiple alleles.

◆ The genotype of an organism showing dominant characteristics can be determined by looking at the offspring produced when it is crossed with an organism showing recessive characteristics. This is called a test cross.

◆ A gene found on the X chromosome but not on the Y chromosome is known as a sex-linked gene.

◆ Monohybrid crosses consider the inheritance of one gene. Dihybrid crosses consider the inheritance of two different genes.

◆ The $\chi^2$ test can be used to find out whether any differences between expected results and observed results of a genetic cross are due to chance, or whether the difference is significant.

◆ Mutation can be defined as an unpredictable change in the base sequence in a DNA molecule (gene mutation) or in the structure or number of chromosomes (chromosome mutation). New alleles arise by gene mutation. Gene mutations include base substitutions, deletions or additions. The sickle cell allele arose by base substitution.

◆ The genotype of an organism gives it the potential to show a particular characteristic. In many cases, the degree to which this characteristic is shown is also affected by the organism's environment.

◆ DNA may be transferred from one species to another by means of gene technology. This technology has been used to produce bacteria that synthesise insulin, and mammalian cells that synthesise human factor VIII.

◆ Genetic engineering may be beneficial, but may have associated hazards.

## CHAPTER 18

# Selection and evolution

**By the end of this chapter you should be able to:**

1 explain how variation is produced in sexually reproducing organisms;

2 explain why variation caused by genes can be inherited, but variation caused by the environment cannot;

3 explain how all organisms can potentially overproduce;

4 describe how different selection pressures may act on individual organisms with different alleles, so affecting allele frequencies in a population;

5 explain, with examples, how environmental factors can act as stabilising or evolutionary forces of natural selection;

6 describe one example of artificial selection;

7 explain the meaning of the term *species*, and explain the roles of natural selection and isolating mechanisms in the evolution of new species.

## Variation

In chapter 17, you have seen how sexual reproduction produces **genetic variation** amongst the individuals in a population. Genetic variation is caused by:

- independent assortment of chromosomes, and therefore alleles, during meiosis;
- crossing over between chromatids of homologous chromosomes during meiosis;
- random mating between organisms within a species;
- random fertilisation of gametes;
- mutation.

The first four of these processes reshuffle alleles in the population. Offspring have combinations of alleles which differ from those of their parents, and from each other. This genetic variation produces phenotypic variation.

Mutation, however, does more than reshuffle alleles that are already present. Mutation can produce completely new alleles. This may happen, for example, if a mistake occurs in DNA replication, so that a new base sequence occurs in a

gene. This is probably how the sickle cell allele of the gene for the production of the β polypeptide of haemoglobin first arose. Such a change in a gene, which is quite unpredictable, is called a **gene mutation**. The new allele is very often recessive, so it frequently does not show up in the population until some generations after the mutation actually occurred, when by chance two descendants of the organism in which the mutation happened mate and produce offspring.

Mutations that occur in body, or **somatic**, cells often have no effects at all on the organism. A malfunctioning cell in a tissue is only one of thousands of similar cells, and it is very unlikely that this cell would cause any problems. Most mutated cells are recognised as foreign by the body's immune systems and are destroyed (see page 85 and chapter 14). Occasionally the mutation may affect the regulation of cell division. If a cell with such a mutation escapes the attack of the immune system, it can produce a lump of cells called a tumour. Tumours often cause little harm, but sometimes the tumour cells are able to spread around the body and invade other tissues. This type of tumour is described as **malignant**, and the diseases caused by such tumours are **cancers**.

Mutations in somatic cells cannot be passed on to offspring by sexual reproduction. However, mutations in cells in the ovaries or testes of an animal, or in the ovaries or anthers of a plant, may be inherited by offspring. If a cell containing a mutation divides to form gametes, then the gametes may also contain the mutated gene. If such a gamete is one of the two which fuse to form a zygote, then the mutated gene will also be in the zygote. This single cell then divides repeatedly to form a new organism, in which all the cells will contain the mutated gene.

*Genetic* variation, whether caused by the reshuffling of alleles during meiosis and sexual reproduction or by the introduction of new alleles by mutation, can be passed on by parents to their offspring giving differences in phenotype. Variation in phenotype is also caused by the *environment* in which organisms live. For example, some organisms might be larger than others because they had access to better quality food while they were growing. This type of variation is *not* passed on by parents to their offspring.

## SAQ 18.1

Explain why variation caused by the environment cannot be passed from an organism to its offspring.

# Overproduction

All organisms have the reproductive potential to increase their populations. Rabbits, for example, produce several young in a litter, and each female may produce several litters each year. If all the young rabbits survived to adulthood and reproduced, then the rabbit population would increase rapidly. *Figure 18.1* shows what might happen.

This sort of population growth actually did happen in Australia in the nineteenth century. In 1859, twelve pairs of rabbits from Britain were released on a ranch in Victoria, as a source of food. The rabbits found conditions to their liking. Rabbits feed on low-growing vegetation, especially grasses, of which there was an abundance. There were very few predators to feed on them, so the number of rabbits soared. Their numbers became so great that they seriously affected the availability of grazing for sheep (*figure 18.2*).

Such population explosions are rare in normal circumstances. Although rabbit populations have the potential to increase at such a tremendous rate, they do not usually do so.

As a population of rabbits increases, various **environmental factors** come into play to keep down their numbers. These factors may be **biotic**, that is caused by other living organisms such as predation, competition for food, or infection by pathogens, or they may be **abiotic**, that is caused by non-living components of the environment such as water supply or nutrient levels in the soil. For example, the increasing number of rabbits eats an increasing amount of vegetation, until food is in short supply. The larger population may allow the populations of predators, such as foxes, stoats and weasels, to increase. Overcrowding may occur, increasing the ease with which diseases such as myxomatosis (*figure 18.3*) may spread. This disease is caused by a virus which is transmitted by fleas. The closer together the rabbits live, the more easily fleas, and therefore viruses, will pass from one rabbit to another.

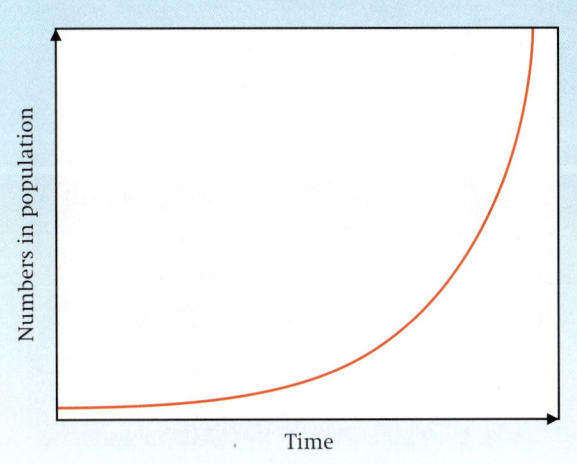

● **Figure 18.1** If left unchecked by environmental factors, numbers in a population may increase exponentially.

● **Figure 18.2** Attempts to control the rabbit population explosion in Australia in the mid to late nineteenth century included 'rabbit drives', in which huge numbers were rounded up and killed. Eventually, myxomatosis brought numbers down.

These environmental factors act to reduce the rate of growth of the rabbit population. Of all the rabbits born, many will die from lack of food, or be killed by predators, or die from myxomatosis. Only a small proportion of young will grow to adulthood

● **Figure 18.3** Rabbits living in dense populations are more likely to get myxomatosis than those in less crowded conditions.

and reproduce, so population growth slows.

If the pressure of the environmental factors is sufficiently great, then the population size will decrease. Only when the numbers of rabbits have fallen considerably will the numbers be able to grow again. Over a period of time, the population will oscillate about a mean level. *Figure 18.4* overleaf shows this kind of pattern in a lemming population over eleven years. The oscillations in lemming populations are particularly marked; in others, they are usually less spectacular!

This type of pattern is shown by the populations of many organisms. The number of young produced is far greater than the number which will survive to adulthood. Many young die before reaching reproductive age.

## Natural selection

What determines which will be the few rabbits to survive, and which will die? It may be just luck. However, some rabbits will be born with a better chance of survival than others. Variation within a population of rabbits means that some will have features which give them an advantage in the 'struggle for existence'.

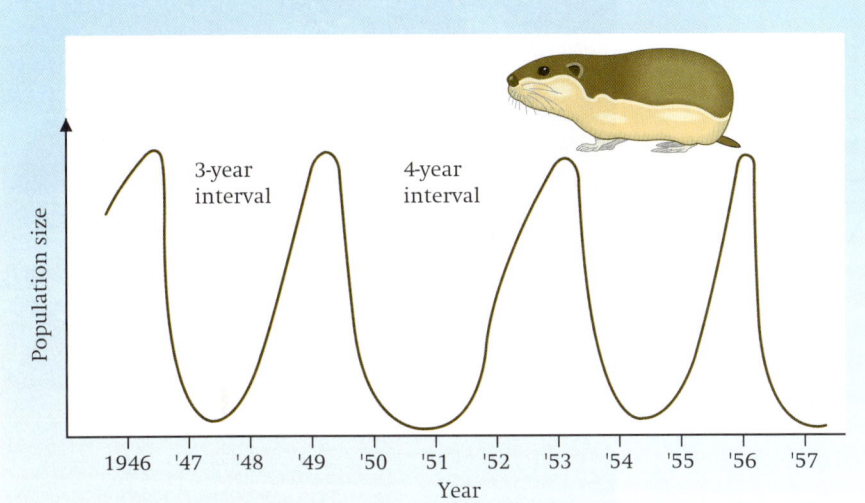

- **Figure 18.4** Lemming populations are famous for their large increases and decreases. In some years, populations become so large that lemmings may emigrate 'en masse' from overcrowded areas. The reason for the oscillating population size is not known for certain, although it has been suggested that food supply or food quality may be the main cause. As the population size rises, food supplies run out, so the population size 'crashes'. Once the population size has decreased, food supplies begin to recover, and the population size rises again.

One feature that may vary is coat colour. Most rabbits have alleles which give the normal agouti (brown) colour. A few, however, may be homozygous for the recessive allele which gives white coat. Such white rabbits will stand out distinctly from the others, and are more likely to be picked out by a predator such as a fox. They are less likely to survive than agouti rabbits. The chances of a white rabbit reproducing and passing on its alleles for white coat to its offspring are very small, so the allele for white coat will remain very rare in the population.

Predation by foxes is an example of a **selection pressure**. Selection pressures increase the chances of some alleles being passed on to the next generation, and decrease the chances of others. In this case, the alleles for agouti coat have a selective advantage over the alleles for white. The alleles for agouti will remain the commoner alleles in the population, while the alleles for white will remain very rare. The alleles for white coat may even disappear completely.

The effects of such selection pressures on the frequency of alleles in a population is called **natural selection**. Natural selection raises the frequency of alleles conferring an advantage, and reduces the frequency of alleles conferring a disadvantage.

## SAQ 18.2

Skomer is a small island off the coast of Wales. Rabbits have been living on the island for many years. There are no predators on the island.

**a** Rabbits on Skomer are not all agouti. There are quite large numbers of rabbits of different colours, such as black and white. Suggest why this is so.

**b** What do you think might be important selection pressures acting on rabbits on Skomer?

# Evolution

Usually, natural selection keeps things the way they are. This is **stabilising selection**. Agouti rabbits are best adapted to survive predation, so the agouti allele remains the most common coat colour allele in rabbit populations. Unless something changes, then natural selection will ensure that this continues to be the case.

However, if a *new environmental factor*, or a *new allele* appears, then allele frequencies may also change. This is called **directional selection** (*figures 18.5* and *18.6*).

## A new environmental factor

Imagine that we are plunged into a new Ice Age. The climate becomes much colder, so that snow covers the ground for almost all of the year. Assuming that rabbits can cope with these conditions, white rabbits now have a selective advantage during seasons when snow lies on the ground, as they are better camouflaged (*figure 18.7*). Rabbits with white fur are more likely to survive and reproduce, passing on their alleles for white fur to their offspring. The frequency of the

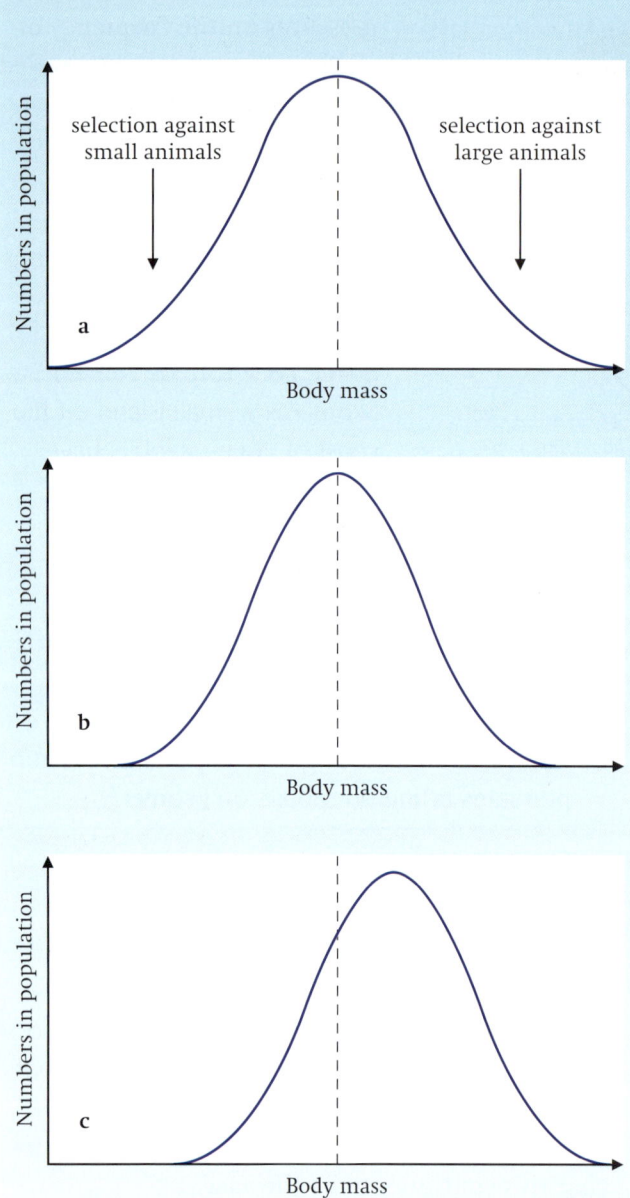

- **Figure 18.6** The tuatara, *Sphenodon punctatus*, is a lizard-like reptile that lives in New Zealand. Fossils of a virtually identical animal have been found in rocks 200 million years old. Natural selection has acted to keep the features of this organism the same over all this time.

allele for white coat increases, at the expense of the allele for agouti. Over many generations, almost all rabbits will come to have white coats rather than agouti.

## A new allele

Because they are random events, most mutations that occur produce features that are harmful. That is, they produce organisms that are less well adapted to their environment than 'normal' organisms. Other mutations may be 'neutral', conferring

- **Figure 18.5** If a characteristic in a population, such as body mass, shows wide variation, selection pressures often act against the two extremes (graph **a**). Very small or very large individuals are less likely to survive and reproduce than those whose size lies nearer the centre of the range. This results in a population with a narrower range of body size (graph **b**). This type of selection, which tends to keep the variation in a characteristic centred around the same mean value, is called **stabilising selection**. Graph **c** shows what would happen if selection acted against smaller individuals but not larger ones. In this case, the range of variation shifts towards larger size. This type of selection, which results in a change in a characteristic in a particular direction, is called **directional selection**.

- **Figure 18.7** The white winter coat of a mountain hare provides excellent camouflage from predators when viewed against snow.

neither an advantage nor a disadvantage on the organisms within which they occur. Occasionally mutations may produce useful features.

Imagine that a mutation occurs in the coat colour gene of a rabbit, producing a new allele which gives a better camouflaged coat colour than agouti. Rabbits possessing this new allele will have a selective advantage. They will be more likely to survive and reproduce than agouti rabbits, so the new allele will become more common in the population. Over many generations, almost all rabbits will come to have the new allele.

Such changes in allele frequency in a population are the basis of **evolution**. Evolution occurs because natural selection gives some alleles a better chance of survival than others. Over many generations, populations may gradually change, becoming better adapted to their environments. Examples of such change are the development of antibiotic resistance in bacteria and industrial melanism in the peppered moth *Biston betularia*.

## Antibiotic resistance

Antibiotics are chemicals produced by living organisms, which inhibit or kill bacteria, but do not normally harm human tissue. Most antibiotics are produced by fungi. The first antibiotic to be discovered was penicillin, which was first used during the Second World War to treat a wide range of diseases caused by bacteria. Penicillin stops cell wall formation in bacteria, so preventing cell reproduction.

If someone takes penicillin to treat a bacterial infection, bacteria which are sensitive to penicillin will die. In most cases, this will be the entire population of the disease-causing bacteria. However, by chance, there may be among them one or more individual bacteria with an allele giving resistance to penicillin. One example of such an allele occurs in some populations of the bacterium *Staphylococcus*, where some individual bacteria produce an enzyme, penicillinase, which inactivates penicillin.

As bacteria have only a single loop of DNA, they have only one copy of each gene, so the mutant allele will have an immediate effect on the phenotype of any bacterium possessing it. These individuals have a tremendous selective advantage. The bacteria without this allele will be killed, while those bacteria with resistance to penicillin can survive and reproduce. Bacteria reproduce very rapidly in ideal conditions, and even if there was initially only one resistant bacterium, it might produce ten thousand million descendants within 24 hours. A large population of a penicillin-resistant strain of *Staphylococcus* would result.

Such antibiotic-resistant strains of bacteria are continually appearing (*figure 18.8*). By using antibiotics, we change the environmental factors which exert selection pressures on bacteria. A constant 'arms race' is on to find new antibiotics against new resistant strains of bacteria.

Alleles for antibiotic resistance often occur on plasmids (see pages 17 and 237). Plasmids are quite frequently transferred from one bacterium to another, even between different species. Thus it is even possible for resistance to a particular antibiotic to arise in one species of bacterium, and be passed on to another. The more we use antibiotics, the greater the selection pressure we exert on bacteria to evolve resistance to them.

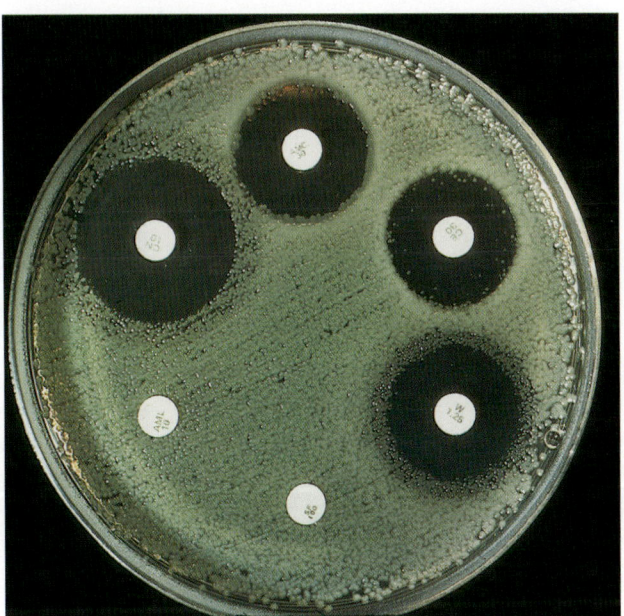

● **Figure 18.8** The grey areas on the agar jelly in this petri dish are colonies of the bacterium *Escherichia coli*. The white discs are pieces of card impregnated with different antibiotics. Where there are clear areas around the disc, the antibiotic has prevented the bacteria from growing. However, you can see that this strain of *E. coli* is resistant to the antibiotics on the discs at the bottom left and has been able to grow right up to the discs.

## SAQ 18.3

Suggest how each of the following might decrease the chances of an antibiotic–resistant strain of bacteria developing:

a  limiting the use of antibiotics to cases where there is a real need;

b  regularly changing the type of antibiotic which is prescribed for a particular disease;

c  using two or more antibiotics together to treat a bacterial infection.

### Industrial melanism

One well-documented case of the way in which changing environmental factors may produce changes in allele frequencies is that of the peppered moth *Biston betularia* (figure 18.9). This is a night-flying moth, which spends the day resting underneath the branches of trees. It relies on camouflage to protect it from insect-eating birds which hunt by sight. Until 1849, all specimens of this moth in collections had pale wings with dark markings, giving a speckled appearance. In 1849, however, a black (melanic) individual was caught near Manchester. During the rest of the nineteenth century, the numbers of black *Biston betularia* increased dramatically in some areas, while in other parts of the country the speckled form remained the more common.

The difference in the black and speckled forms of the moth is caused by a single gene. The normal speckled colouring is produced by a recessive allele of this gene, c, while the black colour is produced by a dominant allele, C. Up until the late 1960s, the frequency of the allele C increased in areas near to industrial cities. In non-industrial areas, the allele c remained the more common allele (figure 18.10).

The selection pressure causing the change of allele frequency in industrial areas was predation by birds. In areas with unpolluted air, tree branches are often covered with grey, brown and green lichen. On such tree branches, speckled moths are superbly camouflaged.

However, lichens are very sensitive to pollutants such as sulphur dioxide, and do not grow on trees near to or downwind of industries releasing pollutants into the air. Trees in these areas therefore have much darker bark, against which the dark moths are better camouflaged. Experiments have shown that light moths have a much higher chance of survival in unpolluted areas than dark moths, while in polluted areas the dark moths have the selective advantage. As air pollution from industry is reduced, the selective advantage swings back in favour of the speckled variety. So we would expect the proportion of speckled moths to increase if we succeeded in reducing the output of certain pollutants. This is, in fact, what has happened since the 1970s. It is predicted that by 2005 there will be hardly any dark individuals left.

It is important to realise that the C allele has probably been present in *B. betularia* populations for a very long time. It has not been produced by pollution. Until the nineteenth century there was such a strong selection pressure against the C allele that it remained exceedingly rare. Mutations of the c allele to the C allele may have occurred quite frequently, but moths with this allele

● **Figure 18.9** Light and melanic forms of peppered moths on light and dark tree bark.

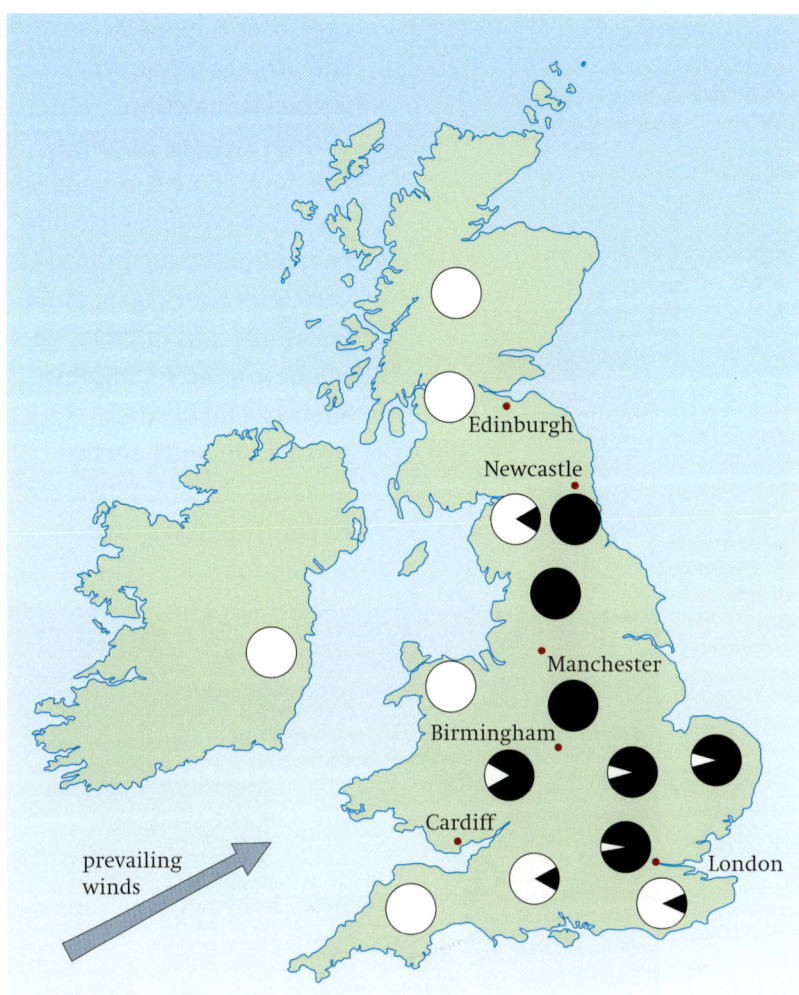

● **Figure 18.10** The distribution of the pale and dark forms of the peppered moth, *Biston betularia*, in the early 1960s. The ratio of dark to light areas in each circle shows the ratio of dark to light moths in that part of the country.

would almost certainly have been eaten by birds before they could reproduce. Changes in environmental factors only affect the likelihood of an allele surviving in a population; they do not affect the likelihood of such an allele arising by mutation.

### Sickle cell anaemia

In chapter 17, we saw how an allele, $H^S$, of the gene which codes for the production of the β polypeptides of the haemoglobin molecule can produce sickling of red blood cells. People who are homozygous for this allele have sickle cell anaemia. This is a severe form of anaemia which is often lethal.

The possession of two copies of this allele obviously puts a person at a great selective disadvantage. People who are homozygous for the sickle cell allele are less likely to survive and reproduce. Until recently, almost everyone with sickle cell anaemia died before reaching reproductive age. Yet the frequency of the sickle cell allele is very high in some parts of the world. In some parts of East Africa, almost 50% of babies born are carriers for this allele, and 14% are homozygous, suffering from sickle cell anaemia. How can this be explained?

The parts of the world where the sickle cell allele is most common are also the parts of the world where malaria is found (*figure 18.12*). Malaria is caused by a protoctist parasite, *Plasmodium*, which can be introduced into a person's blood when an infected mosquito bites (see *figure 18.11* and see pages 167–170). The parasites enter the red blood cells and multiply inside them. Malaria is the major source of illness and death in many parts of the world.

In studies carried out in some African states, it has been found that people who are heterozygous for the sickle cell allele are much less likely to suffer from a serious attack of malaria than people who are homozygous for the normal allele. Heterozygous people with malaria

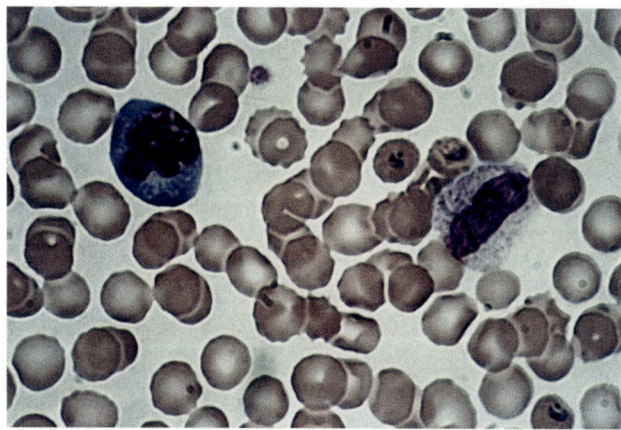

● **Figure 18.11** The purple structure in this micrograph of human blood is a *Plasmodium*, the protoctist which causes malaria (× 1400). At an earlier stage in the life cycle of *Plasmodium*, the organism reproduces inside the red blood cells.

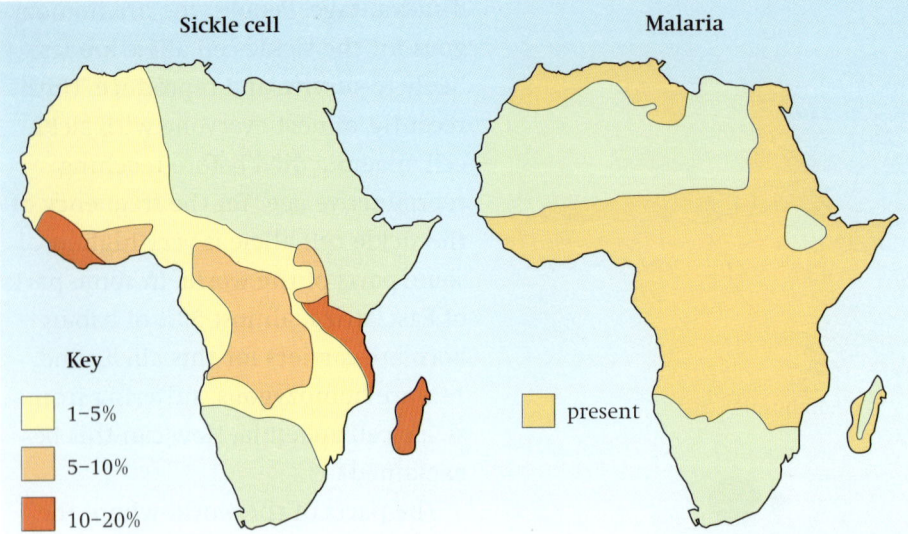

Key
- ☐ 1–5%
- ☐ 5–10%
- ☐ 10–20%

● **Figure 18.12** The distribution of people with at least one copy of the sickle cell allele, and the distribution of malaria, in Africa.

only have about one third the number of *Plasmodium* in their blood as normal homozygotes. In one study, of a sample of 100 children who died from malaria, all except one were normal homozygotes, although within the population as a whole 20% of people were heterozygotes.

There are, therefore, two strong selection pressures acting on these two alleles. Selection against people who are homozygous for the sickle cell allele, $H^SH^S$, is very strong, because they become seriously anaemic. Selection against people who are homozygous for the normal allele, $H^NH^N$, is also very strong, because they are more likely to die from malaria. In areas where malaria is common, heterozygotes, $H^NH^S$, have a strong selective advantage; they do not suffer from sickle cell anaemia and are much less likely to suffer badly from malaria. So both alleles remain in populations where malaria is an important environmental factor. In places where malaria was never present, selection against people with the genotype $H^SH^S$ has almost completely removed the $H^S$ allele from the population.

## Artificial selection

Sometimes, the most important selection pressures on organisms are those applied by humans. When humans purposefully apply selection pressures to populations, the process is known as **artificial selection**.

Consider, for example, the development of modern breeds of cattle. Cattle have been domesticated for a very long time (*figure 18.13*). For thousands of years, people have tried to 'improve' their cattle. Desired features include docility (making the animal easier to control), fast growth

● **Figure 18.13** The original wild cattle from which individuals were first domesticated are thought to have looked very much like **a**, the modern Chillingham White breed. Selective breeding over many centuries has produced many different breeds, such as **b**, the Guernsey. Guernseys have been bred for the production of large quantities of fat-rich milk. Notice the large udder compared with the Chillingham.

rates and high milk yields. Increases in these characteristics have been achieved by **selective breeding**. Individuals showing one or more of these desired features to a larger degree than other individuals have been chosen for breeding. Some of the alleles conferring these features are passed on to their offspring. Again, the 'best' animals from this generation are chosen for breeding. Over many generations, alleles conferring the desired characteristics will increase in frequency, while those conferring characteristics not desired by the breeder will decrease in frequency. In many cases, such 'disadvantageous' alleles are lost entirely.

## The Darwin–Wallace theory of evolution by natural selection

The original theory that natural selection might be a mechanism by which evolution could occur was put forward independently by both Charles Darwin and Alfred Russel Wallace in 1856. They knew nothing of genes or mutations, so did not understand how natural variation could arise or be inherited. Nevertheless, they realised the significance of variation. Their observations and deductions can be summarised briefly as follows:

Observation 1   Organisms produce more offspring than are needed to replace the parents.

Observation 2   Natural populations tend to remain stable in size over long periods.

Deduction 1   There is competition for survival (a 'struggle for existence').

Observation 3   There is variation among the individuals of a given species.

Deduction 2   The best adapted variants will be selected for by the natural conditions operating at the time. In other words, natural selection will occur. The 'best' variants have a selective advantage; 'survival of the fittest' occurs.

As you can see, this theory, put forward well over a century ago, hardly differs from what we now know about natural selection and evolution. The major difference is that we can now think of natural selection as selecting particular *alleles* or groups of alleles.

The title of Darwin's most famous and important book contained the words *On the Origin of Species*. Yet, despite his thorough consideration of how natural selection could cause evolution, he did not attempt to explain how *new species* could be produced. This process is called **speciation**.

## Species and speciation

In this chapter, you have seen how natural selection can act on variation within a population to bring about changes in allele frequencies. Biologists believe that natural selection is the force which has produced all of the different species of organisms on Earth. Yet in the examples of directional selection described on pages 249–251, that is the evolution of antibiotic resistance in bacteria, and changes in the frequency of wing colour in peppered moths, no *new* species have been produced. How can natural selection produce new species? Before we can begin to answer this question, we must answer another: exactly what is a species? This proves to be an extremely difficult question, with no neat answer.

The definition of a species which is most widely accepted by biologists is:

> a group of organisms, with similar morphological, physiological, biochemical and behavioural features, which can interbreed to produce fertile offspring, and are reproductively isolated from other species.

'Morphological' features are structural features, while 'physiological' features are the way that the body works. 'Biochemical' features include the sequence of bases in DNA molecules and the sequence of amino acids in proteins.

Thus all donkeys look and work like donkeys, and can breed with other donkeys to produce more donkeys which themselves can interbreed. All donkeys belong to the same species. Donkeys can interbreed with organisms of another similar species, horses, to produce offspring called mules.

However, mules are infertile, that is they cannot breed and are effectively a 'dead-end'. Thus donkeys and horses belong to different species.

When a decision needs to be made as to whether two organisms belong to the same species or to two different species, they should ideally be tested to find out if they can interbreed successfully, producing fertile offspring. However, as you can imagine, this is not always possible. Perhaps the organisms are dead; they may even be museum specimens or fossils. Perhaps they are both of the same sex. Perhaps the biologist making the decision does not have the time or the facilities to attempt to interbreed them. Perhaps the organisms will not breed in captivity. Perhaps they are not organisms which reproduce sexually, but only asexually. Perhaps they are immature, and not yet able to breed.

As a result of all of these problems, it is quite rare to test the ability of two organisms to interbreed. Biologists frequently rely only on morphological, biochemical, physiological and behavioural differences to decide whether they are looking at specimens from one or two species. In practice, it may only be morphological features which are considered, because physiological and biochemical, and to some extent behavioural, ones are more time-consuming to investigate. Sometimes, however, detailed studies of DNA sequences may be used to assess how similar two organisms are to each other.

It can be extremely difficult to decide when these features are sufficiently similar or different to decide whether two organisms should belong to the same or a different species. This leads to great uncertainties and disagreements about whether to lump many slightly different variations of organisms together into one species, or whether to split them up into many different species.

Despite the problems described above, most biologists would agree that the feature which really decides whether or not two organisms belong to different species is their inability to interbreed successfully. In explaining how natural selection can produce new species, therefore, we must consider how a group of interbreeding organisms, and so all of the same species, can produce another group of organisms which cannot

interbreed successfully with the first group. The two groups must become **reproductively isolated**.

Once again, this is not a question with a neat and straightforward answer. The main difficulty is that this process *takes time*. You cannot, at least not easily, set up a speciation experiment in a laboratory because it would have to run for many years. The evidence which we have for the ways in which speciation can occur is almost all circumstantial evidence. We can look at populations of organisms at one moment in time, that is now, and use the patterns we can see to suggest what might have happened, and might still be happening, over long periods of time.

## Allopatric speciation

One picture which emerges from this kind of observation is that **geographical isolation** has played a major role in the evolution of many species. This is suggested by the fact that many islands have their own unique groups of species. The Hawaiian and Galapagos islands, for example, are famous for their spectacular array of species of all kinds of animals and plants found nowhere else in the world.

Geographical isolation requires a barrier of some kind to arise between two populations of the same species, preventing them from mixing. This barrier might be a stretch of water. We can imagine that a group of organisms, perhaps a population of a species of bird, somehow arrived on one of the Hawaiian islands from mainland America; they might have been blown off course by a storm. Here, separated by hundreds of miles of ocean from the rest of their species on mainland America, the group interbred. The selection pressures on the island were very different from those on the mainland, resulting in different alleles being selected for. Over time, the morphological, physiological and behavioural features of the island population became so different from the mainland population that they could no longer interbreed. A new species had evolved.

You can probably think of many other ways in which two populations of a species could be physically separated. A species living in dense forest, for example, could become split if large areas of forest are cut down, leaving 'islands' of forest in a 'sea' of agricultural land. Very small or immobile organisms can be isolated by even smaller-scale barriers.

Speciation which happens like this, when two populations are separated from each other geographically, is called **allopatric speciation**. 'Allopatric' means 'in different places'. However, it is also possible for new species to arise without the original populations being separated by a geographical barrier. This is known as **sympatric speciation**.

## Sympatric speciation

Perhaps the commonest way in which sympatric speciation can occur is through **polyploidy**.

A polyploid organism is one with more than two complete sets of chromosomes in its cells. This can happen if, for example, meiosis goes wrong when gametes are being formed, so that a gamete ends up with two sets of chromosomes instead of one set. If two such gametes fuse, then the zygote gets four complete sets of chromosomes. It is said to be **tetraploid**.

Tetraploids formed in this way are often sterile. As there are four of each kind of chromosome, all four try to 'pair' up during meiosis I, and get in a terrible muddle. It is very difficult for the cell to divide by meiosis and produce new cells each with complete sets of chromosomes.

However, it may well be able to grow perfectly well, and to reproduce asexually. There is nothing to stop mitosis happening absolutely normally. (Remember that chromosomes do not need to pair up in mitosis – they each behave quite independently.) This does quite often happen in plants but only rarely in animals, largely because most animals do not reproduce asexually anyway.

Just occasionally, this tetraploid plant may manage to produce gametes. They will be diploid gametes. If one of these should fuse with a gamete from the normal, diploid, plant, then the resulting zygote will be triploid. Once again, it may be able to grow normally, but it will certainly be sterile. There is no way in which it can produce gametes, because it cannot share the three sets of chromosomes out evenly between the daughter cells.

So, the original diploid plant and the tetraploid that was produced from it cannot interbreed successfully. They can be considered to be different species. A new species has arisen in just one generation!

The kind of polyploid described here contained four sets of chromosomes all from the same species. It is said to be an **autopolyploid**. ('Auto' means 'self'.) Polyploids can also be formed that contain, say, two sets of chromosomes from one species and two sets from another closely-related species. They are called **allopolyploids**. ('Allo' means 'other' or 'different'.) Meiosis actually happens more easily in an allotetraploid than in an autotetraploid, because the chromosomes from each species are not quite identical. So the two chromosomes from one species pair up with each other, whilst the two chromosomes from the other species pair up. This produces a much less muddled situation than in an autopolyploid where the chromosomes try to get together in fours, so it is much more likely that meiosis can come to a successful conclusion. The allopolyploid may well be able to produce plenty of gametes. It is fertile.

Once again, however, the allopolyploid cannot interbreed with individuals from its parent species, for the same reasons as for the autopolyploid. It is a new species.

One well-documented instance of speciation through allopolyploidy is the cord grass *Spartina anglica*. This is a vigorous grass that grows in salt marshes.

Before 1830, the species of *Spartina* that grew in these places was *S. maritima*. Then, in 1829, a different species called *S. alterniflora* was imported from America. *S. maritima* and *S. alterniflora* hybridised, producing a new species called *S. townsendii*. This is a diploid plant, with one set of chromosomes from *S. maritima* and one set from *S. alterniflora*. It is sterile, because the two sets of chromosomes from its parents cannot pair up, so it cannot undergo meiosis successfully. Nor can it interbreed with either of its two parents, which is what makes it a different species. Although it is sterile, it has been able to spread rapidly, reproducing asexually by producing long underground stems called rhizomes, from which new plants can grow.

At some later time, probably around 1892, faulty cell division in *S. townsendii* somehow produced cells with double the number of chromosomes. A tetraploid plant was produced, probably from the fusion of two diploid gametes from *S. townsendii*. So this tetraploid has two sets of chromosomes that originally came from *S. maritima*, and two sets from *S. alterniflora*. It is an allotetraploid. These chromosomes can pair up with each other, two and two, during meiosis, so this tetraploid plant is fertile. It has been named *S. anglica*. It is more vigorous than any of the other three species, and has spread so widely and so successfully that it has practically replaced them in England.

# SUMMARY

◆ Meiosis, random mating and the random fusion of gametes produce variation amongst populations of sexually reproducing organisms. Variation is also caused by the interaction of the environment with genetic factors, but such environmentally induced variation is not passed on to an organism's offspring. The only source of new alleles is mutation.

◆ All species of organisms have the reproductive potential to increase the sizes of their populations, but, in the long term, this rarely happens. This is because environmental factors come into play to limit population growth. Such factors decrease the rate of reproduction, or increase the rate of mortality so that many individuals die before reaching reproductive age.

◆ Within a population, certain alleles may increase the chance that an individual will survive long enough to be able to reproduce successfully. These alleles are therefore more likely to be passed on to the next generation than others. This is known as natural selection. Normally, natural selection keeps allele frequencies as they are; this is stabilising selection. However, if environmental factors which exert selection pressures change, or if new alleles appear in a population, then natural selection may cause a change in the frequencies of alleles; this is directional selection.

◆ Over many generations, directional selection may produce large changes in allele frequencies. This is how evolution occurs.

◆ Artificial selection involves the choice by humans of which organisms to allow to breed together, in order to bring about a desirable change in characteristics. Thus artificial selection, like natural selection, can affect allele frequencies in a population.

◆ A species can be defined as a group of organisms with similar morphology, behaviour, physiology and biochemistry that are capable of interbreeding to produce fertile offspring. In practice, however, it is not always possible to determine whether or not organisms can interbreed.

◆ New species arise by a process called speciation. In allopatric speciation, two populations become isolated from one another, perhaps by some geographical feature, and then evolve along different lines until they become so different that they can no longer interbreed. In sympatric speciation, new species may arise through polyploidy.

## CHAPTER 19

# Control, coordination and homeostasis

## By the end of this chapter you should be able to:

1 outline the need for communication systems in animals and plants, to respond to changes in the internal and external environment;

2 discuss the need for homeostasis in mammals;

3 define the term *excretion*, and explain the importance of removing nitrogenous waste products and carbon dioxide from the body;

4 describe the gross structure of the kidney and the detailed structure of the nephron with the associated blood vessels;

5 explain the functioning of the kidney in the control of metabolic wastes, using water potential terminology;

6 explain the control of water content of the body as an example of a negative feedback control mechanism;

7 explain what is meant by an *endocrine gland*;

8 describe the cellular structure of an islet of Langerhans from the pancreas, and outline the role of the pancreas as an endocrine gland;

9 explain how the blood glucose concentration is regulated by negative feedback control mechanisms, with reference to insulin and glucagon;

10 explain the advantages of treating diabetics with human insulin produced by genetic engineering;

11 describe the structure of a sensory neurone and a motor neurone, and outline their functions in a reflex arc;

12 describe and explain the transmission of an action potential in a myelinated neurone, including the roles of sodium and potassium ions;

13 explain the importance of the myelin sheath and saltatory conduction, and the refractory period;

14 outline the role of sensory receptors in mammals in converting different forms of energy into nerve impulses;

15 describe the structure and function of a cholinergic synapse, including the role of calcium ions;

16 outline the roles of synapses in the nervous system in determining the direction of nerve impulse transmission, and in allowing the interconnection of nerve pathways;

17 describe the role of auxins in apical dominance;

18 describe the role of gibberellins in stem elongation and in the germination of barley seeds;

19 describe the role of abscisic acid in the closure of stomata and in leaf fall.

ost animals and plants are complex organisms, made up of many millions of cells. Different parts of the organism perform different functions. It is essential that information can pass between these different parts, so that their activities are coordinated. Sometimes, the purpose of this information transfer is to regulate the levels of some substance within the organism, such as the control of blood sugar levels in mammals. Sometimes, the purpose may be to change the activity of some part of the organism in response to some external stimulus, such as moving away from an unpleasant stimulus.

In both animals and plants, chemical messengers called **hormones** (in plants they are sometimes known as **plant growth regulators**) help to transfer information from one part to another and so achieve coordination. In many animals, including mammals, **nerves** transfer information in the form of electrical impulses. We will look at both of these methods of information transfer in this chapter.

First, however, we will look at the need for mammals to maintain a stable environment, and consider how **excretion** (especially by the kidneys) helps to achieve this.

## Homeostasis

A vital function of control systems in mammals is to maintain a stable internal environment. This is called **homeostasis**. 'Internal environment' means the conditions inside the body, in which cells function. For a cell, its immediate environment is the tissue fluid that surrounds it. Many features of the environment affect the functioning of the cell. Three such features are:

- **temperature** – low temperatures slow metabolic reactions, while high temperatures cause denaturation of proteins, including enzymes;
- **amount of water** – lack of water in the tissue fluid causes water to be drawn out of cells by osmosis, causing metabolic reactions in the cell to slow or stop, while too

much water entering the cell may cause it to swell and burst;

- **amount of glucose** – glucose is the fuel for respiration, so lack of it causes respiration to slow or stop, depriving the cell of an energy source, while too much glucose may draw water out of the cell by osmosis.

In general, homeostatic mechanisms work by controlling the composition of blood, which therefore controls the composition of tissue fluid.

Most control mechanisms in living organisms use a **negative feedback** control loop (figure 19.1). This involves a **receptor** (or **sensor**) and an **effector**. The receptor picks up information about the parameter being regulated. This is known as the **input**. This sets off a series of events culminating in some action, by the effector, which is called the **output**. Continuous monitoring of the parameter by the receptor produces continuous adjustments of the output, which keep the parameter oscillating around a particular 'ideal' level, or set point. In negative feedback such as this, a rise in the parameter results in something happening that makes the parameter fall.

There are a few instances of the opposite thing happening in living organisms. For example, if a person breathes air that has a very high carbon dioxide content, this produces a high concentration of carbon dioxide in the blood. This is sensed by carbon dioxide receptors, which cause the breathing rate to increase. So the person breathes faster, taking in even more carbon dioxide, which stimulates the receptors even more, so they breathe faster and faster... This is an example of **positive feedback**. You can see that it cannot play any role in keeping things constant!

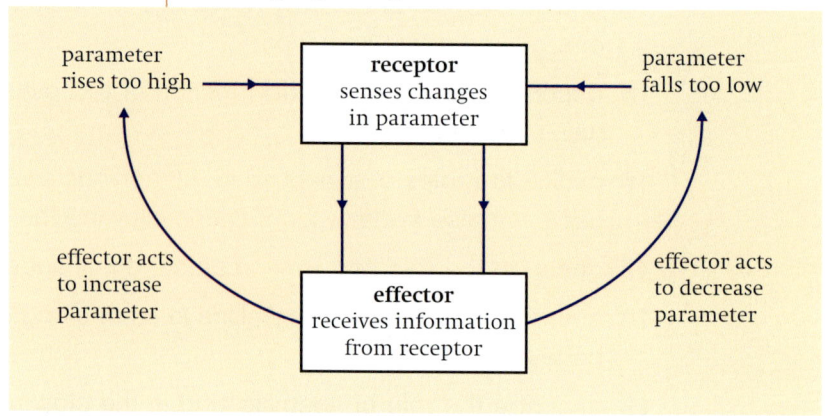

● **Figure 19.1** A negative feedback control loop.

# Excretion

Many of the metabolic reactions occurring within the body produce unwanted substances. Some of these are toxic (poisonous). The removal of these unwanted products of metabolism is known as **excretion**.

Many excretory products are formed in humans, but two are made in much greater amounts than the others. These are **carbon dioxide** and **urea**. Carbon dioxide is produced virtually continuously by almost every cell in the body, by the reactions of aerobic respiration. The waste carbon dioxide is transported from the respiring cells to the lungs, in the bloodstream (see page 111). It diffuses from the blood into the alveoli of the lungs, and is excreted in the air we breathe out (see page 60 and chapter 11).

In contrast, urea is produced in only one organ in the body, that is the **liver**. It is produced from excess amino acids (as described in the next section) and is transported from the liver to the kidneys, in solution in blood plasma. The kidneys remove urea from the blood and excrete it, dissolved in water, as **urine**. Here, we will look more fully at the production and excretion of urea.

## Deamination

If more protein is eaten than is needed, the excess cannot be stored in the body. It would be wasteful, however, simply to get rid of all the excess, because the amino acids contain useful energy. The liver salvages this energy by removing the nitrogen atoms from the amino acids, excreting these in the form of urea, and keeping the rest of each amino acid molecule. The process by which urea is made from excess amino acids is called **deamination**.

*Figure 19.2* shows how deamination takes place. In the liver cells, the amino ($NH_2$) group of an amino acid is removed, together with an extra hydrogen atom. These combine to produce ammonia. The keto acid that remains may become a carbohydrate, which can be used in respiration, or may be converted to fat and stored.

Ammonia is a very soluble and highly toxic compound. In aquatic animals, such as fish, this poses

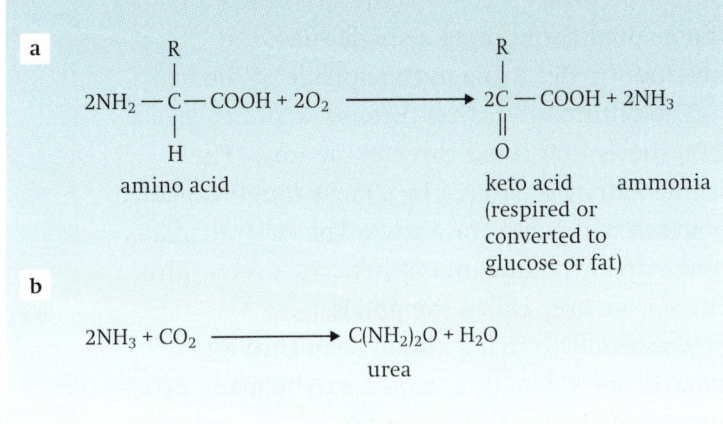

● **Figure 19.2 a** Deamination and **b** urea formation.

no danger as the ammonia can simply dissolve into the water around them. However, in terrestrial animals, such as humans, ammonia would rapidly build up in the blood and cause immense damage. So the ammonia produced in the liver is instantly converted to the less soluble and less toxic compound, **urea**. Urea is made by combining ammonia with carbon dioxide (*figure 19.2b*). An adult produces around 25–30 g of urea per day.

Urea is the main **nitrogenous excretory product** of humans. We also produce small quantities of other nitrogenous excretory products, mainly **creatinine** and **uric acid**. Creatine is made in the liver, from certain amino acids. Much of this creatine is used in the muscles, in the form of creatine phosphate, where it acts as an energy store (see page 200 in chapter 15). However, some is converted to creatinine and excreted. Uric acid is made from the breakdown of nucleic acids, not from amino acids.

The urea made in the liver passes from the liver cells into the blood plasma. All of the urea made each day must be excreted, or its concentration in the blood would build up and become dangerous. As the blood passes through the kidneys, the urea is extracted and excreted. To explain how this happens, we must first look at the structure of a kidney.

## The structure of the kidney

*Figure 19.3* shows the position of the kidneys in the body together with their associated structures. Each kidney receives blood from a **renal artery**, and returns blood via a **renal vein**. A narrow tube, called the **ureter**, carries urine from the kidney to the bladder. From there a single tube, the **urethra**, carries urine to the outside of the body.

A longitudinal section through a kidney (*figure 19.4*) shows that it has three main areas. The whole kidney is covered by a fairly tough **capsule**, beneath which lies the **cortex**. The central area is made up of the **medulla**. Where the ureter joins, there is an area called the **pelvis**.

A section through a kidney seen through a microscope (*figure 19.6*), shows it to be made up of thousands of tiny tubes that are called **nephrons**. *Figure 19.5* shows the position and structure of a single nephron. One end of the tube forms a cup-shaped structure called a **renal (Bowman's) capsule**. The renal capsules of all nephrons are in the cortex of the kidney. From the renal capsule, the tube runs towards the centre of the kidney, first forming a twisted region called the **proximal convoluted tubule**, and then a long hairpin loop in the medulla, the **loop of Henle**. The tubule then runs back upwards into the cortex, where it forms another twisted region called the **distal convoluted tubule**, before finally joining a **collecting duct** which leads down through the medulla and into the pelvis of the kidney. Here the collecting ducts join the ureter.

Blood vessels are closely associated with the nephrons (*figure 19.5c*). Each renal capsule is supplied with blood by a branch of the renal

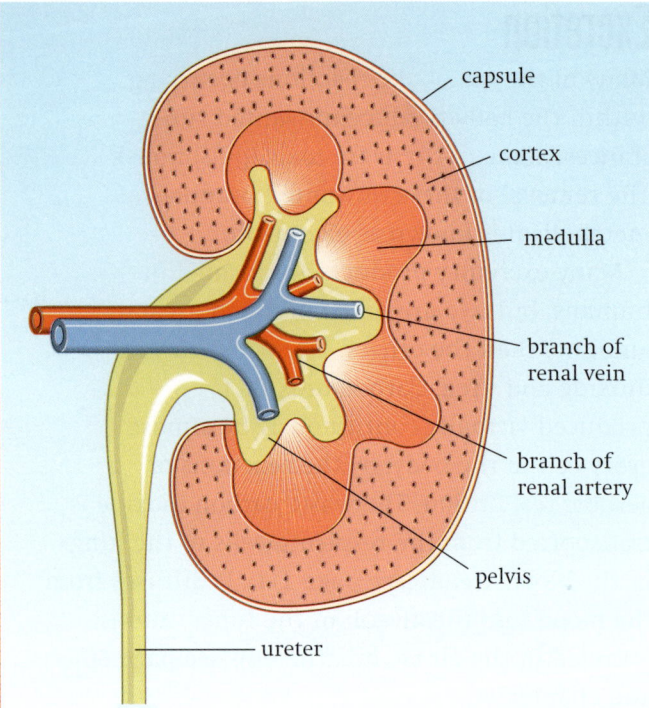

● **Figure 19.4** A kidney, cut in half vertically (LS).

artery, called an **afferent arteriole**, which splits into a tangle of capillaries in the 'cup' of the capsule, called a **glomerulus**. The capillaries of the glomerulus rejoin to form an **efferent arteriole**. This leads off to form a network of capillaries running closely alongside the rest of the nephron, before linking up with other capillaries to feed into a branch of the renal vein.

## Ultrafiltration

The kidney makes urine in a two-stage process. The first stage, **ultrafiltration**, involves filtering small molecules, including urea, out of the blood and into the renal capsule. From here they flow along the nephron towards the ureter. The second stage, **reabsorption**, involves taking back any useful molecules from the fluid in the nephron as it flows along.

*Figure 19.7* shows a section through part of a glomerulus and renal capsule. The blood in the glomerular capillaries is separated from the lumen of the renal capsule by two cell layers and a basement membrane. The first cell layer is the lining, or **endothelium**, of the capillary. Like the endothelium of most capillaries, this has gaps in it, but there are far more gaps than in other capillaries:

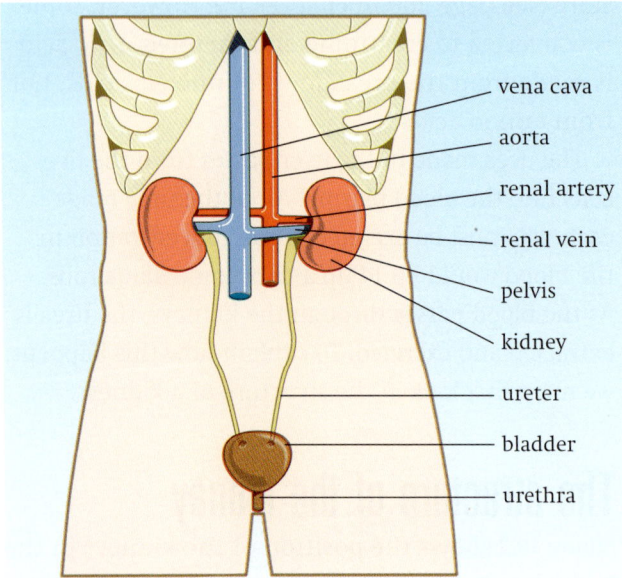

● **Figure 19.3** Position of the kidneys and associated structures in the human body.

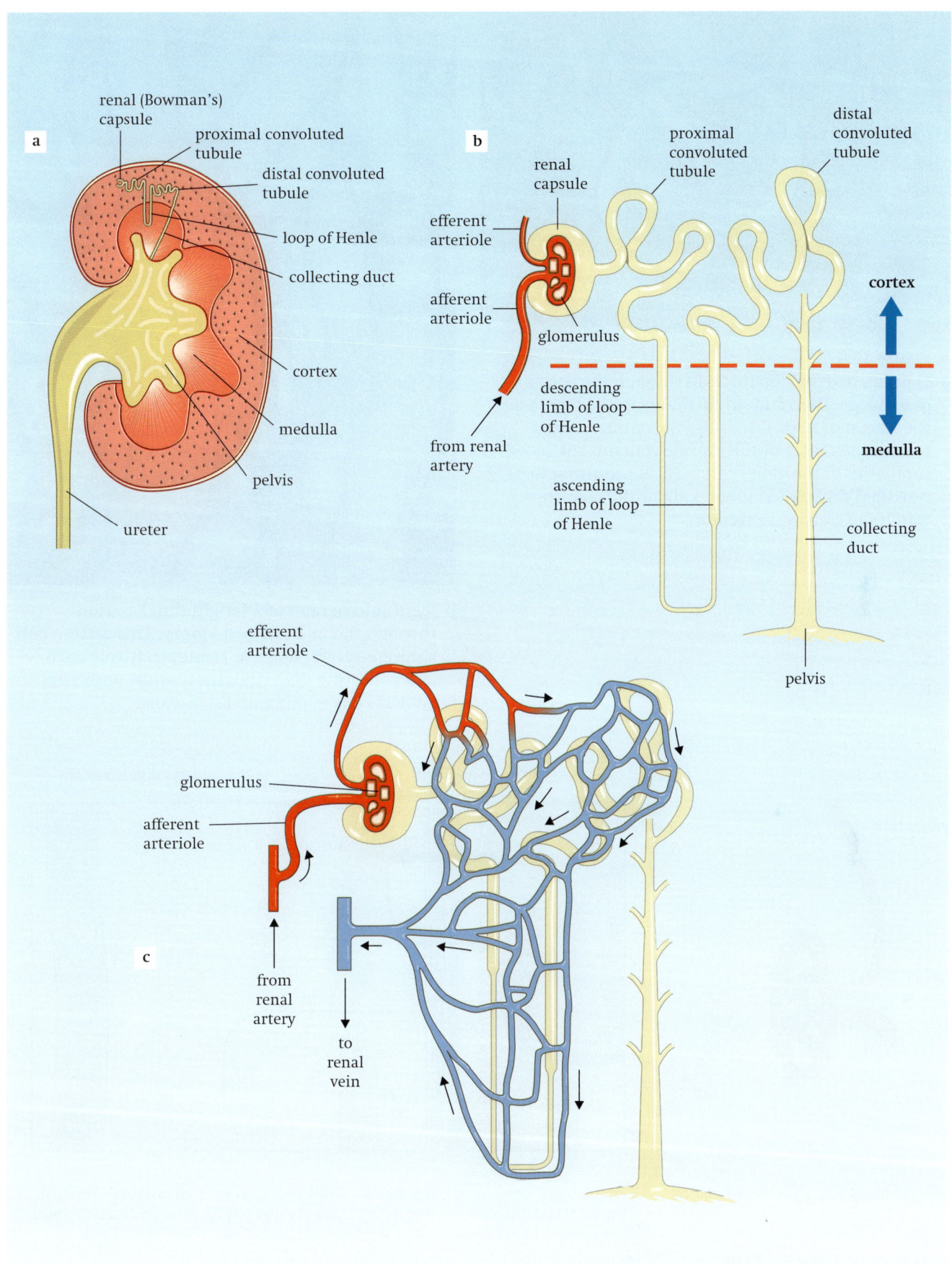

**Figure 19.5 a** Section through the kidney to show the position of a nephron. **b** A nephron.
**c** The blood supply associated with a nephron.

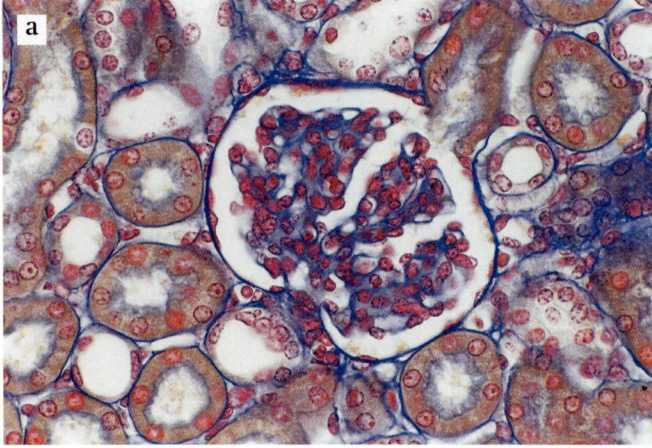

● **Figure 19.6**

**a** Light micrograph of a section through the cortex of a kidney. The white circular area in the centre is the lumen of a renal capsule. The darkly stained area in the centre of the capsule contains the blood capillaries of the glomerulus. There are also several proximal convoluted tubules and distal convoluted tubules in transverse section.

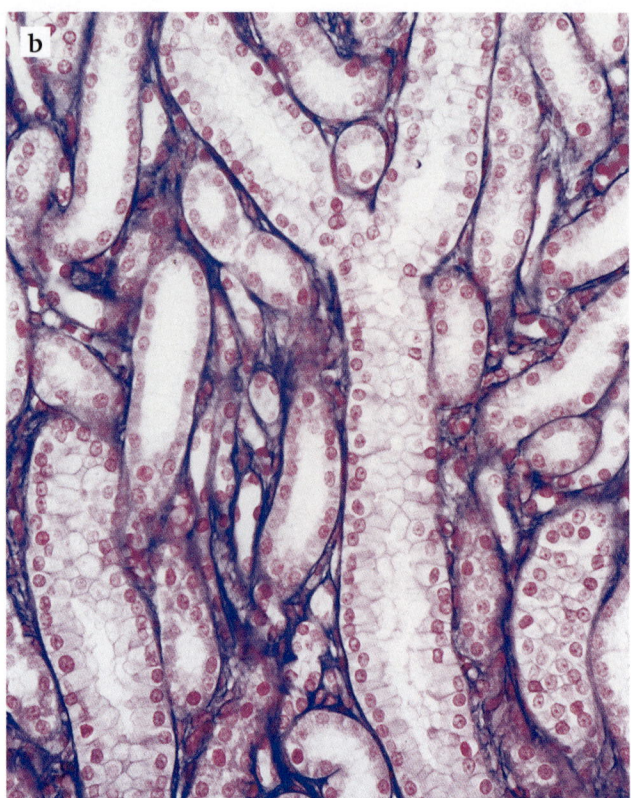

**b** Light micrograph of a longitudinal section through the medulla of a kidney. This section cuts through several loops of Henle (relatively narrow) and collecting ducts (relatively wide, with almost cubical cells making up their walls).

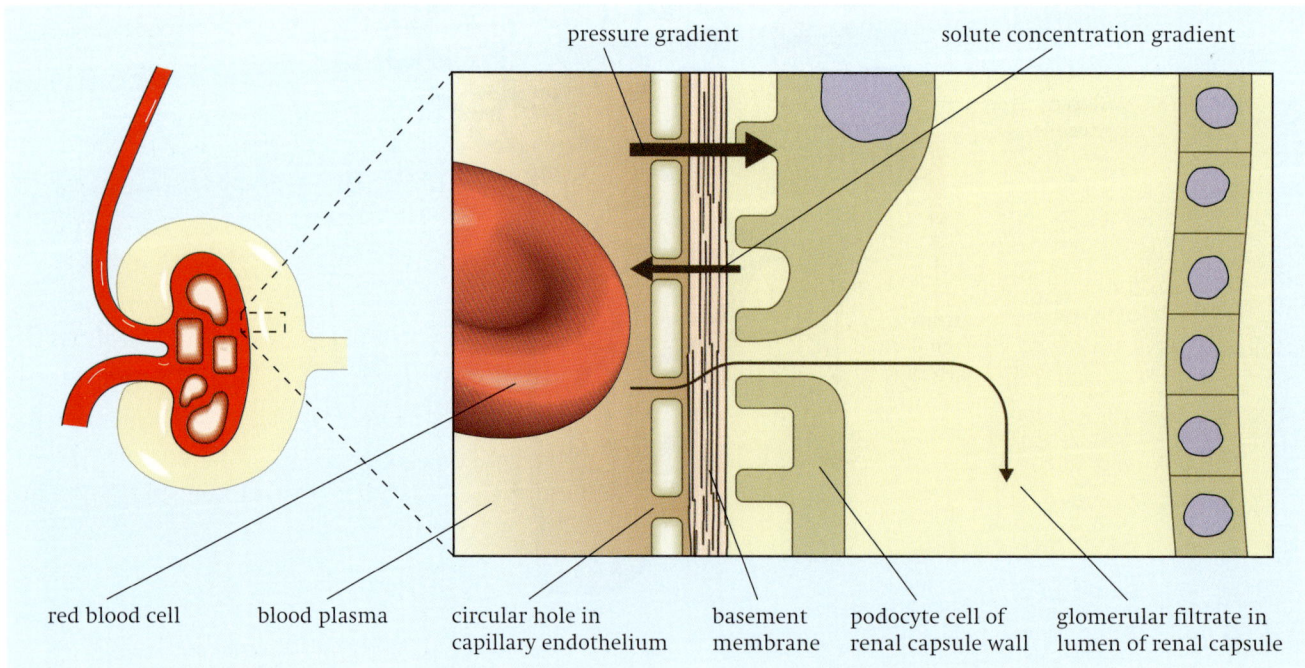

pressure gradient      solute concentration gradient

red blood cell    blood plasma    circular hole in capillary endothelium    basement membrane    podocyte cell of renal capsule wall    glomerular filtrate in lumen of renal capsule

● **Figure 19.7** Detail of the wall of a glomerular capillary and renal capsule. The arrows show how the net effect of higher pressure in the capillary and lower solute concentration in the renal capsule is that fluid moves out of the capillary and into the lumen of the capsule. The basement membrane acts as a molecular filter.

each endothelial cell has thousands of tiny holes in it. Next comes the **basement membrane** which is made up of a network of collagen and glycoproteins. The second cell layer is formed from **epithelial cells** which make up the wall of the renal capsule. These cells have many tiny finger-like projections, with gaps in between them and are called **podocytes**.

The holes in the capillary endothelium, and the gaps in the renal capsule epithelium, are quite large, and make it easy for any substances dissolved in the blood plasma to get through from the blood into the capsule. However, the basement membrane stops large protein molecules from getting through. Any protein molecule with a relative molecular mass of around 69 000 or more cannot pass through the basement membrane, and so cannot escape from the glomerular capillaries. This basement membrane therefore acts as a filter. Blood cells, both red and white, are also too large to pass through this barrier, and so remain in the blood. *Table 19.1* shows the relative concentrations of substances in the blood and in the glomerular filtrate. You will see that glomerular filtrate is identical to blood plasma minus plasma proteins.

### Factors affecting glomerular filtration rate

The rate at which fluid seeps from the blood in the glomerular capillaries, into the renal capsule, is called the **glomerular filtration rate**. In a human, for all the glomeruli in both kidneys, this is about $125 \, cm^3 \, min^{-1}$.

| Substance | Concentration in blood plasma $(g \, dm^{-3})$ | Concentration in glomerular filtrate $(g \, dm^{-3})$ |
|---|---|---|
| water | 900 | 900 |
| proteins | 80.0 | 0.05 |
| amino acids | 0.5 | 0.5 |
| glucose | 1.0 | 1.0 |
| urea | 0.3 | 0.3 |
| uric acid | 0.04 | 0.04 |
| creatinine | 0.01 | 0.01 |
| inorganic ions (mainly $Na^+$, $K^+$ and $Cl^-$) | 7.2 | 7.2 |

● **Table 19.1** Relative concentrations of substances in the blood and in the glomerular filtrate.

What makes the fluid filter through so quickly? This is determined by the differences in **water potential** between the contents of the glomerular capillaries and the renal capsule. You will remember from see page 55 that water moves from a region of high water potential to a region of low water potential, down a water potential gradient. Water potential is lowered by the presence of solutes, and raised by high pressures.

Inside the capillaries in the glomerulus, the blood pressure is relatively high, because the diameter of the afferent arteriole is wider than that of the efferent arteriole, causing a 'traffic jam' inside the glomerulus. This therefore tends to raise the water potential of the blood plasma above the water potential of the contents of the renal capsule (*figure 19.7* again).

However, the concentration of solutes in the blood plasma in the capillaries is *higher* than the concentration of solutes inside the renal capsule. This is because, while most of the contents of the blood plasma filter through the basement membrane and into the renal capsule, the plasma protein molecules are too big to get through, and so stay in the blood. This difference in solute concentration tends to make the water potential in the blood capillaries *lower* than that in the renal capsule.

Overall, though, the effect of differences in pressure outweighs the effect of the differences in solute concentration. Overall, the water potential of the blood plasma in the glomerulus is higher than the water potential of the liquid in the renal capsule. So water continues to move down this water potential gradient, from the blood into the capsule.

## Reabsorption in the proximal convoluted tubule

As you saw in *table 19.1*, the fluid which has filtered through into the renal capsule is virtually identical to blood plasma, except that it does not contain large protein molecules. Many of the substances in the filtrate need to be kept in the body, so they are reabsorbed into the blood as the fluid passes along the nephron. Since only certain substances are reabsorbed, the process is called **selective reabsorption**. Most of the reabsorption takes place in the proximal convoluted tubule.

The basal membranes (the ones nearest the blood and furthest from the lumen) of the cells lining the proximal convoluted tubule actively transport sodium ions out of the cell (*figure 19.8*). The sodium ions are carried away in the blood. This lowers the concentration of sodium ions inside the cell, so that they passively diffuse into it, down their concentration gradient, from the fluid in the lumen of the tubule. However, they do not just diffuse freely through the membrane; they can only enter through special transporter (carrier) proteins in the membrane. There are several different kinds of these, each of which transports something else, such as glucose, at the same time as the sodium. The concentration gradient for the sodium provides enough energy to pull in glucose molecules. Thus glucose is taken up by the cell, and into the blood.

All of the **glucose** in the glomerular filtrate is transported out of the proximal convoluted tubule and into the blood. Normally, no glucose is left in the tubule, so no glucose is present in urine. Similarly, **amino acids**, **vitamins**, and many **sodium** and **chloride ions** are actively reabsorbed here.

The uptake of these substances would decrease the solute concentration of the filtrate. However, **water** can (and does) move freely out of the filtrate, through the walls of the tubule and into the blood. As the substances listed above move into the cells surrounding the tubule, water follows by osmosis. Thus the overall concentration of the filtrate remains about the same. About 65% of the water in the filtrate is reabsorbed here.

### SAQ 19.1

Look back at *figure 19.6*.
**a** Where has the blood in the capillaries surrounding the proximal convoluted tubule come from?
**b** What solutes will this blood contain that are *not* present in the glomerular filtrate?
**c** How might this help in the reabsorption of water from the proximal convoluted tubule?

Surprisingly, quite a lot of urea is reabsorbed too. Urea is a small molecule, which passes easily through cell membranes. Its concentration in the glomerular filtrate is considerably higher than that in the capillaries, so it diffuses passively through the wall of the proximal convoluted tubule and into the blood. About half of the urea in the filtrate is reabsorbed in this way.

The other two nitrogenous excretory products, uric acid and creatinine, are not reabsorbed. Indeed, creatinine is actively **secreted** by the cells of the proximal convoluted tubule into its lumen.

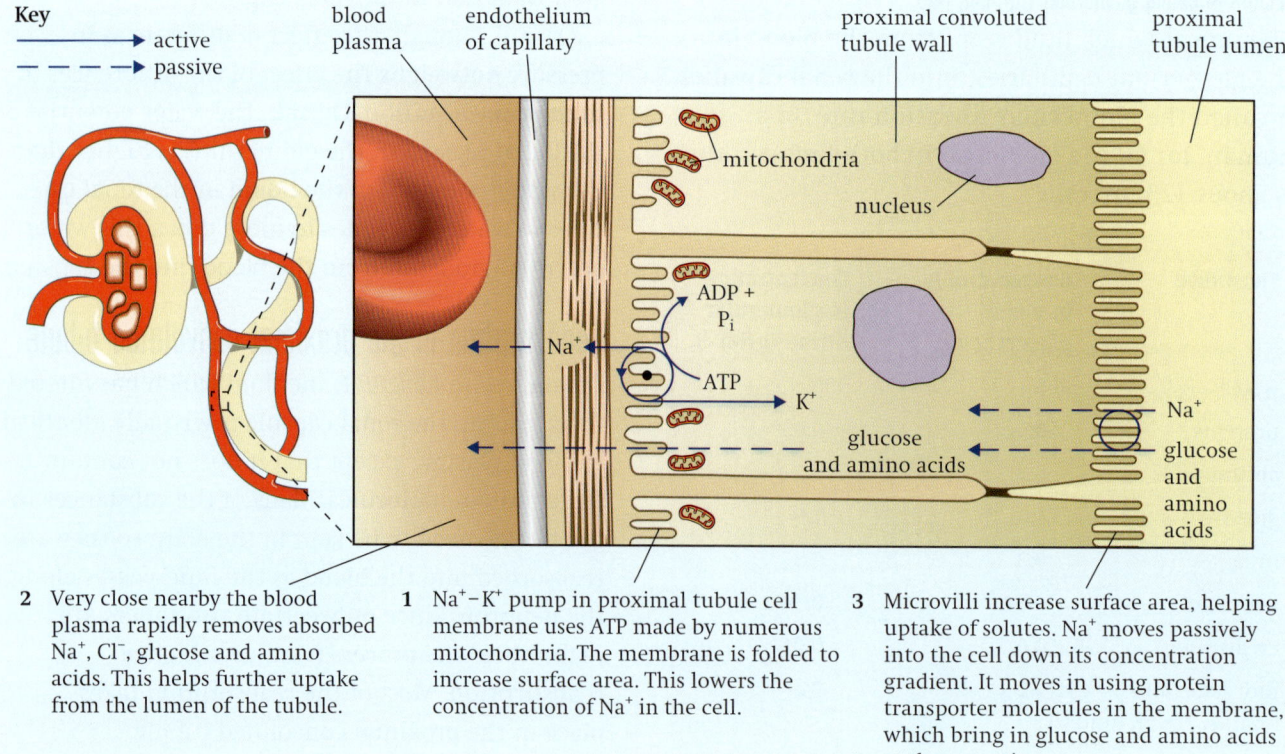

**Key**
→ active
- - → passive

blood plasma
endothelium of capillary
proximal convoluted tubule wall
proximal tubule lumen

mitochondria
nucleus
ADP + $P_i$
$Na^+$
ATP
$K^+$
glucose and amino acids
$Na^+$
glucose and amino acids

**2** Very close nearby the blood plasma rapidly removes absorbed $Na^+$, $Cl^-$, glucose and amino acids. This helps further uptake from the lumen of the tubule.

**1** $Na^+$–$K^+$ pump in proximal tubule cell membrane uses ATP made by numerous mitochondria. The membrane is folded to increase surface area. This lowers the concentration of $Na^+$ in the cell.

**3** Microvilli increase surface area, helping uptake of solutes. $Na^+$ moves passively into the cell down its concentration gradient. It moves in using protein transporter molecules in the membrane, which bring in glucose and amino acids at the same time.

● **Figure 19.8** Reabsorption in the proximal convoluted tubule.

The reabsorption of so much water and solutes from the filtrate in the proximal convoluted tubule greatly reduces the volume of liquid remaining. In an adult human, around 125 cm³ of fluid enter the proximal tubules every minute, but only 45 cm³ per minute are passed on to the next region, the loop of Henle.

## SAQ 19.2

Although almost half of the urea in the glomerular filtrate is reabsorbed from the proximal convoluted tubule, the *concentration* of urea in the fluid in the nephron actually increases as it passes through the proximal convoluted tubule. Explain why this is so.

## Reabsorption in the loop of Henle and collecting duct

The function of the loop of Henle is to create a very high concentration of salts in the tissue fluid in the medulla of the kidney. As you will see, this allows a lot of water to be reabsorbed from the fluid in the collecting duct, as it flows through the medulla. As a result, very concentrated urine can be produced. The loop of Henle therefore allows water to be conserved in the body, rather than lost in urine.

Figure 19.9a shows the loop of Henle. The hairpin loop runs deep down into the medulla of the kidney, before turning back towards the cortex again. The first part of the loop is therefore called the **descending limb**, and the second part the **ascending limb**.

To explain how it works, it is best to begin by looking at what happens in the *second* part of the loop, the ascending limb. The walls of the upper parts of the ascending limb are *impermeable* to water. The cells in the walls of this area actively transport

sodium and chloride ions out of the fluid in the tube, into the tissue fluid between the cells filling the space between the two limbs. This produces a high concentration of sodium and chloride ions around the descending limb. This concentration can be as much as four times greater than the normal concentration of tissue fluid.

The walls of the descending limb are *permeable* to water, and also to sodium and chloride ions. As the fluid flows down this tube, water from it is drawn *out*, by osmosis, into the tissue fluid, because of the high concentration of sodium and chloride ions there. At the same time, sodium and

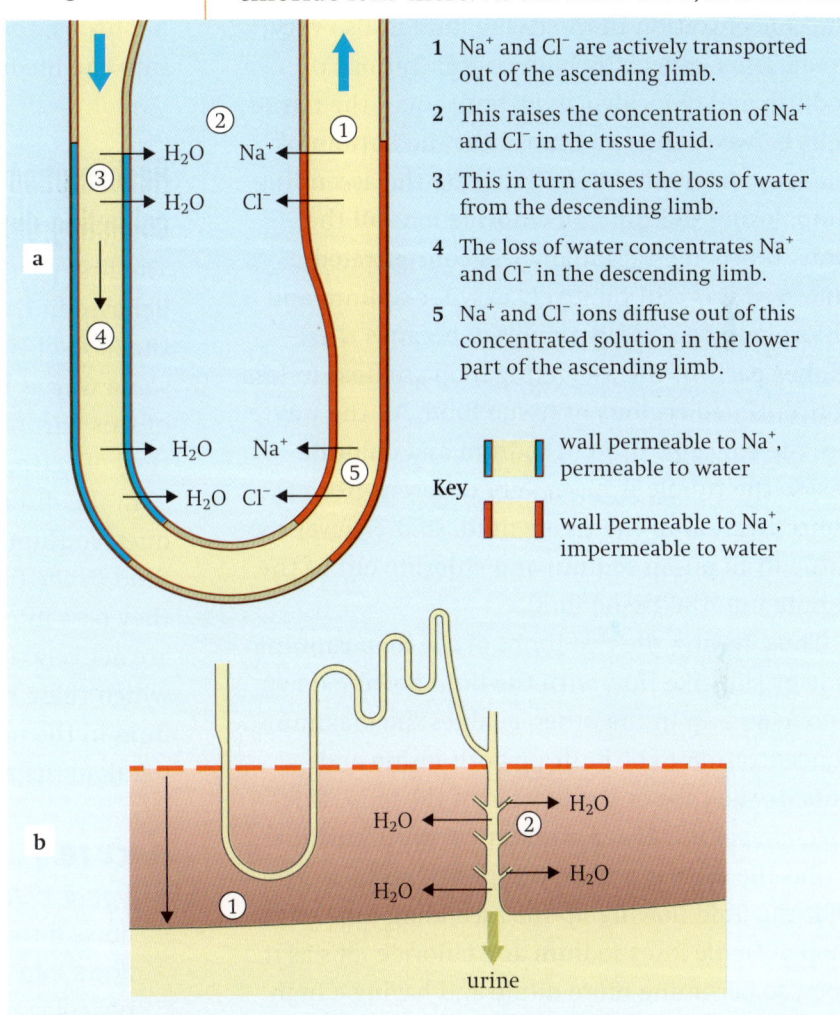

1 Na⁺ and Cl⁻ are actively transported out of the ascending limb.

2 This raises the concentration of Na⁺ and Cl⁻ in the tissue fluid.

3 This in turn causes the loss of water from the descending limb.

4 The loss of water concentrates Na⁺ and Cl⁻ in the descending limb.

5 Na⁺ and Cl⁻ ions diffuse out of this concentrated solution in the lower part of the ascending limb.

**Key**

wall permeable to Na⁺, permeable to water

wall permeable to Na⁺, impermeable to water

1 The tissue in the deeper layers of the medulla contains a very concentrated solution of Na⁺, Cl⁻ and urea.

2 As urine passes down the collecting duct, water can pass out of it by osmosis. The reabsorbed water is carried away by the blood in the capillaries.

● **Figure 19.9** How the loop of Henle allows the production of concentrated urine.   **a** The counter-current mechanism in the loop of Henle builds up a high sodium ion and chloride ion concentration in the tissue fluid of the medulla.   **b** Water can be drawn out of the collecting duct by the high salt concentration in the surrounding tissue fluid.

chloride ions diffuse *into* the tube, down their concentration gradient. So, by the time the fluid has reached the very bottom of the hairpin, it contains much less water and many more sodium and chloride ions than it did at the top. The fluid becomes more concentrated towards the bottom of the loop. The longer the loop, the more concentrated it can become.

This concentrated fluid now turns the corner and begins to flow up the ascending limb. Because the fluid inside the loop is so concentrated, it is relatively easy for sodium and chloride ions to leave it and pass into the tissue fluid, even though the concentration in the tissue fluid is also very great. Thus, especially high concentrations of sodium and chloride can be built up in the tissue cells between the two limbs near the bottom of the loop. As the fluid continues up the ascending limb, losing sodium and chloride ions all the time, it becomes gradually less concentrated. However, it is still relatively easy for sodium and chloride to be actively removed, because these higher parts of the ascending loop are next to less concentrated regions of tissue fluid. All the way up, the concentration of sodium and chloride inside the tubule is never very different from the concentration in the tissue fluid, so it is never too difficult to pump sodium and chloride out of the tubule into the tissue fluid.

Thus, having the two limbs of the loop running side by side like this, with the fluid flowing down in one and up in the other, enables the maximum concentration to be built up both inside and outside the tube at the bottom of the loop. This mechanism is called a **counter-current multiplier**.

But the story is not yet complete! You have seen that the fluid flowing up the ascending limb of the loop of Henle loses sodium and chloride ions as it goes, so becoming more dilute and having a high water potential. However, in *figure 19.9b* you can see that the fluid continues round through the distal convoluted tubule into the **collecting duct**, which runs down into the medulla again. It therefore passes once again through the regions where the solute concentration of the tissue fluid is very high, with a very low water potential. Water therefore moves out of the collecting duct, by osmosis, until the water potential of urine is the same as the

water potential of the tissue fluid in the medulla, which may be much greater than the water potential of the blood. The degree to which this happens is controlled by **antidiuretic hormone, ADH**, and is explained later in this chapter.

The longer the loop of Henle, the greater the concentration that can be built up in the medulla, and the greater the concentration of the urine which can be produced. Desert animals such as kangaroo rats, which need to conserve as much water as they possibly can, have especially long loops of Henle. Humans, however, only have long loops of Henle in about one third of their nephrons, the other two thirds hardly dipping into the medulla at all.

## Reabsorption in the distal convoluted tubule and collecting duct

The first part of the distal convoluted tubule behaves in the same way as the ascending limb of the loop of Henle. The second part behaves in the same way as the collecting duct, so the functions of this part of the distal convoluted tubule and the collecting duct will be described together.

In the distal convoluted tubule and collecting duct, **sodium ions** are actively pumped from the fluid in the tubule into the tissue fluid, from where they pass into the blood. **Potassium ions**, however, are actively transported *into* the tubule. The rate at which these two ions are moved into and out of the fluid in the nephron can be varied, and helps to regulate the amount of these ions in the blood.

### SAQ 19.3
a *Figure 19.10* shows the relative rate at which fluid flows through each part of a nephron. If water flows into an impermeable tube, such as a hosepipe, it will flow *out* of the far end at the same rate that it flows *in*. However, this clearly does not happen in a nephron. Consider what happens in each region, and suggest an explanation for the shape of the graph.
b *Figure 19.11* shows the relative concentrations of four substances in each part of a nephron. Explain the shapes of the curves for (i) glucose, (ii) urea, (iii) sodium ions, and (iv) potassium ions.

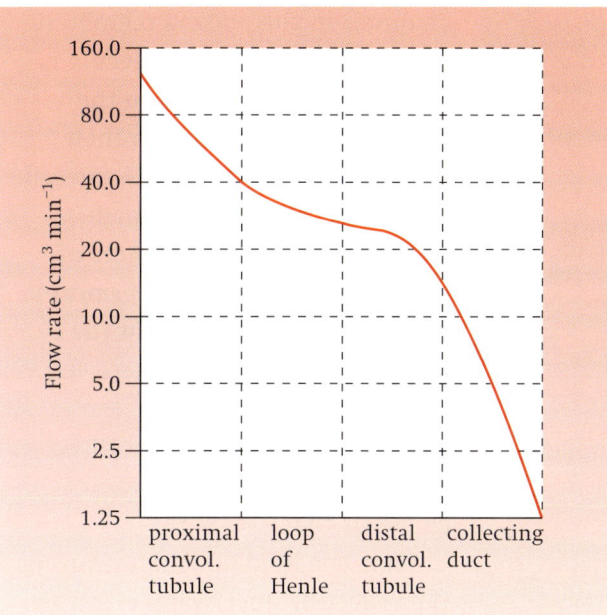

**Figure 19.10** Flow rates in different parts of a nephron.

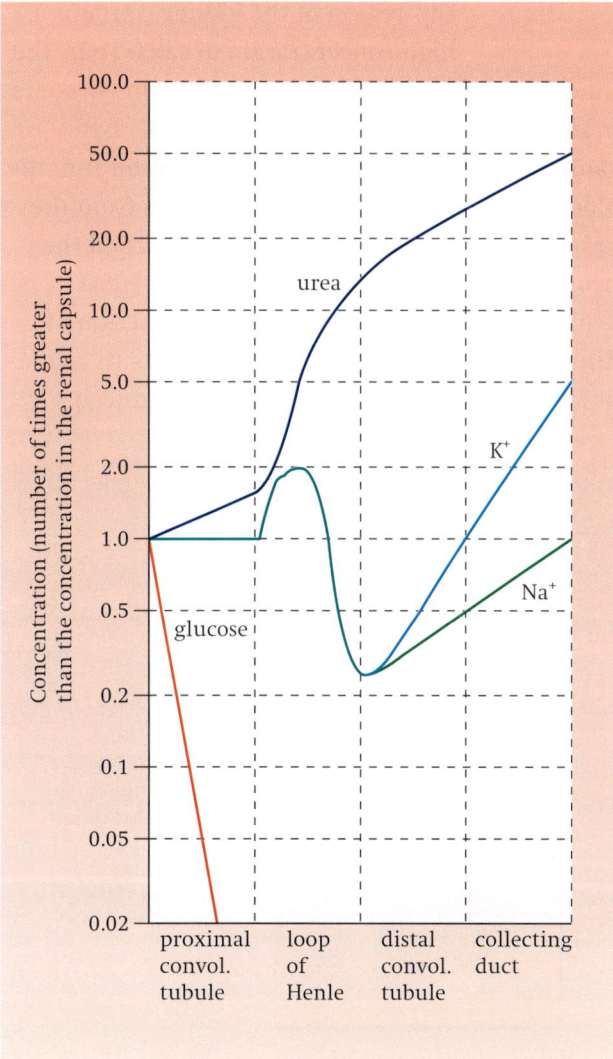

● **Figure 19.11** Relative concentrations of four substances in different parts of a nephron.

# Control of water content

## Osmoreceptors, the hypothalamus and ADH

At the beginning of this chapter, we saw that mammals maintain a relatively stable environment in which their cells can function – a process known as **homeostasis**. The kidneys play an important role in homeostasis by regulating the concentration of water in the body fluids. This is known as **osmoregulation**.

If you look back to *figure 19.1*, you will remember that control mechanisms that keep something relatively constant usually work by a **negative feedback mechanism**. There needs to be a **receptor** that monitors whatever it is that is being controlled, and an **effector** that does something to bring it back to normal if it deviates too far. In the osmoregulation mechanism in mammals, the receptor is cells in the **hypothalamus**, and the effectors are the **pituitary gland** and the walls of the distal convoluted tubules.

The amount of water in the blood is constantly monitored by cells, called **osmoreceptors**, within the hypothalamus (*figure 19.12*). It is not known exactly how these work, but it is probable that differences in water content of the blood cause water to move either into them or out of them by osmosis. When water content of the blood is low, the loss of water from the osmoreceptor cells reduces their volume, which triggers stimulation of nerve cells in the hypothalamus.

These nerve cells are rather different from other nerve cells that will be described later in this chapter, because they produce a chemical called **antidiuretic hormone**, or **ADH**. ADH is a polypeptide made up of just nine amino acids. It is made in the cell bodies of the nerve cells, and passes along them to their endings in the posterior lobe of the pituitary gland. When the nerve cells are stimulated by the osmoreceptor cells, electrical impulses called **action potentials** travel down them (page 280). This causes ADH to be released from their endings into the blood in capillaries in the posterior pituitary gland. From here, it is then carried all over the body.

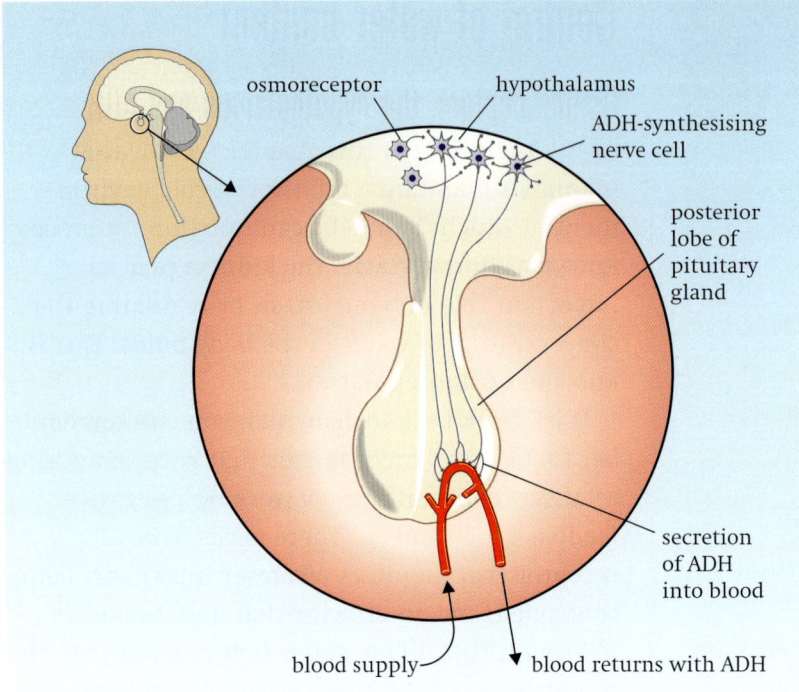

• **Figure 19.12** The secretion of ADH.

## How ADH affects the kidneys

ADH acts on the plasma membranes of the cells making up the walls of the collecting ducts in the kidneys. It makes these membranes more permeable to water than usual (*figure 19.13*).

This change in permeability is brought about by increasing the number of water-permeable channels in the plasma membrane (*figure 19.14*). The ADH molecule is picked up by a receptor on the plasma membrane, which then activates an enzyme inside the cell. Inside the cell are ready-made vesicles surrounded by pieces of membrane full of water-permeable channels. The activation of the enzyme by ADH causes these vesicles to move to, and fuse with, the plasma membrane of the cell, so adding many water-permeable channels to it.

So, as the fluid flows down through the collecting duct, water is free to move out of the tubule and into the tissue fluid, and it does so because this region of the kidney contains a high concentration of salts. Thus, the fluid in the collecting duct loses water and becomes more concentrated. The secretion of ADH has caused the increased reabsorption of water into the blood. The amount of urine which flows from the kidneys into the bladder will be smaller, and the urine will be more concentrated (*figure 19.15*).

The word 'diuresis' means the production of dilute urine. Antidiuretic hormone gets its name because it stops dilute urine being produced.

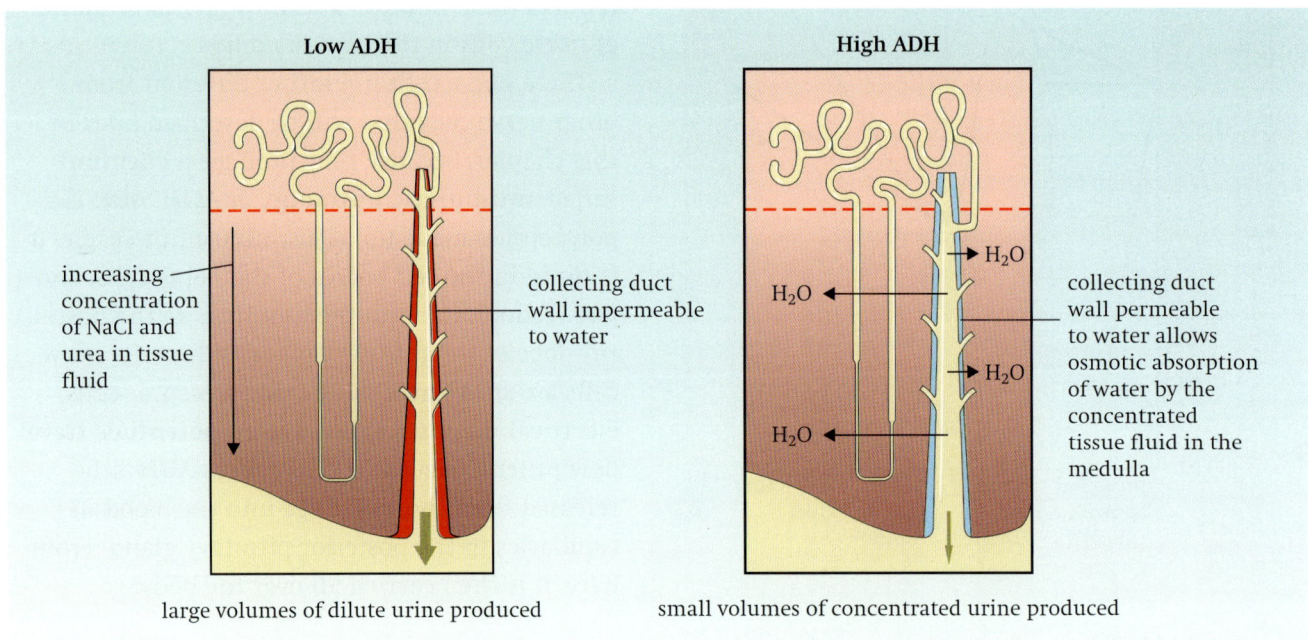

• **Figure 19.13** The effects of ADH on water reabsorption from the collecting duct.

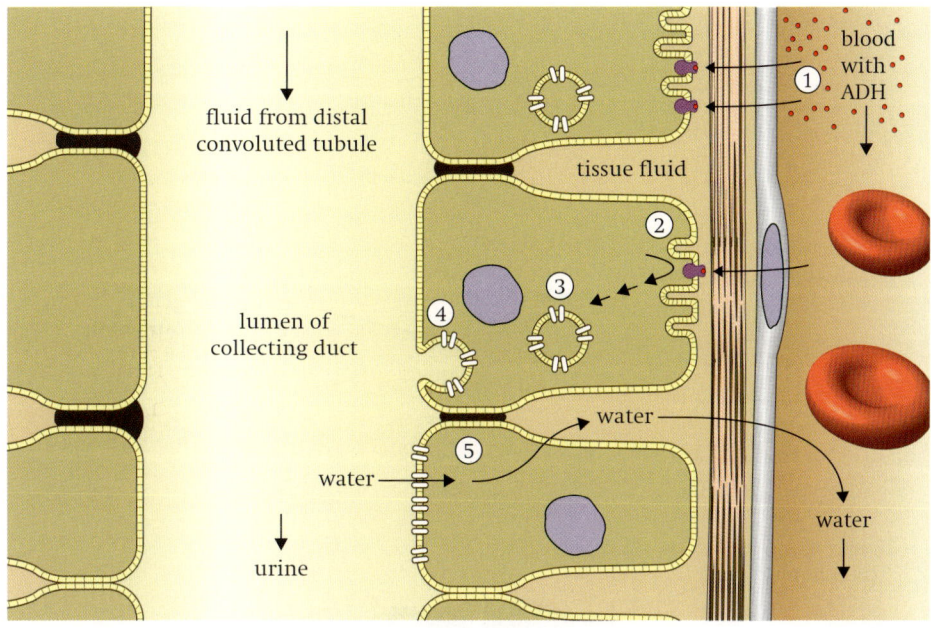

1  ADH binds to receptors in the plasma membrane of the cells lining the collecting duct.
2  This activates a series of enzyme-controlled reactions, ending with the production of an active phosphorylase enzyme.
3  The phosphorylase causes vesicles, surrounded by membrane containing water-permeable channels, to move to the plasma membrane.
4  The vesicles fuse with the plasma membrane.
5  Water can now move freely through the membrane, down its water potential gradient, into the concentrated tissue fluid and blood plasma in the medulla of the kidney.

● **Figure 19.14**  How ADH increases water reabsorption in the collecting duct.

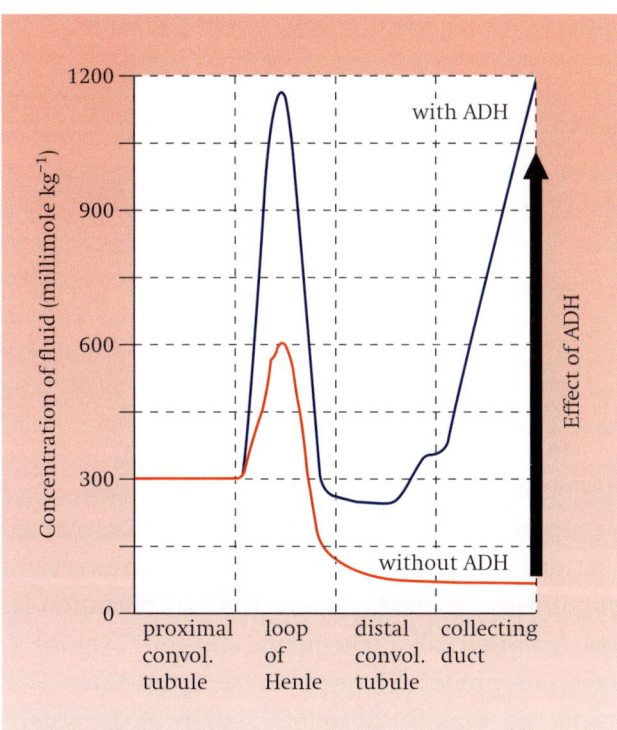

● **Figure 19.15**  The concentration of fluid in different regions of a nephron, with and without the presence of ADH.

## Negative feedback in the control of water content

You have seen how the secretion of ADH, brought about as a result of a low blood water content, causes more water to be absorbed back into the blood from the nephrons in the kidney. Thus the maximum amount of water will be retained in the body.

When the blood water content rises, the osmoreceptors are no longer stimulated, and stop stimulating their neighbouring nerve cells. So ADH secretion slows down. This affects the cells in the walls of the collecting ducts. The water-permeable channels are moved out of the plasma membrane of the collecting duct cells, back into the cytoplasm. Thus, the collecting duct becomes less permeable to water. The liquid flowing down it retains more of its water, flowing into the ureter as a copious, dilute urine.

The collecting duct cells do not respond immediately to the reduction in ADH secretion by the posterior pituitary gland. This is because it takes some time for the ADH already in the blood to be broken down; approximately half of it is destroyed every 15–20 minutes. However, once ADH stops arriving at the collecting duct cells, it takes only 10–15 minutes for the water-permeable channels to be removed from the plasma membrane and taken back into the cytoplasm for storage.

### SAQ 19.4
Construct a flow diagram to show how blood water concentration is controlled. Identify clearly the receptors and effectors, and show how negative feedback is involved.

# Hormonal communication

## Exocrine and endocrine glands

The chemicals, such as ADH, which are used to carry information from one part of a mammal's body to another part are called **hormones**. They are made in **endocrine glands** (*figure 19.16*).

A **gland** is a group of cells which produces and releases one or more substances, a process known as **secretion**. Endocrine glands contain secretory cells which pass their secretions directly into the blood. 'Endocrine' means 'secreting to the inside', a reference to the fact that endocrine glands secrete hormones into blood capillaries inside the gland.

Endocrine glands are not the only type of gland. We have many glands in our digestive system, for example, such as the salivary glands which secrete saliva. These glands are **exocrine glands** (*figure 19.16*). 'Exocrine' means 'secreting to the outside'. The secretory cells of exocrine glands secrete their substances, which are *not* hormones, into a tube or duct, along which the secretion flows. Salivary glands secrete saliva into salivary ducts, which carry the saliva into the mouth.

## Hormones

Mammalian hormones have many features in common. They are usually relatively small molecules. Many hormones, such as insulin, are polypeptides or proteins whereas others, such as testosterone, are steroids.

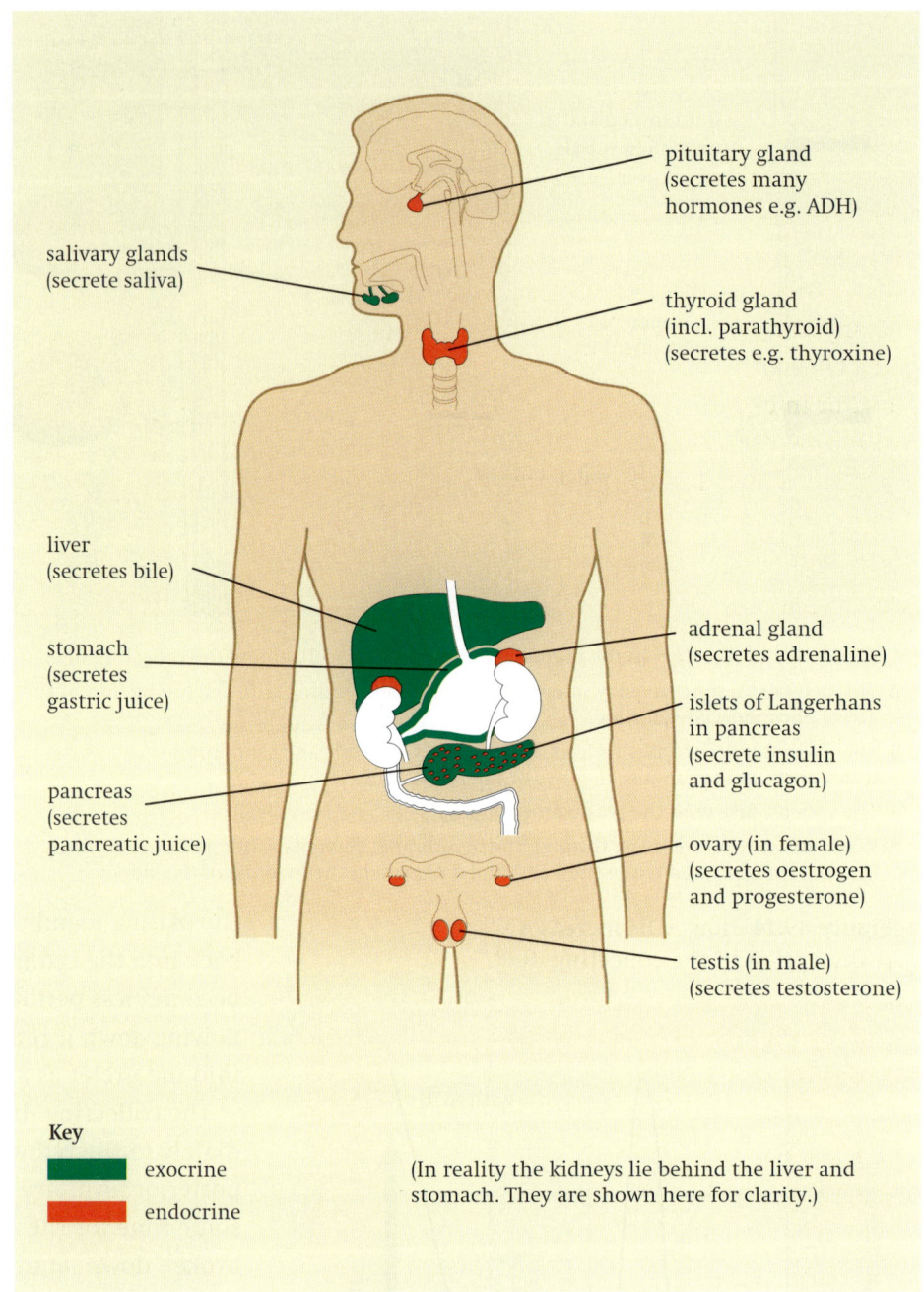

**Key**

▇ exocrine

▇ endocrine

(In reality the kidneys lie behind the liver and stomach. They are shown here for clarity.)

● **Figure 19.16** The positions of some exocrine and endocrine glands in the human body.

After they have been secreted from an endocrine gland, hormones are transported around the body in the blood plasma. The concentrations of hormones in human blood are very small. For any one hormone, the concentration is rarely more than a few micrograms of hormone per $cm^3$ of blood. Their rate of secretion from endocrine glands is also low, usually of the order of a few micrograms or milligrams a day. These small quantities of hormone can, however, have very large effects on the body.

Most endocrine glands can secrete hormones very quickly when an appropriate stimulus arrives. For example, adrenaline, the 'fight or flight' hormone secreted in response to a frightening stimulus, is secreted from the adrenal glands within one second of the stimulus being perceived. This means that the effects of hormones can be 'turned on' quite quickly.

Many hormones have a very short life in the body. They are broken down by enzymes in the blood or in cells, or are lost in the urine. Insulin, for example, lasts for only around 10–15 minutes, while adrenaline lasts for between 1 and 3 minutes. This means that the effects of hormones can also be 'turned off' quite quickly.

## SAQ 19.5

If you are in a frightening situation, adrenaline will be secreted and cause your heart rate to increase. This can go on for several hours. If adrenaline has such a short life-span in the body, how can its effect continue for so long?

Hormones are transported all through the body in the blood. However, each hormone has a particular group of cells which it affects, called **target cells**. These cells, and only these cells, are affected by the hormone because they contain **receptors** specific to the hormone. The receptors for protein hormones, such as insulin, are on the plasma membrane. These hormones bind with the receptors on the outer surface of the membrane, causing a response by the cell without actually entering it. Steroid hormones, however, are lipid-soluble, and so can pass easily through the plasma membrane into the cytoplasm. The receptors for steroid hormones are inside the cell, in the cytoplasm.

## SAQ 19.6

Explain why steroid hormones can pass easily through the plasma membrane, while protein hormones cannot.

## The pancreas

*Figures 19.17, 19.18* and *19.19* show the structure of the pancreas. The pancreas is a very unusual gland, because parts of it function as an exocrine gland, while other parts function as an endocrine gland. The exocrine function is the secretion of pancreatic juice, which flows along the pancreatic duct into the duodenum, where it helps in digestion. The endocrine function is carried out by groups of cells called the **islets of Langerhans**, which are scattered throughout the pancreas.

The islets contain two types of cells. α **cells** secrete **glucagon**, while β **cells** secrete **insulin**. These two hormones, both small proteins, are involved in the control of blood glucose levels (*figure 19.21* on page 274).

### The control of blood glucose

Carbohydrate is transported through the human bloodstream in the form of glucose, in solution in the blood plasma. For storage, glucose can be converted into the polysaccharide **glycogen**, a large, insoluble molecule made up of many glucose units linked together (see page 26), which can be stored inside cells, especially liver and muscle cells (*figure 19.22* on page 274).

In a healthy human, each $100 \text{ cm}^3$ of blood normally contains between 80 and 120 mg of glucose. If blood glucose level drops below this, then cells may run short of glucose for respiration, and be unable to carry out their normal activities. This is

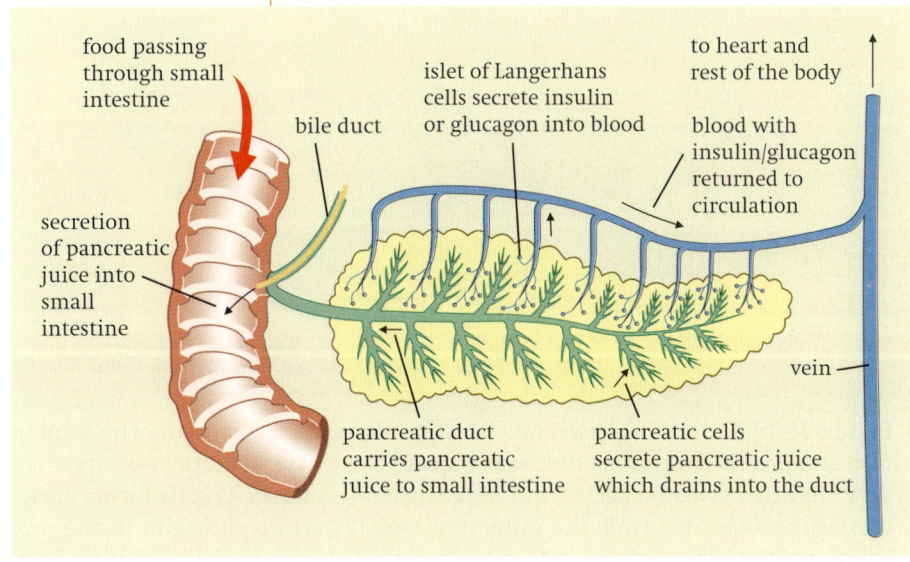

● **Figure 19.17** The pancreas is both an exocrine and endocrine gland.

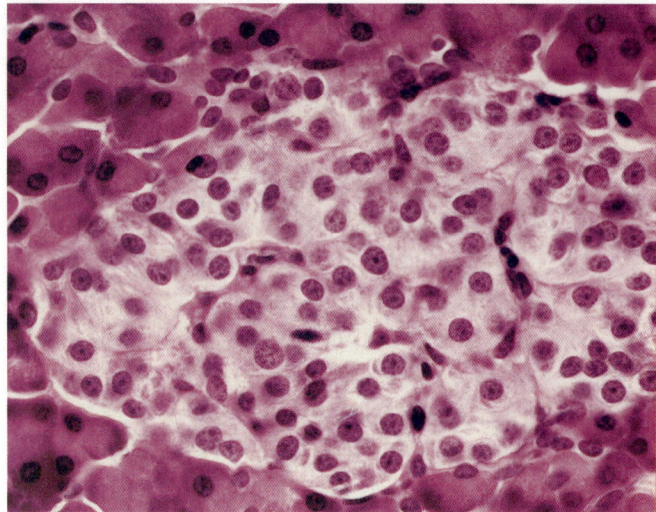

● **Figure 19.18** Light micrograph of pancreas (×500), showing an islet of Langerhans, containing α and β cells.

especially important for cells that can only respire glucose, such as brain cells. Very high blood glucose levels can also cause major problems, again upsetting the normal behaviour of cells.

After a meal containing carbohydrate, glucose from the digested food is absorbed from the small intestine and passes into the blood. As this blood flows through the pancreas, the α and β cells detect the raised glucose levels. The α cells respond

by stopping the secretion of glucagon, while the β cells respond by secreting insulin into the blood plasma. The insulin is carried to all parts of the body, in the blood (see *box 19A* opposite).

Insulin affects many cells, especially those in the liver and muscles. The effects on these cells include:

■ an increased absorption of glucose from the blood into the cells;

■ an increase in the rate of use of glucose in respiration;

■ an increase in the rate at which glucose is converted into the storage polysaccharide glycogen.

All of these processes take glucose out of the blood, so lowering the blood glucose levels. A drop in blood glucose concentration is detected by the α and β cells in the pancreas. The α cells respond by secreting glucagon, while the β cells respond by stopping the secretion of insulin.

The lack of insulin puts a stop to the increased uptake and usage of glucose by liver and muscle cells, although uptake still continues at a more 'normal' rate. The presence of glucagon affects the activities of the liver cells. (Muscle cells are not responsive to glucagon.) These effects include:

■ the breakdown of glycogen to glucose;

■ the use of fatty acids instead of glucose as the main fuel in respiration;

■ the production of glucose from other compounds, such as fats.

As a result, the liver releases glucose into the blood. This blood flows around the body, passing through the pancreas. Here, the α and β cells sense the raised glucose levels, switching off glucagon secretion and perhaps switching on insulin secretion if the glucose levels are higher than normal.

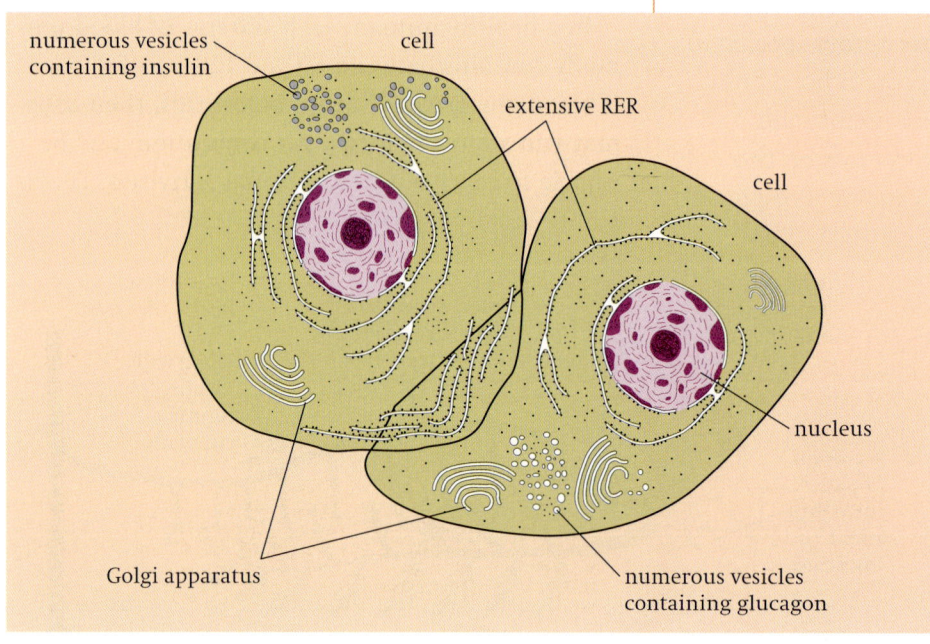

● **Figure 19.19** Diagrams, based on electron micrographs, showing the structure of α and β cells from islets of Langerhans of a rat. The structures of these cells vary slightly between species, but here the insulin in the β cells forms dark, crystalline deposits inside the numerous vesicles, while glucagon in the α cells is less visible. In reality, the cells are almost filled with vesicles, though only a few have been shown here.

## Box 19A The control of insulin secretion

The β cells in the islets of Langerhans release insulin in response to the presence of high levels of glucose in the blood. How do they sense this, and how is the amount of insulin that is released controlled?

The β cells contain several types of channels in their plasma membranes, each of which allows a particular type of ion to pass through (see page 54). These include channels that let $K^+$ ions pass through, and others that let $Ca^{2+}$ pass through.

Normally, the $K^+$ channels are open, leaving the $K^+$ ions free to pass through. These positively charged ions diffuse from inside the cell to the outside. This makes the outside of the cell positively charged compared with the inside. We say that there is a **potential difference** across the membrane. The potential difference across the plasma membrane of a resting β cell – that is one that is not secreting insulin – is about $-70$ mV.

However, when glucose levels around the β cell are high, this starts off a chain of events which alters the situation. The glucose passes into the cell, where it is quickly phosphorylated by the enzyme **glucokinase**. The phosphorylated glucose is then metabolised to produce ATP.

The $K^+$ channels are sensitive to the amount of ATP in the cell, and they respond to this increase in ATP levels by closing. So now the $K^+$ ions cannot diffuse out. As a result, the difference in electrical potential on the inside and the outside of the membrane becomes less. It is now only about $-30$ mV.

Now the $Ca^{2+}$ channels come into the picture. They, unlike the $K^+$ channels, are normally closed. However, they respond to the change in potential across the membrane by opening. $Ca^{2+}$ ions flood into the cell from the tissue fluid outside it.

The $Ca^{2+}$ ions affect the behaviour of the vesicles containing insulin. These vesicles are moved towards the plasma membrane, where they fuse with the membrane and empty their contents outside the cell.

When you have studied the way in which nerve impulses are generated and transmitted (pages 278–287), you will probably be able to pick out several similarities between that mechanism, and the mechanism by which insulin secretion is controlled.

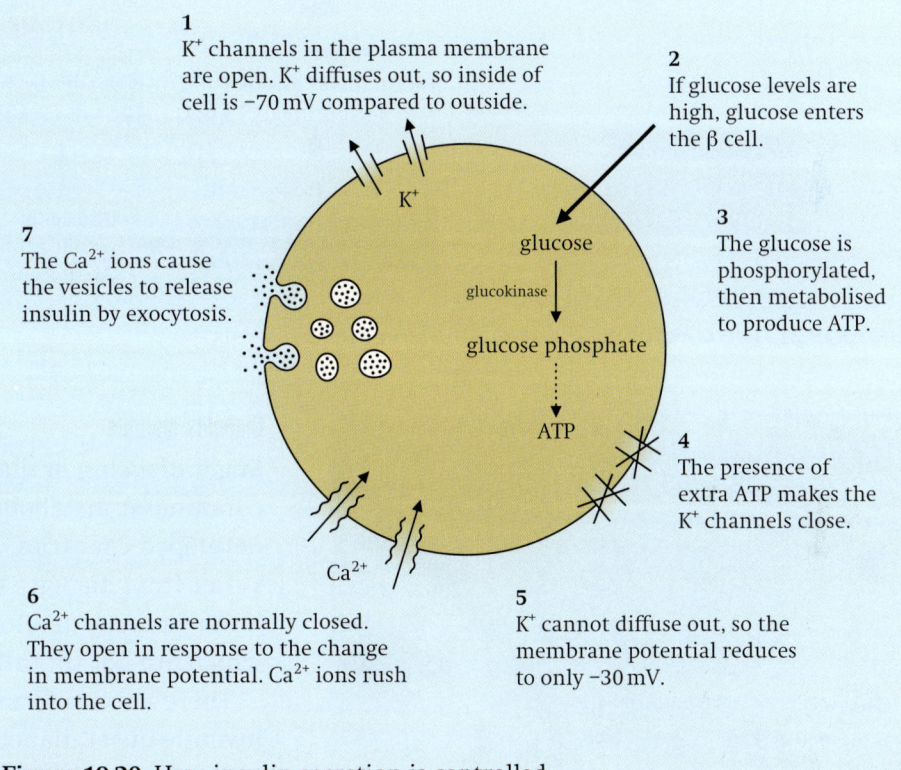

**1** $K^+$ channels in the plasma membrane are open. $K^+$ diffuses out, so inside of cell is $-70$ mV compared to outside.

**2** If glucose levels are high, glucose enters the β cell.

**3** The glucose is phosphorylated, then metabolised to produce ATP.

**7** The $Ca^{2+}$ ions cause the vesicles to release insulin by exocytosis.

**4** The presence of extra ATP makes the $K^+$ channels close.

**6** $Ca^{2+}$ channels are normally closed. They open in response to the change in membrane potential. $Ca^{2+}$ ions rush into the cell.

**5** $K^+$ cannot diffuse out, so the membrane potential reduces to only $-30$ mV.

● **Figure 19.20** How insulin secretion is controlled.

Blood sugar levels never remain constant, even in the healthiest person. One reason for this is the inevitable time delay between a change in the blood glucose level and the onset of actions to correct it. Time delays in control systems result in oscillation, where things do not stay absolutely constant, but sometimes rise slightly above and sometimes drop slightly below the 'required' level.

## SAQ 19.7

The control of blood glucose concentration uses a negative feedback control mechanism.

**a** Explain what is meant by negative feedback.

**b** What are the receptors in this control mechanism?

**c** What are the effectors?

(You may need to look back to page 258.)

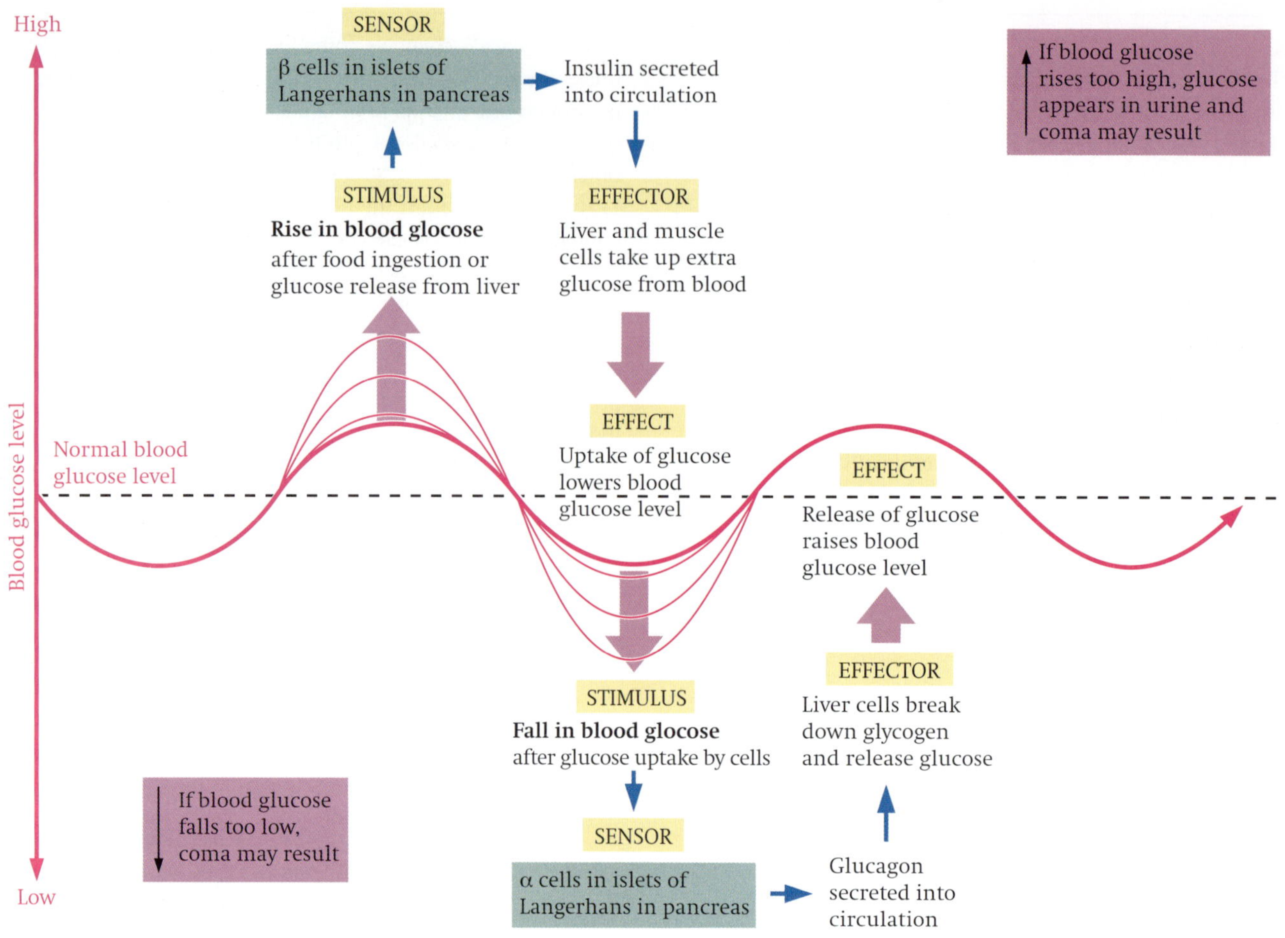

High

Blood glucose level

**SENSOR**

β cells in islets of Langerhans in pancreas → Insulin secreted into circulation

If blood glucose rises too high, glucose appears in urine and coma may result

**STIMULUS**

**Rise in blood glocose** after food ingestion or glucose release from liver

**EFFECTOR**

Liver and muscle cells take up extra glucose from blood

Normal blood glucose level

**EFFECT**

Uptake of glucose lowers blood glucose level

**EFFECT**

Release of glucose raises blood glucose level

**EFFECTOR**

Liver cells break down glycogen and release glucose

**STIMULUS**

**Fall in blood glocose** after glucose uptake by cells

If blood glucose falls too low, coma may result

**SENSOR**

α cells in islets of Langerhans in pancreas → Glucagon secreted into circulation

Low

● **Figure 19.21** The control mechanism for blood glucose levels.

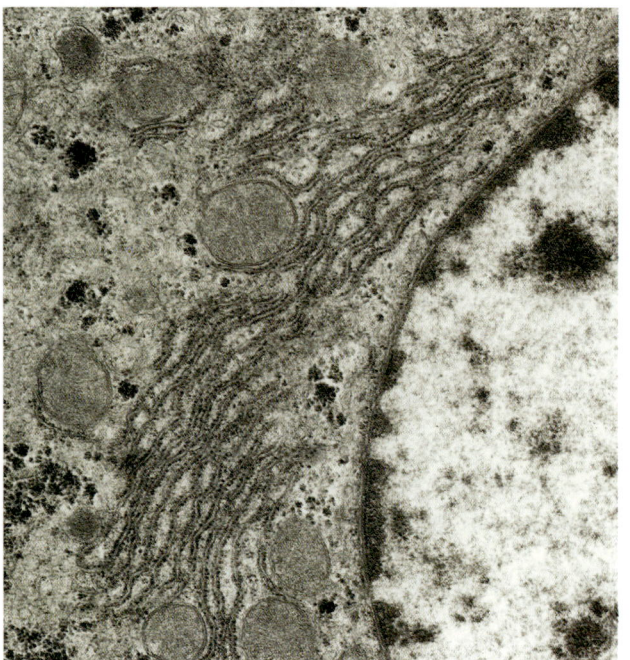

● **Figure 19.22** Electron micrograph of part of a liver cell (× 22 000). The dark spots are glycogen granules in the cytoplasm. Numerous mitochondria can also be seen.

### Diabetes mellitus

Sugar diabetes, or diabetes mellitus, is one of the commonest metabolic diseases in humans. In developed countries, approximately 1% of people suffer from diabetes mellitus. The incidence is lower in many developing countries, for reasons which are not yet fully understood.

There are two forms of sugar diabetes. In **juvenile-onset diabetes**, sometimes called **insulin-dependent diabetes**, the pancreas seems to be incapable of secreting sufficient insulin. It is thought that this might be due to a deficiency in the gene which codes for the production of insulin, or because of an attack on the β cells by the person's own immune system. This form of diabetes, as suggested by its name, usually begins very early in life.

The second form of diabetes is called **non-insulin-dependent diabetes**. In this form, the pancreas does secrete insulin, but the liver and muscle cells do not respond properly to it. It frequently begins relatively late in life and is often associated with obesity.

The symptoms of both types of diabetes mellitus are the same. After a carbohydrate meal, blood glucose levels rise and stay high (*figure 19.23*). Normally there is no glucose in urine, but if blood glucose levels become very high, the kidney cannot reabsorb all the glucose so that some passes out in the urine. Extra water and salts accompany this glucose, and the person consequently feels extremely hungry and thirsty.

In a diabetic person, uptake of glucose into cells is slow, even when blood glucose levels are high. Thus cells lack glucose, and metabolise fats and proteins as an alternative energy source. This can lead to a build-up of substances called keto-acids in the blood, which lowers the blood pH. The combination of dehydration, salt loss and low blood pH can cause coma in extreme situations.

Between meals, when blood glucose levels would normally be kept up by mobilisation of glycogen reserves, the blood glucose levels of a person with untreated diabetes may plummet. This is because there is no glycogen to be mobilised. Once again, coma may result, this time because of a lack of glucose for respiration.

## SAQ 19.8

Explain why people with diabetes mellitus have virtually no glycogen to be mobilised.

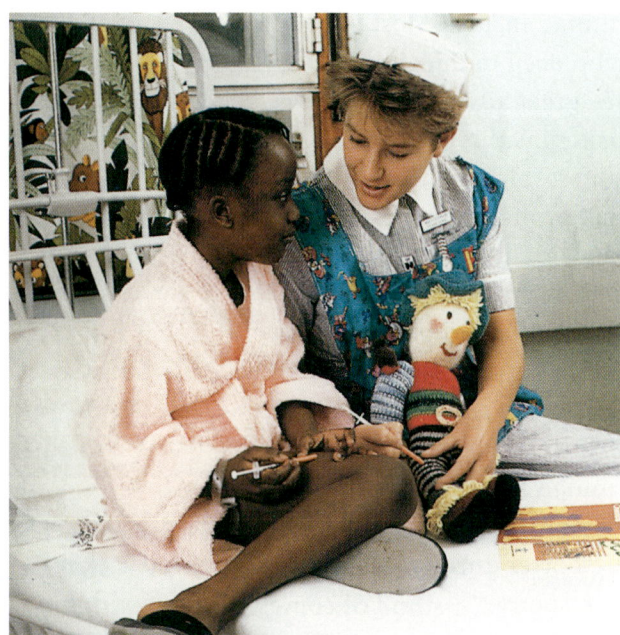

● **Figure 19.24** A nurse teaches a girl with insulin-dependent diabetes to inject insulin. She will have to do this daily, all her life.

In insulin-dependent diabetes, regular injections of insulin, together with a carefully controlled diet, are used to keep blood glucose levels near normal (*figure 19.24*). The person must monitor their own blood glucose level, taking a blood sample several times a day. In non-insulin-dependent diabetes, insulin injections are not normally needed. Control is by diet alone.

## SAQ 19.9

**a** Insulin cannot effectively be taken by mouth. Why is this so?

**b** Suggest how people with non-insulin-dependent diabetes can control their blood glucose level.

Until the early 1980s, all insulin was obtained from animals such as pigs and cattle. In the 1980s, insulin began to be made using bacteria into which the human insulin gene had been inserted (see page 238). This insulin is much cheaper than that obtained from animals, and also has the advantage that it is, of course, human insulin, not pig or cow insulin, which differ slightly from

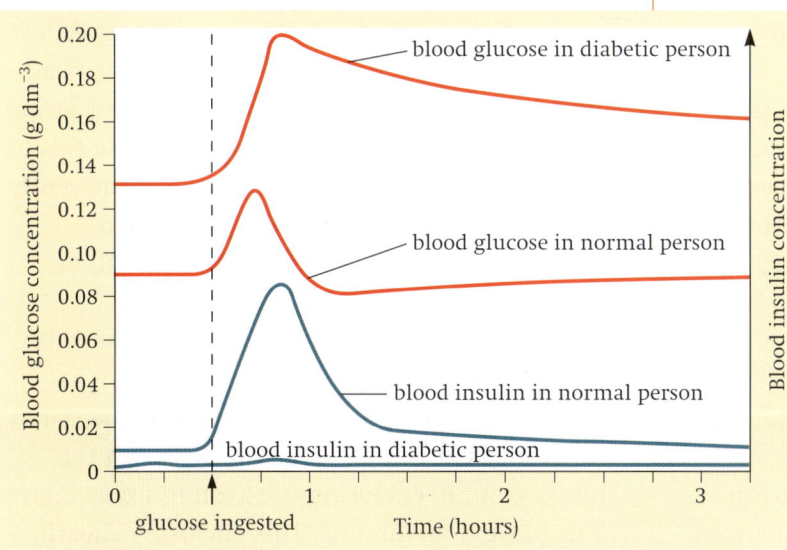

● **Figure 19.23** Blood glucose and insulin levels following intake of glucose in a normal person and a person with insulin-dependent diabetes.

human insulin. Most people who need to inject insulin see great advantages in using insulin from this source, although a few have had problems in making the change from using pig or cattle insulin to using human insulin produced by genetic engineering.

Some of the advantages of using genetically engineered human insulin are:

■ more rapid response;
■ shorter duration of response;
■ less chance of an immune response to the insulin developing;
■ effective in people who have developed a tolerance for animal-derived insulin;
■ more acceptable to people who feel it is unethical to use pig or cattle insulin.

# Nervous communication

## Neurones

So far in this chapter we have looked at the way in which hormones are used by mammals to send messages from one part of the body to another. The hormones are carried in the blood, and so spread all through the body. Mammals also have another method of communication within their bodies. This method is faster and more precise and involves the transmission of electrical signals or impulses along precisely constructed pathways. The cells which carry these signals are called nerve cells or **neurones**.

*Figure 19.25* shows the structure of a mammalian neurone. This is a **motor neurone** which transmits messages from the brain or spinal cord to a muscle or gland.

The cell body of a motor neurone lies within the spinal cord or brain. The nucleus of a neurone is always in its cell body. Often, dark specks can be seen in the cytoplasm. These are groups of ribosomes involved in protein synthesis.

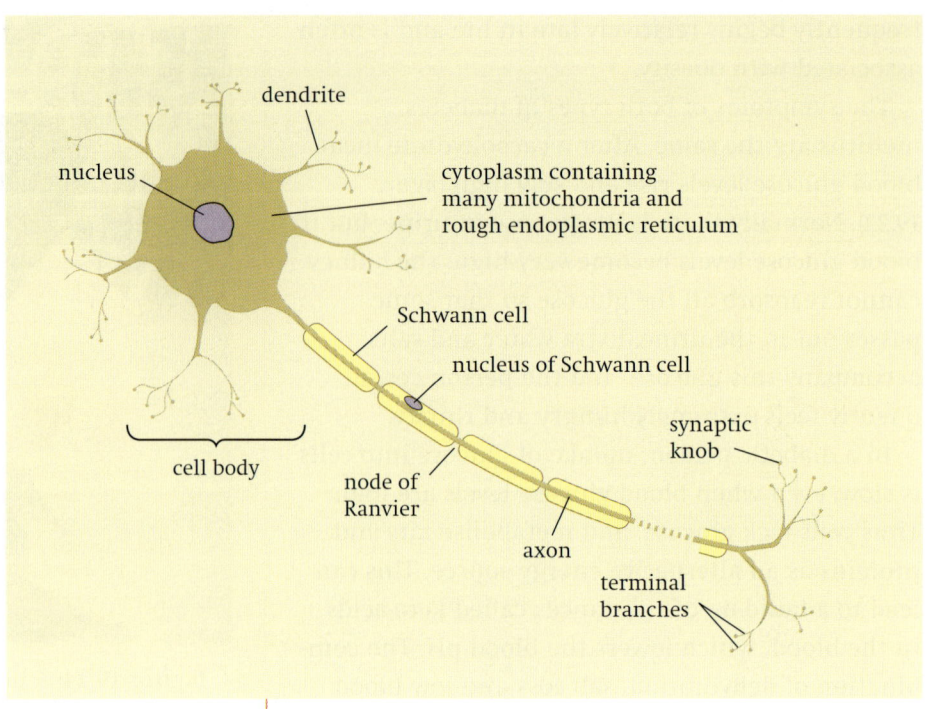

● **Figure 19.25** A motor neurone. The axon may be over a metre long.

Many thin cytoplasmic processes extend from the cell body. In a motor neurone, all but one of these processes are relatively short. They conduct impulses *towards* the cell body, and are called **dendrons** or **dendrites**. One process is much longer, and conducts impulses *away* from the cell body. This is called the **axon**. A motor neurone with its cell body in your spinal cord might have its axon running all the way to one of your toes, so axons may be extremely long. Within the cytoplasm of an axon, all of the usual organelles such as endoplasmic reticulum, Golgi apparatus and mitochondria, are present. Particularly large numbers of mitochondria are found at the tips of the terminal branches of the axon, together with many vesicles containing chemicals called transmitter substances. Their function will be explained later (page 286).

In some neurones, cells called **Schwann cells** wrap themselves around the axon all along its length. *Figure 19.26* shows one such cell, viewed as the axon is cut transversely. The Schwann cell spirals around, enclosing the axon in many layers of its plasma membrane. This enclosing sheath, called the **myelin sheath**, is made largely of lipid, together with some proteins. Not all axons have myelin sheaths. Some invertebrate animals, such

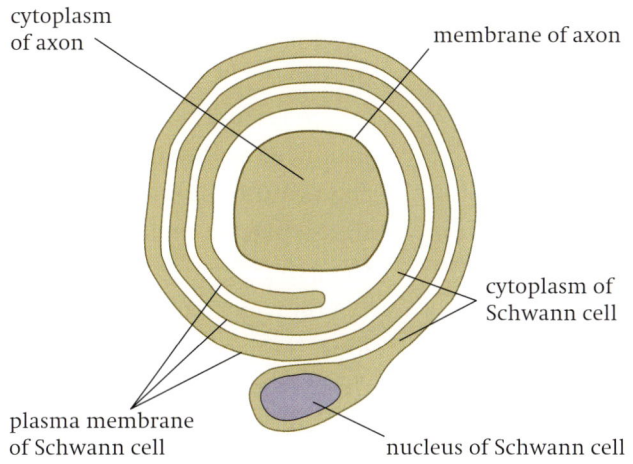

cytoplasm of axon

membrane of axon

cytoplasm of Schwann cell

plasma membrane of Schwann cell

nucleus of Schwann cell

● **Figure 19.26** Transverse section of the axon of a myelinated neurone.

as earthworms, have no myelin sheaths around their neurones. In humans, about one third of our motor and sensory neurones are myelinated. The sheath affects the speed of conduction of the nerve impulse (page 281). The small, uncovered areas of axon between Schwann cells are called **nodes of Ranvier**. They occur about every 1–3 mm in human neurones. The nodes themselves are very small, around 2–3 μm long.

*Figure 19.27* shows the different shapes of neurones with different functions. The basic structure of a **sensory neurone** is the same as

that of a motor neurone, but it has one long dendron and an axon which is often shorter than the dendron. Sensory neurones bring impulses from receptors (cells which pick up stimuli, such as touch or light) to the brain or spinal cord. There, they pass them on to other neurones.

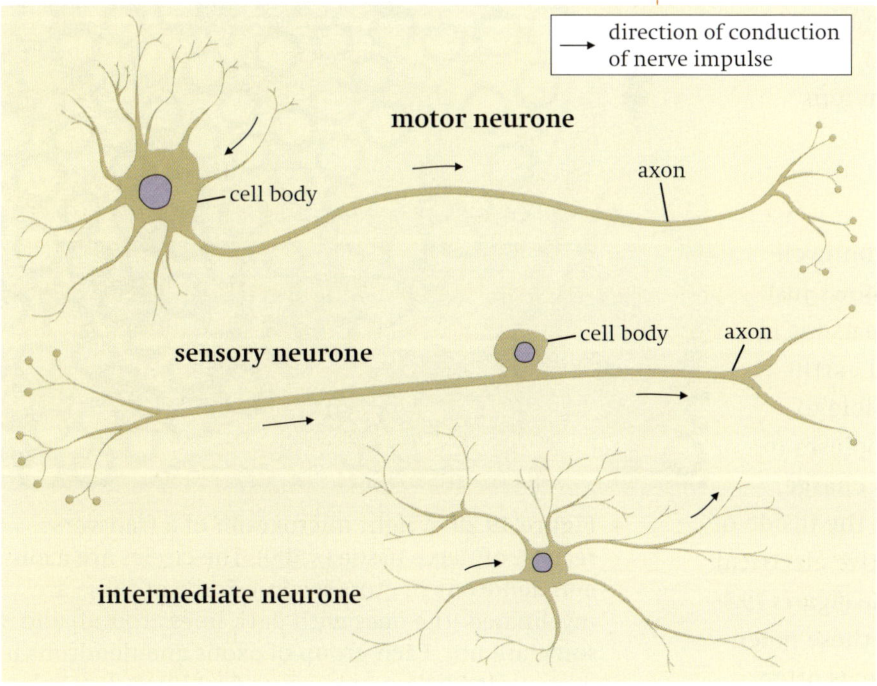

## A reflex arc

In the human body, a sensory neurone and a motor neurone may form a reflex arc. A **reflex arc** is the pathway along which impulses are carried from a receptor to an effector, without involving 'conscious' regions of the brain. *Figure 19.28* shows the structure of a spinal reflex arc in which the impulse is passed from neurone to neurone inside the spinal cord. The neurone between the sensory and motor neurones is called an intermediate neurone. Others may have no intermediate neurone, and the impulse passes directly from the sensory neurone to the motor neurone.

Within the spinal cord, the impulse will also be passed on to other neurones which take the impulse up the cord to the brain. This happens at the same time as the message is travelling along the motor neurone to the effector. The effector therefore responds to the stimulus before there is any voluntary response, involving conscious regions of the brain. This type of reaction to a stimulus is called a **reflex action**. It is a fast, automatic response to a stimulus. Reflex actions are a very useful way of responding to danger signals, such as the touch of a very hot object on your skin or the sight of an object flying towards you.

### SAQ 19.10

Think of three reflex actions other than the two already mentioned. For each action, state the precise stimulus, name the receptor which first detects this stimulus, name the effector which responds to it, and describe the way in which this effector responds.

direction of conduction of nerve impulse

**motor neurone**

cell body

axon

**sensory neurone**

cell body

axon

**intermediate neurone**

● **Figure 19.27** Motor, sensory and intermediate neurones.

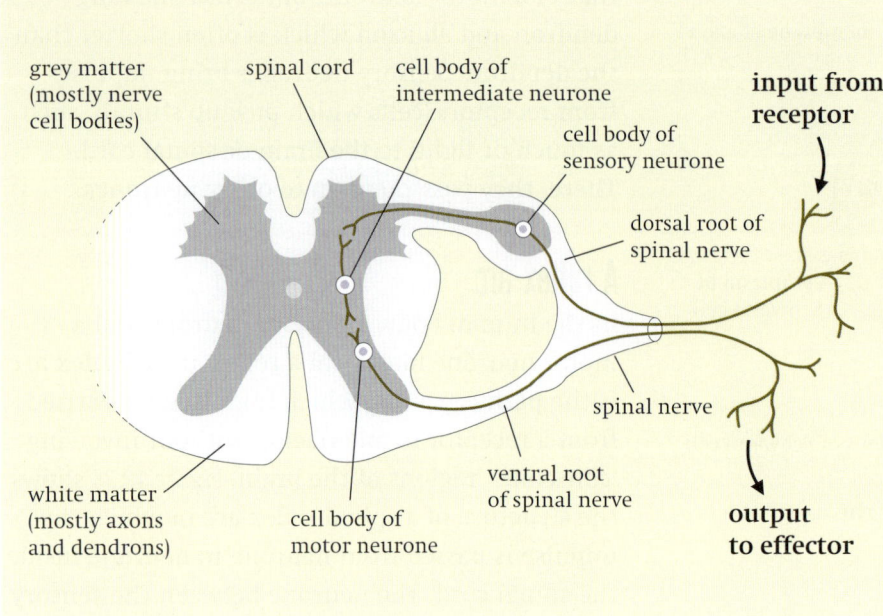

grey matter (mostly nerve cell bodies)

spinal cord

cell body of intermediate neurone

cell body of sensory neurone

**input from receptor**

dorsal root of spinal nerve

spinal nerve

ventral root of spinal nerve

white matter (mostly axons and dendrons)

cell body of motor neurone

**output to effector**

● **Figure 19.28** A reflex arc. The spinal cord is shown in transverse section.

## Transmission of nerve impulses

Neurones transmit impulses as electrical signals. These signals travel very rapidly along the plasma membrane from one end of the cell to the other and are *not* a flow of electrons like an electric current. Rather, the signals are very brief changes in the distribution of electrical charge across the plasma membrane, caused by the very rapid movement of sodium and potassium ions into and out of the axon.

## Resting potential

*Figure 19.29* shows myelinated and nonmyelinated nerve tissue and *figure 19.30* shows just part of a nonmyelinated axon. Some axons in some organisms, such as squids and earthworms, are very wide, and it is possible to insert tiny electrodes into their cytoplasm to measure these changes in electrical charge.

In a resting axon, it is found that the inside of the axon always has a slightly negative electrical potential compared with the outside (*figures 19.30 and 19.31a*). The difference between these potentials, called the **potential difference**, is often around −65 mV. In other words, the electrical potential of the inside of the axon is 65 mV lower than the outside.

The resting potential is produced and maintained by the **sodium–potassium pump** in the plasma membrane of the axon (*figure 19.31b* and *box 15A* on page 200). Sodium ions, $Na^+$, are picked up from the cytoplasm inside the axon by a carrier protein in the membrane and carried to the outside. At the same time, potassium ions, $K^+$, are brought into the cytoplasm from the external fluids. Both of these processes involve moving the ions against their concentration gradients, and so use energy from the hydrolysis of ATP.

The sodium–potassium pump removes three sodium ions from the cell for every two potassium

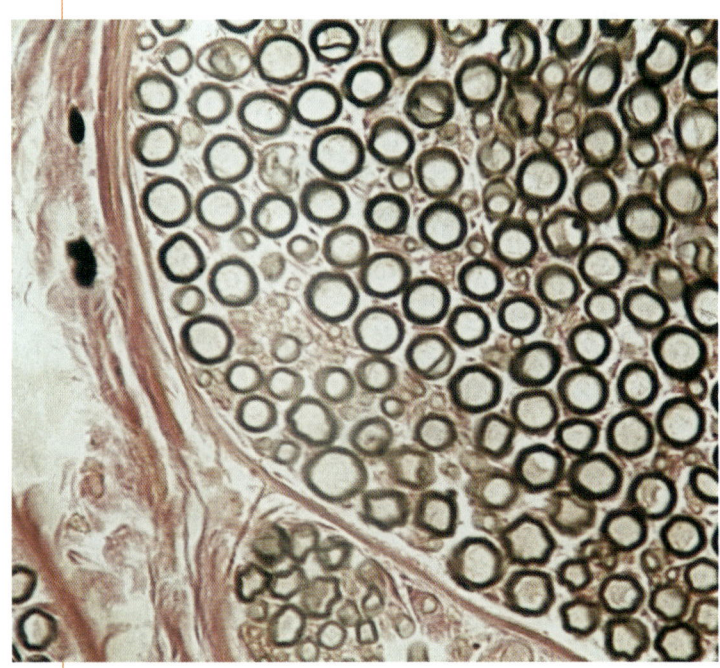

● **Figure 19.29** A light micrograph of a transverse section of nerve tissue (× 500). The circles are axons and dendrons in cross-section. Some of these are myelinated (the ones with dark lines around) and some are not. Each group of axons and dendrons is surrounded by a perineurium (red lines). Several such groups make a complete nerve.

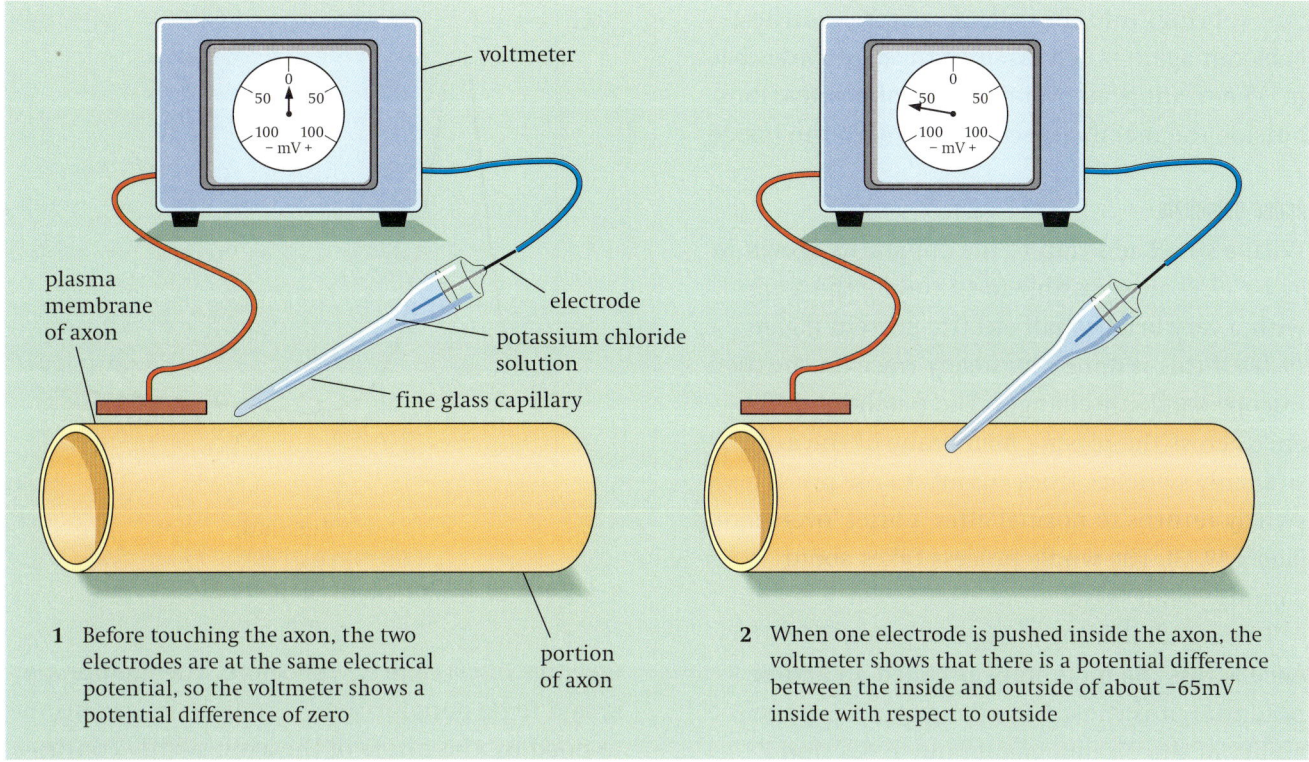

1 Before touching the axon, the two electrodes are at the same electrical potential, so the voltmeter shows a potential difference of zero

2 When one electrode is pushed inside the axon, the voltmeter shows that there is a potential difference between the inside and outside of about −65mV inside with respect to outside

- **Figure 19.30** Measuring the resting potential of an axon.

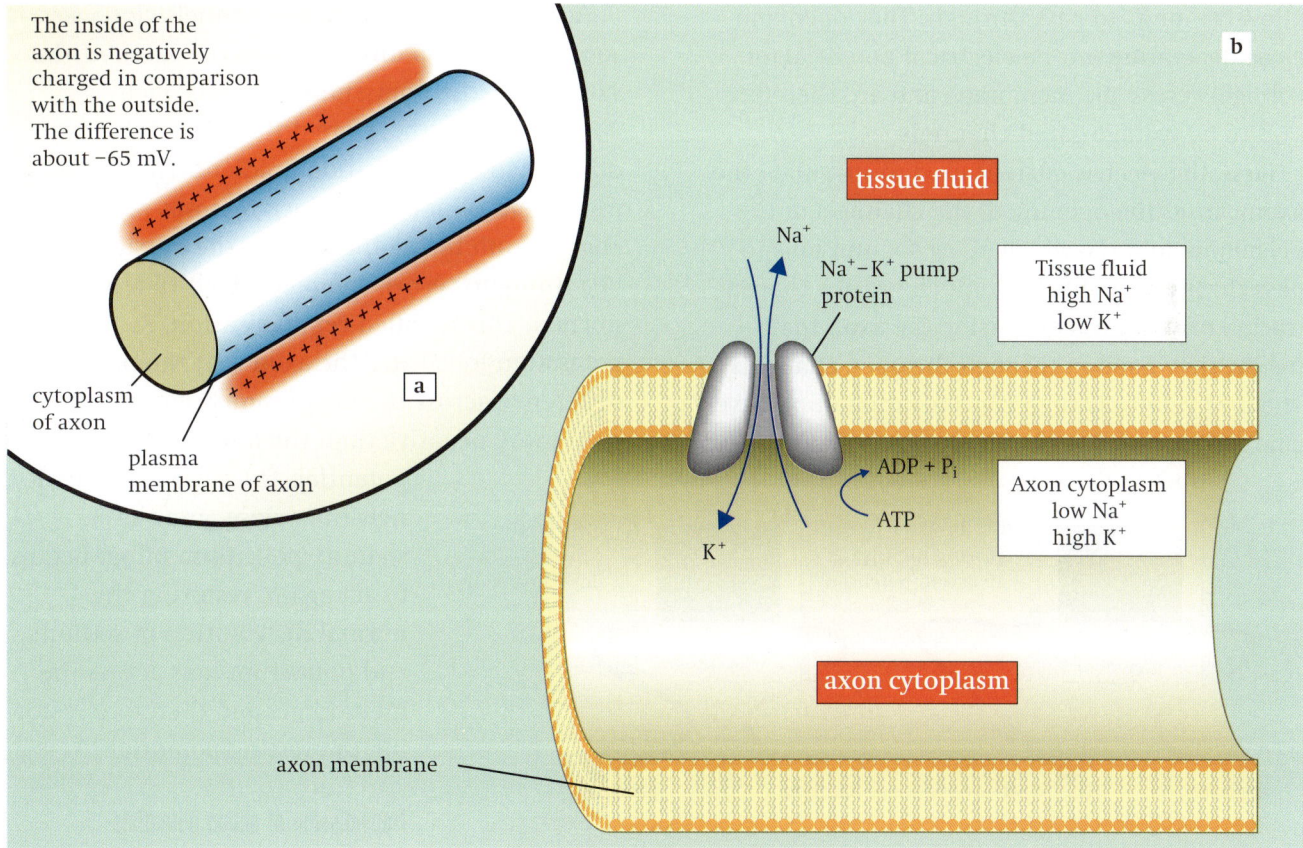

- **Figure 19.31**
a At rest, an axon has negative electrical potential inside.
b The sodium–potassium pump maintains the resting potential by keeping more sodium ions outside than there are potassium ions inside.

ions it brings into the cell. Moreover, K⁺ diffuses back out again much faster than Na⁺ diffuses back in. The result is an overall excess of positive ions outside the membrane compared with the inside.

## Action potentials

With a small addition to the apparatus shown in *figure 19.30*, it is possible to stimulate the axon with a very brief, small electric current (*figure 19.32*). If this is done, the steady trace on the oscilloscope suddenly changes. The potential difference across the plasma membrane of the axon suddenly switches from −65 mV to +40 mV. It swiftly returns to normal after a brief 'overshoot' (*figure 19.33*). The whole process takes about 3 milliseconds (ms).

This rapid, fleeting change in potential difference across the membrane is called an **action potential**. It is caused by changes in the permeability of the plasma membrane to Na⁺ and K⁺.

As well as the sodium–potassium pump, there are other channels in the plasma membrane that allow Na⁺ or K⁺ to pass through. They open and close depending on the electrical potential (or voltage) across the membrane and are therefore said to be **voltage-gated channels**.

First, the electric current used to stimulate the axon causes the opening of the channels in the plasma membrane which allow sodium ions to pass through. As there is a much greater concentration of sodium ions outside the axon than inside, they flood in through the open channels. The now relatively high concentration of positively charged sodium ions inside the axon makes it less

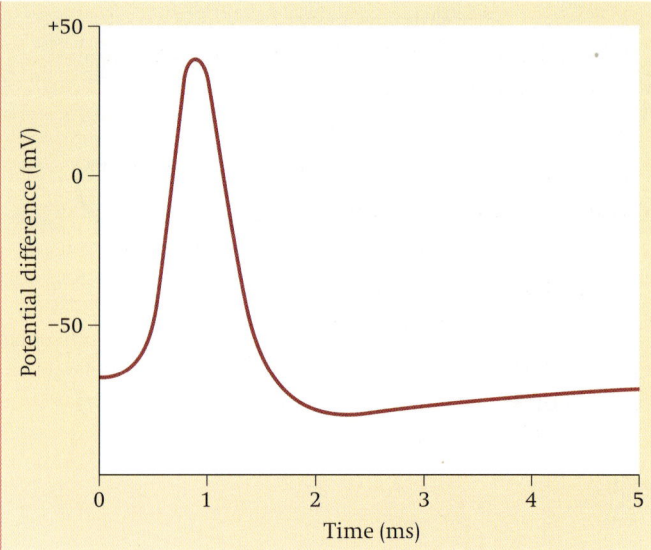

● **Figure 19.33** An action potential.

negative *inside* than it was before. The membrane is said to be **depolarised**. As sodium ions continue to flood in, the inside of the axon swiftly continues to build up positive charge, until it reaches a potential of +40 mV compared with the outside.

At this point, the sodium channels close, so sodium ions stop diffusing into the axon. At the same time, the potassium channels open. Potassium ions therefore diffuse *out* of the axon, down their concentration gradient. The outward movement of potassium ions removes positive charge from inside the axon to the outside, thus beginning to return the potential difference to normal. This is called **repolarisation**. So many potassium ions leave the axon that the potential difference across the membrane briefly becomes even more negative than the normal resting potential. The potassium channels then close, and the sodium–potassium pump begins to act again, restoring the normal distribution of sodium and potassium ions across the membrane, and therefore restoring the resting potential.

## Transmission of action potentials

The description of an action potential above concerns events at one particular point in an axon membrane. However, the

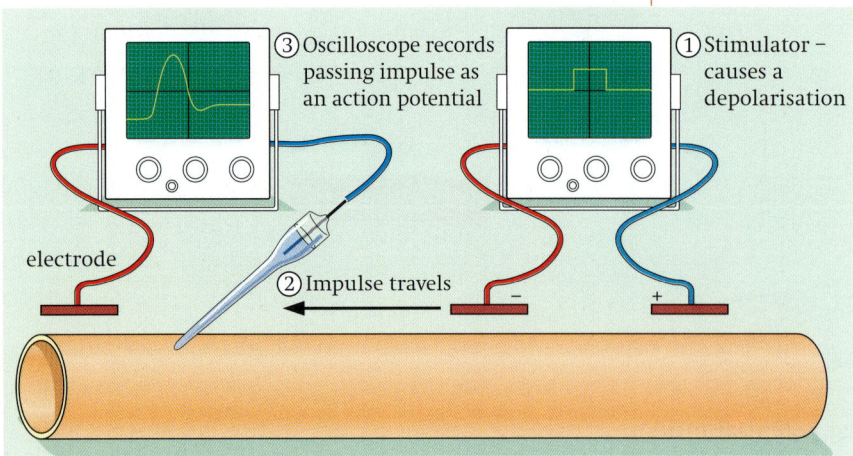

● **Figure 19.32** Recording of an action potential.

function of a neurone is to transmit information *along* itself. How do action potentials transmit information along a neurone?

An action potential at any point in an axon's plasma membrane triggers the production of an action potential in the membrane on either side of it. *Figure 19.34* shows how it does this. The temporary depolarisation of the membrane where the action potential is causes a 'local circuit' to be set up between the depolarised region and the resting regions on either side of it. Sodium ions flow sideways inside the axon, away from the positively charged region towards the negatively charged regions on either side. This depolarises these adjoining regions and so generates an action potential in them.

In practice, if an action potential has been travelling in one direction from a point of stimulation, a 'new' action potential is only generated *ahead* of, and not behind, it. This is because the region behind it will still be recovering from the action potential it has just had and its distribution of sodium and potassium ions will not yet be back to normal. It is therefore incapable of producing a new action potential for a short time. This is known as the **refractory period**.

### How action potentials carry information

Action potentials do not change in size as they travel. However long an axon is, the action potential will continue to reach a value of +40 mV inside all the way along. Moreover, the intensity of

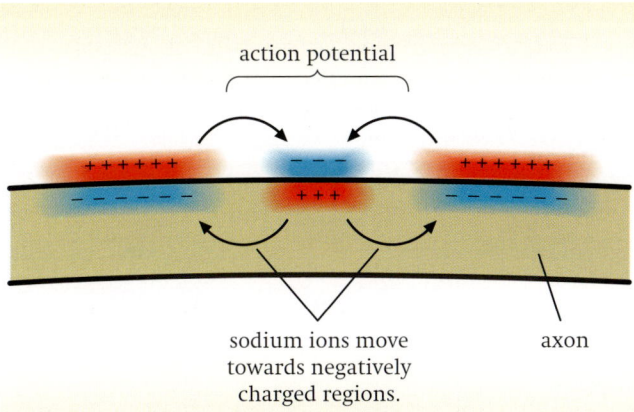

action potential

sodium ions move towards negatively charged regions.

axon

● **Figure 19.34** How action potentials are transmitted along an axon. Local circuits are set up between the region where there is an action potential and the resting regions, causing the resting regions to depolarise.

the stimulus which originally generated the action potential has absolutely no effect on the size of the action potential. A very strong light shining in your eyes will produce action potentials of precisely the same size as a dim light. Nor does the speed at which the action potentials travel vary according to the size of the stimulus. In any one axon, the speed of axon potential transmission is always the same.

What *is* different about the action potentials resulting from a strong and a weak stimulus is their **frequency**. A strong stimulus produces a rapid succession of action potentials, each one following along the axon just behind its predecessor. A weak stimulus results in fewer action potentials per second (*figure 19.35*).

Moreover, a strong stimulus is likely to stimulate more neurones than a weak stimulus. While a weak stimulus might result in action potentials passing along just one or two neurones, a strong stimulus could produce action potentials in many more.

The brain can therefore interpret the *frequency* of action potentials arriving along the axon of a sensory neurone, and the *number* of neurones carrying action potentials, to get information about the *strength* of the stimulus being detected by that receptor. The *nature* of the stimulus, whether it is light, heat, touch or so on, is deduced from the *position* of the sensory neurone bringing the information. If the neurone is from the retina of the eye, then the brain will interpret the information as meaning 'light'. If for some reason a different stimulus, such as pressure, stimulates a receptor cell in the retina, the brain will still interpret the action potentials from this receptor as meaning 'light'. This is why rubbing your eyes when they are shut can cause you to 'see' patterns of light.

### Speed of conduction

In a myelinated human neurone, action potentials travel at up to $100 \, m \, s^{-1}$. In nonmyelinated neurones, the speed of conduction is much slower, being as low as $0.5 \, m \, s^{-1}$ in some cases. Myelin speeds up the rate at which action potentials travel by insulating the axon membrane. Sodium and potassium ions cannot flow through the myelin sheath, so it is not possible for depolarisation or action potentials to occur in parts of the axon which are surrounded by it. They can only occur at the nodes of Ranvier.

*Figure 19.36* shows how an action potential is transmitted along a myelinated axon. The local circuits that are set up stretch from one node to the next. Thus action potentials 'jump' from one node to the next, a distance of 1–3 mm. This is called **saltatory conduction**. It can increase the speed of transmission by up to 50 times that in a nonmyelinated axon of the same diameter.

Diameter also affects the speed of transmission (*figure 19.37*). Thick axons transmit action potentials faster than thin ones. Earthworms, which have no myelinated axons, have a few very thick nonmyelinated ones which run all along their body from head to tail (*figure 19.38*). A bird pecking at an earthworm's head sets up action potentials in these giant axons, which sweep along the length of the body, stimulating muscles to contract. The rapid response which results may enable the earthworm to escape.

## What starts off an action potential?

In the description of the generation of an action potential on page 279, the initial stimulus was a small electric current. In normal life, however, action potentials are generated by a wide variety of stimuli, such as light, touch, sound, temperature or chemicals.

A cell which responds to such stimuli by initiating an action potential is called a **receptor cell**. Receptor cells are often

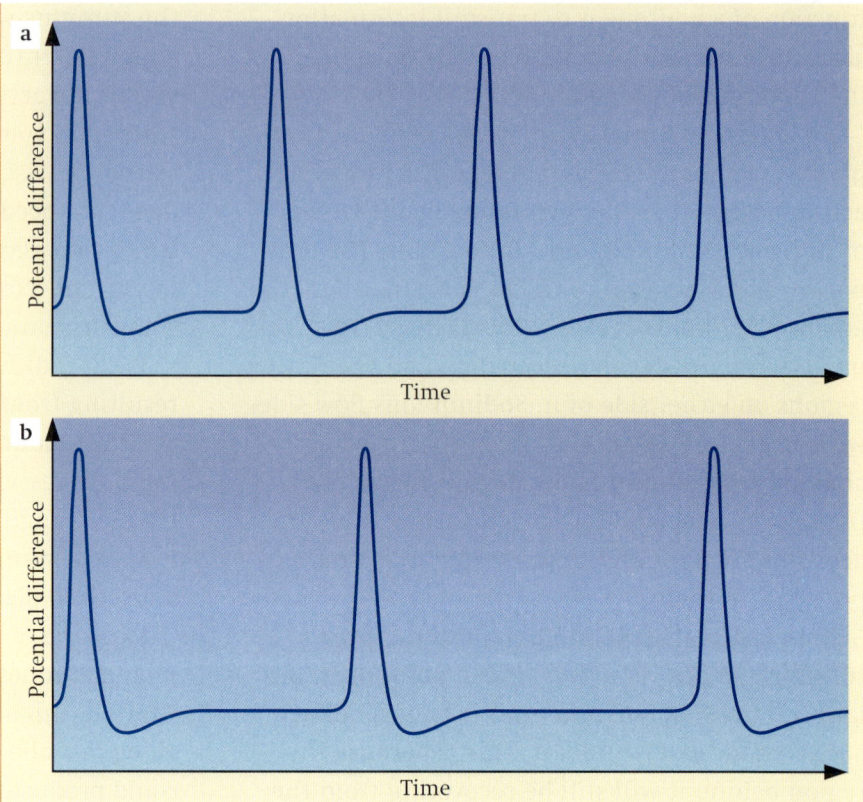

● **Figure 19.35** Action potentials resulting from **a** a strong stimulus and **b** a weak stimulus. Note that the size of each action potential remains the same, only their frequency changes.
**a** A high frequency of impulses is produced when a receptor is given a strong stimulus. This high frequency carries the message 'strong stimulus'.
**b** A lower frequency of impulses is produced when a receptor is given a weaker stimulus. This lower frequency carries the message 'weak stimulus'.

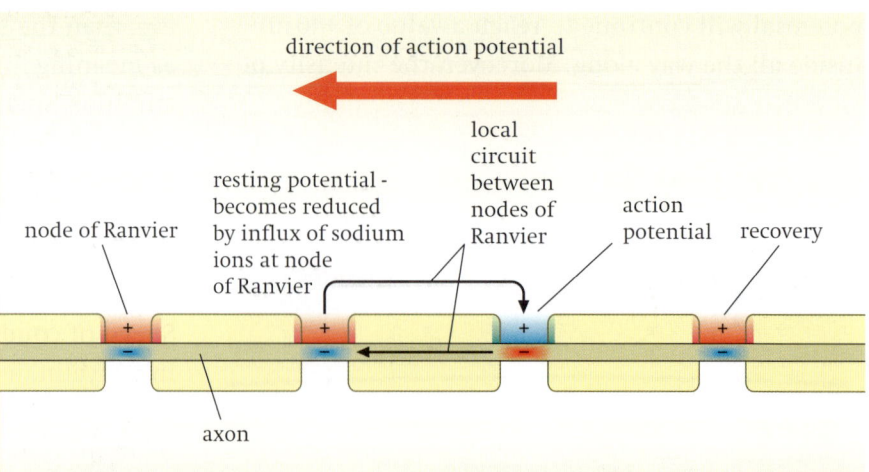

● **Figure 19.36** Transmission of an action potential in a myelinated axon. The myelin sheath acts as an insulator, preventing differences in potential across the parts of the axon membrane surrounded by the sheath. Potential differences can only occur at the nodes of Ranvier. The action potential therefore 'jumps' from one node to the next, travelling much more swiftly than in a nonmyelinated axon.

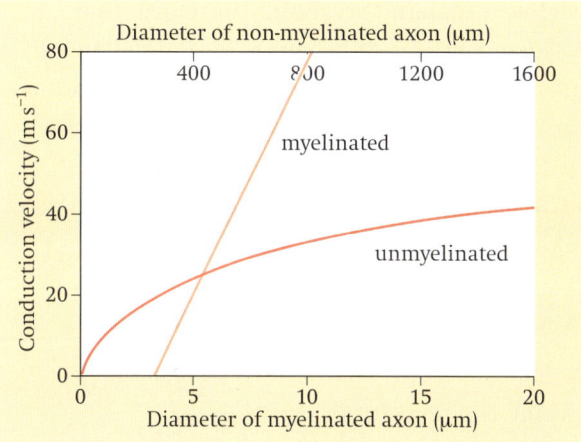

- **Figure 19.37** Speed of transmission in myelinated and nonmyelinated axons of different diameters.

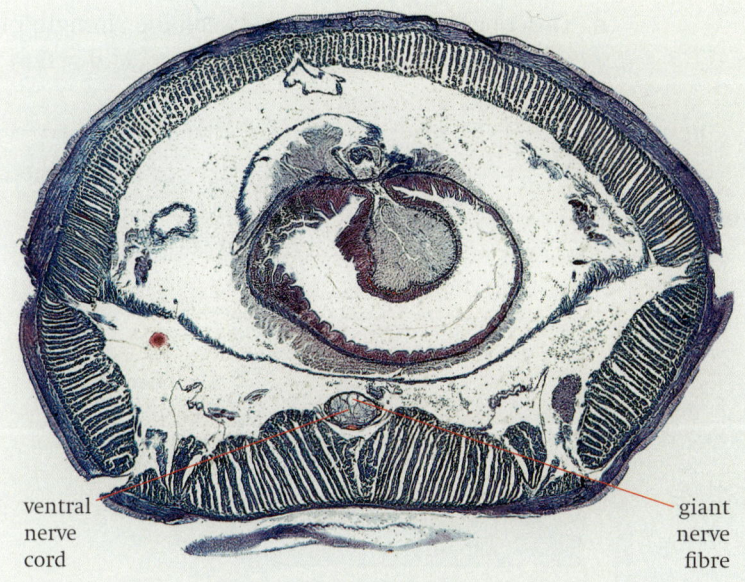

- **Figure 19.38** Transverse section of an earthworm (× 22). The ventral nerve cord contains three giant nerve fibres.

found in sense organs. For example, light receptor cells are found in the eye and sound receptor cells are found in the ear. Some receptors, such as light receptors, are special cells which generate an action potential and send it on to a sensory neurone, while others, such as some kinds of touch receptors, are simply the ends of the sensory neurones themselves. Receptor cells convert energy in one form – such as light, heat or sound – into energy in an electrical impulse in a neurone (*table 19.2*).

One type of receptor found in the dermis of the skin is a **Pacinian corpuscle** (*figure 19.39*). Pacinian corpuscles contain an ending of a sensory neurone, surrounded by several layers of connective tissue, called a **capsule**. The ending of the sensory neurone inside the capsule has no myelin.

When pressure is applied to a Pacinian corpuscle, the capsule is pressed out of shape, and deforms the nerve ending inside it. This deformation causes sodium and potassium channels to open in the cell membrane, allowing sodium ions to flood in and potassium ions to flow out. This depolarises the membrane. The increased positive charge inside the axon is called a **receptor potential**. The harder the pressure applied to the Pacinian corpuscle, the more channels open and the greater the receptor potential becomes. If the pressure is great enough, then the receptor potential becomes large enough to trigger an action potential (*figure 19.40*).

| Receptor | Sense | Form in which energy is received |
|---|---|---|
| rod or cone cells in retina | sight | light |
| taste buds on tongue | taste | chemical potential |
| olfactory cells in nose | smell | chemical potential |
| Pacinian corpuscles in skin | pressure | movement and pressure |
| Meissner's corpuscles in skin | touch | movement and pressure |
| Ruffini's endings in skin | temperature | heat |
| proprioceptors (stretch receptors) in muscles | placement of limbs | mechanical displacement – stretching |
| hair cells in semicircular canals in ear | balance | movement |
| hair cells in cochlea | hearing | sound |

- **Table 19.2** Some examples of energy conversions by receptors. Each type of receptor converts a particular form of energy into electrical energy – that is, a nerve impulse.

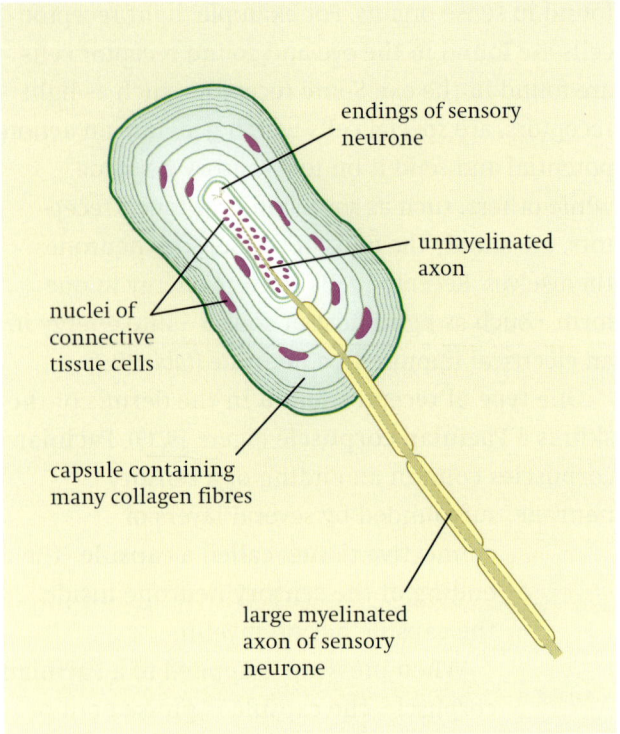

• **Figure 19.39** A Pacinian corpuscle. These corpuscles are found in the dermis, and are sensitive to pressure.

Below a certain threshold, therefore, the pressure stimulus only causes local depolarisation, not an action potential, and therefore no information is transmitted to the brain. Above this threshold, action potentials are initiated. As the pressure increases, the action potentials are produced more frequently.

### SAQ 19.11

Use *figure 19.40* to answer these questions.
**a** What is a receptor potential?
**b** Describe the relationship between the pressure applied to a Pacinian corpuscle and the size of the receptor potential which is generated.
**c** What is the threshold receptor potential?
**d** Describe the relationship between the strength of the stimulus applied and the frequency of action potentials generated.

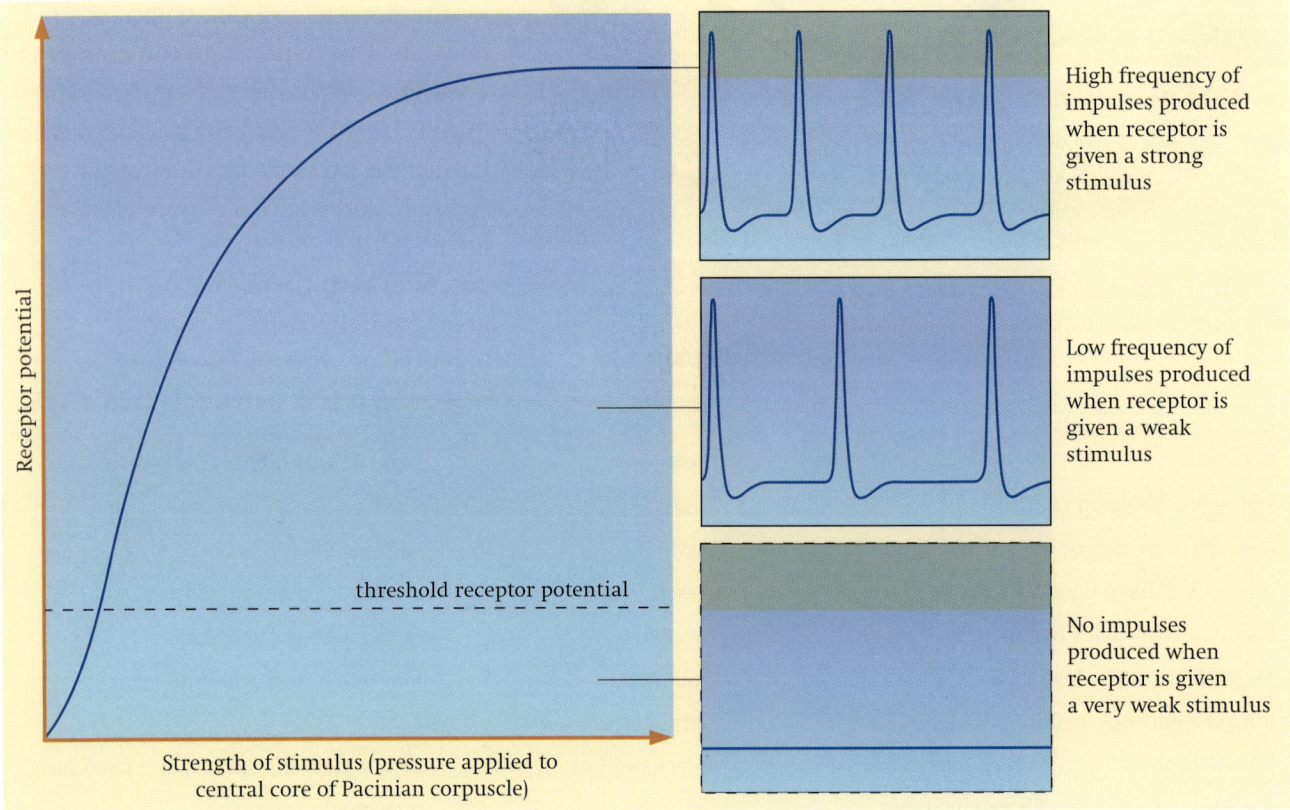

• **Figure 19.40** As pressure is applied to the inner capsule of a Pacinian corpuscle, it produces a depolarisation of the membrane of the sensory nerve ending. This is called the receptor potential. Greater pressures produce greater receptor potentials. If the receptor potential reaches a particular size, called the threshold, then an action potential is triggered.

## Synapses

Where two neurones meet, they do not quite touch. There is a very small gap, about 20 nm wide, between them. This gap is called the **synaptic cleft**. The parts of the two neurones near to the cleft, plus the cleft itself, make up a **synapse** (*figure 19.41*).

### The mechanism of synaptic transmission

Action potentials cannot jump across synapses. Instead, the signal is passed across by a chemical, known as a **transmitter substance**. In outline, an action potential arriving along the plasma membrane of the first, or **presynaptic**, neurone, causes it to release transmitter substance into the cleft. The transmitter substance molecules diffuse across the cleft, which takes less than a millisecond as the distance is so small. This may set up an action potential in the plasma membrane of the second, or **postsynaptic**, neurone.

Let us look at these processes in more detail. The cytoplasm of the presynaptic neurone contains vesicles of transmitter substance (*figure 19.42*). More than 40 different transmitter substances are known; **noradrenaline** and **acetylcholine** are found throughout the nervous system, while

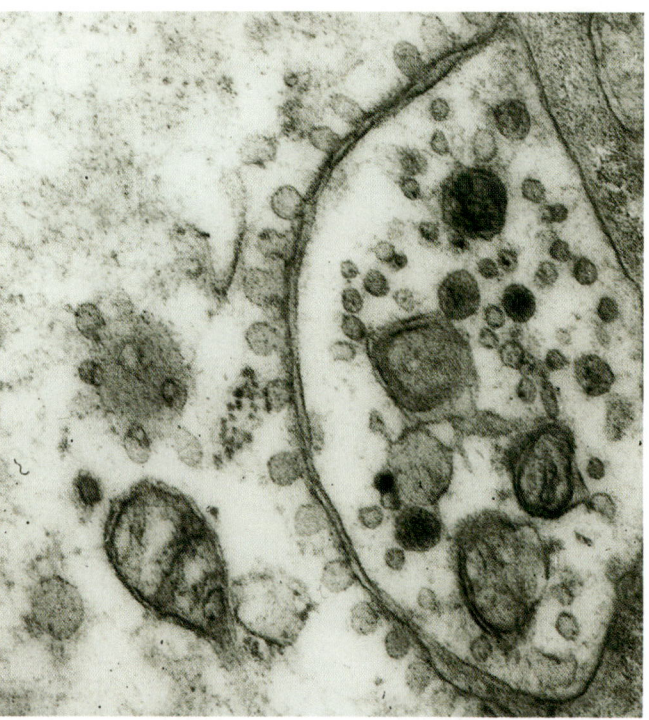

● **Figure 19.42** Electron micrograph of a synapse (×68 000). The presynaptic neurone is to the right of the picture. Several mitochondria and numerous vesicles, which contain transmitter substance, can be seen.

others such as **dopamine** and **glutamic acid** occur only in the brain. We will concentrate on those synapses which use acetylcholine (ACh) as the transmitter substance. These are known as **cholinergic synapses**.

You will remember that, as an action potential sweeps along the plasma membrane of a neurone, local circuits depolarise the next piece of membrane, opening sodium channels and so propagating the action potential. In the part of the membrane of the presynaptic neurone which is next to the synaptic cleft, the arrival of the action potential also causes **calcium channels** to open. Thus, the action potential causes not only sodium ions but also calcium ions to rush in to the cytoplasm of the presynaptic neurone.

This influx of calcium ions causes vesicles of ACh to move to the presynaptic membrane and fuse with it, emptying their contents into the synaptic cleft (*figure 19.43*). Each action potential causes just a few vesicles to do this, and each vesicle contains up to 10 000 molecules of ACh. The ACh diffuses across the synaptic cleft, usually in less than 0.5 ms.

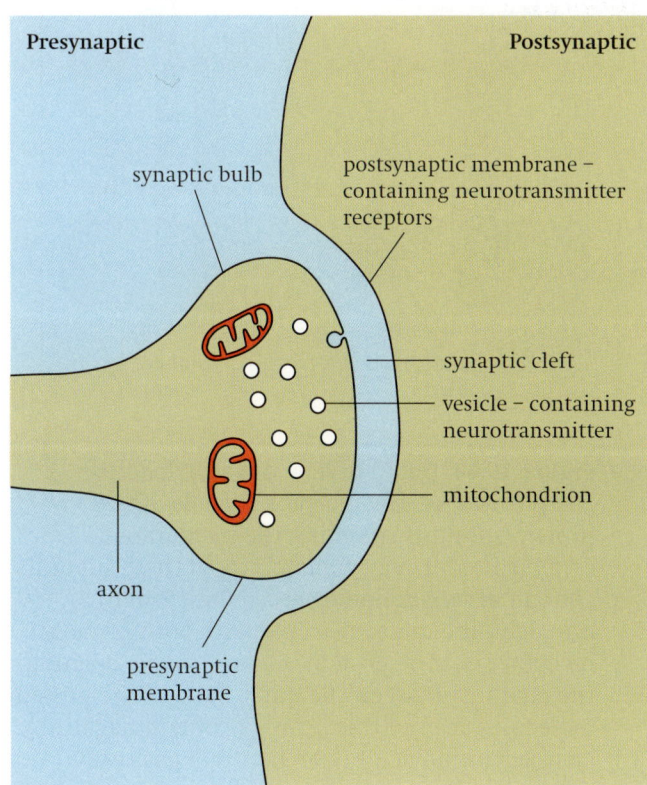

● **Figure 19.41** A synapse.

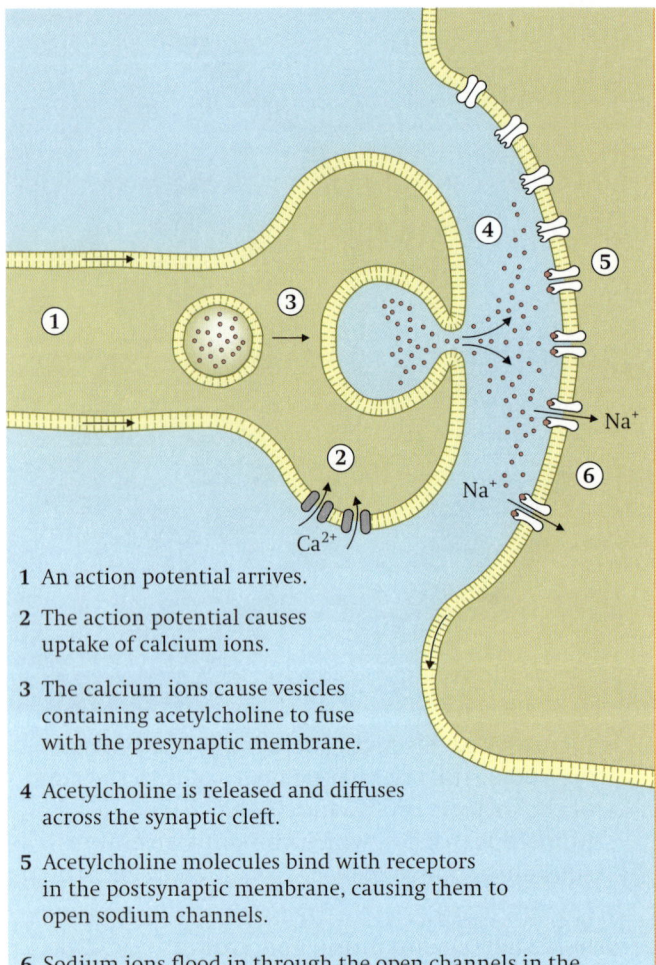

1 An action potential arrives.

2 The action potential causes uptake of calcium ions.

3 The calcium ions cause vesicles containing acetylcholine to fuse with the presynaptic membrane.

4 Acetylcholine is released and diffuses across the synaptic cleft.

5 Acetylcholine molecules bind with receptors in the postsynaptic membrane, causing them to open sodium channels.

6 Sodium ions flood in through the open channels in the postsynaptic membrane. This depolarises the membrane and initiates an action potential.

● **Figure 19.43** Synaptic transmission.

The plasma membrane of the postsynaptic neurone contains **receptor proteins**. Part of the receptor protein molecule has a complementary shape to part of the ACh molecule, so that the ACh molecules can temporarily bind with the receptors. This changes the shape of the protein, opening channels through which sodium ions can pass (*figure 19.44*). Sodium ions rush into the cytoplasm of the postsynaptic neurone, depolarising the membrane and starting off an action potential.

If the ACh remained bound to the postsynaptic receptors, the sodium channels would remain open, and action potentials would fire continuously. To prevent this from happening, and also to avoid wasting the ACh, it is recycled. The synaptic cleft contains an enzyme, **acetylcholinesterase**, which splits each ACh molecule into acetate and choline.

The choline is taken back into the presynaptic neurone, where it is combined with acetyl co-enzyme A to form ACh once more. The ACh is then transported into the presynaptic vesicles, ready for the next action potential. The entire sequence of events, from initial arrival of the action potential to the re-formation of ACh, takes about 5–10 ms.

Much of the research on synapses has been done not at synapses between two neurones, but those between a motor neurone and a muscle.

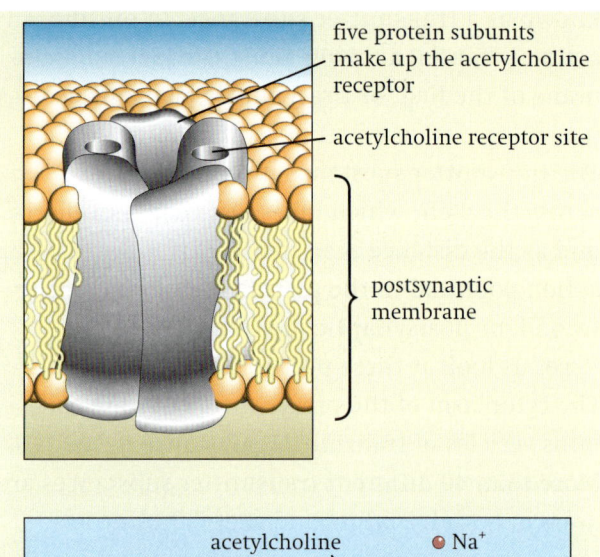

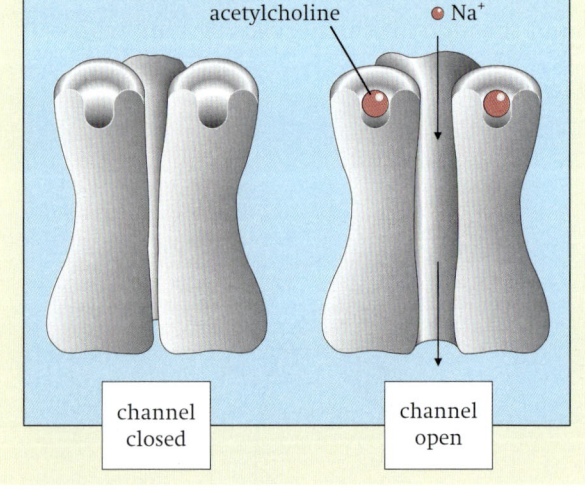

● **Figure 19.44** Detail of how the acetylcholine receptor works. The receptor is made of five protein subunits spanning the membranes arranged to form a cylinder. Two of these subunits contain acetylcholine receptor sites. When acetylcholine molecules bind with both of these receptor sites the proteins change shape, opening the channel between the units. Parts of the protein molecules around this channel contain negatively charged amino acids, which attract positively charged sodium ions and pull them through the channel.

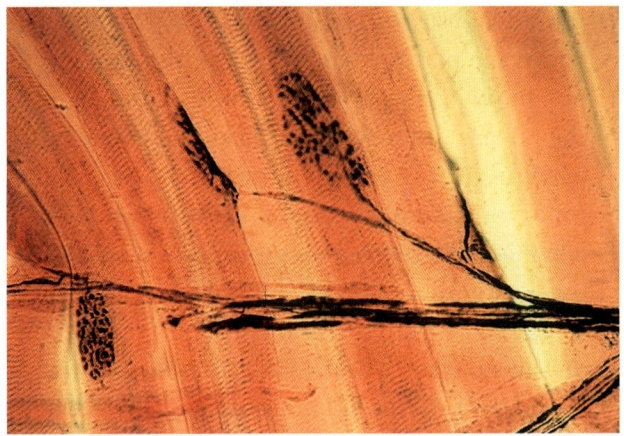

● **Figure 19.45** Light micrograph of neuromuscular junctions (× 320). The red tissue in the background is muscle fibres, while the axons show as dark lines. The axons terminate in a number of branches on the surface of the muscle fibre, forming motor end plates. Action potentials are passed from the axon to the muscle, across a synaptic cleft, at these end-plates.

Here, the nerve forms **motor end plates** and the synapse is called a **neuromuscular junction** (*figure 19.45*). Such synapses function in the same way as described above. An action potential is produced in the muscle, which may cause it to contract (*box 15B on page 201*).

## SAQ 19.12

Suggest why:

**a** impulses travel in only one direction at synapses;

**b** if action potentials arrive repeatedly at a synapse, the synapse eventually becomes unable to transmit the impulse to the next neurone.

### The effects of other chemicals at synapses

Many drugs and other chemicals act by affecting the events at synapses.

You will remember that **nicotine** is one of the main chemicals found in cigarette smoke (see page 151). Part of the nicotine molecule is similar in shape to ACh molecules, and will fit into the ACh receptors on postsynaptic membranes (*figure 19.46*). This produces similar effects to ACh, initiating action potentials in the postsynaptic neurone or muscle fibre. Unlike ACh, however, nicotine is not rapidly broken down by enzymes, and so remains in the receptors for longer than ACh. A large dose of nicotine can be fatal.

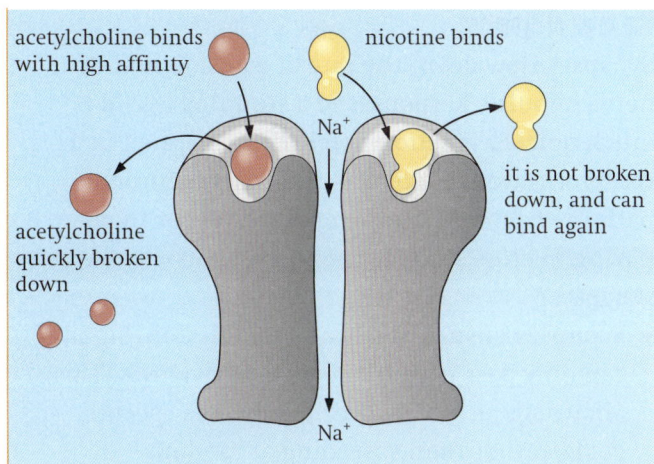

● **Figure 19.46** Nicotine molecules have similarities in shape to acetylcholine molecules, and will fit into some acetylcholine receptor sites, causing sodium channels to open. So nicotine can generate action potentials in postsynaptic neurones. Not all acetylcholine receptors are equally responsive to nicotine; those at neuromuscular junctions have only a low affinity for nicotine.

The **botulinum toxin** is produced by an anaerobic bacterium which occasionally breeds in contaminated canned food. It acts at the pre-synaptic membrane where it prevents the release of ACh. Eating food that contains this bacterium is frequently fatal. However, the toxin does have important medical uses. In some people, for example, the muscles of the eyelids contract permanently, so that they cannot open their eyes. Injections of tiny amounts of the botulinum toxin into these muscles can cause them to relax, so allowing the lids to be raised.

**Organophosphorous insecticides** inhibit the action of acetylcholinesterase, thus allowing ACh to cause continuous production of action potentials in the postsynaptic membrane. Many flea sprays and collars for cats and dogs contain organophosphorous insecticides, so great care should be taken when using them, for the health of both the pet and the owner. Contamination from organophosphorous sheep-dip (used to combat infestations by ticks) has been linked to certain illnesses in farm workers. Several **nerve gases** also act in this way. People involved in the production of these chemicals have regular checks of their free acetylcholinesterase levels to monitor their safety.

## The roles of synapses

Synapses slow down the rate of transmission of a nerve impulse. Responses to a stimulus would be much quicker if action potentials generated in a receptor travelled along an unbroken neuronal pathway from receptor to effector, rather than having to cross synapses on the way. So why have synapses?

■ **Synapses ensure one-way transmission**. Signals can only pass in one direction at synapses. This allows signals to be directed towards specific goals, rather than spreading at random through the nervous system.

■ **Synapses increase the possible range of actions in response to a stimulus**. Synapses allow a wider range of behaviour than could be generated in a nervous system in which neurones were directly 'wired up' to each other. They do this by allowing the interconnection of many nerve pathways. Think for a moment of your possible behaviour when you see someone you know across the street. You can call out to them and walk to meet them, or you can pretend not to see them and hurry away. What decides which of these two responses, or any number of others, you will make?

Your nervous system will receive information from various sources about the situation. Receptors in your eyes will provide details about who the person is, and whether they have seen you or not. Stored away in your brain will be memories about the person: are they a good friend?..is there something you want to talk about with them?..are they boring? Your brain will also have other information to consider: are you in a hurry?..have you time to spend here? All of these pieces of information will produce action potentials in many neurones in your nervous system. As a result of this, action potentials may or may not be sent to the muscles of your legs to make them turn and carry you across the street.

The way in which the 'turn' and 'not turn' decision is reached in your nervous system depends on what happens at synapses. Each neurone within the brain has many, often several thousand, synapses with other neurones. Action potentials arriving at some of these synapses will **stimulate** an action potential in the neurone, as described on page 284. Action potentials arriving at others will cause the release of transmitter substances which, far from producing an action potential in the neurone, will actually make it *more* difficult to depolarise its plasma membrane, and so **inhibit** the production of an action potential. Whether or not an action potential is produced depends on the summed effect of the number and frequency of action potentials arriving at all the stimulatory and inhibitory synapses on that particular neurone. In very simple terms, if the action potentials carrying 'good friend' information outweigh those carrying 'in a hurry' information, you will probably decide to turn towards the person even if it may make you late for your appointment.

The loss of *speed* in this response is more than compensated for in the possible *variety* of responses which can be made. We *do* have, however, some very rapid and very stereotyped responses. These are called **reflex actions**, and involve quick, automatic responses to stimuli. Two examples are blinking when an object speeds towards your eye, or jumping when you hear an unexpected noise. In a reflex action, there are normally only two or three neurones involved: a sensory neurone and a motor neurone, with perhaps an intermediate neurone in between (*figure 19.28*). These actions are ones where the survival value of a very rapid response is greater than the value of a carefully considered one.

■ **Synapses are involved in memory and learning**. Despite much research, little is yet known about how memory operates. However, there is much evidence that it involves synapses. For example, if your brain frequently receives information about two things at the same time, say a sound of a particular voice and a sight of a particular face, then it is thought that *new* synapses form in your brain that link the neurones involved in the passing of information along the particular pathways from your ears and eyes. In future, when you hear the voice, information flowing from your ears along this pathway automatically flows into the other pathway too, so that your brain 'pictures' the face which goes with the voice.

# Plant growth regulators

Plants, like animals, have communication systems that allow coordination between different parts of their bodies. In at least some species, electrical signals rather like action potentials can be detected. For example, in the 'sensitive plant' *Mimosa*, which responds to touch by folding up its leaves, an electrical signal similar to an action potential can be detected, though this travels much more slowly. In the Venus fly-trap (*figure 19.47*), action potentials are set up when an insect touches the sensory hairs in the middle of the trap, and pass into the leaf tissue, causing the trap to close.

However, most communication within plants depends on chemicals. These are known as **plant hormones** or **plant growth regulators**.

Unlike animal hormones, plant growth regulators are not produced in endocrine glands, but in a variety of tissues. They are usually produced in such small quantities that it has proved very difficult to discover exactly where some of them are made. They move in the plant either directly from cell to cell (by diffusion or active transport) or carried in the phloem sap or xylem vessels. Some may not move at all from their site of synthesis.

● **Figure 19.47** The leaves of the Venus fly trap, *Dionaea muscipula*, have a group of stiff, sensitive hairs in their centres. When these are touched, the leaves respond by closing, trapping whatever was crawling over them. Digestive juices are then secreted, and the soluble products absorbed into the leaf cells.

Because they are usually found in only very, very low concentrations it is difficult to determine precisely what some of them do. Moreover, some of them seem to have very different effects when they are present in a relatively low concentration than when they are in a relatively high concentration. They can have different effects in different tissues, in different species, or at different stages of a plant's development. Add to this the fact that two or more plant growth regulators acting together can have very different effects from either of them acting alone, and you can see how difficult it is for plant physiologists to discover exactly what these substances do, where they do it and how.

Here we will look at just one or two roles of three plant growth substances – auxins, gibberellins and abscisic acid.

## Auxins and apical dominance

Plants make several chemicals known as **auxins**, of which the principal one is **IAA** (indole 3-acetic acid, *figure 19.48*). Here, we will refer to this simply as 'auxin' in the singular. Auxin is synthesised in the growing tips of roots and shoots, where the cells are dividing. It is transported from here back down the shoot, or up the root, by active transport from cell to cell, and also to a lesser extent in phloem sap.

Auxin seems to be involved in determining whether a plant grows upwards or whether it branches sideways. When a plant has an active growing point at its apex, this tends to stop buds on the side of the stem, called **lateral buds**, from growing. The plant grows upwards rather than branching out sideways. However, if the bud at the tip of the main shoot – the apical bud – is cut off, then the lateral buds start to grow. Clearly, the presence of the **apical bud** is stopping the lateral buds from growing. This is called **apical dominance**.

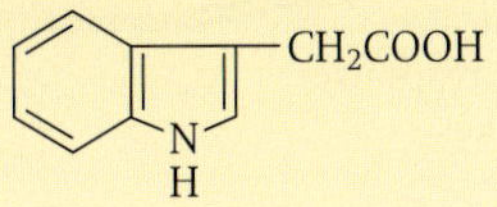

$$\text{indole ring} - CH_2COOH$$

● **Figure 19.48** The molecular structure of indole 3-acetic acid, IAA.

Auxin synthesised in the apical bud is transported down the stem to the lateral buds. One theory to explain apical dominance is that auxin is present in the lateral buds in a concentration that inhibits their growth. Removal of the apical bud causes the concentration of auxin in the lateral buds to drop, so that they can now grow. But the experimental evidence for this is contradictory and uncertain. At the moment, it is not understood how this effect occurs, or exactly what role auxin has in it. It seems likely that other plant growth substances, such as **cytokinins** and **abscisic acid** are also involved (see later).

## Gibberellins and stem elongation

Gibberellins are plant growth regulators that are synthesised in most parts of plants. They are present in especially high concentrations in young leaves and in seeds, and are also found in stems, where they have an important role in determining their growth.

The height of some plants is partly controlled by their genes. For example, tallness in peas is affected by a gene with two alleles; if the dominant allele is present, the plants can grow tall, but plants homozygous for the recessive allele always remain short. The dominant allele of this gene regulates the synthesis of an enzyme that catalyses the synthesis of an active form of gibberellin, $GA_1$. If only the recessive allele is present, then the plant contains only inactive forms of gibberellin. Active gibberellin stimulates cell division and cell elongation in the stem, so causing the plant to grow tall.

Applying active gibberellin to plants which would normally remain short, such as cabbages, can stimulate them to grow tall. As yet, little is known about how gibberellins cause these effects.

## Gibberellins and seed germination

In some seeds, gibberellins are involved in the control of germination. *Figure 19.49* shows the structure of a barley seed. When the seed is shed from the parent plant, it is in a state of **dormancy**; that is it contains very little water and is metabolically inactive. This is useful, as it allows the seed to survive in adverse conditions, such as through a cold winter, only germinating when the temperature rises in spring.

The seed contains an **embryo**, which will grow to form the new plant when the seed germinates. The embryo is surrounded by **endosperm tissue** which is a food store, containing the polysaccharide starch. On the outer edge of the endosperm is a protein-rich **aleurone layer**. The whole seed is covered by a tough, waterproof, protective layer.

When the seed absorbs water, this stimulates the production of gibberellin by the embryo, and the gibberellin in turn stimulates the synthesis of **amylase** by the cells in the aleurone layer. The amylase hydrolyses the starch molecules in the endosperm, converting them to soluble maltose molecules. These are converted to glucose and are transported to the embryo, providing a source of carbohydrate that can be respired to provide energy as the embryo begins to grow.

Gibberellin causes these effects by regulating genes that are involved in the synthesis of amylase. In barley seeds, it has

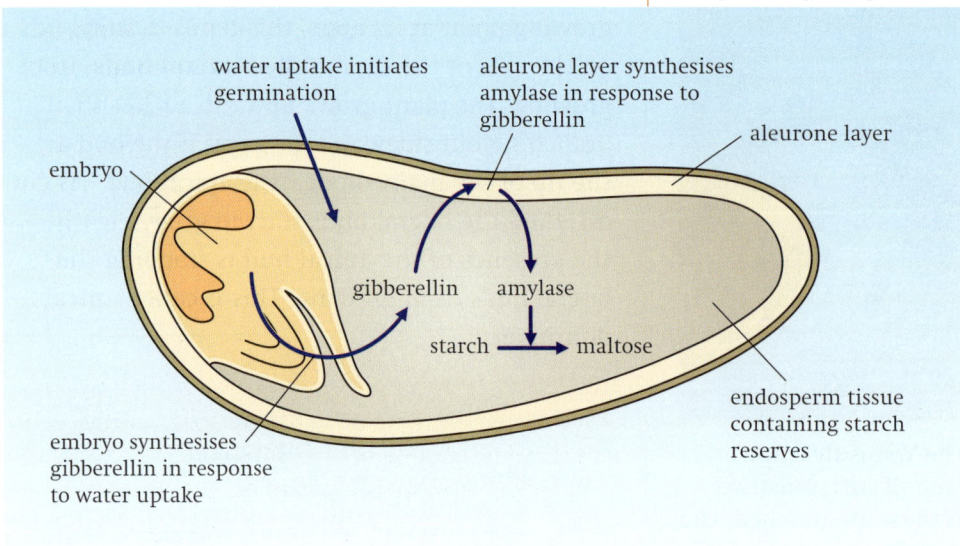

water uptake initiates germination

aleurone layer synthesises amylase in response to gibberellin

aleurone layer

embryo

gibberellin   amylase

starch ⟶ maltose

endosperm tissue containing starch reserves

embryo synthesises gibberellin in response to water uptake

● **Figure 19.49** Longitudinal section through a barley seed, showing how secretion of gibberellins by the embryo results in the mobilisation of starch reserves during germination.

been shown that application of gibberellin causes an increase in the transcription of mRNA coding for amylase.

## Abscisic acid and stomatal closure

Abscisic acid, otherwise known as **ABA**, has been found in a very wide variety of plants, including ferns and mosses as well as flowering plants. It can be found in every part of the plant, and is synthesised in almost all cells that possess chloroplasts or amyloplasts (organelles like chloroplasts, but that contain large starch grains and no chlorophyll).

One role of ABA that is well documented is as a so-called **stress hormone**. If a plant is subjected to difficult environmental conditions, such as very high temperatures, or very reduced water supplies, then it responds by secreting ABA. In a plant in drought conditions, the concentration of ABA in the leaves can rise to 40 times that which would normally be present. This high concentration of ABA causes the stomata to close, which reduces loss of water vapour from the leaf.

In chapter 16, we saw how the guard cells control the opening and closure of stomata (*figure 16.9*). Each guard cell has a relatively thick cell wall on the side next to the stoma (the opening between them), and a relatively thin wall on the opposite side. When the guard cells become turgid, they expand into a curved shape, because the inner, thick cell wall resists expansion more than the outer wall. This leaves a space between them – the open stoma. When the guard cells lose water, they become flaccid and collapse together so that the stoma is closed.

The increase in turgor of the guard cells is brought about by the activities of transporter proteins in their plasma membranes. An ATP-powered 'proton pump' in the membrane actively transports hydrogen ions, $H^+$, out of the guard cells. The lowering of hydrogen ion concentration inside the cells causes potassium channels to open in the plasma membrane, and potassium ions, $K^+$, move into the cell. They do this because the removal of $H^+$ ions has left the inside of the cell negatively charged compared with the outside, and as the $K^+$ ions have a positive charge, they are drawn down an electrical gradient towards the negatively charged region.

The extra $K^+$ ions inside the guard cells lower the solute potential, and therefore the water potential. Now there is a water potential gradient between the outside and the inside of the cell, so water moves in by osmosis. This increases the turgor of the guard cells, and the stoma opens.

It is not known exactly how ABA achieves the closure of stomata, but the fact that the response is very fast indicates that, unlike the effect of gibberellins in seeds, it is not done by regulating the expression of genes. If ABA is applied to a leaf, the stomata close within just a few minutes. It seems that guard cells have ABA receptors on their plasma membranes, and it is possible that when ABA binds with these it inhibits the proton pump. This would stop the hydrogen ions being pumped out, so potassium ions and water would not enter, and the guard cells would become flaccid and close the stomata.

## Leaf abscission

Abscisic acid takes its name from the fact that it was thought to be closely involved in leaf or fruit fall, which is known as **abscission**.

Some trees regularly drop their leaves at certain times in year. In Britain, for example, as in many other temperate countries, deciduous trees such as oak and ash drop their leaves in autumn, as the days grow shorter and cooler.

The leaves fall because the leaf stalk, or petiole, breaks off from the stem (*figure 19.50*). First, useful substances are withdrawn from the leaves and taken back into the stem; this involves the breakdown of some of the pigments in the leaves, changing their green colour to yellow, golds and reds. An **abscission zone** forms where the petiole meets the stem, made up of two layers of cells. Nearest to the leaf is the **separation layer**, which is made of small cells with quite thin cell walls. Nearest to the stem is the **protective layer**, made up of cells whose walls contain **suberin**. (Suberin is a waxy, waterproof substance, also found in the cell walls of cork cells in tree bark.)

Enzymes then break down the cell walls in the separation layer, and the petiole breaks at this point. The protective layer remains, forming a 'scar' on the stem where the leaf used to be. These leaf scars can sometimes be very visible. For example, they form the characteristic horseshoe shapes on twigs of horse chestnut trees.

We still do not know exactly what controls leaf abscission, but it does now seem that abscisic acid has very little to do with it! Abscisic acid does appear to be involved in the senescence (aging) of leaves, but not directly in their falling from the plant. Auxin is a much stronger candidate for this role. Abscission is usually accompanied by a drop in auxin concentration in the leaf, and in many instances abscission can be prevented by applying auxin in the early stages of the process. It is used for this purpose in citrus orchards, where auxin is sprayed onto the trees to prevent the fruit from falling until it can be harvested. Confusingly, however, high concentrations of auxin, applied later, can actually *promote* fruit drop! It is sometimes sprayed in this way onto olive or apple trees to thin the fruit crop, if it looks as though there will be too many small fruits instead of fewer, larger ones.

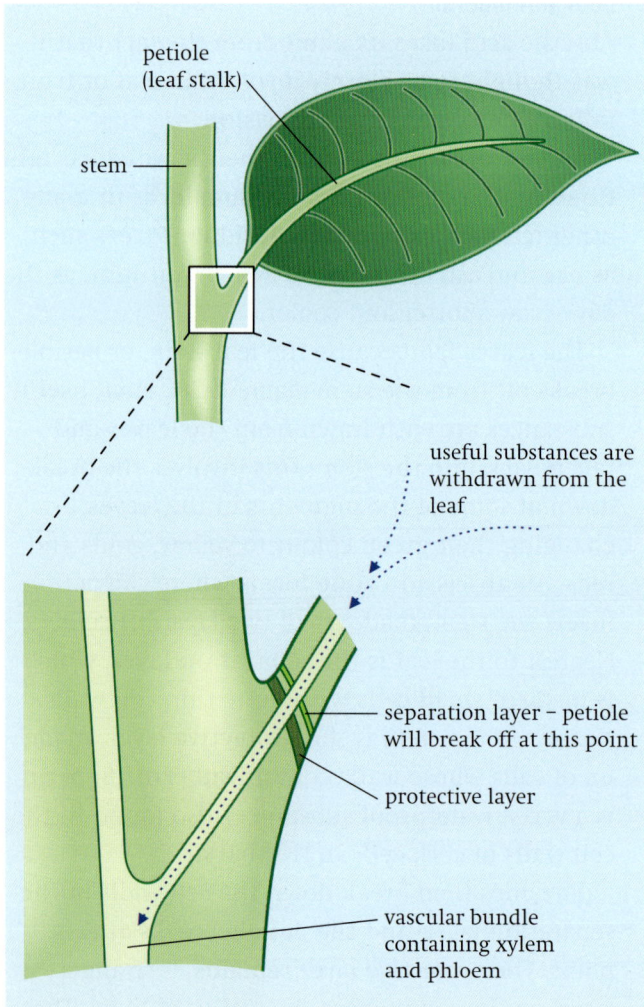

● **Figure 19.50** Leaf fall.

# SUMMARY

♦ Animals and plants have internal communication systems that allow information to pass between different parts of their bodies, and so help them to respond to changes in their external or internal environments.

♦ Mammals keep their internal environment relatively constant, so providing steady and appropriate conditions within which cells can carry out their activities. This is known as homeostasis.

♦ Information transfer and control systems involve receptors and effectors. Regulatory control systems, such as those involved in homeostasis, also involve negative feedback.

♦ Toxic waste products of metabolism, especially carbon dioxide and urea, are removed from the body by the process of excretion. Urea is the main nitrogenous excretory product, formed by the deamination of excess amino acids in the liver. Urea is excreted in solution in water, as urine.

♦ The kidneys regulate the concentration of various substances in the body fluids, by excreting appropriate amounts of them. Each kidney is made up of thousands of nephrons and their associated blood vessels. The kidneys produce urine by ultrafiltration and reabsorption, plus some secretion of unwanted substances. Different regions of a nephron have different functions, and this is reflected in the structure of their walls.

♦ Blood is brought to the glomerulus in the cup of the renal capsule of the nephron in an afferent arteriole. High hydrostatic pressure in the glomerulus forces substances through the capillary walls, the basement membrane and the wall of the renal capsule into the nephron. The basement membrane acts as a filter, allowing only small molecules through. Most reabsorption occurs in the proximal convoluted tubule, by diffusion and active transport, and also in the distal convoluted tubule and collecting duct. The loop of Henle acts as a counter-current multiplier, producing high concentrations of salt in the medulla which can draw out water from the collecting duct and produce a concentrated urine.

♦ The water content of the blood is controlled by changing the amount of water excreted by the urine. This is done by regulating the permeability of the walls of the collecting ducts to water, and hence the amount of water reabsorbed from the collecting ducts into the blood. The permeability is altered by the hormone ADH, which is secreted by the posterior pituitary gland in response to stimulation of osmoreceptors in the hypothalamus.

♦ Hormones are chemicals that are made in endocrine glands and transported in blood plasma to their target cells, where they bind to specific receptors and so affect the behaviour of the cells.

♦ Blood glucose levels are controlled by the action of insulin and glucagon, which are secreted by the islets of Langerhans in the pancreas and affect liver and muscle cells. Negative feedback keeps the blood glucose level from varying too much from the norm.

♦ Neurones are cells adapted for the rapid transmission of electrical signals. Sensory neurones transmit signals from receptors to the central nervous system (brain and spinal cord); motor neurones transmit signals from the central nervous system to effectors; intermediate neurones transmit signals within the central nervous system. In vertebrates, the axons of many neurones are insulated by a myelin sheath which speeds up transmission.

◆ Signals are transmitted as action potentials. A resting neurone has a negative potential inside compared with outside. An action potential is a fleeting reversal of this potential, caused by changes in permeability of the plasma membrane to potassium and sodium ions. Action potentials are always the same size. Information about the strength of a stimulus is given by the frequency of action potentials produced.

◆ Action potentials may be initiated within the brain or at a receptor. Receptors respond to information from the environment. Environmental changes result in permeability changes in the membranes of receptor cells, which in turn produce changes in potential difference across the membrane. If sufficiently great, this will trigger an action potential.

◆ Neurones do not make direct contact with one another, but are separated by a very small gap called a synaptic cleft. Impulses pass across this gap as bursts of transmitter substance, released by the presynaptic neurone when an action potential arrives.

Any one neurone within the central nervous system is likely to have at least several hundred synapses with other neurones, some of which will be stimulatory and some inhibitory. This allows integration within the nervous system, resulting in complex and variable patterns of behaviour, and in learning and memory.

◆ Plants produce several chemicals known as plant growth substances, that are involved in the control of growth and responses to environmental changes. Auxin is synthesised mainly in growing tips of shoots and roots, and appears to be involved in preventing the growth of lateral buds when an intact and active apical bud is present. It is also involved in leaf abscission. Gibberellin is synthesised in young leaves and in seeds. It stimulates growth of stems and germination of seeds. Abscisic acid is synthesised by any cells in a plant that contain chloroplasts or amyloplasts, especially in stress conditions. The presence of large concentrations of abscisic acid in leaves causes stomata to close.

# Amino acid R groups

The general formula for an amino acid is shown in *figure 2.15*. In the list below, only the R groups are shown; the rest of the amino acid molecule is represented by a block.

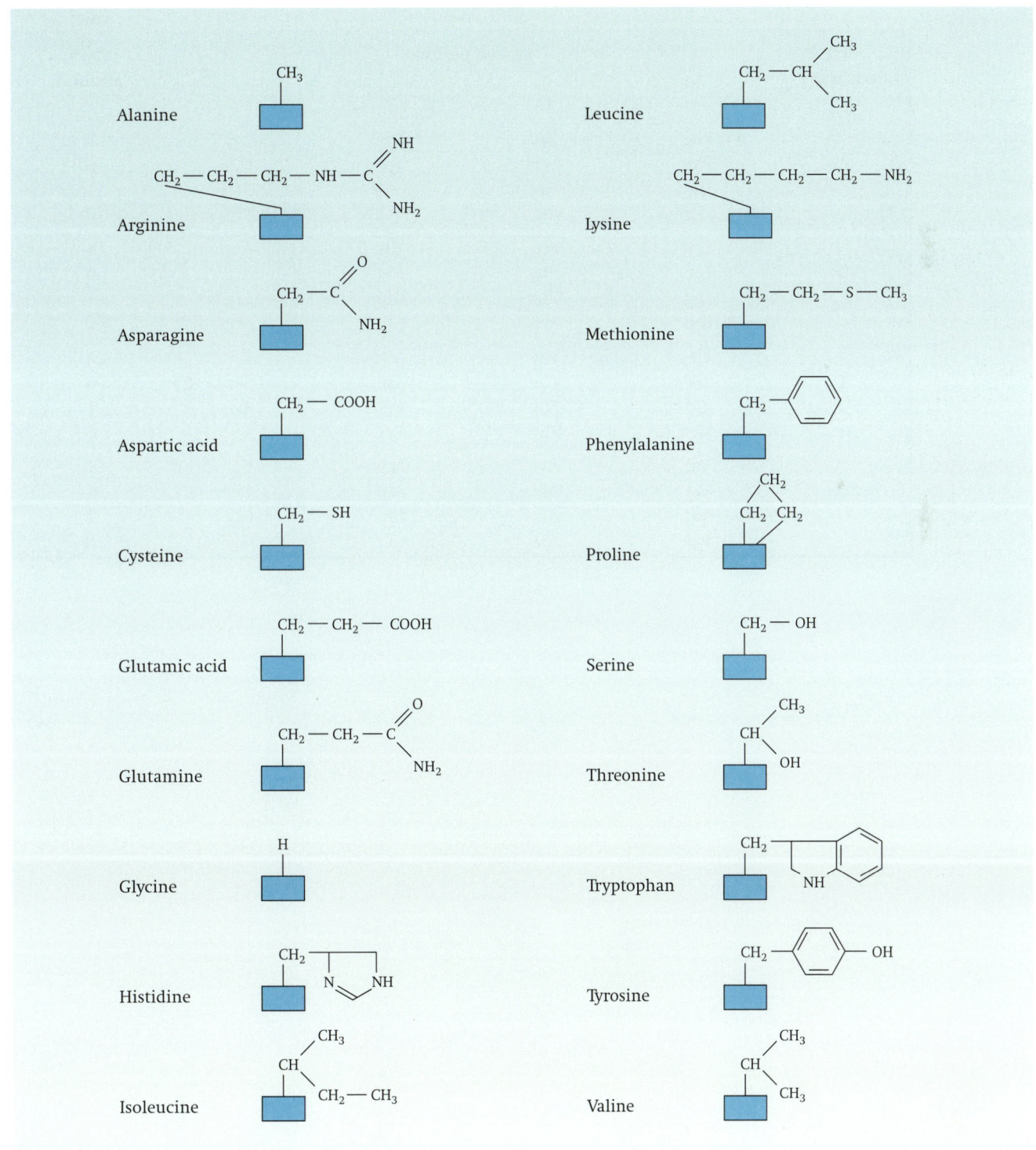

# DNA triplet codes

The table shows all the possible triplets of bases in a DNA molecule and what each codes for. The 3-letter abbreviations for each amino acid are, in most cases, the first three letters of their full name – see Appendix 1.

| first position | Second position | | | | Third position |
|---|---|---|---|---|---|
| | A | G | T | C | |
| A | Phe | Ser | Tyr | Cys | A |
| | Phe | Ser | Tyr | Cys | G |
| | Leu | Ser | STOP | STOP | T |
| | Leu | Ser | STOP | Trp | C |
| G | Leu | Pro | His | Arg | A |
| | Leu | Pro | His | Arg | G |
| | Leu | Pro | Gln | Arg | T |
| | Leu | Pro | Gln | Arg | C |
| T | Ile | Thr | Asn | Ser | A |
| | Ile | Thr | Asn | Ser | G |
| | Ile | Thr | Lys | Arg | T |
| | Met | Thr | Lys | Arg | C |
| C | Val | Ala | Asp | Gly | A |
| | Val | Ala | Asp | Gly | G |
| | Val | Ala | Glu | Gly | T |
| | Val | Ala | Glu | Gly | C |

# t-test

Questions involving an understanding of the use of t- and chi-squared tests may be set in the A2 practical paper (Paper 5). The formulae for the tests will always be provided.

You will find an explanation of the use of the chi-squared test on page 233. An example of the use of the t-test follows below.

## The t-test

The **t-test** allows you to see whether the means of two sets of data differ significantly. Here is a hypothetical example to illustrate how it works. Suppose you count the number of species of lichen found in eight urban churchyards and six rural churchyards within 10 km of your school or college. You might collect data, as follows:

> number of lichen species in urban churchyards:
> 6, 4, 6, 5, 3, 7, 5, 3
> number of lichen species in rural churchyards:
> 6, 6, 8, 5, 9, 7

It looks as though there may be more species of lichen in the rural churchyards but it is difficult to be sure of this simply by looking at the data. How can we investigate this statistically? The first thing to do is to calculate the **means**.

### Calculating means

You will have calculated means before. All you have to do is to add up the individual values and divide by the total number of measurements. So, in this case, the mean number of lichen species in the urban churchyards is:

$(6 + 4 + 6 + 5 + 3 + 7 + 5 + 3) \div 8 = 4.875$

The general formula is:

$\bar{x} = \Sigma x / n$

where:

> $\bar{x}$ is the mean
> $\Sigma$ stands for 'sum of'
> $x$ refers to the individual values in the sample
> $n$ is the total number of individual values in the sample.

Repeating this exercise with the rural lichens, we find that their mean number is 6.833. The next thing we need to do is to calculate **standard deviations**. The standard deviation is a measure of the extent to which individual measurements *vary* around the mean. The greater the variation among the individual measurements, the bigger the standard deviation; the less the variation among the individual measurements, the smaller the standard deviation.

It helps to have a calculator that works out standard deviations. If you *don't* have such a calculator, read the next section, entitled 'Calculating standard deviations'. If you *do* have a calculator that works out standard deviations, you may want to skip this section and go to the section entitled 'Using a calculator to obtain standard deviations'.

### Calculating standard deviations

The standard deviation, $s_x$, is given by the following formula:

$$s_x = \sqrt{\frac{\Sigma x^2 - \frac{(\Sigma x)^2}{n}}{n - 1}}$$

where:

> $\Sigma$ stands for 'sum of'
> $x$ refers to the individual values in the sample
> $n$ is the total number of individual values in the sample.

In the case of our eight urban lichens, the eight individual values, i.e. values of x, are: 6, 4, 6, 5, 3, 7, 5 and 3. The sum of these eight values, i.e. $\Sigma x$, equals 39. (You can forget about any units while actually doing the calculations provided you put them in at the end – the mean and the standard deviation have the same units as the individual values of x.) Using these values of x and $\Sigma x$ in the above formula, we have:

$$s_x = \sqrt{\dfrac{(6^2 + 4^2 + 6^2 + 5^2 + 3^2 + 7^2 + 5^2 + 3^2) - \dfrac{(39)^2}{8}}{7}}$$

$$= \sqrt{\dfrac{205 - 190.125}{7}}$$

$$= 1.458$$

### Using a calculator to obtain standard deviations

You may need to have your calculator in standard deviation mode, and type the data into a memory. Then there will be a key, often labelled x. This key gives you the mean of the data. Another key, often labelled s or $\sigma$, gives you the standard deviation. If you have the key $\sigma_{n-1}$ or $s_{n-1}$ use it in preference to $\sigma_n$ or $s_n$ for working out the standard deviation.

### Using standard deviations to calculate the significance of the difference between two means

If you have a decidedly up-market calculator (or access to a statistical computer package) you may have a key labelled t. This stands for **t-test** because that is the name of the test we are going to carry out. If you do have such a key, use it as instructed by the calculator or statistical package and go to step 8 below. If you don't have a key labelled t, proceed as follows:

1 Work out the means of the two sets of data.
2 Subtract the smaller mean from the larger one.
3 Work out the standard deviation of one set of data. Multiply this number by itself (i.e. square it) and divide it by the number of pieces of data in that set of data.
4 Work out the standard deviation of the other set of data. Multiply this number by itself (i.e. square it) and divide it by the number of pieces of data in that set of data.

5 Add together the figures you calculated in steps 3 and 4.
6 Take the square root of the figure calculated in step 5.
7 Divide the difference between the two means (step 2) by the figure calculated in step 6. This is your t value.
8 Now use the table opposite to see whether your value of t could be expected by chance. For a t-test, the degrees of freedom are simply two less than the total number of individual measurements in the two samples.

- If your value of t is bigger than the critical value – the one that corresponds to a 5% probability that chance could have produced it – in the table you can be at least 95% confident that the difference between the means is significant. Your result is said to be statistically significant and you can reject the null hypothesis that there is no difference between the means.

- If your value of t is smaller than the critical value in the table you are less than 95% confident that the difference between the means is significant. Your result is not statistically significant and you cannot reject the null hypothesis that there is no difference between the means.

To help make this easier to understand here's a worked example using our data on the number of lichen species in rural and urban churchyards.

1 Mean number of urban lichen species = 4.875; mean number of rural lichen species = 6.833.
2 Difference between the means = 1.958.
3 Standard deviation of the number of urban lichens multiplied by itself divided by the number of pieces of data in that set of data = 1.458 × 1.458 ÷ 8 = 0.266.
4 Standard deviation of the number of rural lichens multiplied by itself divided by the number of pieces of data in that set of data = 1.472 × 1.472 ÷ 6 = 0.361.
5 The sum of the figures calculated in steps 3 and 4 = 0.266 + 0.361 = 0.627.
6 The square root of the figure calculated in step 5 = 0.792.
7 The difference between the two means (step 2) divided by the figure calculated in step 6 = 1.958 ÷ 0.792 = 2.47.

| Degrees of freedom | Value of t | | | |
|---|---|---|---|---|
| 1 | 6.31 | 12.7 | 63.7 | 636 |
| 2 | 2.92 | 4.30 | 9.93 | 31.6 |
| 3 | 2.35 | 3.18 | 5.84 | 12.9 |
| 4 | 2.13 | 2.78 | 4.60 | 8.61 |
| 5 | 2.02 | 2.57 | 4.03 | 6.87 |
| 6 | 1.94 | 2.45 | 3.71 | 5.96 |
| 7 | 1.90 | 2.37 | 3.50 | 5.41 |
| 8 | 1.86 | 2.31 | 3.36 | 5.04 |
| 9 | 1.83 | 2.26 | 3.25 | 4.78 |
| 10 | 1.81 | 2.23 | 3.17 | 4.59 |
| 11 | 1.80 | 2.20 | 3.11 | 4.44 |
| 12 | 1.78 | 2.18 | 3.06 | 4.32 |
| 13 | 1.77 | 2.16 | 3.01 | 4.22 |
| 14 | 1.76 | 2.15 | 2.98 | 4.14 |
| 15 | 1.75 | 2.13 | 2.95 | 4.07 |
| 16 | 1.75 | 2.12 | 2.92 | 4.02 |
| 17 | 1.74 | 2.11 | 2.90 | 3.97 |
| 18 | 1.73 | 2.10 | 2.88 | 3.92 |
| 19 | 1.73 | 2.09 | 2.86 | 3.88 |
| 20 | 1.73 | 2.09 | 2.85 | 3.85 |
| 22 | 1.72 | 2.07 | 2.82 | 3.79 |
| 24 | 1.71 | 2.06 | 2.80 | 3.75 |
| 26 | 1.71 | 2.06 | 2.78 | 3.71 |
| 28 | 1.70 | 2.05 | 2.76 | 3.67 |
| 30 | 1.70 | 2.04 | 2.75 | 3.65 |
| 40 | 1.68 | 2.02 | 2.70 | 3.55 |
| 50 | 1.68 | 2.01 | 2.70 | 3.52 |
| 60 | 1.67 | 2.00 | 2.66 | 3.46 |
| 70 | 1.67 | 1.99 | 2.65 | 3.44 |
| 80 | 1.66 | 1.99 | 2.64 | 3.42 |
| 90 | 1.66 | 1.99 | 2.63 | 3.40 |
| 100 | 1.66 | 1.99 | 2.63 | 3.39 |
| Probability that chance could have produced this value of t | 0.10 | 0.05 | 0.01 | 0.001 |
| Confidence level | 10% | 5% | 1% | 0.1% |

● Table of t values.

8 It is clear that 2.47 is greater than the critical value of t, which for a total of 12 degrees of freedom equals 2.18. This means that we are at least 95% confident that the mean number of lichens differs between urban and rural church-yards. Note that our value of t would have had to have been equal to at least 3.06 for us to have been 99% confident of this conclusion.

We can sum up the way to calculate the value of t by these steps as follows:

$$t = \frac{\bar{x} - \bar{y}}{\sqrt{\frac{(s_x)^2}{n_x} + \frac{(s_y)^2}{n_y}}}$$

where:

$\bar{x}$ equals the mean of sample X
$\bar{y}$ equals the mean of sample Y
$s_x$ is the standard deviation of sample X
$s_y$ is the standard deviation of sample Y
$n_x$ is the number of individual measurements in sample X
$n_y$ is the number of individual measurements in sample Y.

The degrees of freedom are equal to $n_x + n_y - 2$.

One final point. The larger your sample sizes, the more likely you are to detect a significant difference – if it exists. You normally need a *minimum* of six individual measurements in each sample.

# Answers to self-assessment questions

## AS Level

### Chapter 1

**1.1** Structures found in both animal and plant cells: nucleus with nucleolus and chromatin; cytoplasm containing mitochondria, Golgi apparatus and other small structures; plasma membrane. Structure found only in animal cells: centriole. Structures found only in plant cells: chloroplasts, large central vacuole, cell wall with middle lamella and plasmodesmata.

**1.2**  **a** Actual diameter = 20 μm (see caption)
Diameter on diagram = 60 mm = 60 000 μm
Magnification = size of image ÷ size of specimen = 60 000 ÷ 20
Therefore magnification = × 3000

**b** Magnification = × 20 000 (see caption)
Length on micrograph = 65 mm = 65 000 μm
Size of specimen = size of image ÷ magnification = 65 000 ÷ 20 000
Therefore actual size of chloroplast = 3.25 μm

**1.3** Resolution of a microscope is limited by the radiation used to view the specimen. Resolution equals half the wavelength of the radiation used. Shortest wavelength of light is 400 nm; therefore resolution of a light microscope is 200 nm. Diameter of ribosome is much smaller than this, namely 22 nm.

**1.4** Detail seen with electron microscope: in the **nucleus**, chromatin can be distinguished; the nucleus is seen to be surrounded by a double membrane with **pores** in it; **mitochondria** have surrounding double membrane, the inner layer forming folds pointing inwards; **endoplasmic reticulum** is extensive throughout cell, some with **ribosomes** and some without; small structures seen under the light microscope can be distinguished as **lysosomes** and **vesicles**; free

**ribosomes** seen throughout cell; **centriole** consists of two structures. (The microvilli seen on *figure 1.12* are not characteristic of all cells.)

**1.5** Details seen with electron microscope: in the **nucleus**, chromatin can be distinguished; **nuclear membrane** can be seen as a double structure, continuous with rough endoplasmic reticulum, and with pores in it; there is extensive **rough** and **smooth endoplasmic reticulum** throughout cell; free **ribosomes** in cytoplasm; **mitochondria** have double membrane, the inner layer having folds into matrix in middle; **chloroplasts** have double outer membrane; **grana** can be seen as stacks of double membrane sacs connected to other grana by longer sacs.

**1.6** **plasma membrane**: essential because it forms a partially permeable barrier between the cell and its environment, regulating movement of materials into and out of the cell. This is necessary to maintain an environment inside the cell which is different from that outside the cell.
**cytoplasm**: site of metabolic activity; contains biochemicals in solution.
**ribosomes**: sites of protein synthesis, an essential activity of all cells. (DNA controls cells by controlling which proteins are made.) Protein synthesis is a complex process involving the interaction of many molecules – the ribosome provides a site where this can happen in an organised way.
**DNA**: the genetic material. Contains the information which controls the activities of the cell. Has the ability to replicate itself, enabling new cells to be formed.
**cell wall** (absent in animal cells): prevents the cell from bursting if it is exposed to a solution of higher water potential.

# Chapter 2

**2.1** **a** $C_3H_6O_3$ or $(CH_2O)_3$ **b** $C_5H_{10}O_5$ or $(CH_2O)_5$

**2.2** **a** To ensure that *all* of the sugar reacts with the Benedict's.

**b** FIrst, carry out the test, using excess Benedict's solution, on a range of samples of known concentration of reducing sugar. Each sample must be of the same volume, and the test carried out in exactly the same way. If you have a colorimeter, take a reading for each concentration and plot reading against concentration a graph. If you do not have a colorimeter, line the tubes up in a rack.

Then carry out the test in exactly the same way on your unknown sample. If using a colorimeter, obtain a reading for it and use the graph to read off the concentration. If not, hold against the row of samples and judge which is the closest match by eye.

Alternatively, you can filter the contents of the tube, and dry and mass the precipitate obtained. Once again, you can compare the results from your unknown solution with those from a range of known ones.

**2.3** First, carry out the test for reducing sugar. If this is negative, then test for non-reducing sugar.

If it is positive, ensure that you have added excess Benedict's reagent, that is, ensure that *all* of the reducing sugar has reacted. Then filter the contents of the tube, and save the filtrate. If there is any non-reducing sugar present, this is where it will be. Test the filtrate for non-reducing sugar.

**2.4** Hydrolysis

**2.5** 1 macromolecules/polymers
2 polysaccharides
3 made from $\alpha$-glucose
4 glucose units held together by 1,4 links (glycosidic bonds formed by condensation)
5 branches formed by 1,6 links

**2.6**

| amylose | cellulose |
|---|---|
| made from $\alpha$-glucose | made from $\beta$-glucose |
| all glucose units have the same orientation | successive glucose units are at 180° to each other |
| molecule is not fibrous – chains not attracted to each other | fibrous molecule – chains held together by hydrogen bonds to form microfibrils and fibres. |

**2.7** At one end of the chain of amino acids; also in the R groups of many of the amino acids in the chain.

**2.8**

| | property | importance |
|---|---|---|
| a | Water requires a relatively large amount of heat energy to evaporate, that is water has a high heat of vapourisation. | Heat energy which is transferred to water molecules in sweat allows them to evaporate from the skin, which cools down, helping to prevent the body from overheating. A relatively large amount of heat can be lost with mimimal loss of water from the body. |
| b | The solid form of water (ice) is less dense than the liquid form and so floats. | Bodies of water such as lakes start to freeze from the top down. Ice insulates the water below it, increasing the chance of survival of organisms in the water. |
| c | High surface tension. | Small insects can land on, or live on, water without drowning. This increases the range of habitats available for feeding, reproduction (especially if aquatic larvae), etc. |
| d | Solvent. | Needed for transport by diffusion or active transport into, out of, or within cells. Also for circulation in blood so that nutrients can reach the sites where they are needed. Chemical reactions take place in aqueous solution. |

# Chapter 3

**3.1** In case of inaccuracy of measurement at 30 seconds. The shape of the curve is more likely to give an accurate value.

**3.2** See figure.

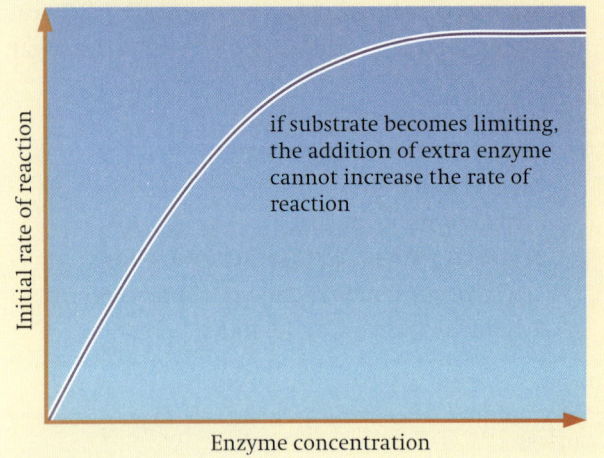

if substrate becomes limiting, the addition of extra enzyme cannot increase the rate of reaction

● **Answer for** SAQ 3.2

**3.3** **a** See figure.

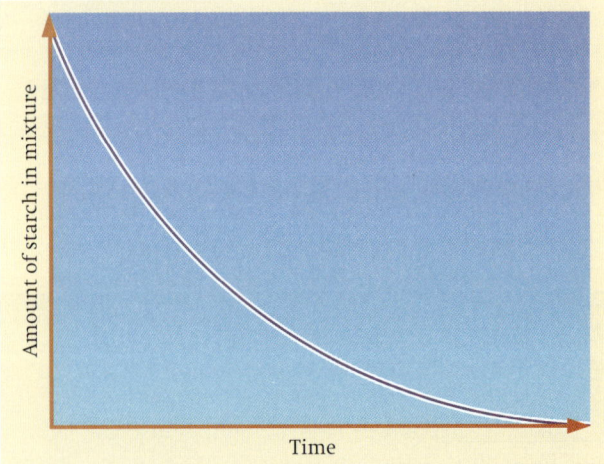

● **Answer for** SAQ 3.3a

**b** Calculate the slope of the curve right at the beginning of the reaction.

**3.4** Measure the volume of oxygen given off over time for several hydrogen peroxide–catalase reactions at different temperatures. In each case, all conditions other than temperature must remain constant. In particular, the same volume of hydrogen peroxide and of catalase solutions should be used each time. Plot total volume of oxygen against time for each reaction. Calculate the slope of the line at the beginning of the reaction in each case to give the initial reaction rate. Then plot initial reaction rate against time.

**3.5** **a** Haemoglobin colours blood stains. Protein-digesting enzymes hydrolyse haemoglobin to amino acids, which are colourless. They are also soluble, so will wash away in water.

**b** Many protein-digesting enzymes have an optimum temperature of around 40 °C.

**c** Other components of washing powders, such as the oil-removing detergents, work best at high temperatures.

**3.6** One possible answer is as follows; other answers might be equally acceptable.

Set up two sets of five tubes containing equal volumes of the same concentration of milk suspension. Make up five buffer solutions of varying pH. Add equal volumes of buffer solution to the milk suspension, two of each pH. To one set of tubes, add equal volumes of trypsin solution. To the other set of tubes, add the same volume of water; these act as controls. Time the disappearance of cloudiness in each tube. Plot rate of reaction (1/time taken) against pH.

# Chapter 4

**4.1** Large number of possible reasons, e.g. gain nutrients, remove waste products, gain oxygen for respiration, secrete hormones, secrete enzymes, maintain constant pH and ionic concentration.

**4.2** Water potentials are equal.

**4.3** **a** The pure water or dilute solution.

**b** The concentrated solution.

**c** The solution with the same concentration as the red cell.

**4.4** **a** From A to B.

**b** Water molecules can move from A to B and from B to A, but more move from A to B in a given time period. Overall therefore, A loses water and B gains water – the overall movement is the net movement.

**c** A has a higher water potential than B (−250 is less negative than −400) and water always moves from regions of higher to lower water potential. Water crosses a partially permeable plasma membrane every time it enters or leaves a cell – this process is called osmosis.

**d** (i) Pure water has a water potential of zero, which is higher than that of cells A and B. There is therefore a net movement of water into cells A and B by osmosis through their partially permeable plasma membranes. As water enters, the volume of the protoplasts will increase, exerting pressure on the cell walls and raising the pressure potential of the cells. This increases the water potential of the cells. This will continue until an equilibrium is reached when the contents of the cell reach the same water potential as the water, namely zero. The cells will then be turgid.

(ii) A 1 mol dm$^{-3}$ sucrose solution has a lower water potential than that of cells A and B. There is therefore a net movement of water out of cells A and B by osmosis through their partially permeable plasma membranes. As water leaves the cells, the protoplasts shrink and the pressure they exert on the cell walls drops; in other words the pressure potential of the cells decreases. This decreases the water potential of the cells. Eventually the pressure potential drops to zero and the cells are at incipient plasmolysis. As shrinkage continues, the protoplasts pull away from the cell walls – this is plasmolysis. The sucrose solution can pass freely through the permeable cell walls and remains in contact with the protoplasts. As water leaves the cells, the contents of the protoplasts get more and more concentrated and their water potential gets lower and lower (more and more negative). Equilibrium is reached when the water potential of the cells equals that of the sucrose solution.

**4.5** The animal cell does not have a cell wall. Plasmolysis is the pulling away of cytoplasm from the cell wall.

**4.6** Five – into and then out of a cell in the alveolar wall, into and then out of a cell in the capillary wall, and then into a red blood cell.

**4.7** Large surface area – increases the number of molecules or ions that can cross the surface in a given time.
Thin barriers across which substances cross – decreases the time taken for the molecules or ions to cross them.

# Chapter 5

**5.1**   a   ATP, which phosphorylates the nucleotides, providing energy to drive the reaction.
DNA polymerase, which catalyses the linkage of adjacent nucleotides once they have correctly base-paired.
    b   The nucleus.

**5.2**   a   The DNA in tube 2 is less dense than that in tube 1. In tube 1, all the N in the DNA molecules is $^{15}$N. In tube 2, each DNA molecule is made up of one strand containing $^{14}$N and one containing $^{15}$N.
    b   One band in the original position (the 'old' DNA containing only $^{15}$N), and another band higher up (the 'new' DNA containing only $^{14}$N).
    c   Assuming that most strands ended up with a mix of $^{14}$N and $^{15}$N – which you would expect if the bits of each kind were scattered randomly – then there would be a single band, like the one shown in tube 2.
    d   Tube 3. If the DNA had replicated dispersively, then instead of two distinct bands there would be a single, wide one because each strand of DNA would contain a mix of $^{14}$N and $^{15}$N. The band would be higher than that in tube 2, because there would now be more $^{14}$N and less $^{15}$N in the DNA molecules.
As it is, the two bands contain molecules with one strand containing $^{14}$N and the other $^{15}$N (the bottom band) and molecules in which all the N is $^{14}$N (the top band).

**5.3**   a   64
    b   For 'punctuation marks', that is for starting or stopping the synthesis of a polypeptide chain. Also, some amino acids could be coded for by two or three different base triplets.
    c   A two-letter code could only code for 16 amino acids.

**5.4** DNA contains the pentose sugar deoxyribose, while RNA contains ribose.
DNA contains the base thymine, while RNA has uracil.
DNA is made up of two polynucleotide strands, whereas RNA has only one.
DNA molecules are much longer than RNA molecules.

**5.5** There are various possible flow diagrams, but a suitable one might be as follows:

DNA unwinds and the two strands separate → complementary mRNA molecule built up against one DNA strand (transcription) → mRNA molecule attaches to ribosome → complementary tRNA loaded with appropriate amino acid pairs with one codon on mRNA (translation) → peptide bond forms between adjacent amino acids

# Chapter 6

**6.1** The chromosomes are arranged in order of size in the karyotype.

**6.2**   a (i) 92 (at this stage the cell is, technically, 4n)
      (ii) 92 (each chromatid contains one)
      (iii) 46 (the diploid number)
      (iv) 92

    b (i)             (ii)

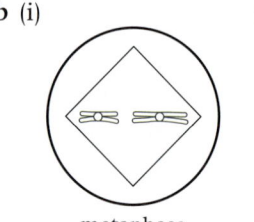 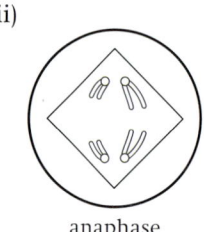

       metaphase           anaphase

    c nucleotides
    d C, H, O, N, P (deoxyribose and bases contain C, H and O; bases also contain N; phosphate contains P and O).
    e To hold chromatids together and to attach chromosomes to spindle.
    f Nine cells out of 75 000 were undergoing mitosis. Therefore mitosis occupies 9/75 000 of the cell cycle. Mitosis lasts 1 hour. Therefore cell cycle = 75 000/9 hours long, = 8333 hours = 8333/24 days = 347 days. (Cell cycles vary in length in adult animals from less than 8 hours to more than 1 year.)

# Chapter 7

**7.1** These animals are homeothermic, generating heat within their cells to keep their body temperature constant. This body temperature is normally above the environmental temperature, and so large quantities of heat are lost from their bodies.

**7.2**   a $38\,000\,\text{kJ}\,\text{m}^{-2}\,\text{year}^{-1}$, $31\,000\,\text{kJ}\,\text{m}^{-2}\,\text{year}^{-1}$
    b $54\,000\,\text{kJ}\,\text{m}^{-2}\,\text{year}^{-1}$
    c For example, higher light intensity, higher temperatures and higher rainfall allow photosynthesis to take place at a greater rate; there are few seasonal variations in these factors, so photosynthesis can continue all the year round; coniferous trees in the pine forest must be adapted to withstand cold and water shortage in winter, so have narrow needles, limiting maximum rates of light absorption even when environmental conditions are ideal for photosynthesis; tropical rain forest has a greater density of plants.

    d Alfalfa plants are young and growing, so much of the carbon they fix in photosynthesis is incorporated into new cells rather than being respired. In the rain forest, the trees are mostly mature and amounts of growth will be small. Alfalfa is a nitrogen-fixer and this, together with the probable application of fertiliser to the crop, could allow greater rates of growth than in the rain forest or pine forest.

# Chapter 8

**8.1**   a Size is important, but is not the only factor. Microscopic organisms such as *Paramecium* do not have transport systems, whereas all large organisms such as green plants, fish and mammals do. However, cnidarians do not have transport systems even though some of them are considerably larger than insects, which do.
    b Surface area to volume ratio is important. Small organisms have large surface area to volume ratios, and as explained in a, these generally do not have a transport system. Organisms with branching bodies, such as plants, can have large surface area to volume ratios even if they are large; they do have transport systems, but (as you will see) these are not used for transporting gases, and they do not have pumps.
    c Level of activity is important. Animals such as fish and mammals have a transport system containing a pump; plants, most of which are less active than most animals, do not have a pump. Insects have pumps in their transport system, even though they are smaller than the less active cnidarians, which do not have a pump.

**8.2**   a The fish has a single circulatory system, whereas the mammal has a double circulatory system. In the fish, blood leaves the heart and travels to the gills, where it picks up oxygen, before continuing around the body. In the mammal, the blood returns to the heart after picking up oxygen at the lungs, and is then pumped around the body.
    b Oxygenated blood can be pumped around the body at a higher pressure, and therefore faster, in a mammal than in a fish, because pressure is lost in the capillaries in the gills. This can provide a more efficient oxygen supply to mammalian cells than to fish cells.

**8.3** a Elastic fibres allow the artery to stretch and recoil as blood pulses through. Nearer the heart, the pressure changes between systole and diastole will be greater, and the maximum systolic pressure greater, than anywhere else in the circulatory system. Thus more elastic fibres are needed to cope with these large pressures and pressure changes.

**8.4** Blood cells, and haemoglobin in red blood cells, would cause scattering and absorption of light before it reached the retina. The aqueous humour supplies the cornea with its requirements.

**8.5** a Gravity pulls blood downwards. Normally, contraction and relaxation of leg muscles squeezes in on leg veins; valves in them ensure blood moves upwards and not downwards. When standing to attention, these muscles are still, so blood accumulates in the feet.

  b As thoracic volume increases, pressure inside the thorax decreases. This decreases the pressure in the blood vessels in the thorax. The effect is very small in the arteries, but more significant in the veins. The relatively low pressure of the blood in the veins in the thorax, compared with the pressure in veins elsewhere in the body, produces a pressure difference causing blood movement towards the thorax.

**8.6** Answers to this may be found in the text.

**8.7** Answers should include reference to: the fluctuating pressure in arteries; why the fluctuations become gradually less as the blood passes through the arterial system; the rapid drop in pressure as the blood flows along the arterioles and capillaries and reasons for this; the rise of pressure as blood enters the pulmonary circulation via the right-hand side of the heart, but not so high as the pressure in the aorta, and reasons for this.

**8.8** a The larger the relative molecular mass, the lower the permeability.

  b Net diffusion for glucose would be into the muscle. Respiration within the muscle requires glucose, so that its concentration within the muscle cells is lower than in the blood plasma.

  c Albumin in the blood plasma raises its solute concentration (osmotic pressure), thus helping to draw water back from the tissue fluid into capillaries. If albumin could diffuse out of

capillaries into tissue fluid, more water would accumulate in the tissue fluid. (This is called oedema.)

**8.9** a Protein in tissue fluid comes from the cells making up the tissues, many of which secrete proteins.

  b If plasma protein concentrations are low, then, as explained in SAQ 8.8c, water will not be drawn back into capillaries from tissue fluid.

**8.10** $2.1 \times 10^{11}$

**8.11** a Protein synthesis – no; there is no DNA, so no mRNA can be transcribed.

  b Cell division – no; there are no chromosomes, so mitosis cannot occur, nor are there centrioles for spindle formation.

  c Lipid synthesis – no; this occurs on the smooth endoplasmic reticulum, and there is none.

  d Active transport – yes; this occurs across the plasma membrane, and can be fuelled by ATP produced by anaerobic respiration.

**8.12** a $195\,cm^3$

  b $25\,cm^3$

**8.13** a (i) 96.5%

    (ii) $1.25\,cm^3$

  b (i) 24.0%

    (ii) $0.31\,cm^3$

**8.14** Less oxygen would enter the blood by diffusion, and therefore less oxygen would be carried to the body cells. The percentage saturation of haemoglobin will be only about 30% saturated (*figure 8.13*).

**8.15** At these heights, the percentage saturation of the haemoglobin is relatively low. If the number of red blood cells is increased, then the number of haemoglobin molecules is also increased. Even though the percentage saturation of the haemoglobin is low, the fact that that there is more of it can increase the actual quantity of oxygen carried in the blood.

**8.16** Spending a length of time at high altitude stimulates the body to produce more red blood cells. When the athlete returns to sea level, these 'extra' red blood cells remain in the body for some time, and can supply extra oxygen to muscles enabling them to work harder and for longer than they would otherwise be able to do.

## Chapter 9

**9.1** **a** (i) 0.7–0.8 seconds

(ii) 60/0.8 = 75 beats per minute

For **b**, **c**, **d**, **e** and **f**, see figure below.

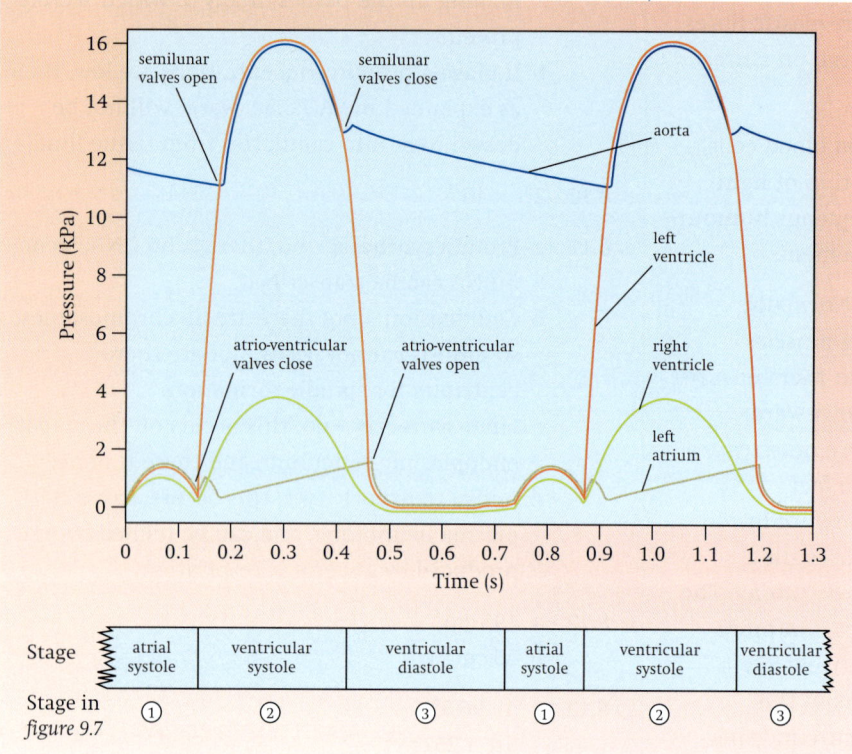

| Stage | atrial systole | ventricular systole | ventricular diastole | atrial systole | ventricular systole | ventricular diastole |
|---|---|---|---|---|---|---|
| Stage in figure 9.7 | ① | ② | ③ | ① | ② | ③ |

● **Answer for** SAQ 9.1

**9.2** **a** 1 beat = about 20 mm on the grid.

25 mm on the grid represents 1 second.

So 20 mm represents $\frac{20}{25}$ seconds = 0.8 seconds.

If one beat lasts 0.8 seconds, then in 1 second there are $\frac{1}{0.8}$ beats.

So in 1 minute there are $\frac{1 \times 60}{0.8}$ = 75 beats.

**b** (i) This is the time during which the ventricles are contracting.

(ii) On the grid, the distance betweeen Q and T is about 7 mm.

This represents $\frac{7}{25}$ = 0.28 seconds.

**c** (i) This is the time when the ventricles are relaxed, and are filling with blood.

(ii) On the grid, the distance between T and Q is about 13 mm.

This represents $\frac{13}{25}$ = 0.52 seconds.

A quicker way of working this out is to subtract your answer to b(ii) from 0.8 seconds.

**d** (i) By performing varying levels of exercise.

(ii) See figure on page 307.

(iii) As heart rate increases, contraction time remains constant, but filling time decreases. This indicates that the increase in heart rate is produced by a shorter time interval between ventricular contractions, rather than by a faster ventricular contraction.

The more frequent contractions increase the rate of circulation of blood around the body, providing extra oxygen to exercising muscles. If this was done by shortening the time over which the ventricles contract, much of the advantage would be lost, as less blood would probably be forced out by each contraction. By shortening the time *between* contractions, the amount of blood pumped out of the heart per unit time is increased.

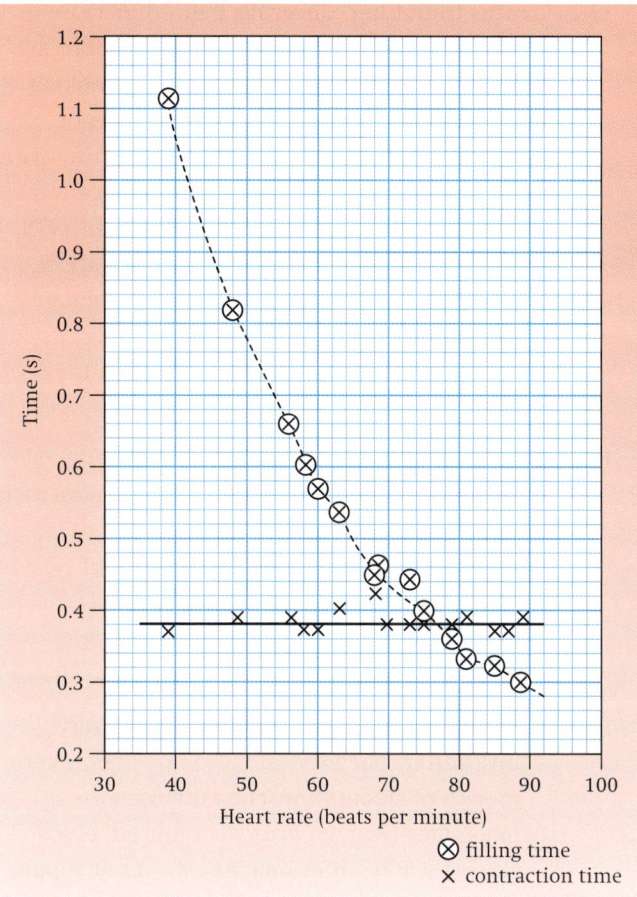

filling time
× contraction time

• **Answer for** SAQ 9.2d(ii)

# Chapter 10

**10.1 a** Mutualism.
**b** The relationship between *Rhizobium* and leguminous plants. The bacterium stimulates the roots to form nodules, in which the bacteria live and where they fix nitrogen. The plant uses the fixed nitrogen to make amino acids and proteins. The bacteria obtain organic nutrients and a source of energy from the plant.

**10.2** Spiral, annular or reticulate walls can all stretch and lengthen as the plant grows. The walls of pitted vessels cannot do this.

**10.3 a** Increased wind speed moves water vapour away from the leaf more rapidly, thus maintaining a steeper water potential gradient between the air spaces of the leaf and the surrounding air.
**b** Rise in temperature increases the kinetic energy of water molecules which move, and therefore diffuse, more rapidly. High temperatures may also decrease the humidity of the air (as warm air can hold more water), so increasing the diffusion gradient.

**10.4** Mammals use the evaporation of water in sweat for cooling purposes; the water evaporates from the skin surface, absorbing heat energy. Thus the two processes are very similar.

**10.5 a** In plants, osmosis is involved in the uptake of water from soil into the root hair. It may also be involved in movement from cell to cell across the root, but only if the water moves through the cell surface membranes of the cells. If it travels by the apoplast pathway, or via plasmodesmata between cells, then osmosis is not involved. Movement across the root endodermis does involve crossing cell surface membranes, so osmosis is again involved here. In the leaf, osmosis is involved if water moves into the cytoplasm of a cell across the cell surface membrane, but not if it moves by the apoplast pathway or plasmodesmata, as in the root.

In animals, osmosis is involved in the movement of water from the ileum and colon into the cells lining these parts of the alimentary canal, and then out of these cells into the blood. The water again moves by osmosis when leaving the blood and entering cells in body tissues. Osmosis is also involved in the reabsorption of water from nephrons.
**b** In plants, mass flow is involved in the movement of water up the xylem vessels.

In animals, mass flow is involved in the movement of blood plasma (containing water) along the blood vessels.

**10.6 a** The total lack of cell contents provides an uninterrupted pathway for the flow of water.
**b** Lack of end walls also provides an uninterrupted pathway for the flow of water.
**c** The wider the diameter, the more water can be moved up through a xylem vessel per unit time. However, if the vessels are too wide, there is an increased tendency for the water column to break. The diameter of xylem vessels is a compromise between these two requirements.
**d** The lignified walls provide support, preventing the vessels from collapsing inwards.
**e** Pits in the walls of the vessels allow water to move into and out of them.

**10.7** Sucrose, amino acids, ATP and plant growth substances.

**10.8** Sources: storage tissue of a potato tuber when the buds are beginning to sprout.
Sinks: nectary, developing fruit, developing potato tuber.

**10.9** All the required contents of this comparison table are in the text on pages 128–139. Care should be taken that equivalent points are kept opposite each other.

# Chapter 11

**11.1** Mouth/nostril; nasal cavity; pharynx; trachea; bronchus; terminal and respiratory bronchioles; alveolar duct; alveolus; epithelium; connective tissue; endothelium of capillary; plasma; red blood cell.

**11.2** Large surface area; thin epithelium, therefore short diffusion distance between air and blood; well supplied with many blood capillaries.

**11.3** During exercise the bronchioles are wider to allow more air to reach the alveoli to supply the large quantities of oxygen needed during exercise and to remove carbon dioxide.

**11.4**  a  tidal volume = $0.5\,dm^3$
vital capacity = $3.75\,dm^3$
b  (i)  12 breaths per minute
(ii)  $6.0\,dm^3\,min^{-1}$
(iii)  $0.38\,dm^3\,min^{-1}$

**11.5** Exercise; smoking; excitement; release of adrenaline; sleep; fear; meditation.

**11.6** With age, the arteries do not stretch as well and there is an increased resistance to the flow of blood. The heart needs to beat harder to overcome this resistance.

# Chapter 12

**12.1** Bronchitis: enlargement of mucus glands in airways; increased secretion of mucous; narrowing/obstruction of airways; inflammation; severe coughing; coughing-up of phlegm.
Emphysema: digestion by phagocytes of pathways through alveolar walls; loss of elastin; overextension and bursting of alveoli to form large air spaces; decrease in surface area for gaseous exchange; lack of recoiling of air spaces when breathing out; decrease in volume of air forced out from lungs; shortness of breath.

**12.2** Deaths from lung cancer lag behind increase in cigarette smoking by some 20 years or more in both men and women. Women began smoking later than men, so rise in death rate did not begin until later. More men than women smoke, and more men than women die of lung cancer. The male death rate started to decrease in 1975, roughly twenty years after the fall in cigarette consumption. Fall in consumption by women began about 1975, but by 1991 this had yet to be reflected in a decrease in mortality from lung cancer.

**12.3** Tar: paralyses or destroys cilia; stimulates oversecretion of mucus; leads to the development of bronchitis and emphysema (see SAQ 12.1). Carcinogens: cause changes in the DNA of cells in the bronchial epithelium, leading to the development of bronchial carcinoma – lung cancer.

**12.4**  a  A break in the wall of the coronary artery; invasion by phagocytes; build-up of cholesterol; growth of smooth muscle; atherosclerosis; blood flow through artery is reduced; blood clots at site of atheroma; blood cannot supply oxygen and nutrients to heart muscle; heart muscle dies; myocardial infarction (heart attack).
b  As blood does not supply oxygen and nutrients to part of the brain, the nerve cells in that area die. This may be fatal; if not, the part of the brain affected may not function properly or at all. As a result, a stroke may lead to loss of e.g. memory, movement or the ability to speak.

**12.5** Carbon monoxide: combines with haemoglobin to form stable compound carboxyhaemoglobin, with the result that less oxygen is transported in the blood.
Nicotine: raises blood pressure; raises heart rate; makes platelets sticky and therefore increases likelihood of thrombosis; decreases blood supply to the extremities.

**12.6**  a  Cardiovascular diseases are degenerative diseases; deaths in this age range are premature deaths.
b  (i) The death rates are much higher in North Karelia than Catalonia for all cardiovascular diseases in both men and women. They are higher in North Karelia than the national figures for Finland, but lower in Catalonia than the national figures for Spain.

(ii) CHD accounts for over half of all deaths in this category in North Karelia, less than half in Catalonia.

(iii) Strokes are responsible for the lowest number of deaths, but the death rate is still more than twice as high in North Karelia than in Catalonia.

c men

**12.7 a** total number of deaths attributable to smoking: men = 55 508; women = 39 975.

**b** percentages: men = 17.5%; women = 11.74%; total = 14.5%.

# Chapter 13

**13.1** Faeces from infected person contain *Vibrio cholerae*. These bacteria are transmitted to uninfected people in drinking water, contaminated food (e.g. vegetables irrigated with raw sewage or food prepared by a symptomless carrier), or when washing or bathing in contaminated water.

**13.2** $\dfrac{10^{13}}{10^{6}}$ = 10 million

**13.3** Refugees rarely have access to proper sanitation, clean water, or uncontaminated food.

**13.4** The visitor can drink bottled or boiled water; and avoid eating salads and raw vegetables.

**13.5** When a female *Anopheles* mosquito bites an infected person she takes up some gametes of the parasite. These develop into infective stages which enter an uninfected person when the same mosquito takes another blood meal.

**13.6** The resistance of mosquitoes to insecticides such as DDT and dieldrin; the difficulty of controlling the breeding of mosquitoes because they lay eggs in small bodies of water; the resistance of some strains of *Plasmodium* to anti-malarial drugs such as chloroquine.

**13.7** People can avoid being bitten by mosquitoes, sleep under nets impregnated with insecticide, use repellents; and use anti-malarial drugs as prophylactics (but not those to which *Plasmodium* is resistant).

**13.8** Many cases of AIDS are not diagnosed or reported.

**13.9** Condoms can split when in use or may not be put on correctly.

**13.10** Practise safer sex (e.g. use condoms); do not use unsterile needles; have one sexual partner; do not donate blood if at risk of HIV infection; do not use prostitutes (male or female); have a blood test to find out if you are HIV+.

**13.11** HIV is a blood-borne virus; blood donations may not be screened or heat-treated for HIV.

**13.12** It is important for people to know whether they are HIV+ so that they can make sure they reduce the chances of transmitting the virus to others.

**13.13** Highest incidence/prevalence in South-East Asia, sub-Saharan Africa, countries of old Soviet Union, India, China, Mexico, Peru, Bolivia. Poor nutrition; HIV infection; poor disease control; large cities with poor housing and homeless people; countries with limited health facilities and large numbers of displaced people (e.g. migrants and refugees).

**13.14** B and E. These have inhibition zones larger than the minimum required to be in the sensitive range. These antibiotics could be used together.

# Chapter 14

**14.1** Antigen: any large molecule (e.g. protein) recognised by the body as foreign.
Antibody: a protein made by the immune system in response to the presence of an antigen and targeted specifically at it.
Immune response: lymphocytes respond to the presence of a pathogen in the body by producing antibodies.

**14.2 a** The lymphocyte nucleus takes up most of the cell; there is very little cytoplasm. Neutrophils have a lobed nucleus, with a larger amount of cytoplasm. The neutrophil is larger.

**b** 10 μm.

**14.3** B lymphocytes originate and mature in bone marrow. T lymphocytes originate in bone marrow, but mature in the thymus gland.

**14.4** By puberty, T cells have matured and left the thymus gland. The thymus has no further use, so decreases in size.

**14.5** The cytoplasm of plasma cells is full of rough endoplasmic reticulum where protein is made. There are also Golgi bodies for packaging antibodies in vacuoles for secretion.

**14.6** The plasma cells will be identical to the original B cell and will therefore all produce exactly the same antibody molecules. Also the memory cells will be identical so that the same antibody molecules are produced during any subsequent immune response to the same antigen.

**14.7** Polysaccharides are made from only a small number of different sugars unlike proteins that are made from twenty different amino acids. Polysaccharides would not give the same huge number of different molecular shapes as is achieved with proteins in the variable region of antibodies.

**14.8** Only some B and T cells have receptors of the correct specificity.

**14.9** Immunity to one strain does not provide immunity to all of them as they do not all share the same antigens.

**14.10** The primary response to an antigen is slow. It takes several weeks to produce enough antibody molecules to fight the infection effectively. During this time we usually show the symptoms of the disease concerned.

**14.11** B lymphocytes with antibody receptors specific to the invading antigen divide by mitosis to form plasma cells and memory cells. The short-lived plasma cells secrete antibody molecules. T helper lymphocytes and killer T lymphocytes with T cell receptors specific to the invading antigen also divide by mitosis. The T helper cells secrete cytokines to activate appropriate B lymphocytes to divide and macrophages to carry out phagocytosis. The killer T cells search for any infected cells in the body and kill them.

**14.12** Active: antigens are introduced into the body by injection or by mouth, and stimulate an immune response by B and T cells. This provides long-term immunity but is not immediate as the immune response takes several weeks to become effective. Passive: antibodies are injected into the body to give immediate protection against a pathogen or toxin. Antibodies are soon removed from circulation and no immune response has occurred, so this is a temporary form of immunity.

**14.13** Maternal IgG increases during pregnancy as it crosses the placenta; it decreases after birth as it is removed from the circulation. This is natural passive immunity. The fetus does not produce its own antibodies because it does not have any mature T or B cells and develops in a sterile environment. The infant produces its own antibodies shortly after birth as it begins to encounter infections.

**14.14** The infant is protected against diseases which are endemic and which the mother has caught or been vaccinated against. For example, measles is a serious childhood infection; the infant is protected for several months by its mother's antibodies. (Note that the infant will not gain passive immunity to any diseases the mother has not encountered.)

**14.15** Antibodies are proteins. If children have protein energy malnutrition they may not have the ability to produce many antibodies or develop T and B cell clones during an immune reponse.

**14.16** Every time the parasite changes its antigens a new primary response will be activated. As soon as there are some antibodies in the blood, the parasite exposes different antigens, so making the antibodies ineffective.

**14.17 a** *Mycobacterium tuberculosis, M. bovis*
**b** HIV

# A Level

## Chapter 15

**15.1** $ADP + P_i (H_3PO_4) + 30.5\,kJ \rightarrow ATP + H_2O$

**15.2** In polynucleotides, the bases are linked by covalent bonds between phosphate groups and sugars. In NAD, the link is by a covalent bond between two phosphate groups.

**15.3** By decarboxylation, citrate, a six-carbon compound, can be converted to five-carbon and four-carbon compounds, finally giving oxaloacetate which can act as an acceptor for an incoming two-carbon unit from acetyl coenzyme A giving citrate again.

**15.4** Reduced NAD per glucose:

| | |
|---|---|
| from glycolysis | 2 |
| from the link reaction (1 × 2) | 2 |
| from the Krebs cycle (3 × 2) | 6 |
| Total | 10 |

Reduced FAD per glucose:

| | |
|---|---|
| from the Krebs cycle (1 × 2) | 2 |

Remember that two molecules of pyruvate go through the link reaction, and that there are two turns of the Krebs cycle for each molecule of glucose respired.

**15.5** Each reduced NAD produces 2.5 ATP in oxidative phosphorylation; each reduced FAD produces 1.5 ATP.

Oxidative phosphorylation gives 28 ATP per molecule of glucose, as follows:

| | |
|---|---|
| via 2 reduced NAD from glycolysis (2 × 2.5) | 5 |
| via 2 reduced NAD from the link reaction (2 × 2.5) | 5 |
| via 2 reduced FAD from the Krebs cycle (2 × 1.5) | 3 |
| via 6 reduced NAD from the Krebs cycle (6 × 2.5) | 15 |

**15.6** Only 2 ATP (1 ATP per turn) are made directly in the Krebs cycle.

Hydrogens are lost at four different stages of each turn of the cycle. Once these have been taken up by hydrogen carriers they can be transferred to the reactions of oxidative phosphorylation to give much more ATP.

**15.7** Points should include: the link reaction and Krebs cycle take place in the liquid matrix where enzymes and substrates can freely interact; mitochondria in active tissues are large and have many cristae; the large surface area of cristae for the layout of the sequences/'production lines' of carriers needed for electron transfer; the importance of the membranes and the intermembrane space for building up a hydrogen ion gradient in chemiosmosis; the role of ATP synthase.

**15.8** $C_{18}H_{36}O_2 + 26O_2 \rightarrow 18CO_2 + 18H_2O + energy$

$$RQ = \frac{CO_2}{O_2} = \frac{18}{26} = 0.69$$

**15.9** Take readings of oxygen consumption at one temperature, say 15 °C, including replicate readings to give a mean value. Increase the temperature to, say, 25 °C. Leave the organisms at that temperature for about 10 minutes for the rate of respiration to reach an equilibrium. Take readings as before. Repeat at other temperatures.

# Chapter 16

**16.1** **a** Both chlorophyll *a* and *b* have absorption peaks in the blue (400–450 nm) and red (650 nm) wavelengths. The carotenoids also have two peaks in the blue wavelengths. The action spectrum also peaks in the blue and red wavelengths. It is these absorbed wavelengths that provide energy for photosynthesis.

**b** Different pigments have different absorption spectra. The maxima of the action spectrum do not closely match the absorption spectrum of any single pigment. Although the peaks at the two ends of the absorption spectrum are of similar height, the action spectrum has a larger peak at 650 nm. The action spectrum does not perfectly match the absorption spectra since not all absorbed light is used in photosynthesis.

**16.2** **a** The chloroplasts absorb light and split water (photolysis) generating hydrogen ions. This reduces DCPIP from blue to colourless, so that the colorimeter reading falls.

**b** The chloroplasts in light reduce DCPIP at a steady rate. The chloroplasts in the dark for five minutes do not reduce DCPIP during that time. When placed in the light, reduction occurs at a slightly slower rate. (A possible reason for this is gradual loss of activity by isolated chloroplasts because of damage.)

**16.3** The Hill reaction shows that chloroplasts have 'reducing power' necessary to reduce fixed carbon dioxide to carbohydrate. They produce hydrogen ions. This is seen by their reduction of a coloured redox agent (blue DCPIP) to colourless.

**16.4** Thylakoid membranes provide a large surface area for many pigments, enzymes and electron carriers and for light absorption.

The arrangement of pigments into photosystems provides efficient light absorption.

Grana with ATP synthase allow ATP synthesis.

The stroma bathes all the membranes and holds the enzymes, reactants and products of the Calvin cycle.

**16.5** **a** Experiments 1 and 2 differ only in temperature and show the limiting effect of temperature. The photosynthetic rate is approximately doubled by each $10\,°C$ temperature rise, both in initial increase in light intensity and at light saturation. The effect is via the light-independent stage since increased temperature increases the rate of these reactions.

**b** Experiments 1 and 3 differ in carbon dioxide concentration and show limiting effect of that concentration. A tenfold increase of external carbon dioxide concentration produces an approximate doubling of the rate of photosynthesis. The limiting effect is not only external carbon dioxide concentration but the rate at which the leaf can be supplied with carbon dioxide. This depends on the steepness of the diffusion gradient and the permeability of the leaf.

# Chapter 17

**17.1** **a** prophase I (in fact, they pair before this, during interphase but can only be seen for the first time during prophase)

**b** prophase I

**c** anaphase I

**d** anaphase II

**e** telophase I

**17.2** Meiosis could not take place in a triploid, 3n, cell, because there is an odd number of each chromosome so they will not be able to pair up.

In theory, meiosis can take place in a tetraploid, 4n, cell because there is an even number of each kind of chromosome, so they can each find a partner to pair up with. In practice, meiosis is often very difficult in a 4n cell because, if there are four homologous chromosomes present, they all tend to join up with each other. Crossing over between chromatids of different chromosomes results in an inextricable tangle, so meiosis cannot proceed effectively.

**17.3** 6

**17.4** 3 homozygous, 3 heterozygous

**17.5** **a** Symbols should use the same capital letter, with a different superscript for each allele. For example:
$C^R$ to represent the allele for red coat
$C^W$ to represent the allele for white coat

**b** $C^RC^R$ red coat
$C^RC^W$ roan coat
$C^WC^W$ white coat

**c** (i) Red Poll × roan gives $C^RC^R$ (red coat) and $C^RC^W$ (roan coat) in a ratio of $1:1$.
(ii) Roan × roan gives $C^RC^R$ (red coat), $C^RC^W$ (roan coat) and $C^WC^W$ (white coat) in a ratio of $1:2:1$.

**17.6** Symbols should use the same letter of the alphabet, using the capital letter to represent the dominant allele, and the small letter to represent the recessive allele. For example:
B to represent the allele for black eyes.
b to represent the allele for red eyes.

The cross would be expected to produce Bb (black-eyed mice) and bb (red-eyed mice) in a ratio of $1:1$.

**17.7** If a cross between an unspotted and spotted plant can sometimes produce offspring which are all unspotted, then unspotted must be the dominant allele. Suitable symbols could be:
A to represent the dominant unspotted allele;
a to represent the recessive spotted allele.
(U and u or S and s are not good choices, as they are difficult to distinguish.)

An unspotted plant could therefore have either the genotype AA or Aa.
A spotted plant could only have the genotype aa.
Therefore, a cross between spotted and unspotted could either be:

*Parents*    AA × aa
*Offspring*    Aa

or it could be:

*Parents*    Aa × aa
*Offspring*    Aa and aa in a ratio of $1:1$.

**17.8** She may be right, but not necessarily. It is true that if her bitch were homozygous for the dominant allele for black spots all of her eggs would contain this dominant allele, and therefore all of her offspring would be black no matter what the genotype of the male parent. If the bitch was heterozygous, it might be expected that a mating with a homozygous recessive dog would produce black-spotted and brown-spotted offspring in a ratio of 1:1. However, as only three puppies were born, it may just be chance that no brown-spotted puppy was born. The breeder would need to produce more litters from the bitch before she could be sure of her dog's genotype.

**17.9** The child with blood group O must have the genotype $I^oI^o$. Therefore, each parent must have one $I^o$ allele. The genotypes are therefore:

Man and the child with blood group B     $I^BI^o$
Woman and the child with blood group A     $I^AI^o$
Child with blood group O     $I^oI^o$

**17.10** Suitable symbols for these four alleles could be:

$C^A$     to represent the agouti allele
$C^g$     to represent the chinchilla (grey) allele
$C^h$     to represent the Himalayan allele
$C^a$     to represent the albino allele

**a** $C^AC^A$     agouti
$C^AC^g$     agouti
$C^AC^h$     agouti
$C^AC^a$     agouti
$C^gC^g$     chinchilla
$C^gC^h$     chinchilla
$C^gC^a$     chinchilla
$C^hC^h$     Himalayan
$C^hC^a$     Himalayan
$C^aC^a$     albino

**b** (i) The genotype of the albino parent must be $C^aC^a$, as the allele $C^a$ is recessive to everything else. Each of the offspring will therefore get a $C^a$ allele from this parent. The offspring are all chinchilla, so their genotypes must all be $C^gC^a$. This means that the chinchilla parent must have given a $C^g$ allele to each offspring, so the chinchilla parent almost certainly has the genotype $C^gC^g$. If it had any other allele in its genotype, this would be expected to show in the phenotype of its offspring.

When the $C^gC^a$ offspring are crossed, they will produce genotypes of $C^gC^g$, $C^gC^a$ and $C^aC^a$ in a ratio of 1:2:1, that is a ratio

of chinchilla to albino in a ratio of 3:1. This is close enough to the actual ratio of 4 chinchilla to 2 albino, as with these very small numbers it is unlikely for ratios to work out exactly.

(ii) Following similar reasoning to that in (i) above, the agouti rabbit probably has the genotype $C^AC^h$ and the Himalayan parent the genotype $C^hC^h$. This would produce agouti and Himalayan offspring in a ratio of 1:1.

(iii) As chinchilla rabbits are produced in the first generation, at least one of the agouti parents must carry a chinchilla allele. As a Himalayan rabbit is produced in the second generation, one of the original parents must carry a Himalayan allele. The first cross is therefore:

| Parental phenotype | agouti | | agouti | |
|---|---|---|---|---|
| Parental genotype | $C^AC^h$ | | $C^AC^g$ | |
| Gametes | $C^A$ or $C^h$ | | $C^A$ or $C^g$ | |
| Offspring | $C^AC^A$ | $C^AC^h$ | $C^AC^g$ | $C^hC^g$ |
| | agouti | agouti | agouti | chinchilla |

The second cross, between two chinchilla rabbits with the genotype $C^h C^g$, would be expected to produce offspring with the genotypes $C^hC^h$, $C^hC^g$ and $C^gC^g$ in a ratio of 1:2:1, so giving the phenotypic ratio of 3 chinchilla:1 Himalayan.

**17.11**

| Parental phenotype | female | | male |
|---|---|---|---|
| Parental genotype | XX | | XY |
| Gametes | all $X$ | | $X$ or $Y$ |
| Offspring | XX | or | XY |
| | female | | male |

**17.12 a** No. The son will receive a Y chromosome from his father, which cannot carry a haemophilia allele.

**b** Yes. The man could pass on his haemophilia allele to a daughter, who could then pass it on to a son.

**17.13 a** Suitable symbols could be:

$X^N$ allele for normal colour vision

$X^n$ allele for red-green colour blindness

$X^N X^N$ normal female

$X^N X^n$ carrier (normal) female

$X^n X^n$ female with colour blindness

$X^N Y$ normal male

$X^n Y$ male with colour blindness

**b** 

| | normal | | normal | |
|---|---|---|---|---|
| *Parental phenotypes* | | | | |
| *Parental genotypes* | $X^N$ $Y$ | | $X^N$ $X^n$ | |
| *Gametes* | $X^N$ $Y$ | | $X^N$ $X^n$ | |

*Offspring genotypes and phenotypes*

*Genotypes of eggs*

| | | $X^N$ | $X^n$ |
|---|---|---|---|
| *Genotypes of sperm* | $X^N$ | $X^N X^N$ normal female | $X^N X^n$ carrier female |
| | $Y$ | $X^N Y$ normal male | $X^n Y$ male with colour blindness |

This can happen if the woman is heterozygous. The affected child will be male.

**c** Yes, if the mother has at least one allele for colour blindness, and the father has colour blindness.

**17.14 a** Male cats cannot be tortoiseshell because a tortoiseshell cat has two alleles of this gene. As the gene is on the X chromosome, and male cats have one X chromosome and one Y chromosome, then they can only have one allele of the gene.

**b**

| | orange male | tortoiseshell female |
|---|---|---|
| *Parental phenotypes* | | |
| *Parental genotypes* | $X^{C^O} Y$ | $X^{C^O} X^{C^B}$ |
| *Gametes* | $X^{C^O}$ $Y$ | $X^{C^O}$ $X^{C^B}$ |

*Offspring genotypes and phenotypes*

*Genotypes of eggs*

| | | $X^{C^O}$ | $X^{C^B}$ |
|---|---|---|---|
| *Genotypes of sperm* | $X^{C^O}$ | $X^{C^O} X^{C^O}$ orange female | $X^{C^B} X^{C^O}$ tortoiseshell female |
| | $Y$ | $X^{C^O} Y$ orange male | $X^{C^B} Y$ black male |

The kittens would therefore be expected to be in the ratio of 1 orange female : 1 tortoiseshell female : 1 orange male : 1 black male.

**17.15 a** Independent assortment results from the random alignment of the pairs of homologous chromosomes on the equator during metaphase I. It ensures that the chromosomes carrying the A or a alleles behave quite independently from those carrying the D or d alleles. This means that allele A can end up in a gamete with either D or d, and allele a can also end up in a gamete with either D or d. Thus independent assortment is responsible for the fact that each parent can produce four different types of gamete, AD, Ad, aD and ad.

**b** Random fertilisation means that it is equally likely that any one male gamete will fuse with any female gamete. This results in the 16 genotypes shown in the square. Thus random fertilisation is responsible for the different genotypes of the offspring, and the ratios in which they are found.

**17.16 a** AaBb and Aabb in a ratio of 1:1

**b** GgHh, Gghh, ggHh and gghh in a ratio of 1:1:1:1

**c** All TtYy

**d** EeFf, Eeff, eeFf and eeff in a ratio of 1:1:1:1

**17.17 a** They would all have genotype GgTt, and phenotype grey fur and long tail.

**b** Grey long, grey short, white long, white short in a ratio of 9:3:3:1.

**17.18** Let T represent the allele for tall stem, t the allele for short stem, $L^G$ the allele for green leaves and $L^W$ the allele for white leaves.

| Parents | $TTL^GL^G$ | $ttL^GL^W$ |
|---|---|---|
| Gametes | $(TL^G)$ | $(tL^G)$ or $(tL^W)$ |
| Offspring | $TtL^GL^G$, $TtL^GL^W$ in a ratio of 1 : 1. | |

**17.19 a** If B represents the allele for black eyes and b represents the allele for red eyes, L represents the allele for long fur and l represents the allele for short fur, then the four possible genotypes of an animal with black eyes and long fur are

BBLL, BbLL, BBLl and BbLl.

**b** Perform a test cross, that is breed the animal with an animal showing both recessive characteristics. If the offspring show one of the recessive characteristics, then the 'unknown' genotype must be heterozygous for that characteristic.

**17.20** The expected ratio would be 9 grey long : 3 grey short : 3 white long : 1 white short.

The total number of offspring is 80, so we would expect 9/16 of these to be grey long, and so on.

Expected numbers:
$9 \div 16 \times 80 = 45$ grey long
$3 \div 16 \times 80 = 15$ grey short
$3 \div 16 \times 80 = 15$ white long
$1 \div 16 \times 80 = 5$ white short

Now complete a table like the one on page 234.

Now look at *table 17.3* on page 234. We have four classes of data, so there are 3 degrees of freedom. Looking along this line, we can see that our value for $\chi^2$ is much greater than any of the numbers there, and certainly well above the value of 7.82, which is the one indicating a probability of 0.05 that the difference between the observed and expected results is due to chance. Our value is way off the right hand end of the table, so we can be certain that there is a **significant difference** between our observed and expected results. Something must be going on that we had not predicted.

(Note: if you study *Applications of Genetics* later in your course, you will find out what may be causing these rather unexpected results!)

**17.21** The potential risk is greatest with bacteria, because genes may be passed from modified bacteria by horizontal transmission into other bacteria. Such transfers could have unforeseen effects.

**17.22** Virus treatment reduces caterpillar damage compared with the untreated control plants. Plants treated with genetically modified viruses showed less damage than those treated with unmodified viruses. Treatment with viruses at the higher dose rate reduces leaf damage.

In comparison with the control plants, damage is reduced by low doses of unmodified virus to 93%, and genetically modified virus to 65%. Damage is reduced by high doses of unmodified virus to 59%, and of genetically modified virus to 47%.

| Phenotypes of animals | grey, long | grey, short | white, long | white, short |
|---|---|---|---|---|
| Observed number (O) | 54 | 4 | 4 | 18 |
| Expected ratio | 9 | 3 | 3 | 1 |
| Expected number (E) | 45 | 15 | 15 | 5 |
| O − E | +9 | −11 | −11 | +13 |
| $(O - E)^2$ | 81 | 121 | 121 | 169 |
| $(O - E)^2/E$ | 1.8 | 8.1 | 8.1 | 33.8 |

$\Sigma(O - E)^2/E = 51.8$
$\chi^2 = 51.8$

This is a huge value for $\chi^2$.

# Chapter 18

**18.1** Characteristics are passed from parents to offspring in their genes. Variation caused by the environment does not change the DNA of an organism.

**18.2** **a** There seems to be no selection pressure against unusual colours, as there are no predators.
**b** Possibilities include ability to cope with a limited food or water supply, ability to cope with the limited breeding space, and susceptibility to disease such as myxomatosis if this is present on the island.

**18.3** **a** The more frequently antibiotics are used, the more frequently resistant bacteria will be selected for. If antibiotic use is infrequent, then other selection pressures will be more important in bacterial populations, decreasing the likelihood of resistant bacteria surviving.
**b** Changing the antibiotic changes the selection pressure. Different strains of bacteria will be selected for when a different antibiotic is used, decreasing the likelihood of a resistant strain for each antibiotic becoming widespread.
**c** It is far less likely that any individual bacterium will be resistant to two antibiotics than to any single antibiotic, so decreasing the chance of any bacteria surviving in an environment where two antibiotics are used together.

# Chapter 19

**19.1** **a** From the glomerulus.
**b** Proteins.
**c** It will increase the solute concentration of the blood plasma therefore lowering its water potential and increasing the water potential gradient between the filtrate and the blood.

**19.2** A large percentage (60%) of the water in the fluid is reabsorbed in the proximal convoluted tubule; thus the amount of water in which the urea is dissolved decreases. This increases the concentration of urea in the fluid.

**19.3** **a** Flow rate is highest at the beginning of the proximal convoluted tubule, where fluid is entering via filtration into the renal capsule. As the fluid flows along the proximal convoluted tubule, a large percentage of it is reabsorbed, thus decreasing its volume. There is thus less fluid to flow, so less passes a given point in unit time: in other words, its flow rate decreases.

This reabsorption continues all along the nephron, which is why the flow rate continues to drop. The rate of flow decreases rapidly in the collecting duct, as a high proportion of water may be reabsorbed here.
**b** (i) Glucose concentration drops rapidly to zero as the fluid passes through the proximal convoluted tubule, because all of it is reabsorbed into the blood at this stage.
(ii) Urea concentration increases because water is reabsorbed from the tubule.
(iii) The concentration of sodium ions remains constant in the proximal convoluted tubule, as, although some sodium is reabsorbed here, this is balanced by the reabsorption of water. In the loop of Henle, the counter-current multiplier builds up sodium ion concentration in the lower parts of the loop; the concentration drops as you pass up the ascending limb towards the distal convoluted tubule, as sodium ions are lost from the tubule. In the distal convoluted tubule, sodium ions are actively pumped out of the tubule, so you might expect their concentration to drop. However, this is counterbalanced by the continued removal of water from the tubule, which results in an increasing concentration.
(iv) As for sodium ions, until the distal convoluted tubule, where potassium ions are actively transported into the tubule, so increasing their concentration more than that of sodium.

**19.4** There are many possible ways in which this flow diagram could be constructed. It should show the following: input (change in blood water concentration) to sensor (osmoreceptor cells); resulting in secretion of ADH from posterior pituitary if water concentration low; producing output (change in rate of water reabsorption) by effector (walls of collecting duct); and negative feedback to sensor.

**19.5** The adrenal glands may continue to secrete adrenaline over a long period, for as long as the stimulus to do so continues.

**19.6** The plasma membrane is a bilayer of phospholipids. Steroids can dissolve in these lipids, and so pass through.

**19.7 a** In negative feedback, a change initiates a response that brings things back to normal. So, in the case of blood glucose concentration, a rise in the concentration brings mechanisms into play that reduce it.

**b** The α and β cells in the islets of Langerhans in the pancreas.

**c** The effectors are the α and β cells themselves (because they respond by secreting or not secreting glucagon and insulin), and also the cells in the liver and muscles.

**19.8** The lack of insulin, or the lack of response to insulin, means that cells either do not take up extra glucose when it is in excess, or they do not convert it to glycogen stores.

**19.9 a** Insulin is a protein. Its molecules would be hydrolysed to amino acids in the digestive system.

**b** People with non-insulin-dependent diabetes are encouraged to test their blood or urine regularly for glucose. They can adjust their diet accordingly. They should eat small amounts of carbohydrate fairly regularly, rather than large quantities at any one time. High-sugar foods, such as confectionary, should be avoided, as these may result in a rapid and dangerous rise in blood glucose levels.

**19.10** A wide variety of answers is possible, some of which are suggested below:

**19.11 a** A receptor potential is an electrical potential generated in a receptor such as a Pacinian corpuscle. It is produced by the opening of sodium channels which results in a depolarisation of the nerve ending, that is a less negative potential inside the axon than when it is at rest.

**b** Increasing pressure produces an increasing receptor potential. At low levels of pressure, a small increase in pressure results in a relatively large increase in receptor potential. At higher levels of pressure, the increase in receptor potential is less. (The functional significance of this pattern, which is found in most receptors, could be discussed; it results in a relatively high level of sensitivity to low-level stimuli as long as they are above the critical threshold).

**c** The threshold receptor potential is the smallest receptor potential at which an action potential is generated.

**d** The greater the strength of the stimulus applied, the greater the frequency of action potentials generated.

**19.12 a** At the synapse, vesicles of transmitter substance are only present in the presynaptic neurone, not in the postsynaptic neurone.

**b** Repeated action potentials may cause the release of transmitter substance into the cleft at a greater rate than it can be replaced in the presynaptic neurone.

| Stimulus | Receptor | Effector | Response |
|---|---|---|---|
| sudden loud sound | hair cells in cochlea of ear | various muscles especially in legs | rapid contraction producing movement |
| smell of food cooking | chemoreceptors in nose | salivary glands | secretion of saliva |
| sharp tap on knee | stretch receptors | thigh muscle | contraction, causing lower leg to be raised |

# Past examination questions and answers

Past University of Cambridge Local Examination Syndicate (UCLES) examination questions
(drawn from 9700 AS and A Level Biology papers and their predecessors (8700, 9266 and 9263))

## Chapter 1

**1.1** Fig. 1.1 is an electron micrograph of a plasma cell from a mammal.

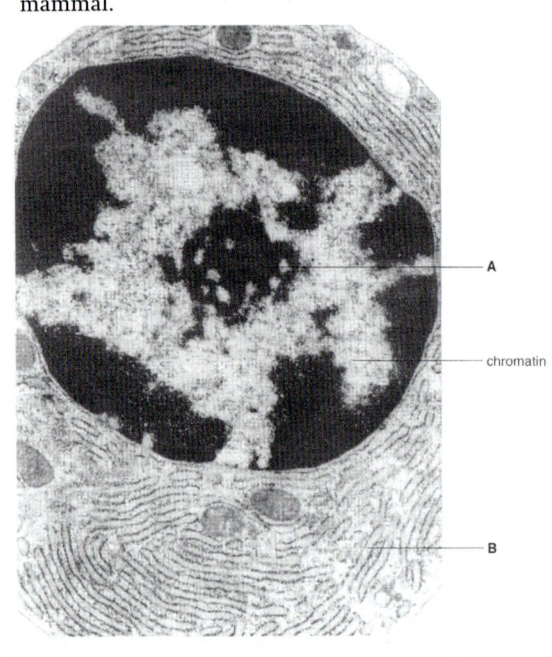

16μm

Fig. 1.1

(a) (i) Name the structures **A** and **B**. [2]
  (ii) State the functions of A and B. [2]
(b) Describe the change in appearance of the chromatin when the nucleus is about to divide. (ref. Chapter 6) [1]
(c) Calculate the magnification of the nucleus in Fig. 1.1. [1]

*9266/03 June 1998 Q1* Total [6]

## Chapter 2

**2.1** Fig. 2.1 is a diagram of part of an α helix of a polypeptide chain commonly found in many types of protein.

(a) (i) Name the repeating monomer of a polypeptide chain. [1]
  (ii) Explain what would happen to the α helix if the polypeptide chain was heated to a temperature above 60°C. [2]
(b) In globular proteins, the polypeptide chain bends and folds to give a more compact shape. This is the tertiary structure of the protein.
  Name **three** types of bond that help to maintain the tertiary structure. [3]
(c) Monosaccharides can also be linked together to form long chain molecules called polysaccharides.
  State **two** ways, other than the names of the monomers present, in which the structure of a polysaccharide differs from that of a polypeptide chain. [2]
(d) The fibrous protein collagen and the polysaccharide cellulose both posses considerable tensile strength.
  List **two** features that contribute to the strength of collagen; [2]
  and the strength of cellulose. [2]

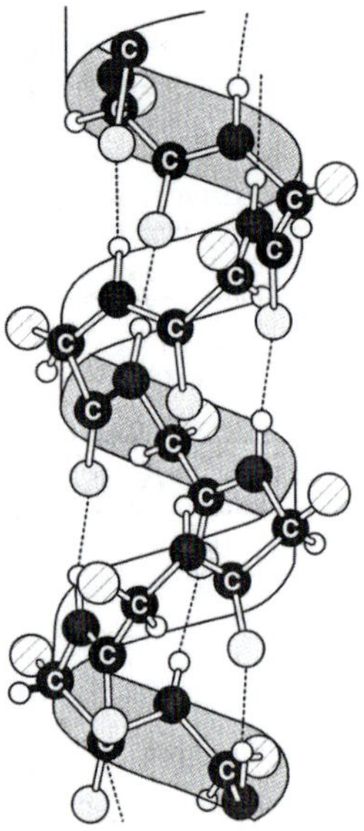

*8700/02 Nov 2001 Q3* Total [12]

Fig. 2.1

**2.2** Fig. 2.2 shows the reaction to form triglycerides.

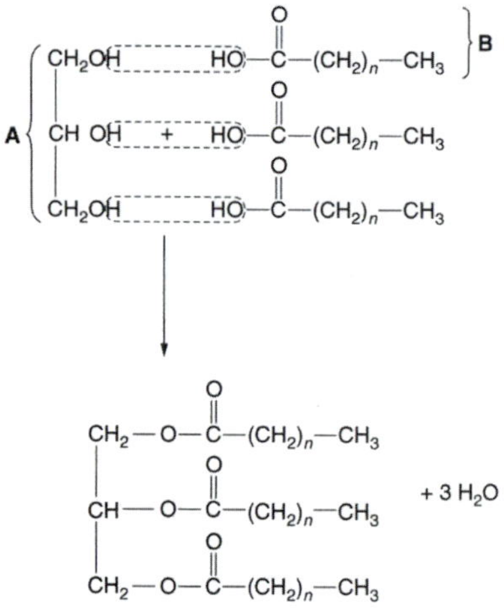

$$
A\begin{cases} CH_2OH \\ CH\ OH \\ CH_2OH \end{cases} +\ \begin{aligned} &HO-\overset{\overset{O}{\|}}{C}-(CH_2)_n-CH_3 \\ &HO-\overset{\overset{O}{\|}}{C}-(CH_2)_n-CH_3 \\ &HO-\overset{\overset{O}{\|}}{C}-(CH_2)_n-CH_3 \end{aligned}\Bigg\} B
$$

$$
\begin{aligned} &CH_2-O-\overset{\overset{O}{\|}}{C}-(CH_2)_n-CH_3 \\ &CH-O-\overset{\overset{O}{\|}}{C}-(CH_2)_n-CH_3 \\ &CH_2-O-\overset{\overset{O}{\|}}{C}-(CH_2)_n-CH_3 \end{aligned} \quad +\ 3\ H_2O
$$

Fig. 2.2

(a) With reference to Fig. 2.2,
  (i) name the molecules **A** and **B**; [2]
  (ii) state the name of the reaction shown. [1]
(b) Animals and plants store triglycerides as energy reserves.
  Explain the advantages of storing triglycerides as energy reserves rather than carbohydrates, such as starch. [2]

*9700/02 Nov 2002 Q2 (part)* Total [5]

## Chapter 3

**3.1** Fig. 3.1 shows the effect of increasing substrate concentration on the rate of a particular reaction in the presence and absence of an enzyme.

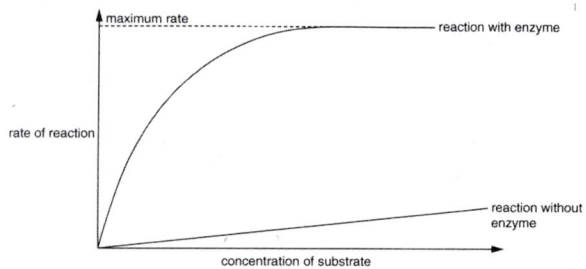

Fig. 3.1

(a) With reference to Fig. 3.1 explain the difference between the two graphs. [3]

(b) (Draw out Fig. 3.1) On your drawing of Fig. 3.1, draw two labelled curves to show the effect on the rate of the enzyme catalysed reaction of the addition of a competitive inhibitor; and a non-competitive inhibitor. [2]

(c) Explain the effect of a competitive inhibitor on the rate of enzyme activity. [3]

*9266/03 Nov 2000 Q4* Total [8]

**3.2** Fig. 3.2 shows the energy changes during the course of a reaction with and without an enzyme catalyst.

(a) State the name given to the energy needed before the reaction can occur. [1]

(b) State why an enzyme catalysed reaction is more likely to occur. [1]

(c) Describe the ways in which an enzyme interacts with its substrate. [3]

(d) Explain how increasing the temperature of an enzyme-catalysed reaction will increase the rate of the reaction. [3]

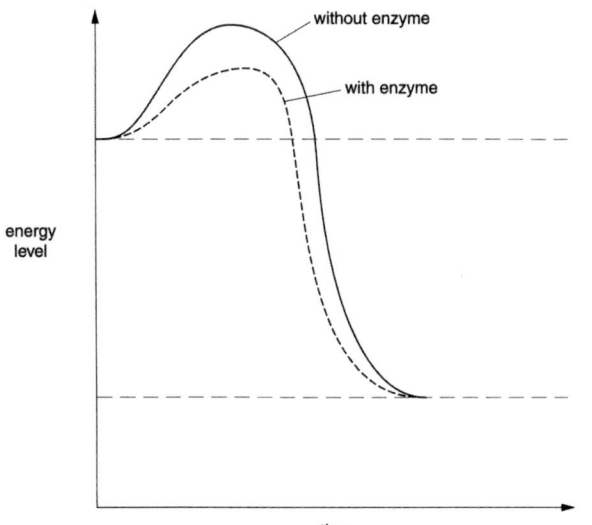

Fig. 3.2

*9266/03 June 2001 Q3* Total [8]

## Chapter 4

**4.1** Fig. 4.1 is a drawing made from an electron micrograph showing a cross-section of an alveolus and two adjacent capillaries.

(a) Calculate the magnification of Fig. 4.1. Show your working and express your answer to the nearest whole number. (ref. Chapter 1) [2]

(b) With reference to Fig. 4.1, describe the process of gas exchange in the alveolus. [4]

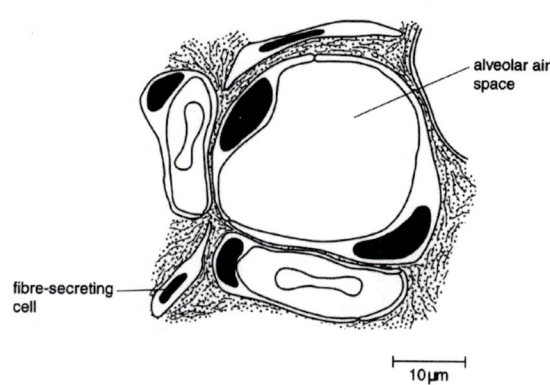

Fig. 4.1

*9700/02 Nov 2002 Q1 (part)* Total [6]

**4.2** T Lymphocytes have protein receptors in their cell surface membranes. These T cell receptors are very similar in structure to antibody molecules. Each type of T cell receptor binds specifically to one type of antigen. Fig. 4.2 shows part of a cell surface membrane of a T cell with an antigen bound to a T cell receptor. (ref. Chapter 14)

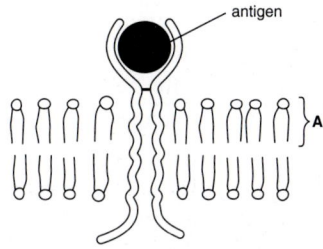

Fig. 4.2

(a) With reference to Fig. 4.2,

   (i) name the molecule labelled **A**; [1]

   (ii) describe the structure of molecule **A**; [3]

   (iii) explain why membranes are described as *fluid mosaic* in structure. [2]

(b) Explain, in terms of protein structure, how it is possible for each type of T cell receptor to bind specifically to one type of antigen. (ref. Chapters 2 and 14) [3]

*9700/02 June 2002 Q3* Total [9]

## Chapter 5

**5.1** (a) List **three** ways in which transcription differs from translation in protein synthesis. [3]

Fig. 5.1 represents a polyribosome with several translation sites.

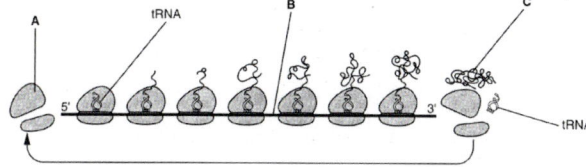

Fig. 5.1

(b) Name the structures labelled **A**, **B** and **C**. [3]

(c) Name two molecules, in addition to the molecules shown in Fig. 5.1, which are required to complete translation. [2]

(d) Describe two structural features that adapt tRNA to its role in translation. [2]

*9266/03 June 2000 Q4* Total [10]

5.2 Fig. 5.2 shows the process of translation occurring at a ribosome in a cell that synthesises an enzyme that is secreted out of the cell to carry out its function.

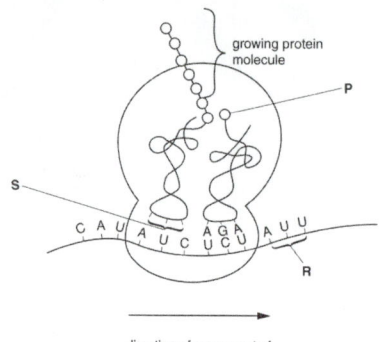

growing protein molecule

P

S

R

direction of movement of ribosome along mRNA

Fig. 5.2

Table 5.2 shows some triplet base sequences of mRNA and the amino acids for which they code.

| mRNA | amino acid |
|------|-----------|
| AUU | isoleucine |
| AUC | isoleucine |
| AUG | methionine |
| UUU | phenylalanine |
| UCU | serine |
| CAU | histidine |

Table 5.2

(a) With reference to Fig. 5.2 and Table 5.2,
  (i) name the amino acid **P**; [1]
  (ii) state the base sequence at **S**; [1]
  (iii) state the name given to the triplet base sequence on mRNA; [1]
  (iv) describe the change that would occur to the protein if the base sequence at **R** was UUU instead of AUU. [1]
(b) Describe what happens to the enzyme molecule after it has left the ribosome until it leaves the cell. [4]

*8700/02 Nov 2001 Q4* Total [8]

5.3 Table 5.3 shows the relative amounts of the bases Adenine, Thymine, Guanine and Cytosine in DNA from different organisms.

| Source | Adenine | Thymine | Guanine | Cytosine |
|--------|---------|---------|---------|----------|
| Bacterium | 23.8 | 23.1 | 26.8 | 26.3 |
| Maize | 26.8 | 27.2 | 22.8 | 23.2 |
| Fruit fly | 30.7 | 29.5 | 19.6 | 20.2 |
| Chicken | 28.0 | 28.4 | 22.0 | 21.6 |
| Human | 29.3 | 30.0 | 20.7 | 20.0 |

Table 5.3

(a) Explain the importance of the ratios of A to T and G to C to the structure of DNA. [3]
(b) In about half an A4 page, draw and fully label a short length of a DNA molecule. [4]
  The bacteriophage virus φX-174 has single stranded DNA with the four bases present in the following relative amounts.
  Adenine    Thymine    Guanine    Cytosine
  24.0       31.2       23.3       21.5
(c) Suggest why the ratios of A to T and C to G for the virus do not correspond to the ratios found in living organisms. [1]

*9266/03 Nov 1998 Q1* Total [8]

## Chapter 6

6.1 Fig. 6.1 shows the life cycle of a species of brown seaweed.

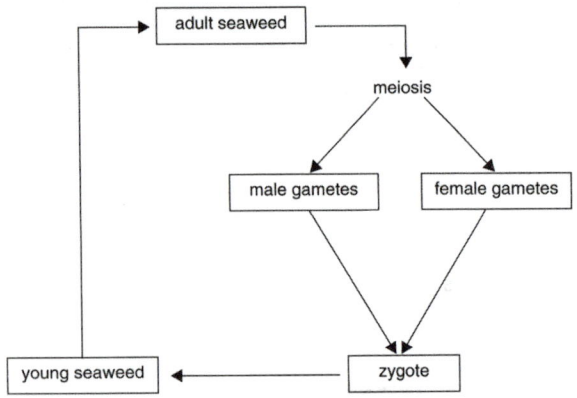

Fig. 6.1

(a) Indicate on Fig. 6.1, with the letter **M**, one stage where mitosis occurs. [1]
(b) DNA replication occurs in cells during interphase before they divide by mitosis. Explain why it is important that replication occurs before mitosis. [2]
(c) Explain why DNA replication is described as semi-conservative. (ref Chapter 5) [2]
(d) Explain why meiosis occurs in the life cycle of this seaweed. [3]

*9700/02 June 2002 Q1* Total [8]

6.2 The diploid number for a certain species of animal is 12. Fig. 6.2 shows the appearance of a pair of homologous chromosomes in a cell from this animal during mitosis.

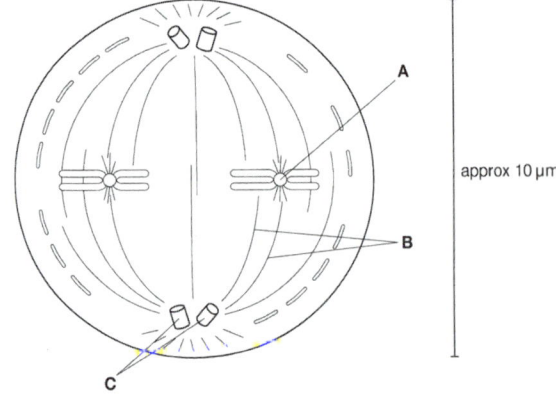

A

B

C

approx 10 μm

Fig. 6.2

(a) Name the stage of mitosis shown in Fig. 6.2. [1]
(b) Name the structures **A**, **B** and **C** and describe their functions in mitosis. [3]
(c) Describe the events that occur during mitosis immediately following the stage shown in Fig. 6.2 until the formation of the nuclear membrane. [2]
(d) Explain why the chromosomes shown in Fig. 6.2 are described as homologous. [2]
(e) State, in terms of chromosome number, how a gamete would differ from the cell shown in Fig. 6.2. [1]

*8700/02 June 2001 Q1* Total [9]

6.3 Movements within a cell during mitosis are plotted in Fig. 6.3 in three curves that show changes in distance between:
  1  the centromeres of the chromosomes and the poles of the spindle,

2  the centromeres of sister chromatids,
3  the poles of the spindle.
On the time scale, 0 marks the time when chromosomes line up on the equator.
(a)  With reference to Fig. 6.3,
  (i)  identify the two main stages of mitosis when these movements occur;  [2]
  (ii)  describe what happens to the chromatids after 15 minutes.  [2]
(b)  Describe and explain the changes in curve 3.  [2]

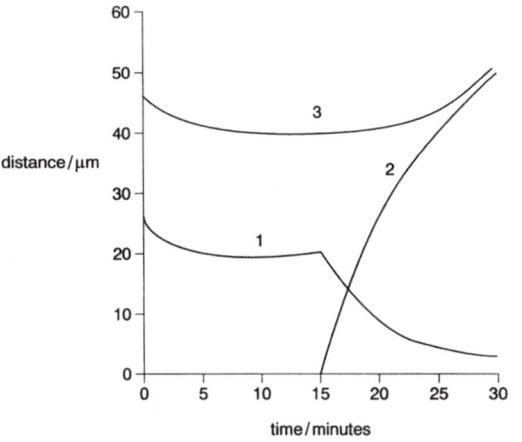

Fig. 6.3

*9266/03 Nov 1999 Q3 (part)*  Total [6]

## Chapter 7

**7.1**  Fig. 7.1 shows part of a food web for a marine ecosystem. The food web is based around a species of starfish.

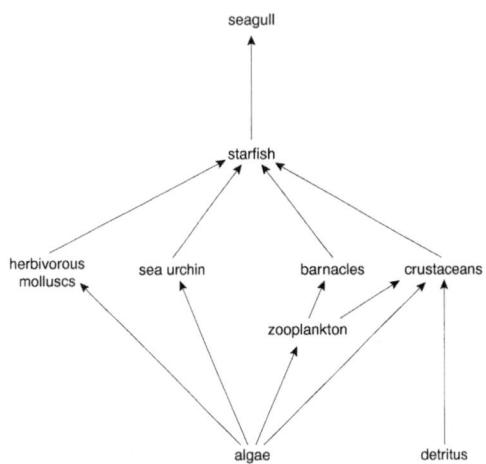

Fig. 7.1

Starfish feed at different trophic levels.
(a)  Write out **one** food chain in which the starfish feeds at the secondary consumer trophic level.  [1]
(b)  Explain why there is less energy available to the starfish compared to the zooplankton in this food web.  [3]
(c)  Suggest an advantage to populations of starfish given by feeding on a variety of prey species.  [2]
(d)  State the meaning of the term ecosystem.  [1]

*8700/02 June 2001 Q7*  Total [7]

**7.2**  Fig. 7.2 shows part of the nitrogen cycle.

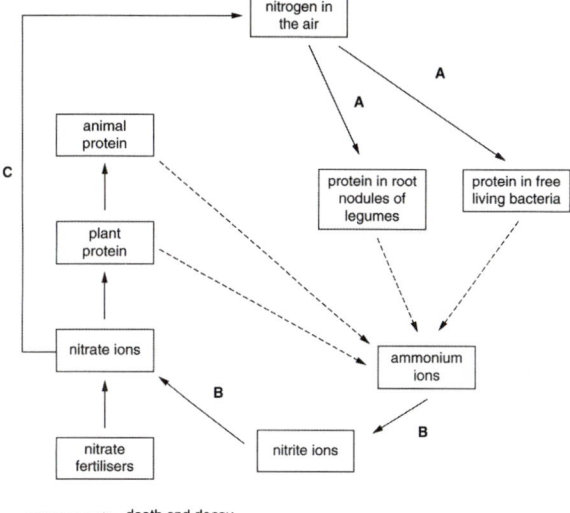

Fig. 7.2

(a)  With reference to Fig. 7.2,
  (i)  name the processes **A** to **C**;  [3]
  (ii)  explain how a farmer may maintain the fertility of soil without using nitrogen fertilisers.  [3]

*9700/02 June 2002 Q6*  Total [6]

## Chapter 8

**8.1**  Fig. 8.1 is a light micrograph of a transverse section through part of a healthy human artery.

Fig. 8.1

(a)  With reference to Fig. 8.1,
  (i)  name the tissues **A** to **C**;  [3]
  (ii)  state two ways in which a transverse section of a vein would be different;  [2]
  (iii)  suggest why the tissue **A** is folded.  [1]
(b)  Explain how smoking may affect the arteries supplying the cardiac muscle, resulting in coronary heart disease.  [3]

*9266/03 Nov 1998 Q2 (modified)*  Total [9]

**8.2**  Fig. 8.2a shows a wireframe model of a molecule of haemoglobin.

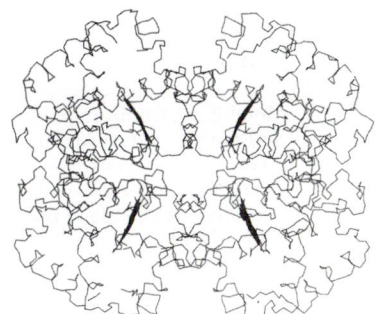

Fig. 8.2a

(a) (i) On a sketch of Fig. 8.2a, clearly indicate, using arrows, the positions of iron in the molecule. [1]

(ii) State two other features of the molecular structure of haemoglobin that are visible in Fig. 8.2a. (ref. Chapter 2) [2]

Fig. 8.2b shows the oxygen dissociation curves for haemoglobin at two concentrations of carbon dioxide ($CO_2$).

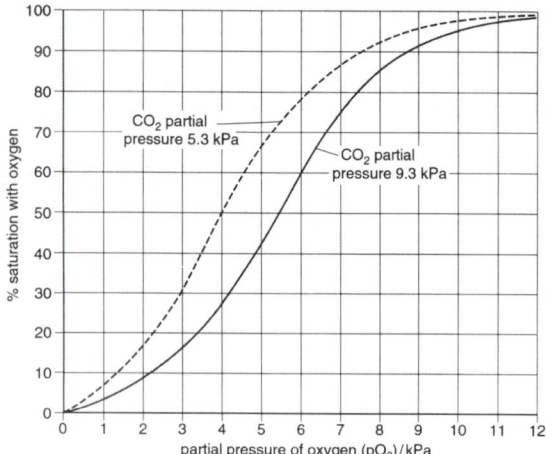

Fig. 8.2b

The partial pressure of oxygen ($pO_2$) in tissues is approximately 3.5 kPa.

(b) With reference to Fig. 8.2b,
(i) state the difference in % saturation of haemoglobin between the two concentrations of $CO_2$ at this partial pressure of oxygen; [1]
(ii) explain the physiological importance of this difference in the % saturation of haemoglobin. [3]

*9266/03 June 1998 Q5*      Total [7]

## Chapter 9

**9.1** Fig. 9.1 shows the left side of the human heart at a particular phase in the cardiac cycle.

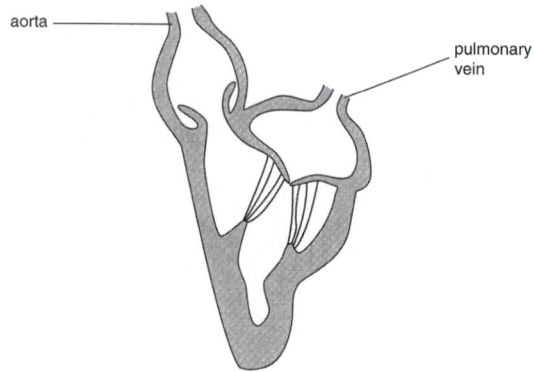

Fig. 9.1

(a) Describe the events occurring in the phase of the cardiac cycle shown in Fig. 9.1. [2]
(b) Explain how the structure of the heart ensures that blood flows from the pulmonary vein to the aorta and not in the reverse direction. [3]
(c) Blood pressure is usually measured by placing a cuff over an artery in the arm.
Explain why the blood pressure in the left ventricle falls to zero in the cardiac cycle, but the lowest pressure recorded in the arteries is about 10 kPa. (ref. Chapter 8) [3]

(d) Coronary heart disease and stroke are two forms of cardiovascular disease. Smoking is recognised as a contributory factor to these diseases.
Explain how smoking may contribute to the development of cardiovascular diseases, such as coronary heart disease and stroke. (ref Chapter 12) [3]

*9700/02 June 2002 Q4*      Total [11]

**9.2** Fig. 9.2 shows a section through the heart with structures involved in the initiation and transmission of heart action.

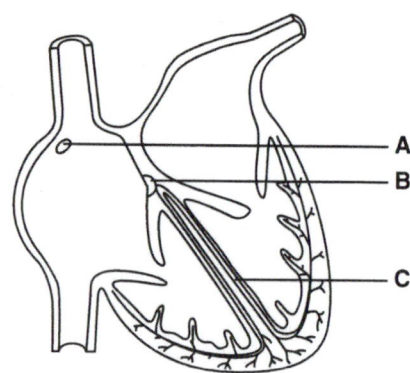

Fig. 9.2

(a) Name the structures labelled **A**, **B** and **C**. [3]
(b) Describe how the contraction of the atria and ventricles occur in sequence. [2]
(c) Suggest why structure **C** is required for the transmission of depolarisation through the ventricles but a similar structure is not required for the atria. [2]

*9266/03 Nov 2000 Q2 (part)*      Total [7]

## Chapter 10

**10.1** (a) Explain why transpiration is the inevitable consequence of gaseous exchange in land plants. [3]
Fig. 10.1 shows some of the cells from the lower part and under surface of a leaf. The water potentials of three cells, A, B and C are shown.

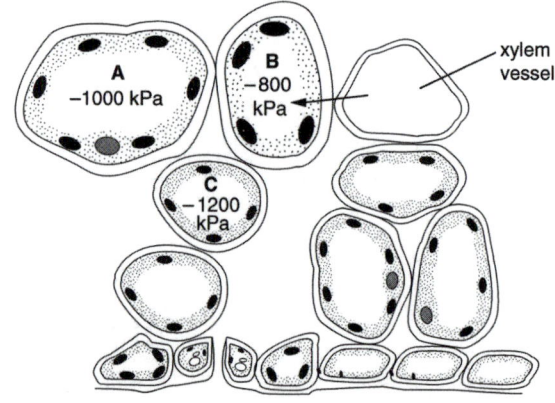

Fig. 10.1

(b) Explain how water moves from the xylem vessel to cell **B**. [3]
(c) Sketch Fig. 10.1 and draw labelled arrows on the sketch to show the direction in which:
(i) water flows between the cells A, B and C; [2]
(ii) water vapour diffuses. [1]
(d) State two features of xerophytic plants that help to reduce the loss of water by transpiration from their leaves. [2]

*9700/02 Nov 2002 Q3*      Total [11]

**10.2** Fig. 10.2 shows a three-dimensional diagram of part of a plant stem.

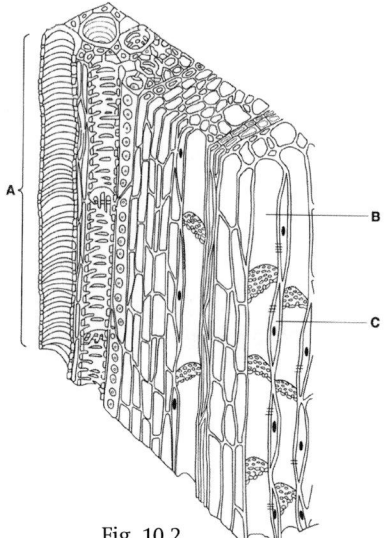

Fig. 10.2

(a) Identify the structure labelled **A** and state one function it performs in the stem. [1]
(b) State three main structural features that help **A** to carry out this function. [3]
(c) The cell labelled **B** is involved in the transport of the products of photosynthesis up and down the stem. Identify the cell labelled **C** and explain how it is involved in this process. [3]

*8700/02 Nov 2001 Q1* Total [7]

## Chapter 11

**11.1** (a) Explain the terms tidal volume and vital capacity. [4]
(b) Fig. 11.1 shows two different epithelial cells from the human gas exchange system.

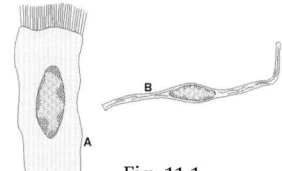

Fig. 11.1

(i) State two differences that can be seen between epithelial cell **A** and cell **B**. [2]
(ii) Describe the distribution and function of epithelial cells like **A** in the human gas exchange system. [3]
(iii) Explain how the cell contents of cells like cell **B** contribute to reduction in lung volume during exhalation. [2]

*Exemplar* Total [11]

## Chapter 12

**12.1** Fig. 12.1 shows diagrams of two alveoli from the lungs of a smoker, A, and a non-smoker, B, after exhalation is complete.

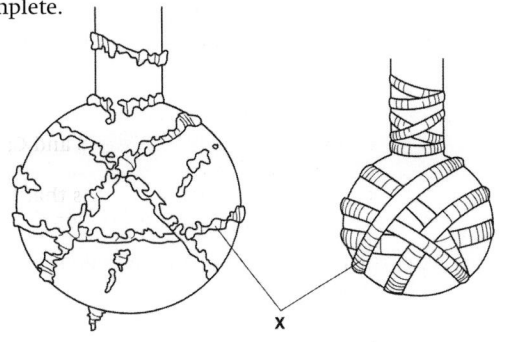

Fig. 12.1

(a) Identify the fibres labelled **X**, in Fig. 12.1 that surround the alveoli. [1]
(b) With reference to Fig. 12.1, explain why people who have smoked heavily for a long time often have difficulty in breathing and in obtaining sufficient oxygen. [3]
(c) State two pieces of epidemiological evidence and one piece of experimental evidence to link smoking with lung cancer. [3]

*8700/02 Nov 2001 Q2* [Total 7]

## Chapter 13

**13.1** (a) Complete Table 13.1 to show which of the four statements about the transmission and control of disease apply to cholera, malaria, HIV/AIDS and tuberculosis (TB). In each box, use a tick (✓) to show that the statement applies and a cross (✗) if it does not.

| Statement | Cholera | Malaria | HIV/AIDS | TB |
|---|---|---|---|---|
| causative organism is a bacterium | | | | |
| causative organism is water-borne | | | | |
| transmitted by an insect vector | | | | |
| sexually transmitted | | | | |

[4]

Table 13.1

(b) Sickle cell anaemia is an inherited disease that is common in areas of the world where malaria is endemic.
Explain why sickle cell anaemia is common in areas where malaria is endemic. [3]

*8700/02 Nov 2001 Q6* Total [7]

**13.2** (a) Describe how the malarial parasite is normally transmitted from an infected person to an uninfected person. [2]
Fig. 13.2 is drawn from an electron micrograph of a red blood cell taken from a person suffering from malaria.

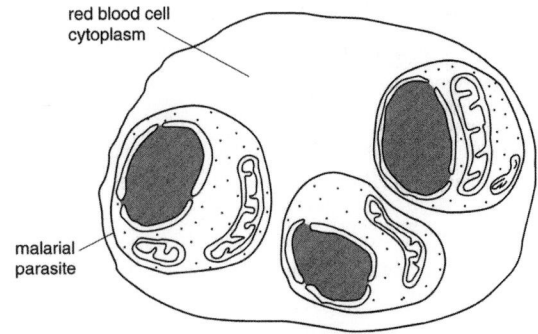

Fig. 13.2

(b) State two features, visible in Fig. 13.2, that indicate that the malarial parasite is eukaryotic. [2]
(c) Outline the likely effects on the body of the presence of malarial parasites in red blood cells. [3]

*9700/02 Nov 2002 Q5* Total [7]

## Chapter 14

**14.1** Phagocytes are involved in removing bacteria from infected tissues.

Fig. 14.1 is a drawing made from an electron micrograph of a phagocyte that has engulfed some bacteria.

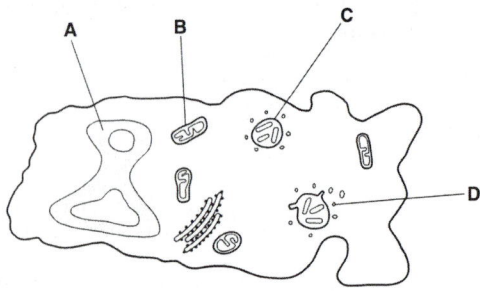

Fig. 14.1

(a) Name **A** to **D**. [2]
(b) Explain why it would not be possible to see the same detail of a phagocyte, as shown in Fig. 14.1, using a light microscope. (ref. Chapter 1) [3]
(c) Explain how phagocytes engulf and destroy invading organisms, such as bacteria. [4]

*8700/02 Nov 2001 Q5* Total [9]

**14.2** Both phagocytes and lymphocytes are involved in the body's defence against bacterial infections.

(a) (i) State the site of origin of phagocytes and lymphocytes. [1]
(ii) Name the organ where T lymphocytes mature. [1]
(b) Describe how phagocytes remove bacteria from a site of infection.
(c) Describe two ways in which active immunity differs from passive immunity. [2]

*8700/02 Nov 2001 Q5* Total [7]

## Chapter 15

**15.1** (free response question)
(a) Describe the main stages of glycolysis. [7]
(b) Explain the role of NAD and oxygen in oxidative phosphorylation. [8]

*9266/03 Nov 1998 Q7* Total [15]

**15.2** ATP is described as the energy 'currency' in all living organisms.
State the role of ATP in:
(a) glycolysis; [2]
(b) the light independent stage of photosynthesis; [2]
(c) protein synthesis; [2]
(d) the replication of DNA. [2]

*9266/03 Nov 2001 Q5* Total [8]

**15.3** Fig. 15.3 shows the Krebs cycle and the reactions preceding it.

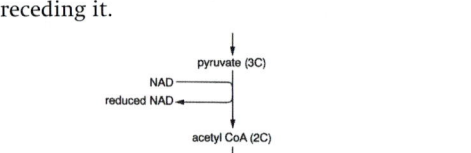

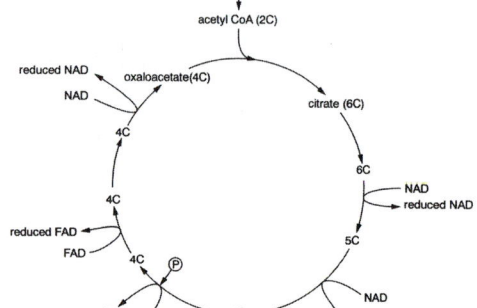

Fig. 15.3

(a) State precisely where the Krebs cycle occurs in cells. [1]
(b) Draw Fig. 15.1, then label on your drawing, all the stages where:
(i) decarboxylation reactions occur. Label these with a large letter C. [2]
(ii) dehydrogenation reactions occur. Label these with a large letter H. [2]
(c) Explain the role of NAD in the Krebs cycle. [3]

*9266/03 Nov 1999 Q5* Total [8]

## Chapter 16

**16.1** Fig. 16.1 shows the changes in concentration of a 3C compound, glycerate phosphate, GP, and a 5C compound, ribulose bisphosphate, RuBP, extracted from samples taken from actively photosynthesising green algae in an experimental chamber when the light source was turned off.

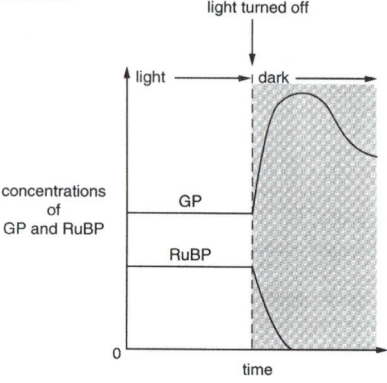

Fig. 16.1

(a) With reference to Fig. 16.1, describe what happens, after the light source was turned off, to the concentration of:
(i) GP; [2]
(ii) RuBP. [1]
(b) Explain, with reference to the Calvin cycle, the reasons for these observed changes in:
(i) GP; [2]
(ii) RuBP. [2]
(c) State the two products of photophosphorylation that drive the Calvin cycle. [2]

*9700/04 Nov 2002 Q1* Total [9]

**16.2** Fig. 16.2 shows the follow of electrons in cyclic and non-cyclic photophosphorylation.

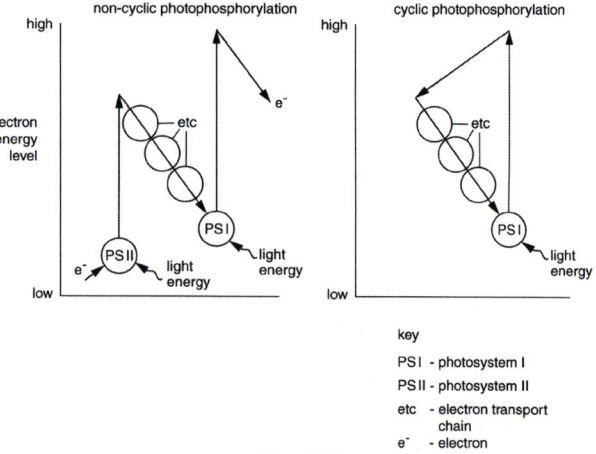

key

PSI - photosystem I
PSII - photosystem II
etc - electron transport chain
e⁻ - electron

Fig. 16.2

(a) State the precise location of photophosphorylation in a chloroplast. [2]
(b) Describe the role of light in photophosphorylation. [2]
(c) Explain how non-cyclic photophosphorylation differs from cyclic photophosphorylation. [4]

(d) Paraquat is a herbicide that prevents the flow of electrons from photosystem 1 and reduces oxygen to a chemically reactive superoxide radical. This results in severe damage to chloroplasts. It is now possible to make a crop plant resistant to such herbicides. Suggest a use for such plants. [1]

*9700/04 June 2002 Q1* Total [9]

**16.3** (free response question)
(a) Describe the role of light in photosynthesis. [7]
(b) Explain the main stages of the Calvin cycle. [8]

*9266/03 June 1998 Q7* Total [15]

## Chapter 17

**17.1** Enzymes are globular proteins that catalyse specific reactions.
(a) Restriction enzymes cut DNA into fragments. They cut at specific sites determined by the sequence of bases. Fig. 17.1 shows the base sequences cut by three restriction enzymes and a section of DNA cut by one of those enzymes.
  (i) Identify the restriction enzyme that has cut the section of DNA shown in Fig. 17.1. [1]
  (ii) State the name given to the unpaired base sequences that remain after DNA has been cut by the three restriction enzymes shown in Fig. 17.1. [1]

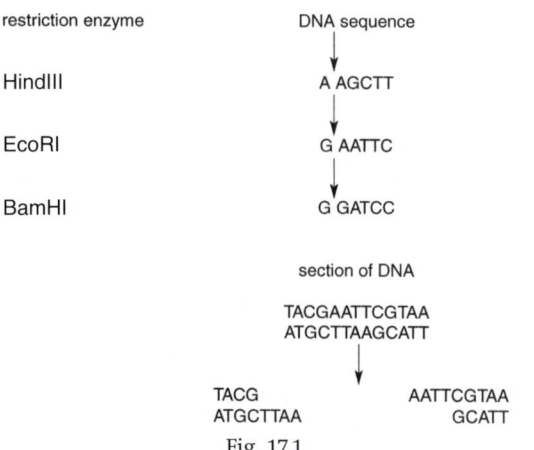

Fig. 17.1

(b) Human genes may be cloned by inserting lengths of DNA into bacteria. This may be carried out by inserting the DNA into a plasmid.
Explain how lengths of DNA, cut by restriction enzymes, are inserted into plasmids. [3]

*9700/02 Nov 2002 Q4 (part)* Total [5]

**17.2** Fig. 17.2 shows the results of an animal cell that has undergone meiosis I.
(a) State which stage of meiosis II is shown. [1]
(b) Describe the next stage of meiosis II. [3]
(c) Describe what has happened before the start of meiosis to: the nuclear membrane; [1] and the centrioles. [2]
(d) Name and explain **two** ways in which meiosis can lead to variation. [4]

*9700/04 Nov 2002 Q4* Total [11]

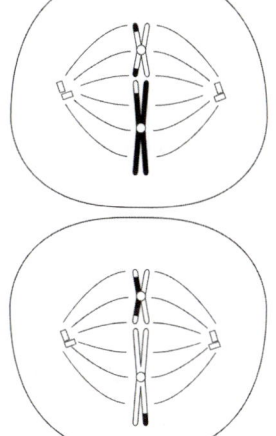

Fig. 17.2

**17.3** Pure-breeding pea plants with round, yellow seeds were crossed with pure-breeding pea plants with wrinkled, green seeds. The offspring all had round, yellow seeds. These seeds were grown and the resulting plants allowed to self-pollinate.
This produced 1112 offspring with the following characteristics.

> 630 round, yellow seeds
> 202 round, green seeds
> 216 wrinkled, yellow seeds
> 64 wrinkled, green seeds

(a) Using symbols R for round, r for wrinkled, B for yellow and b for green, draw a genetic diagram to explain these results [4]
(b) Explain why the wrinkled, green seeds produced pure-breeding offspring, whilst the round, yellow seeds did not. [3]
(c) A ratio of 9:3:3:1 was expected.
A chi squared test was carried out to test the significance of the differences between the observed and expected results. This gave a value of 0.47.

| probability | 0.99 | 0.98 | 0.95 | 0.90 | 0.50 | 0.10 | 0.05 | 0.02 | 0.01 |
|---|---|---|---|---|---|---|---|---|---|
| at 3 degrees of freedom | 0.12 | 0.19 | 0.35 | 0.58 | 2.4 | 6.3 | 7.8 | 9.8 | 11.3 |

With reference to the table of probabilities, explain how the value for the chi squared test supports the hypothesis that these are two pairs of segregating alleles at two loci. [2]

*9700/04 June 2001 Q5* Total [9]

## Chapter 18

**18.1** In Lake Tanganyika in Africa, there are six species of fish of the genus *Tropheus* and a much larger number of distinctly coloured subspecies of each of the six species. *Tropheus* species are small fish that are confined to isolated rocky habitats around the shores of Lake Tanganyika.
  Recent research has compared DNA sequence data from these various species and subspecies and linked this with geological data on the lake. This suggests that some 1.25 million years ago, when the lake was first filled, the six species evolved during the primary radiation phase. They arose from river dwelling ancestors and then filled all available niches. Secondary radiations into the many subspecies occurred during the last 200 000 years. Sometime during this period, the water level in the lake fell, resulting in the formation of three separate lake basins. These basins persisted for many thousands of years before the water level rose again.
  Fig. 18.1 shows an outline map of the lake and the location of the three temporary basins caused by lowering of lake levels.

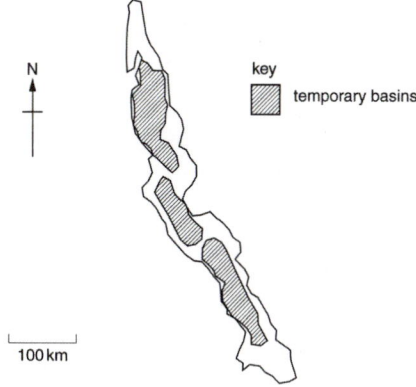

Fig. 18.1

(a) Define the terms species [3] and niche [2].

(b) Explain how natural selection could have caused the evolution of the six closely related species in the primary radiation. [3]

(c) Suggest how the lowering of the water level in the lake to form three separate lake basins could have caused the evolution of so many subspecies. [1]

*9266/03 June 2001 Q4*                    Total [9]

## Chapter 19

**19.1** (a) List three functions of the mammalian kidney [3]

Fig. 19.1 is a diagram of a section through the proximal (first) convoluted tubule of a kidney nephron showing details of cell structure, as seen with an electron microscope.

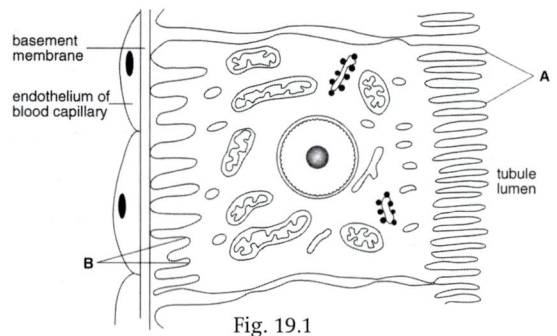

Fig. 19.1

(b) Name structures **A** and **B**. [2]

(c) State two ways in which the cells of the proximal convoluted tubule are adapted for the reabsorption of substances from the glomerular filtrate. [2]

(d) About 80% of the glomerular filtrate is reabsorbed in the proximal convoluted tubule. Explain the mechanism of reabsorption in the kidney. [4]

*9266/03 Nov 2001 Q1 (part)*                    Total [11]

**19.2** (free response question)

(a) Explain what is meant by an endocrine gland, with reference to the islets of Langerhans in the pancreas. [7]

(b) Outline how the blood glucose concentration is regulated by insulin and glucagon. [8]

*9263/03 Nov 2000 Q8*                    Total [15]

**19.3** (free response question)

(a) Outline the need for, and differences between, communication systems within mammals and flowering plants. [7]

(b) Describe the roles of auxins, gibberellins and abscissic acid in control of plant activities. [8]

*Exemplar*                    Total [15]

**19.4** Fig. 19.4 shows an action potential.

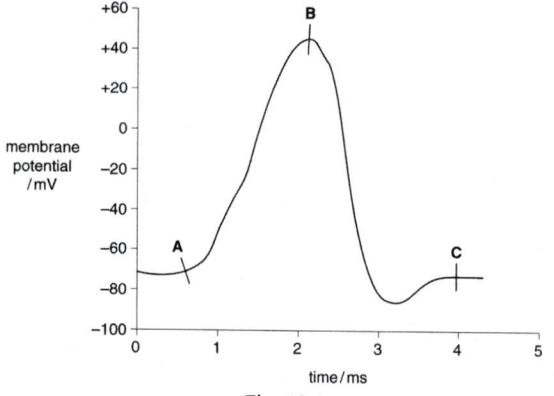

Fig. 19.4

(a) Describe what is happening to $Na^+$ and $K^+$ channels in nerve axons between:

  (i) A and B; [2]

  (ii) B and C. [2]

---

(b) Describe the role of $Ca^{2+}$ ions in the passage of impulses across a synapse. [3]

(c) Explain why the concentration of $Ca^{2+}$ ions of the synaptic knob remains low, despite frequent opening of channels. [2]

(d) Describe how the transmitter causes depolarisation on the postsynaptic membrane. [2]

*9266/03 June 2001 Q2*                    Total [11]

## Answers

**Mark Scheme disclaimer**

The answers provided in the mark schemes are the ones looked for by examiners and, therefore, the most useful for teaching purposes. However they are not necessarily the only answers given credit on the occasion of a particular examination. Alternative wording to that given can often be acceptable.

UCLES/CIE will not enter into discussions or correspondence in connection with these extracts from their mark schemes.

| | |
|---|---|
| ; = end of mark point | / = alternatives |
| AW = accept similar idea | A = accept, R = reject |
| Underlining = essential word(s) | AVP = additional valid point |
| ref. = makes contextual reference to… | () = context or A or R |

**Chapter 1**

**1.1** (a) (i) A = nucleolus / nucleoli; B = <u>rough</u> endoplasmic reticulum / RER; [2]

  (ii) A = RNA / subunit / ribosome synthesis; B = protein / polypeptide synthesis / transport, / translation / mRNA to protein; [2]

(b) condense / thicken / denser / chromosomes become visible; [1]

(c) observed length $\frac{65mm}{16\,\mu m} = \frac{65000\,\mu m}{16\,\mu m} = \times\, 4062$

depending on precise length of bar [1]

**Chapter 2**

**2.1** (a) (i) amino acid; [1]

  (ii) hydrogen bonds break; loses shape / uncoils / disrupted / becomes straight chain; denatured; [max 2]

(b) disulphide / sulphur bridges; van der Waals / AW; hydrogen; ionic; [max 3]

(c) (polysaccharide) may be branched; glycosidic links; no R groups; one monomer / repeating unit structure; AVP; [max 2]

(d) (collagen) 3 polypeptides twisted round each other; hydrogen bonding between polypeptides; covalent links between chains; every 3rd amino acid = glycine which is small so tightly coiled; lay parallel to form fibrils; ends staggered / fibrils overlap, so strong; [max 2]

(cellulose) many -OH groups; hydrogen bonds within molecule; hydrogen bonds between molecules; straight chain / not helix / unbranched / linear; 60–70 chain associated to form microfibrils; arranged in bundles to form macrofibrils / fibres; [max 2]

**2.2** (a) (i) A = glycerol; B = fatty acid;

  (ii) condensation / esterification / ester bond formation; [3]

(b) more energy released / stored, per gram / unit / given mass (R per mole); 37 v 17 kJ (or equivalent); fats more highly reduced; more hydrogens / fewer oxygens / higher carbon to hydrogen ratio / more CH bonds; release / yield more energy when respired / oxidised; [max 2]

**Chapter 3**

**3.1** (a) enzyme lowers activation energy; enzyme concentration limiting; all available active sites occupied at any one time / AW; without enzyme rate of reaction increases steadily (as substrate increases); [max 3]

(b)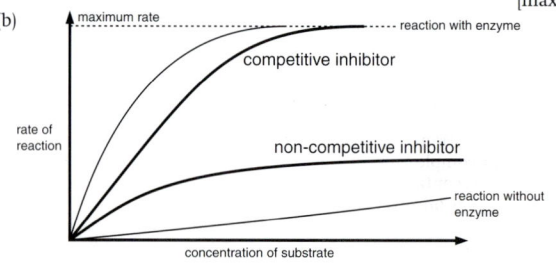

[2]

(c) competitive inhibitor similar shape to substrate / complementary shape to active site; competitive inhibitor attaches to active site; temporarily; prevents substrate from fitting / entering active site;

ref. to relative concentration of substrate / competitive inhibitor on degree of inhibition; [max 3]

**3.2** (a) activation energy; [1]
(b) increases number of substrate molecules with the required energy / the reaction proceeds faster; [1]
(c) ref. to binding with substrate; ionic interactions / attraction / ref. to charged groups; hydrogen bonding; hydrophobic/hydrophilic interactions; ref. to formation of enzyme/substrate complex; ref. to induced fit; ref. to distortion of / strain placed on substrate; ref. to temporary; [max 3]
(d) increase kinetic energy of substrate / molecules; molecules move faster; increase collision rate; increase number of molecules with activation energy; $Q_{10} = 2$; ref. to improved fit of substrate; [max 3]

**Chapter 4**

**4.1** (a) correct measurement of scale bar used as basis for finding magnification (A 9–10 mm) with appropriate working e.g. $9 \times 1000/10$; Correct answer (A 900 to 1000); [2]
(b) movement of oxygen into alveolus; concentration gradient (of $CO_2$ or $O_2$ between air and blood) / AW; oxygen dissolves in liquid film / surfactant film; ref. diffusion; $O_2$ and $CO_2$ exchanged / idea of; squamous / alveolar epithelium; endothelium (of capillary); red blood cell; short diffusion distance / 2–3 μm; [max 4]

**4.2** (a) (i) phospholipid; [1]
(ii) glycerol; 2 fatty acids / hydrocarbon (tails/chains); phosphate (R phosphorus); [3]
(iii) phospholipids are, fluid / liquid / move about / diffuse, within their own monolayer / layer; proteins (are separate pieces), floating / moving about, in liquid / AW; ref. to pattern / arrangement of proteins; [max 2]
(b) variable region / AW; different / particular, sequences of amino acids / primary structure; different, 3D shape / conformation / folding / tertiary structure; ref. to R groups / side chains; complementary to / matches, shape of antigen (R similar / same) [max 3]

**Chapter 5**

**5.1** (a) (transcription) DNA code used / ref. DNA triplets / (translation) mRNA code used ref. codons; (transcription) in nucleus / (translation) in cytoplasm / ribosomes / RER; (transcription) produces mRNA / (translation) produces polypeptide / protein; [3]
(b) A = ribosome; B = mRNA / messenger RNA; C = polypeptide / protein; [3]
(c) ATP; amino acids; [2]
(d) ref. to partly single / double stranded / folded so it fits in a ribosome; receptor site for specific amino acid; ref. to anticodon / 3 bases sticking out; [max 2]

**5.2** (a) (i) P = serine;
(i) UAG;
(iii) codon;
(iv) phenylalanine; [4]
(b) secondary structure / α-helix / pleated sheet; further folding / tertiary structure; ref. named bonds; transported in ER; Golgi; packaged / sugars added / trimmed; vesicles; exocytosis; [max 4]

**5.3** (a) 1:1 / equal amounts; complementary / (base) pairing / A with T & C with G; occupy central position in molecule; ref. copying of strands / replication; ref. role of hydrogen bonding maintaining particular pairing / holding two strands together; ref. mutation when pairing ratio lost; [max 3]
(b) nucleotides; deoxyribose; phosphate; [max 2 if only 1 strand shown]
phosphate correctly linked to deoxyribose / phosphodiester link / bond; ref. to strands antiparallel / ↑↓ on diagram / 3' to 5' and 5' to 3'; hydrogen bonds between bases labelled; ref. to correct number of hydrogen bonds between bases A=T and G≡C; [max 4]
(c) ref. to no base pairing (as single stranded); [1]

**Chapter 6**

**6.1** (a) M between zygote and young seaweed / between young seaweed and adult; [1]
(b) each chromosome then contains two chromatids; (genetically) identical / exact replica; daughter cells receive a copy of, each chromosome / DNA molecule / same genetic material / AW; [max 2]
(c) each strand / polynucleotides act as template / (A sense strand); for complementary, strand / (poly)nucleotides / base pairing; new DNA contains half old, half new; [max 2]
(d) reduction division; gametes / sex cells / eggs and sperms, have half the chromosome number / haploid / n; zygote is diploid / has full number of chromosomes / 2n; chromosome number remains same / does not increase with each generation; [max 3]

**6.2** (a) metaphase;
(b) A = centromere, attaches chromosomes to spindle; B = spindle (A microtubules), for moving / pulling chromosomes;

C = centriole, centre for formation of spindle microtubules / forms poles of spindle; [3]
(c) chromatids separate / split; move to opposite poles; chromosomes (A chromatids) start to uncoil; anaphase; spindle starts to disintegrate; [max 2]
(d) the homologous chromosomes both have the same genes / (sets of) loci; length / shape / position of centromeres; A pair in meiosis / form bivalents; [max 2]
(e) half the number of chromosomes / 6 chromosomes / haploid (not diploid); [1]

**6.3** (a) (i) metaphase; anaphase; [2]
(ii) centromeres split / divide / chromatids separate; chromatids become chromosomes; pulled apart by spindle fibres / microtubules; pulled towards poles (of spindle) / asters / centrioles; [max 2]
(b) initially move closer (during first 5 minutes); as spindle forms / due to spindle formation; then move further apart (after 20 minutes); ref. to pole to pole fibres / microtubules; ref. to sliding / action of fibres / microtubules; [2 max]

**Chapter 7**

**7.1** (a) algae → herbivorous mollusc / sea urchin / crustacean → starfish; [1]
(b) energy lost at each trophic level; any 2 reasons from respiration (in zooplankton / primary consumers); movement; indigestible material / food not absorbed / passed out as faeces; material not eaten; material passed to decomposers; [max 3]
(c) constant supply of food; some prey species vary in number; some prey species may become locally extinct; less competition with other predators; wider niche; do not have to migrate for food; AVP; [max 2]
(d) area with living / biotic and non-living / abiotic components or community (A populations) interacting together and with the physical environment; [1]

**7.2** (a) (i) A = nitrogen fixation; B = nitrification; C = denitrification; [3]
(ii) legumes / green manure / compost / organic matter / humus / manure / AW ploughed / dug into soil; decay / decomposition to release ammonium (ions) / ammonification; nitrification / form nitrates; grow legumes for nitrogen fixation; crop rotation / intercropping / alley cropping; ref. improve aeration / drainage (in context of favouring nitrification / reducing denitrification); [max 3]

**Chapter 8**

**8.1** (a) (i) A = endothelial epithelium / lining epithelium / tunica intima; B = elastic fibres and smooth muscle / tunica media; C = collagen (fibres) / tunica adventitia / tunica externa; [3]
(ii) vein would have thinner muscle layer / tunica media; fewer elastic fibres; larger lumen / lumen to wall ratio; thinner wall; greater proportion of collagen fibres / thicker tunic externa / adventitia; A vasa vasoria / blood vessels penetrate closer to inner surface of wall; [max 2]
(iii) ref. to expansion / stretching / pulse / blood pressure / systole / A ref. to state of contraction of muscles in wall; [1]
(b) (nicotine / CO) damage endothelium (easing penetration by fat); encourage platelets to stick to endothelium (initiating blood clots); CO, combines with haemoglobin / forms carboxyhaemoglobin (reducing oxygen supply to myocardium); (nicotine) causes vasoconstriction of coronary arteries / increases blood pressure / increases heart rate; encourages fibrinogen production / inhibits clot buster enzymes / AW; increased risk of atherosclerosis / thrombus; raises blood fat levels; [max 3]

**8.2** (a) (i) 4 arrows pointing to centres of each of 4 subunits; [1]
(ii) 4 subunits (2 α and 2 β) / chains / parts; globular / folded protein / ref. tertiary structure; α helix present; 4 haem / (A porphyrin) / prosthetic / non-protein molecules; ref. to quaternary structure; ref. to symmetry; [max 2]
(b) (i) 19 ± 1%; [1]
(ii) faster respiration / most active tissues, produce most $CO_2$ / during activity $CO_2$ levels increase; (increase in $CO_2$) reduces affinity of haemoglobin for oxygen (A $O_2$) / (oxyhaemoglobin) dissociates more rapidly; oxygen unloaded to those tissues respiring fastest / most needing oxygen; to complete aerobic respiration; producing more / faster ATP; [max 3]

**Chapter 9**

**9.1** (a) ventricle contracts / ventricular systole; blood forced through semi-lunar valve; into aorta; atrio-ventricular / bicuspid / mitral (R tricuspid) valve closed; [max 2]
(b) ref. to unidirectional valves / valves prevent backflow; blood pressure greater on one side of valve than the other; atrial systole forces open bicuspid / AW valve; ventricular systole closes bicuspid / AW valve; semi lunar valve opens during ventricular systole / as pressure in ventricles exceeds that in arteries; semi lunar valve closes after ventricular systole / as ventricular pressure drops; ref. role of tendons; ref. role of papillary muscles; [max 3]

(c) *pressure falls to zero because* all blood expelled from ventricle / ventricle completely empties / AW; *blood pressure falls to 10 kPa* because elastic fibres absorb and then release energy / artery walls elastic / elastic recoil of artery wall; ref. to relevant role of <u>smooth</u> muscle; narrow diameter of capillaries / arterioles (R arteries unless it says small); high resistance to flow;    [max 3]

(d) nicotine causes contraction of arteries / arterioles; nicotine raises blood pressure; nicotine causes blood platelets to become 'sticky' forming a blockage / 'clot'; nicotine / carbon monoxide damages artery lining / endothelium; increased risk of atherosclerosis / thrombus; detail of atherosclerosis (e.g. ref. to deposition of lipid in artery wall); carbon monoxide reduces oxygen transport / levels;    [max 3]

9.2 (a) A = sinoatrial node; B = atrioventricular node; C = bundle of His / Purkinje / Purkyne fibres;    [3]

(b) ring of fibrous tissue / no connection, between atria and ventricles; ref. to delay at atrioventricular node (so ventricles contract later than atria); ref. to fast transmission down bundle of His / Purkinje fibres (so ventricles contract from bottom up);    [max 2]

(c) ref. thick ventricular wall / thin atrial wall; ref. need for ventricular muscle to contract simultaneously; ref. to wave of contraction starting at bottom of ventricles; papillary muscles contract first to stabilise valves;    [max 2]

## Chapter 10

10.1 (a) stomata (open) for gas exchange / $CO_2$ / $O_2$ uptake / release (for photosynthesis and respiration; large surface area / many cell surfaces; in <u>spongy mesophyll</u>; (so) evaporation from (damp) cell walls / membranes (into air spaces); (and) diffusion / loss down concentration gradient, of water <u>vapour</u>; to air / atmosphere via stomata;    [max 3]

(b) ref. cohesion / tension (in context of xylem); hydrogen bonds (between water molecules); through (freely permeable) cell walls / apoplast pathway; through partially permeable membrane / AW (of cell B); ref. osmosis; down water potential / $\Psi$ gradient / from high / less negative to low / more negative water potential / $\Psi$ AW;    [max 3]

(c) (i) B to both A and C; A to C;    [2]
(ii) water vapour from cell surfaces through air through stomata;    [1]

(d) small leaves / small surface area / reduction of leaf surfaces / needles shaped leaves (R spines); rolled / curled leaves (R folded); shed / lose leaves; sunken stomata / stomata in pits / crypts / grooves; stomata surrounded by hairs / hairy leaves; waxy / impermeable / thick, <u>cuticle</u>;    [max 2]

10.2 (a) xylem vessel to transport water / ions;    [1]

(b) no cross / end walls; wide lumen / AW; strengthened / lignified / thick cell wall; no cytoplasm / no cell contents / no organelles / hollow / empty (R dead unqualified); pits (allow water to move sideways);    [max 3]

(c) C = companion cell; movement of sucrose / (A amino acids / phosphate / $K^+$ ions); into / out of, sieve tubes / loading / unloading; through plasmodesmata; provides energy; ref. active transport;    [max 3]

## Chapter 11

11.1 (a) *tidal volume* = air breathed in / out during one breath / gas exchanged during one breath in and out; when not trying consciously to breath / lowest at rest / increases during exercise;[2] *vital capacity* = largest breath / maximum (forced) ventilation; (measured by) maximum exhalation (after maximum inhalation) / maximum inhalation (after maximum exhalation) / maximum volume of gas exchanged during one breath in and out;    [2]

(b) (i) cilia / brush border (on A); columnar (A) / squamous / flattened / thin / pavement (B); elastic fibres in B;    [max 2]
(ii) trachea and bronchi and bronchioles; [1] cilia move mucus; containing trapped dust / bacteria; up into throat to be swallowed;    [max 2]
(iii) elastic fibres; recoil / AW; pulling lung tissue inwards / forcing air out / AW;    [max 2]

## Chapter 12

12.1 (a) elastin / elastic fibres;

(b) no recoil / alveoli do not recoil; air not forced out / air not refreshed in alveoli; concentration of oxygen falls / not maintained; less oxygen diffuses into blood;    [max 3]

(c) *epidemiological evidence* increase in smoking followed about 20 years later by increase in cases / deaths from lung cancer; when more men (than women) smoke, more of them get lung cancer; lung cancer rare / unknown before smoking became common; lung cancer rare in non-smokers; lower incidence in smokers who do not inhale; ref. to national differences in smoking and cancer; ref. to studies involving British doctors; ref. to evidence from passive smoking; AVP;    [max 2]

*experimental evidence* experimental animals get tumours like human lung cancer / carcinogens have been isolated from tar / isolated carcinogens painted on animal skin induce tumours;    [1]

## Chapter 13

13.1 (a) ✓ ✗ ✗ ✓; ✓ ✗ ✗ ✗; ✗ ✓ ✓ ✗; ✗ ✗ ✓ ✗;    [one mark per row] [4]

(b) people who are heterozygous / carriers; resistant to malaria; malaria only attacks healthy RBC; children of heterozygotes; 1 in 4 chance of inheriting sickle cell anaemia; A further detail of inheritance (e.g. Punnett square);    [max 3]

13.2 (a) <u>female Anopheles mosquito</u> sucks <u>blood</u> from (infected person (R bites); parasites / plasmodia / pathogens reproduce / multiply / form gametes (in mosquito); injects / inserts / pumps in saliva with parasites / transmits/transfers parasite as it feeds/in saliva;    [max 2]

(b) malarial parasite has nucleus / nuclear membrane / nuclear envelope; mitochondria; membranous organelles; R ribosomes R nucleolus    [max 2]

(c) fewer red blood cells / number of RBCs reduced; due to bursting / rupturing of RBCs / parasite destroys RBCs; less haemoglobin; less oxygen transported / reduced ability to transport oxygen; waste excreted / toxins released (by parasite); e.g. of symptom (A one from: anaemia / fatigue / tiredness / muscular pain / headaches / nausea / fever / high temperature and sweating / inability to control temperature / shivering)    [max 3]

## Chapter 14

14.1 (a) A = nucleus; B = mitochondrion; C = phagosome / vacuole / secondary lysosome; D = lysosome;    [half mark each rounded up] [2]

(b) lower resolution in light microscope; ref. wavelength of light; maximum is half wavelength; 200 nm; many organelles / named e.g. smaller than this;    [max 3]

(c) attachment to cell membrane; receptors; role of antibodies / opsonins; folding inwards of cell membrane / endocytosis; formation of vacuole / phagosome; lysosomes fuse with vacuole; toxins kill bacteria; enzymes digest bacteria;    [max 4]

14.2 (a) (i) origin – bone marrow;
(ii) thymus;    [2]

(b) move towards site of infection / chemotaxis; bacteria attach to cell membrane / receptors; may be marked by antibodies; engulfed into vacuole; lysosomes fuse with vacuole; bacteria digested;    [max 3]

(c) active immunity: immune response occurs; antibodies made within body; antigen, encountered / enters body; permanent / ref. to memory cells;    [max 2]

## Chapter 15

15.1 (a) glucose phosphorylated / activated; by ATP; to fructose (6) phosphate (A valid alternatives, R glucose); fructose (1, 6) bisphosphate / 6C sugar diphosphorylated; splits into <u>2 ×</u>, 3C / triose phosphate / TP (A valid alternatives); oxidised / dehydrogenated / formation of reduced NAD (A valid alternatives); ATP formed; net gain 2 ATP; pyruvate formed (A valid alternatives); ref. <u>enzymes</u> involved; e.g. named enzyme;    [max 7]

(b) hydrogens / 2H / hydrogen atoms (R hydrogen molecules / $H_2$); high energy / excited electrons / $e^-$ and $H^+$ / protons; from Krebs cycle; reduced NAD (A valid alternatives); transferred to electron transport chain / ETC / chain of electron carried / AW; flow of electrons along ETC; ref. energy levels; creates $H^+$ proton gradient; ref. stalked particles / ATPase / backflow of $H^+$ / protons; results in ATP synthesis; 3 ATP per reduced NAD; oxygen combines with 2H / hydrogens / hydrogen atoms / $H^+$ + $e^-$ (R hydrogen molecules / $H_2$); to form water; ref. <u>aerobic respiration</u>;    [max 8]

15.2 (a) *glycolysis* add phosphate to hexose; activation of glucose; second phosphorylation of fructose phosphate to bisphosphate;    [max 2]

(b) *light-independent stage of photosynthesis* ref. to Calvin cycle; GP to TP (for energy); regeneration of RuBP (for phosphorylation);    [max 2]

(c) *protein synthesis* formation of mRNA; ref. to formation of peptide bond; activation of amino acids by attachment to tRNA;    [max 2]

(d) *replication of DNA* ref. linking nucleotides / phosphodiester bonds; ref. to nucleoside triphosphates / ATP / TTP / GTP / CTP;    [2]

15.3 (a) <u>matrix of mitochondria</u>;    [1]

(b) C marked 3 times, between pyruvate and acetyl CoA, between 6C and 5C, between 5C and 4C on diagram;    [2 marks, −1 for each error] H marked 5 times, each place where NAD or FAD reduced;    [2 marks, −1 for each error] [4]

(c) coenzymes; remove / carry protons / $H^+$ (R hydrogen / $H_2$); and high energy / excited electrons / $e^-$ (from Krebs cycle); (A hydrogens / hydrogen atoms / 2H / H for 1 mark) ref. to electron transfer chain / ETC (A valid alternatives); ref. inner membrane / cristae (of mitochondria);    [max 3]

**Chapter 16**

**16.1** (a) (i) increase; rapid / sharp / steep; then decrease; does not drop to original value; [max 2]
  (ii) decrease to 0 / all used up; [1]
  (b) (i) GP continues to be formed from RuBP; (until) all RuBP used up; then GP falls as converted to TP / hexose / glucose; [max 2]
  (ii) in dark RuBP not regenerated / converted to GP (R used up); requires, products from light reaction / ATP / reduced NADP from light reaction / photophosphorylation (to be occurring)
  (c) ATP; reduced NADP (A valid alternatives); [2]

**16.2** (a) thylakoid (membrane) / lamellae; in grana; [2]
  (b) source of energy; to excite electrons / AW; for synthesis of ATP (from ADP); and reduced NADP (A valid alternatives); ref. photolysis; [max 2]
  (c) electron lost from photosystem / PSII / non cyclic has PSII and PSI / cyclic only has PSI; replaced by electron from water in non-cyclic; ref. photolysis of water / oxygen produced in non-cyclic; non-cyclic produces reduced NADP (A valid alternatives); in cyclic, excited electron returns to chlorophyll / photosystem 1; in cyclic, electron not taken from water / hydroxyl ion; [max 4]
  (d) idea that paraquat can be used to kill weeds around growing crop; [1]

**16.3** (a) light energy converted to chemical energy; ref. to wavelength of light absorbed; by chlorophyll / ref. to photosystem / PS traps light energy; electrons excited / AW; harnessed / used in electron transport chain / ETC; synthesis of ATP; synthesis of reduced NADP (A valid alternatives); (ATP and reduced NADP) used to drive light independent reaction / Calvin cycle; photolysis of water / dissociation of water; electron from hydroxyl ion; forming oxygen; non-cyclic (photophosphorylation) produced ATP and reduced NADP; cyclic only produces ATP; [max 7]
  (b) carbon dioxide combines with RuBP / AW; (RuBP) 5C (compound); ref. enzyme / RuBP carboxylase / Rubisco; forming unstable 6C compound that immediately splits to form; two molecules of glycerate (3) phosphate / GP (A valid alternatives); (then) using hydrogens / H (R hydrogen molecules / H₂) from reduced NADP (A valid alternatives); (and) using ATP / ATP → ADP + Pi (as an energy source); forming triose phosphate / TP (A valid alternatives); TP may be respired / converted to glucose; TP may be converted to amino acids / lipids / carboxylic acids; TP may be used to regenerate RuBP; using ATP (as a source of phosphate); [max 8]

**Chapter 17**

**17.1** (a) (i) EcoR1; (ii) sticky ends; [2]
  (b) plasmid DNA cut with same restriction enzyme / endonuclease; DNA and plasmid mixed together / AW (R inserted); ref. complementary base pairing (A C and G on sticky ends pair up); ref. to hydrogen bonding; ligase forms bonds between sugar and phosphate / phosphodiester bonds; [max 3]

**17.2** (a) metaphase; [1]
  (b) centromeres divide / split (R break); chromatids separate; move to opposite poles / centrioles; by spindle fibres / microtubules; ref. mechanism of movement; [max 3]
  (c) nuclear membrane breaks down / disperses; [1]
   centrioles divide / replicate; to form 2 pairs; move to (opposite) poles (of cell; [max 2]
  (d) independent assortment / random alignment (A description); different mix of maternal and paternal chromosomes (A chromatids); different gametes produced (allow once); crossing over / chiasmata formation exchange of genetic material; between chromatids of homologous chromosomes; breaks up linkage groups / mixes maternal and paternal alleles; different gametes produced (allow once); [max 4]

**17.3** (a) correct parental genotype (both RrBb); correct gametes (both RB Rb rB rb) (A from Punnett square); correct genotypes of offspring (in Punnett square, e.g. below); correct offspring phenotypes linked to genotypes (in Punnett square, or by a key, or by a table); [4]

| gametes | RB | Rb | rB | rb |
|---|---|---|---|---|
| RB | RRBB * | RRBb * | RrBB * | RrBb * |
| Rb | RRBb * | RRbb # | RrBb * | Rrbb # |
| rB | RrBB * | RrBb * | rrBB + | RrBb + |
| rb | RrBb * | Rrbb # | rrBb + | Rrbb – |

Key * = round yellow, # = round green, + = wrinkled yellow – = wrinkled green
  (b) wrinkled green seeds homozygous / double recessive; round yellow seeds variety of different genotypes / may be heterozygotes; some round yellow seeds will breed true / be homozygous dominant; [3]
  (c) probability of as large a deviation as 0.47 is between 0.9 and 0.95; so there is a close fit to the expected results / no significant difference from the expected result; [2]

**Chapter 18**

**18.1** (a) species similar morphology, behaviour and physiology; interbreed to produce fertile offspring; reproductively isolated from other species; ref. common gene pool; ref. chromosome number; ref. common niche; ref. to problems; [max 3]
   niche set of conditions within which an organism lives; role in the community; ref. realised and potential niche; [max 2]
  (b) colonisation of new / unoccupied habitat; adapt to varied selective pressures in different parts of the lake; reproductive / geographical isolation; ref. to specific selective pressures; [max 3]
  (c) created geographical barriers preventing gene flow / interbreeding / ref. creation of new habitats; [1]

**Chapter 19**

**19.1** (a) excretion of metabolic wastes / urea; osmoregulation / control of water balance / ionic balance of urine / selective reabsorption of ions / control of ion content of urine; maintaining acid-base balance of body; maintaining ionic balance of body; AVP, e.g. secretion of renin; [max 3]
  (b) A = microvilli / brush border; B = infolding of membrane / basement channels; [2]
  (c) increased surface area of microvilli for absorption; increase surface area of basement channels; numerous mitochondria; closeness of blood capillaries; cells separated / space between cells; many transport proteins in cell surface membranes; [max 2]
  (d) diffuse in; active transport out; by carrier protein; glucose by facilitated diffusion; into spaces between cells / basal channels; diffuse into blood capillaries; movement creates concentration / diffusion gradient; ref. role of Na⁺ / K⁺ pumps [max 4]

**19.2** (a) ductless; specialised / secretory cells; (small clumps of cells) within pancreas; ref. to set point / norm reference / homeostasis; negative feedback; ref. to control of secretion / receptor / detection (of blood glucose); synthesis / secretion of a hormone; in small quantities; into blood stream / plasma; ref. role of blood capillaries; to target cells / tissues; [max 7]
  (b) glucagon from α cells; insulin from β cells; blood glucose 80–120 mg per 100 cm⁻³ / 0.1% glucose / 4 mmols dm⁻³; insulin lowers blood glucose; (binds to receptors on) target cells / liver / muscles; ref. to vesicles containing glucose carrier proteins; conversion of glucose to glycogen / glycogenesis; increase (in rate of) uptake of glucose / increased permeability to glucose of cells; increased fatty acid / lipid synthesis / glucose to lipid; increased oxidation / respiration of glucose; glucagon raises blood glucose; binds to receptors (on target cells / liver); conversion of glycogen to glucose / glycogenolysis; conversion of amino acids to glucose; ref. lipids to glucose; [max 8]

**19.3** (a) all organisms need to respond to environmental change; ref. irritability / sensitivity; ref. adaptive responses to stimuli; mammals need to respond fast; need to control cycles; need to control long term-term processes; flowering plants need to coordinate flowering / seed germination / growth; [max 4 on need]
   mammals have slow and fast communication systems / nerves and hormones / electrical and chemical; plants have slow only / plant growth substances (A hormones) / chemical; up to 2 details of: mammalian nervous system; mammalian endocrine system; flowering plant growth substances; [max 4 on differences] [in total max 7]
  (b) auxin ref. positive phototropism of shoots; ref. promotion of root growth by low concentrations; inhibition of roots by high conc.; experimental detail of phototropism; ref. cellular detection; ref. gravitropism; ref other roles e.g. fruit development in strawberry; [max 4 from auxin]
   gibberellins growth promoters; ref. internode extension; ref. dormancy breaking in seeds; ref. role in cereal / barley germination; [max 3 from gibberellins]
   abscisic acid growth inhibitor; closes stomata; ref. seed / bud dormancy; abscission of cotton fruit (R abscission of leaves); [max 3 from abscisic acid] [max 8]

**19.4** (a) (i) Na⁺ channels open; K⁺ channels closed; ref. to delay in opening; [max 2]
  (ii) Na⁺ channels closed; ref. to Na⁺ channels inactive / refractory period; K⁺ channels open; [max 2]
  (b) opened by depolarisation; allows Ca²⁺ ions in; causes vesicles containing neurotransmitter to fuse with presynaptic membrane; releases transmitter into synaptic cleft (where it diffuses across); ref. exocytosis; [max 3]
  (c) ref. to Ca²⁺ pump / carrier protein; Ca²⁺ ions pumped out; using ATP; against concentration gradient; [max 2]
  (d) transmitter / acetylcholine binds to receptor; Na⁺ channels opened / increase permeability to Na⁺ ions; Na⁺ ions diffuse in depolarising post-synaptic membrane; action potential set up (A impulse transmitted); [max 2]
   [A metal ion (e.g. sodium ions) / ionic symbols (e.g. Na⁺) R metal (e.g. sodium) / atomic symbol (e.g. Na)]

# Glossary

**α cell** a cell in the islets of Langerhans in the pancreas that senses when blood glucose levels are low and secretes glucagon in response.

**abiotic factor** the physical characteristics of a habitat, such as temperature, light intensity and soil pH.

**abscisic acid (ABA)** a plant growth regulator that causes closure of stomata in dry conditions, and inhibits seed germination.

**abscission** the dropping of leaves or fruits from a plant.

**absorption spectrum** a graph of the absorbance of different wavelengths of light by a compound such as a photosynthetic pigment.

**accessory pigment** a pigment that is not essential to photosynthesis but which absorbs light of different wavelengths and passes the energy to chlorophyll *a*.

**acetylcholine** a transmitter substance found, for example, in the presynaptic neurone at neuromuscular junctions.

**acetylcholinesterase** an enzyme that rapidly breaks down acetylcholine at synapses.

**action potential** a fleeting reversal of the resting potential across the plasma membrane of a neurone, which rapidly travels along its length.

**action spectrum** a graph of the rate of photosynthesis at different wavelengths of light.

**activation energy** the energy that must be provided to make a reaction take place; enzymes reduce the activation energy required for a substrate to change into a product.

**active immunity** immunity gained when an antigen enters the body, an immune response occurs and antibodies are produced by plasma cells.

**active site** an area on an enzyme molecule where the substrate can bind.

**active transport** the movement of molecules or ions through transport proteins across a cell membrane, against their concentration gradient, involving the use of energy from ATP.

**ADP** adenosine diphosphate.

**adrenaline** a hormone secreted by the adrenal glands in times of stress or excitement.

**afferent** leading towards, e.g. the afferent blood vessel leads towards a glomerulus.

**alcoholic fermentation** anaerobic respiration in which glucose is converted to ethanol.

**aleurone layer** a layer of tissue around the endosperm in a cereal seed that synthesises amylase during germination.

**allele** a particular variety of a gene.

**allergen** an otherwise harmless substance that sensitises the immune system to give an immune response.

**allopatric speciation** speciation that takes place as a result of two populations living in different places and having no contact with each other.

**allopolyploid** possessing more than two sets of chromosomes, where the chromosomes come from two different species.

**anabolic reaction** synthesis of complex substances from simpler ones.

**antibiotic** a substance produced by a living organism that is capable of destroying or inhibiting the growth of a microorganism.

**antibody** a protein (immunoglobulin) made by plasma cells derived from B lymphocytes, secreted in response to an antigen; the variable region of the antibody molecule is complementary in shape to its specific antigen.

**antidiuretic hormone (ADH)** a hormone secreted from the pituitary gland that increases water reabsorption in the kidneys and therefore reduces water loss in urine.

**antigen** a substance that is foreign to the body and stimulates an immune response.

**apical dominance** the tendency for the apical bud of a plant to grow, while the lateral buds do not; removal of the apical bud stimulates growth of the lateral buds.

**artery** a blood vessel with a relatively thick wall containing large amounts of elastic fibres and that carries blood away from the heart.

**artificial immunity** immunity gained either by vaccination (**active**) or by injecting antibodies (**passive**).

**artificial selection** the selection by humans of organisms with desired traits.

**assimilates** substances, such as sucrose, that have been made within a plant.

**atherosclerosis** progressive build-up of fatty material in the lining of arteries.

**ATP** adenosine triphosphate – the universal energy currency of cells.

**ATP synthase** the enzyme catalysing the phosphorylation of ADP to ATP.

**atrio-ventricular node** a patch of tissue in the septum of the heart, through which the wave of electrical excitation is passed from the atria to the Purkyne tissue.

**autopolyploid** possessing more than two sets of chromosomes, where all the sets are from the same species.

**autosomes** all the chromosomes except the X and Y (sex) chromosomes.

**autotroph** an organism that can trap an inorganic carbon source (carbon dioxide) using energy from light or from chemicals.

**auxin** a plant growth substance synthesised in the growing regions of shoots and roots.

**axon** a long cytoplasmic process of a neurone, that conducts action potentials away from the cell body.

**β cell** a cell in the islets of Langerhans in the pancreas that senses when blood glucose levels are high and secretes insulin in response.

**β galactosidase** an enzyme which catalyses the hydrolysis of lactose to glucose and galactose.

**B lymphocyte** a type of lymphocyte that gives rise to plasma cells, which secrete antibodies.

**base pairing** the pairing, held by hydrogen bonds, between the nitrogenous bases cytosine and guanine, and between thymine and adenine or uracil, that occurs in the polynucleotides DNA and RNA.

**Benedict's test** a test for the presence of reducing sugars; the unknown substance is heated with Benedict's reagent, and a change from a clear blue solution to the production of a yellow or red precipitate indicates the presence of reducing sugars such as glucose.

**biotic factor** a factor which affects a population or a process, that is caused by other living organisms; examples include competition, predation and parasitism.

**biuret test** a test for the presence of amine groups, and thus for the presence of proteins; biuret reagent is added to the unknown substance, and a change from pale blue to purple indicates the presence of proteins.

**Bohr effect** the decrease in affinity of haemoglobin for oxygen that occurs when carbon dioxide is present.

**Bowman's capsule** *see* renal capsule.

**bronchitis** a disease in which the airways in the lungs become inflamed and congested with mucus; chronic bronchitis is often associated with smoking.

**calorimeter** the apparatus in which the energy value of a compound can be measured by burning it in oxygen.

**Calvin cycle** a closed pathway of reactions in photosynthesis in which carbon dioxide is fixed into carbohydrate.

**cancer** a disease, often but not always treatable, that results from a breakdown in the usual control mechanisms that regulate cell division; certain cells divide uncontrollably and form tumours, from which cells may break away and form secondary tumours in other areas of the body (metastasis).

**capillary** the smallest type of blood vessel, whose function is to facilitate exchange of substances between the blood and the tissues; capillary walls are made up of a single layer of squamous epithelium, and their internal diameter is only a little larger than that of a red blood cell.

**carcinogen** a substance that can cause cancer.

**cardiac cycle** the sequence of events taking place during one heart beat.

**cardiovascular diseases** degenerative diseases of the heart and circulatory system, for example coronary heart disease, stroke.

**carotenoid** a yellow, orange or red plant pigment used as an accessory pigment in photosynthesis.

**cell** a structure bounded by a plasma membrane, containing cytoplasm and organelles.

**cell cycle** the sequence of events that takes place from one cell division until the next; it is made up of interphase, mitosis and cytokinesis.

**chemiosmosis** the synthesis of ATP using energy stored as a difference in hydrogen ion concentration across a membrane in a chloroplast or mitochondrion.

**chi-squared ($\chi^2$) test** a statistical test that can be used to determine if any difference between observed results and expected results is significant, or due to chance.

**chlorophyll** a green pigment responsible for light capture in photosynthesis in algae and higher plants.

**chloroplast** the photosynthetic organelle in eukaryotes.

**cholinergic synapse** a synapse at which the transmitter substance is acetylcholine.

**chromatid** one of two identical parts of a chromosome, held together by a centromere, formed during interphase by the replication of the DNA strand.

**chromosome** a structure made of DNA and histones, found in the nucleus of a eukaryotic cell. The term **bacterial chromosome** is now commonly used for the circular strand of DNA present in a prokaryotic cell.

**chromosome mutation** a random and unpredictable change in the structure or number of chromosomes in a cell.

**chronic obstructive pulmonary disease** a disease of the lungs characterised by bronchitis and emphysema.

**closed circulation** a circulatory system in which the blood is always contained within vessels, as in mammals.

**codominance** alleles are said to be codominant when both alleles have an effect on the phenotype of a heterozygous organism.

**collecting duct** the last section of a nephron, from which water can be absorbed back into the blood stream before the urine flows into the ureter.

**community** all of the living organisms, of all species, that are found in a particular habitat at a particular time.

**companion cell** a cell with an unthickened cellulose wall and dense cytoplasm that is found in close association with a phloem sieve element to which it is directly linked via many plasmodesmata.

**competition** when a particular resource is in short supply, organisms requiring that resource are said to compete with each other.

**consumer** a heterotrophic organism; an organism that obtains its food in organic form, either directly or indirectly from that which has been synthesised by producers.

**coronary heart disease** a disease of the heart caused by damage to the coronary arteries, often as a result of atherosclerosis.

**counter-current multiplier** an arrangement in which fluid in adjacent tubes flows in opposite directions, allowing relatively large differences in concentration to be built up.

**creatinine** a nitrogenous excretory substance produced from the breakdown of creatine.

**crista** (*pl.* **cristae**) a fold of the inner membrane of the mitochondrial envelope on which are found stalked particles of ATP synthase.

**crossing over** an event that occurs during meiosis I, when chromatids of two homologous chromosomes break and rejoin so that a part of one chromatid swaps places with the same part of the other.

**cytokinins** plant growth regulators that stimulate cell division.

**deamination** the breakdown of excess amino acids in the liver, by the removal of the amine group; ammonia and eventually urea are formed from the amine group.

**dendrite** a short cytoplasmic process of a neurone, that conducts action potentials towards the cell body.

**dendron** a long cytoplasmic process of a neurone, that conducts action potentials towards the cell body.

**depolarisation** the reversal of the resting potential across the plasma membrane of a neurone or muscle cell, so that the inside becomes positively charged compared with the outside.

**diabetes** an illness in which the pancreas does not make sufficient insulin, or where cells do not respond appropriately to insulin.

**diastolic blood pressure** the minimum pressure of blood in the arteries when the ventricles of the heart are relaxing; it is usually about 80 mmHg (10.5 kPa).

**diffusion** the net movement, as a result of random motion of its molecules or ions, of a substance from an area of relatively high concentration to an area of relatively low concentration.

**dihybrid cross** a genetic cross in which two different genes are considered.

**diploid cell** one that possesses two complete sets of chromosomes; the abbreviation for diploid is 2n.

**directional selection** a type of selection in which the most common varieties of an organism are selected against, resulting in a change in the features of the population.

**disease** a form of ill-health or illness with a set of symptoms.

**diuresis** the production of large volumes of dilute urine.

**DNA** Deoxyribonucleic acid, a polynucleotide that contains the pentose sugar deoxyribose.

**dominance** an allele is said to be dominant when its effect on the phenotype of a heterozygote is identical to its effect in a homozygote.

**dormancy** a state of 'suspended animation', in which metabolism is slowed right down, enabling survival in adverse conditions.

**double circulation** a circulatory system in which the blood travels twice through the heart on one complete circuit of the body; the pathway from heart to lungs and back to the heart is known as the pulmonary circulation, and that from heart to the rest of the body and back to the heart as the systemic circulation.

**ecosystem** all of the living organisms of all species, and all of the non-living components, that are found together in a defined area and that interact with one another.

**effector** an organ or tissue that responds to a stimulus, for example a muscle.

**efferent** leading away from.

**electron transport chain** chain of adjacently arranged carrier molecules in the inner mitochondrial membrane along which electrons pass by redox reactions.

**emphysema** a disease in which alveoli are destroyed, giving large air spaces and decreased surface area for gaseous exchange; it is often associated with chronic bronchitis.

**endocrine gland** a gland that secretes its products, which are always hormones, directly into the blood.

**endocytosis** the movement of bulk liquids or solids into a cell, by the indentation of the plasma membrane to form vesicles containing the substance; endocytosis is an active process requiring ATP.

**endosperm** a tissue made of triploid cells that stores food in some seeds, such as cereal grains.

**endothelium** a tissue that lines the inside of a structure, such as the inner surface of a blood vessel or a nephron.

**enzyme** a protein produced by a living organism that acts as a catalyst in a specific reaction by reducing activation energy.

**enzyme specificity** the ability of an enzyme to catalyse reactions involving only a single type of substrate; specificity results from the need for the substrate molecule to bind with the active site of the enzyme, and only substrates with particular shapes are able to bind.

**epithelium** a tissue that covers the outside of a structure.

**eukaryotic cell** a cell containing a nucleus and other membrane-bound organelles.

**excretion** the removal of toxic or excess products of metabolism from the body.

**exocrine gland** a gland that secretes substances into a duct, for example the salivary glands.

**exocytosis** the movement of bulk liquids or solids out of a cell, by the fusion of vesicles containing the substance with the plasma membrane; exocytosis is an active process requiring ATP.

**facilitated diffusion** the diffusion of a substance through protein channels in a cell membrane; the proteins provide hydrophilic areas that allow the molecules or ions to pass through a membrane that would otherwise be less permeable to them.

**factor VIII** one of several substances that must be present in blood in order for clotting to occur.

**fibrous protein** a protein whose molecules have a relatively long, thin structure that are generally insoluble and metabolically inactive, and whose function is usually structural, e.g. keratin and collagen.

**fixation, of nitrogen** the conversion of gaseous nitrogen, $N_2$, into a more reactive form such as nitrate or ammonia.

**gaseous exchange** the movement of gases between an organism and its environment, e.g. the intake of oxygen and the loss of carbon dioxide; gaseous exchange often takes place across a specialised surface such as the alveoli of the lungs.

**gene** a length of DNA that codes for a particular protein or polypeptide.

**gene mutation** a change in the base sequence in part of a DNA molecule.

**genetic engineering** a procedure by which one or more selected genes are removed from one organism and inserted into another, which is then said to be transgenic.

**genotype** the alleles possessed by an organism.

**gibberellin** a plant growth regulator that increases growth in genetically dwarf plants, and that stimulates germination.

**globular protein** a protein whose molecules are curled into a relatively spherical shape and that is often water soluble and metabolically active, e.g. insulin and haemoglobin.

**glomerular filtration rate** the rate at which fluid passes from the glomerular capillaries into the renal capsules in the kidneys.

**glomerulus** a knot of capillaries in the 'cup' of a renal capsule.

**glycosidic bond** a C–O–C link between two monosaccharide molecules.

**glucagon** a small peptide hormone secreted by the α cells in the islets of Langerhans in the pancreas that brings about an increase in the blood glucose level.

**glycogen** a polysaccharide, made of many glucose molecules linked together, that stores glucose in liver and muscle cells.

**glycolysis** the splitting (lysis) of glucose.

**granum** (*pl.* **grana**) a stack of thylakoids in a chloroplast.

**guard cell** a sausage-shaped epidermal cell found in pairs bounding a stoma and controlling its opening or closure.

**habitat** the place where an organism, a population or a community lives.

**haemoglobin** the red pigment found in red blood cells whose molecules contain four iron ions within a globular protein made up of four polypeptides and that combines reversibly with oxygen.

**haemophilia** a genetic disease in which there is an insufficient amount of a clotting factor, such as factor VIII, in the blood.

**haploid cell** one that possesses one complete set of chromosomes; the abbreviation for haploid is n.

**heterotroph** an organism needing a supply of organic molecules as its carbon source.

**heterozygous** having two different alleles of a gene.

**homeostasis** maintaining a constant environment for the cells within the body.

**homologous chromosomes** a pair of chromosomes in a diploid cell that have the same structure as each other, with the same genes (but not necessarily the same alleles of those genes) at the same loci, and that pair together to form a bivalent during the first division of meiosis.

**homozygous** having two identical alleles of a gene.

**hormone** a substance secreted by an endocrine gland, that is carried in blood plasma to another part of the body where it has an effect.

**hydrogen bond** a relatively weak bond formed by the attraction between a group with a small positive charge on a hydrogen atom and another group carrying a small negative charge, e.g. between two $-O^{\delta-}H^{\delta+}$ groups.

**hypertension** abnormally high blood pressure.

**IAA** a type of auxin.

**immune response** the action of lymphocytes in response to the entry of an antigen into the body.

**immune system** the body's defence system.

**immunity** protection against infectious diseases, gained either actively or passively.

**immunological memory** the ability of the immune system to respond quickly to antigens that it recognises as having entered the body before.

**independent assortment** the way in which different alleles of genes on different chromosomes may end up in any combination in gametes, resulting from the random alignment of bivalents on the equator during meiosis I.

**industrial melanism** an increase in the frequency of dark individuals in a population, for example the peppered moth, as a result of selection pressures favouring these forms in areas polluted by industrial emissions.

**infectious disease** a disease caused by an organism such as a bacterium or virus.

**inhibitor, competitive** a substance that reduces the rate of activity of an enzyme by competing with the substrate molecules for the enzyme's active site. Increasing the concentration of the substrate reduces the degree of inhibition.

**inhibitor, non-competitive** a substance that reduces the rate of activity of an enzyme, but where increasing the concentration of the substrate does not reduce the degree of inhibition. Many non-competitive inhibitors bind to areas of the enzyme molecule other than the active site itself.

**insulin** a small peptide hormone secreted by the β cells in the islets of Langerhans in the pancreas that reduces blood glucose levels.

**intermediate neurone** a neurone whose cell body and many dendrites are all within the brain or spinal cord; it receives action potentials from a sensory neurone and transmits action potentials to a motor neurone

**iodine in potassium iodide solution test** a test for the presence of starch; the solution is added to the unknown substance, and a change from brown to blue-black indicates the presence of starch.

**islets of Langerhans** groups of cells in the pancreas which secrete insulin and glucagon.

**Krebs cycle** a closed pathway of reactions in aerobic respiration in a mitochondrion in which hydrogens pass to hydrogen carriers for subsequent ATP synthesis and some ATP is synthesised directly.

**lactate** (*or* **lactic acid**) the end product of anaerobic respiration, often produced by muscles during exercise.

**lamina** the blade of a leaf.

**limiting factor** the one factor, of many affecting a process, that is nearest its lowest value and hence is rate-limiting.

**link reaction** decarboxylation and dehydrogenation of pyruvate and formation of acetyl coenzyme A, linking glycolysis with the Krebs cycle.

**loop of Henle** the part of the nephron between the proximal and distal convoluted tubules; in humans, about 14% of the loops of Henle are long and reach down into the medulla of the kidney.

**lymph** an almost colourless fluid, very similar in composition to blood plasma but with fewer plasma proteins, that is present in lymph vessels.

**lymphocyte** a type of white blood cell that is involved in the immune response; unlike phagocytes they become active only in the presence of a particular antigen that 'matches' their specific receptors or antibodies.

**magnification** the number of times greater that an image is than the actual object. Magnification = image size ÷ object size.

**meiosis** the type of cell division that results in a halving of chromosome number and a reshuffling of alleles; in humans, it occurs in the formation of gametes.

**memory cells** lymphocytes which develop during an immune response and retain the ability to respond quickly when an antigen enters the body on a second, or any subsequent, occasion.

**mesophyll** the internal tissue of a leaf blade with chloroplasts for photosynthesis and consisting of an upper layer of palisade cells (the main photosynthetic tissue) and a lower layer of spongy mesophyll with large air spaces for gas exchange.

**mitochondrion** the organelle in eukaryotes in which aerobic respiration takes place.

**mitosis** the division of a nucleus such that the two daughter cells acquire exactly the same number and type of chromosomes as the parent cell.

**monohybrid cross** a cross in which the inheritance of one gene is considered.

**motor end plate** the ending of an axon of a motor neurone, where it forms a synapse with a muscle.

**motor neurone** a neurone whose cell body is in the brain or spinal cord, and that transmits action potentials to an effector such as a muscle or gland.

**multiple alleles** the existence of three or more alleles of a gene, as, for example, in the determination of A,B,O blood groups.

**mutagen** a substance that can cause mutation.

**mutation** an unpredictable change in the structure of DNA, or in the structure and number of chromosomes.

**myelin** a substance that surrounds many axons and dendrons, made up of many layers of the plasma membranes of Schwann cells.

**natural immunity** immunity gained by being infected (**active**) or by receiving antibodies from the mother across the placenta or in breast milk (**passive**).

**natural selection** the way in which individuals with particular characteristics have a greater chance of survival than individuals without those characteristics, which are therefore more likely to breed and pass on the genes for these characteristics to their offspring.

**negative feedback** a process in which a change in some parameter, such as blood glucose level, brings about processes which move its level back towards normal again.

**nephron** a kidney tubule.

**nerve** a bundle of numerous axons and dendrons of many different neurones, surrounded by a sheath called the perineurium.

**neuromuscular junction** a synapse between the axon of a motor neurone and a muscle.

**neurone** a nerve cell; a cell which is specialised for the conduction of action potentials.

**niche** the role of an organism in an ecosystem.

**nicotine** a chemical found in tobacco smoke that can bind with acetylcholine receptors on the postsynaptic membrane of cholinergic synapses.

**nitrogenous excretory product** an unwanted product of metabolism that contains nitrogen, for example ammonia, urea or uric acid.

**node of Ranvier** a short gap in the myelin sheath surrounding an axon.

**non-infectious disease** a disease that is not caused by an organism.

**noradrenaline** a neurotransmitter substance.

**organ** a structure within a multicellular organism that is made up different types of tissues working together to perform a particular function, e.g. the stomach in a human or a leaf in a plant.

**organelle** a functionally and structurally distinct part of a cell, for example a ribosome or mitochondrion.

**organic molecule** a compound containing carbon and hydrogen.

**osmoreceptor** a receptor cell that is sensitive to the water potential of the blood.

**osmoregulation** the control of the water content of the fluids in the body.

**osmosis** the net movement of water molecules from a region of high water potential to a low water potential, through a partially permeable membrane, as a result of their random motion.

**oxidative phosphorylation** synthesis of ATP from ADP and $P_i$ using energy released by the electron transport chain in aerobic respiration.

**oxygen debt** the volume of oxygen that is required at the end of exercise to metabolise lactate that accumulates as a result of anaerobic respiration in muscles.

**palisade cell** *see* mesophyll.

**pancreas** an organ lying close to the stomach that functions both as an exocrine gland (secreting pancreatic juice) and an endocrine gland (secreting insulin and glucagon).

**passive immunity** immunity gained without an immune response; antibodies are injected (artificial) or pass from mother to child across the placenta or in breast milk (natural).

**pathogen** an organism that causes infectious disease.

**peptide bond** a C–N link between two amino acid molecules.

**petiole** a leaf stalk.

**phagocyte** a type of cell, some of which are white blood cells, that ingests and destroys pathogens or damaged body cells.

**phenotype** the characteristics of an organism, often resulting from an interaction between its genotype and its environment.

**phenylketonuria (PKU)** a genetic disease resulting from a mutation in a gene that codes for an enzyme involved in the metabolism of phenylalanine.

**phloem tissue** tissue containing phloem sieve tubes and other types of cell, responsible for the translocation of assimilates such as sucrose through a plant.

**phospholipid** a substance whose molecules are made up of a glycerol molecule, two fatty acids and a phosphate group; a bilayer of phospholipids forms the basic structure of all cell membranes.

**phosphorylation** the transfer of a phosphate group to an organic compound.

**photolysis** the splitting of water using light energy: $H_2O \rightarrow 2H^+ + 2e^- + \frac{1}{2}O_2$

**photophosphorylation** the synthesis of ATP from ADP and $P_i$ using light energy in photosynthesis.

**photosynthesis** the fixation of carbon from carbon dioxide into organic molecules using light energy.

**photosystem** a cluster of light-harvesting accessory pigments surrounding a primary pigment or reaction centre.

**plant growth regulator** the chemicals produced in plants that affect their growth and development; they include auxins, gibberellins, cytokinins and abscisic acid.

**plaque** fatty material in the lining of an artery.

**podocyte** one of the cells which make up the endothelium of a renal capsule.

**polyploidy** possessing more than two complete sets of chromosomes.

**population** all of the organisms of the same species present in the same place and at the same time that can interbreed with one another.

**positive feedback** a process in which a change in some parameter brings about processes that move its level even further in the direction of the initial change.

**postsynaptic neurone** the neurone on the opposite side of a synapse to the neurone in which the action potential arrives.

**presynaptic neurone** a neurone at a synapse from which neurotransmitter is secreted when an action potential arrives.

**primary pigment** *see* reaction centre.

**primary structure** the sequence of amino acids in a polypeptide or protein.

**producer** an autotrophic organism; an organism that obtains its food from inorganic sources by photosynthesis or chemosynthesis.

**prokaryotic cell** a cell that does not contain a nucleus or any other membrane-bound organelles; bacteria are prokaryotes.

**Purkyne tissue** an area of tissue in the septum of the heart that conducts the wave of excitation from the atria to the base of the ventricles.

**quaternary structure** the three-dimensional arrangement of two or more polypeptides, or of a polypeptide and a non-protein component such as haem, in a protein molecule.

**reabsorption** taking back into the blood some of the substances in a kidney nephron.

**reaction centre** a molecule of chlorophyll *a* that receives energy from the light absorbed by surrounding accessory pigments in a photosystem.

**receptor potential** a change in the normal resting potential across the membrane of a receptor cell, caused by a stimulus.

**receptor** a cell which is sensitive to a change in the environment that may generate an action potential as a result of a stimulus.

**recessive** an allele is said to be recessive if it is only expressed when no dominant allele is present.

**redox reaction** an oxidation–reduction reaction involving the transfer of electrons from a donor to an acceptor.

**reflex action** a fast, automatic response to a stimulus; reflex actions may be innate (inborn) or learned (conditioned).

**reflex arc** the pathway taken by an action potential leading to a reflex action; the action potential is generated in a receptor, passes along a sensory neurone into the brain or spinal cord and then along a motor neurone to an effector.

**refractory period** a period of time during which a neurone is recovering from an action potential, and during which another action potential cannot be generated.

**regulator gene** part of a DNA molecule that codes for a protein that controls the expression of another gene.

**renal capsule** the cup-shaped part at the beginning of a nephron; sometimes known as Bowman's capsule.

**repolarisation** getting the resting potential of a neurone back to normal after an action potential has passed.

**reproductive isolation** the inability of two groups of organisms to breed with one another, for example because of geographical separation, or because of behavioural differences.

**resolution** the ability to distinguish between two objects very close together; the higher the resolution of an image, the greater the detail that can be seen.

**respiration** enzymatic release of energy from organic compounds in living cells.

**respiratory quotient (RQ)** the ratio of the volume of carbon dioxide given out in respiration to that of oxygen used.

**respirometer** the apparatus for measuring the rate of oxygen consumption in respiration or for finding the RQ.

**resting potential** the difference in electrical potential that is maintained across a neurone when it is not transmitting an action potential; it is normally about $-60\,\text{mV}$ inside and is maintained by the sodium–potassium pump.

**RNA** ribonucleic acid, a polynucleotide that contains the pentose sugar ribose.

**saltatory conduction** conduction of an action potential along a myelinated axon or dendron, in which the action potential jumps from one node of Ranvier to the next.

**Schwann cell** a cell which is in close association with a neurone, whose plasma membrane wraps round and round the axon or dendron of the neurone to form a myelin sheath.

**secondary structure** the structure of a protein molecule resulting from the regular coiling or folding of the chain of amino acids, for example an alpha helix or beta pleated sheet.

**secretion** the release of a useful substance from a cell or gland.

**selection pressure** an environmental factor that confers greater chances of survival and reproduction on some individuals than on others in a population.

**selective breeding** choosing only organisms with desirable features, from which to breed.

**semi-conservative replication** the method by which a DNA molecule is copied to form two identical molecules, each containing one strand from the original molecule and one newly synthesised strand.

**sensory neurone** a neurone that transmits action potentials from a receptor to the central nervous system.

**sex chromosomes** the pair of chromosomes that determine the gender of an individual; in humans, they are the X and Y chromosomes.

**sex linked gene** a gene that is carried on an X chromosome but not on a Y chromosome.

**sexual reproduction** reproduction involving meiosis, gametes, fertilisation and the production of zygotes.

**sickle cell anaemia** a genetic disease caused by a faulty gene coding for haemoglobin, in which haemoglobin tends to precipitate when oxygen concentrations are low.

**sickle trait** a person who is heterozygous for the sickle cell allele is said to have sickle trait; there are normally no symptoms, except occasionally in very severe conditions of oxygen shortage.

**sieve tube element** a cell found in phloem tissue, with non-thickened cellulose walls, very little cytoplasm, no nucleus and end walls perforated to form sieve plates, through which sap containing sucrose is transported.

**silent mutation** a mutation in which the change in the DNA has no discernible effect on an organism.

**sinoatrial node** a patch of muscle in the wall of the right atrium of the heart, whose intrinsic rate of rhythmic contraction is faster than that of the rest of the cardiac muscle, and from which waves of excitation spread to the rest of the heart to initiate its contraction during the cardiac cycle.

**sodium–potassium pump** a membrane protein (or proteins), that moves sodium ions out of a cell and potassium ions into it, using ATP.

**speciation** the production of new species.

**stabilising selection** a type of natural selection in which the status quo is maintained because the organisms are already well adapted to their environment.

**stimulus** a change in the environment that is detected by a receptor, and which may cause a response.

**stoma** a pore in the epidermis of a leaf bounded by two guard cells.

**stroke** damage to the brain caused by bursting or blockage of an artery.

**stroma** the matrix of a chloroplast in which the light-independent reactions of photosynthesis occur.

**suberin** a waxy, waterproof substance found in some plant cell walls, for example cells in bark.

**sympatric speciation** the emergence of a new species from another species where the two are living in the same place; it can happen, for example, as a result of polyploidy.

**synapse** a point at which two neurones meet but do not touch; the synapse is made up of the end of the presynaptic neurone, the synaptic cleft and the end of the postsynaptic neurone.

**synaptic cleft** a very small gap between two neurones at a synapse.

**systolic blood pressure** the maximum blood pressure in an artery when the ventricles of the heart contract; at rest this is usually about 120 mmHg (15.8 kPa).

**T lymphocyte** a lymphocyte that does not secrete antibodies; T helper cells stimulate the immune system to respond during an infection, and killer T cells destroy human cells that are infected with pathogens, such as bacteria and viruses.

**target cell** a cell that is affected by a hormone; target cells have receptors with which the hormone is able to bind.

**tertiary structure** the structure of a protein molecule resulting from the three-dimensional coiling of the already-folded chain of amino acids.

**test cross** a genetic cross in which an organism showing a characteristic caused by a dominant allele is crossed with an organism that is homozygous recessive; the phenotypes of the offspring can be a guide to whether the first organism is homozygous or heterozygous.

**tetraploid** possessing four complete sets of chromosomes.

**thylakoid** a flattened membrane-bound, fluid-filled sac, which is the site of the light-dependent reactions of photosynthesis in a chloroplast.

**tidal volume** the volume of air breathed in or out during a single breath.

**tissue** a layer or group of cells of similar type, which together perform a particular function.

**tissue fluid** the almost colourless fluid that fills the spaces between body cells; tissue fluid forms from the fluid that leaks from blood capillaries, and most of it eventually collects into lymph vessels where it forms lymph.

**translocation** the transport of assimilates such as sucrose through a plant, in phloem tissue; translocation requires the input of metabolic energy.

**transmission** the transfer of a pathogen from one person to another.

**transmitter substance** a chemical that is released from a presynaptic neurone when an action potential arrives that then diffuses across the synaptic cleft and may initiate an action potential in the postsynaptic neurone.

**transpiration** the loss of water vapour, by diffusion down a water potential gradient, from a plant to its environment; most transpiration takes place through the stomata on the leaves.

**triglyceride** a lipid whose molecules are made up of a glycerol molecule and three fatty acids.

**triploid** possessing three complete sets of chromosomes.

**trophic level** the level in a food chain at which an organism feeds.

**t-test** a statistical procedure used to determine whether the means of two samples differ significantly

**ultrafiltration** filtration on a molecular scale, for example the filtration that occurs as fluid from the blood passes through the basement membrane of a renal capsule.

**urea** a nitrogenous excretory product produced in the liver from the deamination of amino acids.

**ureter** a tube that carries urine from a kidney to the bladder.

**urethra** a tube that carries urine from the bladder to the outside.

**uric acid** the main nitrogenous excretory product of birds; some uric acid is also excreted by humans.

**vaccination** giving a vaccine containing antigens for a disease either by injection or by mouth; vaccination confers artificial active immunity.

**vein** a blood vessel with relatively thin walls, and containing valves, that carries blood back towards the heart.

**vital capacity** the maximum volume of air that can be breathed out after breathing in as deeply as possible.

**voltage-gated channel** a protein channel through a cell membrane that opens or closes in response to changes in electrical potential across the membrane.

**water potential** the tendency of a solution to lose water; water moves from a solution with high water potential to one with low water potential. Water potential is decreased by the addition of solute, and increased by the application of pressure. Symbol is $\psi$.

**work** all energy transfers other than those involving thermal energy.

**xerophyte** a plant adapted to survive in conditions where water is in short supply.

**xylem tissue** tissue containing xylem vessels and other types of cells, responsible for support and the transport of water through a plant.

**xylem vessel** a dead, empty vessel with lignified walls and no end walls, through which water is transported.

**zygote** a cell formed by the fusion of two gametes; if the gametes are haploid, then the zygote is diploid.

# Index

Terms shown in **bold** also appear in the glossary (see page 330). Pages in *italics* refer to figures.